# Animal Learning
# and Conditioning

# ANIMAL LEARNING AND CONDITIONING

## GRAHAM DAVEY

*Department of Social Science and Humanities*
*The City University*

**UNIVERSITY PARK PRESS**
Baltimore

*First published 1981 by*
THE MACMILLAN PRESS LTD
*Published in North America by*
UNIVERSITY PARK PRESS
*233 East Redwood Street*
*Baltimore, MD 21202*

*Printed in Hong Kong*


Davey, Graham
  Animal learning and conditioning.
  1. Conditioned response
  2. Learning in animals
  I. Title
  156′.3′15  BF319

*ISBN: 0-8391-4149-1*

For my parents
Ray and Betty

# Contents

11, 38, 467

vii

CONTENTS

# Preface

In recent years the field of learning and conditioning has seen a number of important developments. First, in animal learning the traditional laws of classical and operant conditioning are no longer considered sacrosanct; what an animal learns during conditioning not only depends on what contingencies exist in the environment, but more often on *what* that organism is, and what species-specific behaviours it brings with it to the learning task. Our expanding knowledge of what are popularly called 'constraints on learning' has dictated that theorists take a radically new look at the principles of associative learning. The second development has involved a growing belief that animal learning concerns much more than simple associative learning; studies on concept learning, language learning and problem solving suggest that the behaviour of some animals can transcend the limits of simple association, while the growing domains of perceptual learning and animal memory point to important 'contextual' effects in conditioning and learning. Finally, perhaps the greatest furore of all has centred around the application of conditioning theory to the behaviour of man. Whether human behaviour is indeed governed by the principles of conditioning has traditionally been a source of heated argument, but with the advent in the last ten years of an increasingly applied behavioural technology based on conditioning theory, the debate has acquired added dimensions of intensity.

Basically, these are the three issues around which the book revolves and I have constructed the presentation so as to make the book suitable as an advanced undergraduate text on both animal learning and in particular, conditioning theory. This being the case, the book commences with a historical overview of the subject and a fairly detailed and non-jargonised discussion of the principles and concepts of traditional conditioning theory; this should provide a suitable introduction for newcomers to the topic. In later chapters I have assumed that the reader is familiar with the terminology and jargon explained in chapters 1 and 2, but as a further aid I have also compiled a glossary of technical terms which should help the uninitiated reader. Each chapter is intended to be as comprehensive as possible in its coverage of a particular topic, both from the factual viewpoint and in terms of the current status of theoretical interpretations of these facts. Because of this I have often had to refer the reader to alternative sources of information when elaboration of a particular point would prove tangential or just unnecessarily 'bulky'. Hence, if a particular issue captures the reader's interest there should be adequate guidance for further reading. Finally, it is conceivable that the student who is encountering the subject for the first time may feel distinct dissatisfaction with the inconclusiveness of many of the theoretical discussions. There are two reasons for this. First, I have tried to present as unbiased a picture as possible of the theoretical accounts of learning phenomena by presenting the *pros* and *cons* of each theory in some detail. Secondly, as things stand at the present time there is often evidence for and against a number of accounts of a particular phenomenon without there being any definitive evidence which separates them. In these cases one can only summarise rather than conclude.

Although the three sections of this book reflect very different aspects of learning and conditioning, I have tried to integrate them as much as possible. In general this is achieved simply by cross-referencing so that each section, if necessary, can stand on its own as an independent unit. However, conditioning theory has now diversified so greatly, especially when we consider its applications to animals and man respectively, that some topics have little in common apart from sharing similar terminology. In many cases – especially Section III – I have had to stress the differences rather than similarities between conditioning approaches to certain topics. This, I feel, is especially important because it is an aspect which is regularly overlooked. Certainly many workers in the field of animal associative learning would have little sympathy for the way their theories are utilised to explain human behaviour; and conversely most radical behaviourists merely adopt the conceptual framework of animal associative learning without adopting its principles *in toto*. It is this diversity of development that I feel it is important the student should understand.

Finally, the planning and writing of this book required not only my own concentration, but also the patience and understanding of many other people. I must apologise to all those students in the Social Science Department at The City University for the many times they knocked on my office door only to find that I was not in, and I must also thank my friends in the Psychology Department of the University of Sheffield for allowing me to use their facilities during the course of my writing. Special thanks must go to Paul Dean and Dave Oakley for drawing my attention to omissions and inconsistencies in initial drafts of this book, and also to Fergus Lowe for supplying me with prints of certain figures. Marita McLaughlin was responsible for translating my illegible handwriting (yes, it really was illegible!) into typescript for which I am extremely grateful. However, despite all this advice and assistance, responsibility for errors and inconsistencies of course rests solely with the author.

G. C. L. Davey

*City University*

# Section One
# Conditioning in Animals

# 1 Historical Perspective

From classical times right through to the present century animal behaviour has been a rich source of debate for philosophers of Nature. Whether it be the attempts to identify the 'life essences' which differentiated living organisms from inert matter, or to compare the mental processes of man and animals, there has been popular discussion centring around some aspect of animal behaviour in almost every age since the time of Aristotle.

Of the classical philosopher-scientists, the Greeks concerned themselves with two basic problems. The first was related to the causes of behaviour – 'What makes animals behave the way they do?'. The second problem was concerned with the development of different life forms – 'What is the origin of life?'. It is interesting to note that in somewhat modified form these questions still motivate the majority of modern biological research. Nowadays, however, the questions have become more sophisticated ones: 'What are the important mechanisms of animal learning?'. 'Under what conditions can biologically active material be manufactured?'. 'What is the evidence for phylogenetic determinants of behaviour?'. Nevertheless, these problems still embody the 'Hows?' and 'Whys?' of animal behaviour which occupied the classical philosophers.

The scientific study of animal learning has a relatively short history. In the late nineteenth century, when interest began to shift from the study of similarities in the behaviour of man and animals to the study of how animals behaved under controlled conditions, two distinct methods of study evolved. These have come to be known as the comparative and analytical approaches. By studying as wide a variety of organisms as possible the former attempted to throw light on the influences that evolution has had on the learning capacities of animals; the latter concentrated on identifying in some detail the mechanisms of learning in a limited number of species. Although the comparative approach is historically much older than the analytical method, the latter has tended to dominate animal learning since the turn of the century. None the less, both disciplines have contributed substantially to our knowledge of animal learning and both have influenced modern attitudes to research in animal learning.

## HISTORICAL PERSPECTIVES

Man first began to study animal behaviour in a serious fashion in the Grecian era, at a time when the chief interest of the philosophers centred around the problem of the structure and laws of the universe. The study of animals was essentially incidental to this colossal task, and very few thinkers over-indulged their curiosity in animal behaviour. Early attempts at detailed consideration of living organisms can be subdivided into what are loosely called the dualist and monist schools. The monists, among them Leucippus and Democrites (460–370 B.C.) believed that the atoms which comprised matter were endowed with the power of movement. Organisation of atoms was explained merely in terms of physical principles. The dualists, however, such as Anaxagoras (500–428 B.C.) considered that material objects were reducible to physical elements, these being fire, earth, air and water. Organisation and movement of these elements, such as is found in living organisms, was accounted for by a nonphysical principle which Anaxagoras called Mind. Plato (427–347 B.C.) elaborated the dualistic argument and developed a theory of the soul that enabled distinctions to be made between the behaviour of man, animals, and plants. The soul was divided into a rational and irrational part, the latter being further divided into a 'passionate' and 'nutritive' part. Plants possessed only the nutritive soul, animals possessed the passionate and nutritive souls, and man possessed both rational and irrational attributes, the rational soul being located in the head, the passionate soul in the heart and the nutritive soul in the abdomen. Differences in behaviour between plants, animals and man reflected these different attributes. Later, however, Aristotle considered that animals (and man) differed from plants by possessing the general function of 'sensing', as manifested in its most primitive form by touch. Aristotle was perhaps the first to make what approached scientific observation of animals and recorded them in his book *Historia Animalium*. Here he discusses the differences in physical form, reproduction, rearing of young, and locomotion between different species of animals and even makes a preliminary attempt at comparing behavioural phenomena in animals and man. In fact he even makes the far-sighted suggestion that all animals seek pleasure in accordance with that which nature holds appropriate for them – a statement which smacks of the Thorndikian law of effect!

Attempts to define the essential elements which were characteristic of different forms of matter continued right up to the Middle Ages. For Thomas Aquinas there was a multiplicity of substances that had matter in their nature. Four types of matter could be distinguished – non-living bodies, plants, animals, and men. Inanimate objects had material activities alone, plants showed material activities and vegetable activities (that is, the need for nutrition), animals exhibited these two plus sensory activities, and man showed rationality in addition to these three activities. However, from this time right up to the time of Darwin the study of behaviour was taken off the objective footing established by the classical philosophers and was engulfed by the religious dogma of the Roman Catholic Church. This dictated that at the outset of any philosophical or experimental enquiry man was to be considered as qualitatively different from animals: he possessed both free will, and a soul which was immortal. This teaching emphasised that all human behaviour was a result of the 'free' decisions of forces within the person, whilst animals were mere automata. In fact, as a general rule of thumb, the church refused to accept any statement relating to natural philosophy unless it contained a reference to the Bible!

The conception of animals as automata lasted right up to what is popularly known as the Darwinian period. Following the publication of the *Origin of Species* (1859), Charles Darwin (1809–1882) attempted to apply the principles of evolution developed in this book to the physical and psychological development of man. His conclusions are published in *The Descent of Man* (1872). As Waters points out

> . . . (Darwin) argued that differences in all animals present a series of fine graduations and that all differences are quantitative rather than qualitative in nature. In support of this claim he presented the following general theses: Man and animals possess similar senses, similar instinctive action, the same fundamental emotions and higher mental processes of attention, memory, imagination, deliberation, and choice. He even argued that animals show a trace of what, if the animal's intelligence equalled that of man, would be called language, a sense of beauty, a belief in spirits.
>
> (Waters, 1953)

As a result of Darwin's theses, research with animals accelerated rapidly and took as its major problem the relating of different forms of animal behaviour to the phylogenetic scale. In the early stages this was conceived of as the contrast between instinct and reason, and certainly in *The Descent of Man* Darwin emphasises the similarity of reasoning in man to what appeared to be similar processes in animals, suggesting that many of man's mental capacities appeared to have rudimentary equivalents among lower animals. Thus, the study of animal behaviour had been given re-emphasis – the search for evidence of 'reasoning' in animals. Much of the early work involved collecting anecdotal evidence about animals, mainly accounts of casual observations of the behaviour of animals upon which workers chanced, or had searched out by questioning such people as farmers, zoo-keepers and animal breeders. This enthusiastic, but amateur, approach was certainly unsatisfying to the scientific mind, mainly because simple observation could not help in unequivocably identifying the causes of behaviour; it inevitably fostered the desire to anthropomorphosise in Darwinians whose aim it was to narrow the traditional chasm in mental abilities between man and animals. However, comparative psychology can be said to have been born out of the Darwinian revolution. Because of the emphasis on the phylogenetic history implicit in the theory of evolution, it was considered by many Darwinians that no account of behaviour was adequate unless it took these factors into consideration – a tenet still characteristic of the comparative approach to animal learning championed by the ethologists, and which we will come across frequently in later discussions.

At the turn of the century the experimental study of animal behaviour was beginning to emerge, and perhaps the first animal studies which had any flavour of true experimental control about them were performed by Lloyd Morgan (1852–1936). Although many of the situations were elaborate and connived, and although his experiments were not carried out in the strictly controlled environment of a laboratory, the rudiments of experimental method were apparent. The following passage give some idea of the types of 'experiment' Morgan carried out.

I watched Tony's [his dog] behaviour when a solid india-rubber ball was thrown gently towards a wall standing at right angles to its course. At first he followed it close up to the wall and then back as it rebounded. So long as it travelled with such velocity as to be only just ahead of him he pursued the same course. But when it was thrown more violently so as to rebound as he ran towards the wall, he learnt that he was thus able to seize it as it came towards him. And profiting by the experience thus gained, he acquired the habit – though at first with some uncertainty of reaction – of slowing off when the object of his pursuit reached the wall so as to await its return.

When the ball was thrown so as to glance off at a wide angle from a surface, at first – when the velocity was such as to keep it just ahead of him – he followed its course. But when the speed was increased he learnt to take a short cut along the third side of the triangle so as to catch the object some distance from the wall.

A third series of experiments was made where an angle was formed by the meeting of two faces of rock. One side of the angle, the left, was dealt with for a day or two. At first the ball was directly followed. Then a short cut was taken to meet its deflected course. On the fourth day this procedure was well established. On the fifth, the ball was thrown so as to strike the other (the right-hand) side of the angle, and thus be deflected in the opposite direction. The dog followed the old course (the short cut to the left) and was completely non-plussed, searching that side, and then more widely, and not finding the ball for eleven minutes. On repeating the experiment thrice, similar results were on that day obtained.

On the following day the ball was thrown just ahead of him, so as to strike to the right of the angle, and was followed and caught. This course was pursued for three days, and he then learnt to take a short cut to the right. On the next day the ball was sent, as at first, to the left, and the dog was again non-plussed. I did not succeed in getting him to associate a given difference of initial direction with a resultant difference of deflection.

(Lloyd Morgan, 1930)

In addition to executing experiments of this kind, which involved the controlled manipulation of variables, he adopted an attitude to the explanation of behaviour which was in strict reaction to contemporary Darwinian approaches which attributed higher mental faculties to animals.

In no case may we interpret an action as the outcome of the exercise of a higher psychical faculty, if it can be interpreted as the exercise of one which stands lower in the psychological scale.

(Lloyd Morgan, 1894)

This has since become known as Lloyd Morgan's canon, and the principle it embodies demands that notions such as 'purpose', 'reasoning' and 'deliberation', that at the time were being applied to animal behaviour, should be relegated in favour of explanatory notions more directly related to the observed behaviours.

As it was, reductionist accounts of animal learning, which did not directly involve the need for talking about higher mental faculties, were becoming prevalent. This stemmed from the upsurge in controlled experimental studies of animal behaviour, particularly animal learning and problem solving. The experimental paradigm allowed behaviour to be studied in relation to strictly defined antecedent and consequential environmental events, and these environmental factors could be manipulated systematically to produce factual statements about the environmentally observable correlates of behaviour.

Two of the earliest experimentalists in this respect were Thorndike and Pavlov. Thorndike (1874–1949) was, at the turn of the century, using laboratory facilities to study learning in a variety of animals including cats, chicks, dogs and monkeys. Similarly, Pavlov, a Russian, and a physiologist by training, neatly emphasises the necessity of controlled experimentation in the following passage.

The more complex the phenomenon (and what could be more complex than life?), the greater the need for experiment. Experiment alone crowns the efforts of medicine, experiment limited only by the natural range of powers of the human mind. Observation discloses in the animal organism numerous phenomena existing side by side and interconnected now profoundly, now indirectly, or accidentally. Confronted with a multitude of different assumptions the mind must *guess* the real nature of these connections. Experiment, as it were, takes the phenomena in hand, sets in motion now one of them, now another, and thus, by means of artificial, simplified combinations, discovers the actual connection between the phenomena. To put it in another way, observation collects all that which nature has to offer, whereas experiment takes from her that which it desires. And the power of biological experimentation is truely colossal.

Since important aspects of the work of both Thorndike and Pavlov are to be discussed in later chapters, suffice it to say here that their work was important (1) as an experimental source from which the important and influential school of Behaviourism was to flourish, and (2) for initiating an integration of the study of animal behaviour into experimental psychology. Prior to this time, psychology had been the science that investigated the facts of consciousness and sensation (along the lines initiated by the work of Wundt, Fechner, Ebbinghaus and Galton, for example). Although Pavlov's experimental findings were related in the initial stages solely to the salivary responses of dogs, the concepts he developed to deal with the phenomena he observed were soon to be employed in analysing both complex animal behaviour and diverse aspects of human behaviour. The emergence of principles of conditioning from studies of animal learning, and their subsequent application to human behaviour was primarily led by the eminent behaviourist, Watson. The result of this application was a reappraisal of both the scientific outlook of psychology and its achievements up to that time. Watson writes

> Behaviouristic psychology attempts to formulate, through systematic observation and experimentation, the generalizations, laws and principles which underly man's behaviour. When a·human being acts – does something with arms, legs or vocal cords – there must be an invariable group of antecedents serving as a 'cause' of the act. For this group of antecedents the term situation or stimuli is a convenient term . . . Psychology is thus confronted with two problems – the one of predicting the possible causal situation or stimulus giving rise to the response; the other given the situation, of predicting the probable response.
>
> (Watson, 1919)

Such an approach came to be known as 'empty organism' psychology because of its concentration on the observable environmental antecedents of behaviour to the detriment of the role played in learning by cognitive mechanisms. However, it can be argued that classical Behaviourists such as Watson, although helping to establish psychology as a natural science, may have placed too great an emphasis on the prediction and control of behaviour to the detriment of the identification and description of the mechanisms responsible for adaptive behaviour. To the classical behaviourists, behaviour was to all intents and purposes 'explained' when its immediate antecedent causes had been identified since this enabled them to fulfil, to a large degree, their aims of prediction and control.

However, although the behaviourist school dominated the study of animal learning in the first half of this century, at about the time that the work of Pavlov and his associates was becoming widely publicised, another 'school' of psychology was developing. The Gestalt movement, pioneered in Germany by the work of Wertheimer, Köhler and Koffka, provided an approach to the explanation of animal behaviour which contrasted sharply with contemporary explanations in terms of association (conditioning). The important notion in Gestalt psychology, at least as far as learning was concerned, is *insight* (Köhler, 1925). This involves the perception of relations between the elements of a learning situation with the implication that the organism acts as intelligently as it can, interrelating the elements of the problem until the solution is hit upon. Important implications of this approach are that (1) it endows the organism with some form of reasoning (that is, the ability to cognitively 'shuffle' the elements of a problem until the correct solution is found), and (2) perception or organisation of the world into a good gestalt (obtaining a state of pragnanz) can be as good a reinforcer as the reduction of any physiological need state. Needless to say, the stark contrast in interpretations led to much argumentative dialogue between the Gestalt and behaviourist camps, the latter claiming that in all situations in which 'insight' was supposedly involved, learning could more simply be explained by the reinforcement of the correct response, that response being generated in a 'trial-and-error' fashion. Despite the attractive approach that Gestalt psychology brought to the study of animal learning, it was soon pushed into the background by the weight of knowledge being generated by studies of animal conditioning and also by the unverifiable nature of 'insight' as an explanatory notion, dependent for its acceptance 'not upon demonstrable fact, but upon faith in the validity of intuitive judgements' (Thorpe, 1956).

Towards the 1930s and 1940s, with studies of animal learning generating a substantial amount of data, animal learning was going through what can only be described as an 'Age of Theories' (Koch, 1964). Principal theorists of the time were Guthrie, Hull and Tolman, each of whom proposed different, but influential, accounts of animal learning based on the principles of conditioning. In fact, the conceptual framework provided by conditioning studies had become so prevalent in animal learning that conception of the animal had to all intents and purposes reverted to the 'brute' or 'automaton' initially described by Descartes – an organism whose behaviour was only to be explained mechanistically. Certainly, the elaborate formulations of Hull smacked of the internal engineering of a machine, and the simple stimulus–response (S–R) associations of Guthrie presented the stimulus as a 'goad' that 'forced' the organism into action like a cue knocks a billiard ball across the table. Only Tolman credited the animal with any higher mental faculties, suggesting that the animal, through experience of learning situations, formed 'expectancies' of the outcomes of its behaviour in these situations and on future occasions acted in accordance with these 'expectancies'. Nevertheless, such unobservable cognitive notions were all related to observable events, and they were only of a summarising nature, concerned with the complex interactions between stimuli and responses. Accounts of animal learning at this time were rarely to stray from the strict positivist approach fostered by the conditioning paradigm and the principles of behaviourism.

However, with the development of the behaviourist tradition during the twentieth century, the type of questions being asked by students of animal learning and animal behaviour changed. Whereas in pre-Thorndikian and pre-Pavlovian times the important problems had been related to the question 'Do animals exhibit behaviour patterns characteristic of the higher mental processes found in man?', the advent of conditioning and related experimental analyses of behaviour helped to turn this question on its head, making the study of animal learning a discipline in its own right. So much so that the possibility was being avidly explored that man's higher intellectual faculties might be explained in terms of the principles of conditioning discovered from studies of animal learning. This was Lloyd Morgan's canon aimed at every kind of organism no matter how complex the behaviour it exhibited or how privileged its place in Nature. A variety of writers have attempted to account for human behaviour in terms of its interaction with the environment in much the same way that the associative principles of conditioning attempt to explain animal learning. Although such accounts have remained largely unverified, their influence on psychology in general cannot be overestimated.

Perhaps the originator of modern reductionist accounts was Watson. He took as the basis for learning the unconditioned reflex, and it was conditioned reflexes which served as the basic units out of which 'habits' were formed. In fact Watson is taken as the archetypal reductionist and Behaviourist, claiming that human behaviour can be moulded into almost any form by environmental contingencies which operate through the principles of conditioning. He writes

> . . . Give me a dozen healthy infants, well-formed, and my own specified world to bring them up in and I'll guarantee to take any one at random and train him to become any type of specialist I might select – doctor, lawyer, artist, merchant-chief and yes, even beggarman and thief, regardless of his talents, penchants, tendencies, abilities, vocations, and race of his ancestors.

> (Watson, 1924)

The implication of this statement is that heredity or innate factors played little if any role in determining behaviour – a Shakespeare or a Beethoven could theoretically be moulded out of any physiologically complete human being. In fact, for Watson, the principal role of unlearnt activity was merely 'to initiate the process of learning' – and the rules of learning were the same for all organisms, human or not.

Another influential figure who has attempted to relate the findings of animal learning to human behaviour is Skinner. The Skinnerian argument states that all animals obey universal laws of behaviour – for example, all animals, including man, learn as a result of the experienced consequences of their behaviour. Skinner's behaviourism differed from that of his predecessors in that he moved away from the traditional notion of stimuli as mechanical 'elicitors' of behaviour, drawing a distinction between operant behaviour, which was 'emitted' by the organism, and classically conditioned behaviour which was 'elicited'. He claimed that most behaviour of any psychological importance was operant (emitted) behaviour, which suggested that, rather than being a passive mechanism goaded into action by stimuli – the traditional S–R approach – an animal was capable of *initiating* behaviour. The important fact of operant psychology, as this branch of Behaviourism has come to be known, is that behaviour is modified (either in form or frequency) by its consequences. Those consequences of a behaviour which increase its future frequency are known as reinforcers, and those which decrease its future frequency are known as punishers. In order to study these principles of operant psychology, Skinner designed an experimental environment which has since become known as the *Skinner-box*. A rat, usually deprived of food, is trained to press a lever in the chamber, which, according to a particular schedule of reinforcement, will release a small food pellet. This technique, and hence the Skinner-box, has become extremely popular amongst students of animal behaviour since it allows the behaviour of an organism to be studied in a 'free-flowing stream' rather than in the more artificial discrete-trials method necessary with mazes or alleyways.

As well as accounting for the actions of rats in limited experimental environments, Skinner has extended his principles of operant psychology in an attempt to explain many facets of human behaviour. These include, for example, the acquisition of language (Skinner 1957b), problem solving and social interactions in general, and the *en masse* control of societies using behavioural technology derived from conditioning principles (*Skinner*, 1971). Not all psychologists have been ready to extrapolate the findings of animal research to human behaviour in this way, and attempts to explain man's actions in a fashion which tends to present him as a machine has been excellent fuel for the emotional debate on behaviour and freedom of the will. None the less, principles derived from Skinner's work have been successfully applied to the modification of human behaviour in both the clinical and natural environments, and – moral issues aside for the moment – they often seem to be effective. However, whether one can obtain a greater understanding of the more complex aspects of human behaviour in terms of operant principles is open to question. Much of Skinner's theorising in this respect is still unproven and in some instances his analyses of complex human behaviour extend beyond the limits of experimental verification. Nevertheless, his writings illustrate the extent to which knowledge obtained from studies of animal learning have been applied to human behaviour, and are exemplary of the behaviourist's desire to put the study of behaviour on a totally objective footing.

One other topic of major controversy to befall animal learning since the turn of the century has been concerned with the learning–instinct dichotomy. Initially the relatively simple associative mechanisms which, conditioning studies suggested, could explain learning (for example, S–R associationism, the principle of reinforcement, etc.) appeared to lend themselves admirably to an explanation of adaptive behaviour in all ranges of organisms from man right down to creatures with the most primitive nervous systems. Thorndike was perhaps the first to suggest that one law of learning might be applicable in all situations to all species. He wrote

If my analysis is correct the evolution of behaviour is a rather simple matter. Formally the crab, fish, turtle, dog, cat, monkey and baby have very similar intellects and characters. All are systems of connections subject to change by the laws of exercise and effect.

(Thorndike, 1911)

Although this conclusion went far beyond the experimental evidence available at the time, later data appeared in general to support it. None the less, the increasing dependence, in studies of animal learning, on data generated by animals selected merely because they were easy to breed, easy to handle, and cheap to maintain, may have oversimplified our conceptions of animal learning. Studying in depth the behaviour of a single species has been defended vigorously by workers who claim that basic principles will emerge from a detailed analysis of learning in a single species such as the rat (Mackintosh, 1974; Skinner, 1953). Skinner writes that in the experimental analysis of behaviour

. . . species differences in sensory equipment, in effector systems, in susceptibility to reinforcement, and in possibly disruptive repertoires are minimized. The data then show an extraordinary uniformity over a wide range of species. For example, the processes of extinction, discrimination and generalization, and the performances generated by various schedules of reinforcement are reassuringly similar. . . . Although species differences exist and should be studied, an exhaustive analysis of the behaviour of a single species is as easily justified as the study of the chemistry or microanatomy of nerve tissues in one species.

(Skinner, 1969)

Although the detailed study of a limited number of species can without doubt be said to have advanced our knowledge of animal learning, as more species are studied in a wider variety of learning situations the assumed basic principles of associative learning have needed to be extended and qualified. It appears that not all stimuli have the equal capacity to control behaviour, not all reinforcers work in all situations, and the selection of the response for study is not an arbitrary matter. For these reasons such fundamental notions as stimulus control, generalisation, discrimination, and even reinforcement have been cast in a critical light. This development has fused the analytic study of learning processes – as exemplified by those concerned with conditioning and associative learning – with the comparative approach to animal behaviour exemplified by the broader-based studies of the ethologists. These were two disciplines which, until the 1960s had evolved in parallel and, except for the odd brief and cursory interchange of dogma, had developed in almost total disregard for each others findings. One benefit of this convergence of disciplines is that the 'learnt–innate' dichotomy now becomes less of a dilemma and less a point of theoretical hegemony. The fact that certain organisms possess adaptive mechanisms reflective of specialised adaptations to phylogenetic pressures, and that these mechanisms can interfere with (or even facilitate) associative learning, redirects attention away from the sterility of classing behaviours as 'innate' or 'learnt' to a consideration of the ways in which phylogenetically determined response tendencies and associative learning mechanisms interact.

This convergence of ethology and conditioning is perhaps typical of present day interplay between approaches to animal learning. Although most ethologists still use observation as their main source of data they nowadays use techniques of operant conditioning to study such phenomena as species-specific learning and early perceptual learning. Meanwhile, conditioning is being studied in an ever growing variety of animals – not necessarily in order to extract any general principles of learning, but in order to assess the constraints on learning that different species bring to arbitrary learning situations.

## SUMMARY AND OVERVIEW

Having briefly looked at the historical development of animal learning and conditioning, the next step is to tackle in some depth the current state of our knowledge of animal learning. There are many phenomena which parade under the name of 'learning', and for the sake of clarity these have been grouped under a number of convenient headings. The first section of this book deals with the most basic learning processes, namely processes of conditioning or associative learning. Traditionally, conditioning has been divided into classical and operant, usually on the basis of methodology and historical development, but with some importance being attached to supposed qualitative differences between the two conditioning processes (chapter 2). However, recent evidence has tended to erode the theoretical basis of this dichotomy, and because of this, has modified the learning theorist's views of the nature of the learning processes inherent in conditioning phenomena (chapters 3, 4 and 7). Similarly, the study of conditioning in an ever increasing number of species has illuminated a variety of different species-specific constraints on learning (chapters 5 and 6). Basic conditioning processes, it appears, do not act in isolation of species peculiar behavioural tendencies. Apart from highlighting such interactions, these findings have also emphasised the need to modify traditional principles of conditioning which were the legacy of the intense study of a limited number of species during the 1930s and 1950s. Nevertheless, many basic conditioning phenomena can be identified in the adaptive behaviour of organisms under both appetitive and aversive conditions, and the principles derived from the study of these phenomena have provided the basis for the modern technologies of human learning we shall encounter in the final section. The second section takes a look at more complex and diverse forms of animal learning, centring on the processes by which animals tackle what appear to us to be relatively complex learning situations. Apart from discussing what we might loosely term 'intelligent' abilities such as concept learning, insight and language learning (chapter 11), this section also deals with processes of imitation learning (chapter 9), perceptual learning (chapter 8) and animal memory (chapter 10) – processes which are all essential for effective behavioural adaptation and hence the survival of most animals. The final section deals with the sticky but intriguing ground which relates some of the processes of learning encountered in the first two sections to the behaviour of man. There are many phenomena of, for instance, conditioning which have been demonstrated in man and it is important that the reader should be familiar with this hard body of facts before considering the philosophical, moral, theoretical and technological issues which this topic inevitably raises. In anticipation of this requirement chapters 12 and 13 bring together the experimental evidence on the developmental aspects of conditioning in humans and also the performance of human subjects on schedules of reinforcement. This should allow consideration from a knowledgeable viewpoint of the relative status of technologies of behaviour modification (chapter 14) and what have been popularly described as 'chimpomorphic' or 'rat-man' views of human behaviour (chapter 15).

The relationship of the study of animal learning to human psychology is, however, less well-definied today than it was in the earlier eras of animal study when the sole purpose of studying lower organisms was to make explicit comparison with human behaviour. The present-day rule of thumb appears to be 'let him extrapolate who will!' and while some writers see the benefit in trying to find comparisons between human and animal behaviour (Skinner, 1971, 1974), or even in extrapolating directly from principles of animal learning to human behaviour (notably Skinner), the course of the study of animal learning will not be as substantially influenced by these excursions as it might have been at the turn of the century. This is not to underestimate the theoretical and practical contributions that the study of human behaviour has derived from principles of animal learning; but animal learning is now a legitimate topic of study in its own right.

# 2 Classical and Operant Conditioning

In the late nineteenth century the study of psychological phenomena had been essentially concerned with the understanding of sensory and perceptual 'experience' in man – a study which involved analysing the content of the 'conscious mind' and discovering principles which related the elements of conscious experience. The main investigative tool adopted to achieve these ends was that of *introspection*, but the lack of objective criteria by which theoretical disputes could be settled led to many bitter quarrels between the eminent schools of psychology. However, at the turn of the century the roots of a viable alternative methodology could be discerned, and in 1913 a paper entitled 'Psychology as the Behaviourist views it' not only set out the methodological tenets of this new approach but also defined in strict terms what should and should not be considered as the legitimate subject matter of psychology. This paper was written by an American psychologist, John Broadus Watson, and paved the way for *Behaviourism* to become the most influential school in both human and animal psychology for the next fifty years.

Behaviourism developed not only out of the need for a more objective methodology in human psychology, but also out of the need for a theoretical framework in which so-called 'psychic' phenomena in animals could be explained. The fact that dogs salivated at the sight of food seemed at that time to reflect some aspect of the animal's psychical activity: the secretion of the salivary glands reflected an inner state of the organism which was unavailable for investigation by scientific methods. This was one of the first problems encountered by the Russian physiologist Ivan Petrovich Pavlov during his studies of glandular secretion in the digestive tract. Many of his contemporaries felt that 'psychical' phenomena such as that exhibited by the salivating dog were impossible to study since the internal experiences of the animal were inaccessible. Nevertheless, Pavlov was convinced that the concept of 'psychical secretion' could be replaced by objective scientific concepts; it was the relationship between the organism's behaviour and his environment that was the important link, not the relationship between 'ideas' or 'sensations' wholly within the organism. Pavlov held three factors to be important in achieving an understanding of behaviour: (1) It was necessary to have a detailed knowledge of the role of the environment in shaping adaptive behaviour; (2) This could be achieved by adopting the experimental method in the tackling of behavioural problems; and (3) It was also necessary to carry out the intensive study of physiologically intact and normal animals just as much as it was to study surgically treated animals. Although there had been many eminent Russian physiologists before Pavlov, he was perhaps the first to develop the notion of the adaptive character of physiological phenomena and also the first to bring together the contemporarily popular doctrine of associationism with reflexology, thus giving birth to the concept of *conditioning*.

This drift from introspectionism to the methodology of the natural sciences was taken almost to its logical extreme by Watson. He suggested

> Why don't we make what we can *observe* the real field of psychology? Let us limit ourselves to things that can be observed, and formulate laws concerning only those things.
>
> (Watson, 1924, p. 6)

Watson believed that, as subjects, animals offered the experimenter the advantage of being able to achieve almost complete control of the experimental conditions. It was possible not only to control carefully the animal's history of environmental conditions such as rest, diet, activity, etc., but also to control the animal's history of environmental stimulation. Also, one of the main cornerstones of Behaviourism was the belief that, as adaptive organisms, human beings did not differ in any important way from other animals; this appeared to be further justification for intensifying the study of animal learning and incorporating it into a unified science of behaviour (see also chapter 15, pp. 392ff.). However, the newly discovered process of conditioning was seen by Watson as the methodological tool *par excellence*. It enabled the psychologist to analyse the relationships between discrete environmental events (stimuli) and the organism's observable reactions (responses). It will become clear to the reader as we progress through this section that this analysis has become fundamental to our understanding of associative learning. Defining the environment in terms of stimuli, or behaviour in terms of discrete response units enabled investigators to discard those 'psychical' concepts such as ideas, intentions etc. which had hitherto proved unhelpful in understanding adaptive behaviour. As Watson put it

> The rule, or measuring rod, which the behaviourist puts in front of him always is: Can I describe this bit of behaviour I see in terms of 'stimulus and response?' By stimulus we mean any object in the general environment or any change in the tissues themselves due to the physiological condition of the animal such as the change we get when we keep an animal from sex activity, when we keep it from feeding, when we keep it from building a nest. By response we mean anything the animal does – such as turning toward or

away from a light, jumping at a sound, and more highly organized activities such as building a skyscraper, drawing plans, having babies, writing books and the like.

(Watson, 1924, p. 6)

Although the writings of Watson put the study of behaviour on a firm scientific footing – and indeed, had now made the study of animal learning itself quite respectable – the experimental work of Pavlov and an American contemporary called Thorndike provided the main directions in which studies of animal conditioning have since travelled. The experimental and theoretical work of Pavlov provided the basis for what we now know as *classical* or *Pavlovian* conditioning, and the animal studies of Thorndike formed a basis from which *operant* or *instrumental* conditioning has developed.

# PAVLOV – THE ORIGINS OF CLASSICAL CONDITIONING

Pavlov first encountered the phenomenon he was later to call 'conditioning' in the course of studies on the reactions of the digestive tract to certain natural stimuli such as bread, meat or milk. (Initially he called conditioned responses 'psychic reflexes'.) These important observations led him to conduct the first of his prototypical conditioning experiments.

## Sham feeding

Pavlov observed that when food was placed into the mouth of a hungry dog, but – by being channelled through an oesophageal fistula – failed to reach the stomach, gastric secretion still occurred as it does in normal feeding. This suggested that the taste or feel of food in the mouth leads to a kind of 'psychic secretion' in the stomach; that is, the animal is able to discriminate that food in the mouth will soon mean food in the stomach, and the stomach is thus prepared to receive it. Although on the face of it this appears to be a perfectly unexceptional physiological finding, Pavlov noted the adaptive significance of this phenomenon and constructed some rather crude, but insightful, follow-up experiments.

## The conditioning of salivation in the dog

Pavlov had earlier observed that the dog's innate response of salivation to food would often occur at the sight of food or to the sound of the handler's footsteps. Realising that this might be the result of the frequent association of the ingestion of food with these sights and sounds, Pavlov carried out a number of simple experiments where the dog would be shown a piece of bread which it was later allowed to eat. Sure enough, salivation occurred even at the sight of this food.

Following these unstructured observations, Pavlov went about designing a number of controlled experiments where the salivary response could be precisely measured and stimuli more objectively defined. In the classical studies, a hungry dog was subjected to several pairings of a bell with food whilst being restrained in a harness: first the bell was sounded and at its termination a meat powder would mechanically be introduced into the dog's mouth. Secretion of saliva was collected in a fistula which had been surgically attached to the parotid gland of the dog. Before this procedure had been repeated many times, the dog would begin to salivate during the sounding of the bell and prior to the presentation of food. This was a reaction which 'appeared quite regular, could be produced at will again and again as usual physiological phenomena, and was capable of being definitely systematized.' (Pavlov, in Koshtoyants, 1955). From these findings Pavlov set out a number of principles concerning the adaptive nature of physiological phenomena; these principles have since become collectively known as *classical* or *Pavlovian conditioning*. He considered that there were two types of reflexes: inborn or *unconditioned reflexes*, and individually acquired, or *conditioned reflexes*. This latter group of responses was acquired during the lifetime of the organism through the pairing of previously 'neutral' environmental events (stimuli) with the stimuli which naturally produced these reflexive responses.

The importance of Pavlov's work in human and animal psychology cannot be underestimated. It led not only to the inclusion of what were previously considered to be 'psychic phenomena' in animals into a scientific animal psychology, but also to an emphasis on experimentation and objectivity in the study of behaviour.

## THORNDIKE – THE ORIGINS OF OPERANT CONDITIONING

Like Pavlov, Thorndike too laid great store by the experimental method in psychology. Apart from formulating some extremely influential laws of adaptive behaviour, he can perhaps be credited with introducing the modern laboratory experiment into animal psychology. At the turn of the century Thorndike had developed two important techniques for studying problem-solving behaviour in animals.

## The puzzle-box

Here a hungry kitten is placed in a 'strong' box, one side of which is open with escape being prevented by a wire-grill door. Escape can only be effected by the manipulation of a particular mechanism by the kitten. The learning tasks involved such manipulations as pulling a string, turning a button, and pressing a lever. Once this had been accomplished the door would open and the animal would be allowed to eat a morsel of food left outside the box. The kitten was continually placed in this situation until it had effectively mastered the particular task. Thorndike found that in these situations behaviour went through a number of fairly well-defined stages: (1) the kitten would move erratically around the box, clawing and scratching at the bars; (2) during these wild cavortings the animal would accidentally operate the mechanism and obtain the food reward; (3) after a number of trials the frantic behaviour was eliminated and operation of the release mechanism occurred earlier and earlier; and (4) the animal would eventually be able to operate the mechanism as soon as it was placed in the box. Thorndike characterised this process by naming it *trial and error learning*.

## Maze learning

Thorndike also pioneered the use of mazes in the study of animal learning. Instead of having to learn to escape by operating a mechanism, escape from the box could be achieved by traversing a particular pathway bounded by walls and covered on top by wire-netting. In his studies with mazes Thorndike found that the course of learning followed many of the same stages he had discovered with the puzzle boxes. After a number of trials the correct route is traversed and this eventually becomes dominant with the animal learning to eliminate errors.

From the results of experiments such as these Thorndike formulated two quite fundamental laws of learning: the 'Law of Effect' and the 'Law of Exercise'. The *law of effect* states that any response in a given situation which results in a 'satisfying' state of affairs becomes associated with that situation and is thus likely to be repeated when the animal is next in that situation. Conversely, responses which have 'annoying' consequences result in the subsequent elimination of those responses. A statement of this kind may seem fairly obvious and appears at first sight to be just a further restatement of well-known pleasure-pain or hedonistic principles. However, the law of effect is historically important since it stressed the importance of the *consequences* of behaviour in learning. This was the principle on which later formulation of operant and instrumental conditioning were based. The *law of exercise* consisted of two parts, the laws of use and disuse. The former merely states that there is a strengthening of connections or associations with practice, and the latter that there is a weakening of connections when practice is discontinued. This appeared to be a principle loosely moulded from the old law of association, but again it is important since it appealed not to the association of 'ideas', but to the association of strictly observable events couched in scientifically acceptable terms.

# CLASSICAL CONDITIONING

## Basic concepts

The terminology of classical conditioning contains four basic notions which refer to the stimuli and responses identifiable in the classical conditioning procedure:

– (1) *The unconditioned stimulus (UCS)*   This is a stimulus or event which at the outset of the conditioning procedure gives a measurable response; that is, there is no previous history of the organism having *learnt* to respond to it, and the relationship between this stimulus and the response appears to be 'innate'. Examples of unconditioned stimuli (UCSs) often used in classical conditioning experiments include meat (producing salivation in the dog), a puff of air to the cornea (producing an eye-blink), or mild electric shock to the leg of a harnessed dog (producing leg flexion). These types of stimuli were originally called 'unconditioned' by Pavlov because the resulting reflexive response appeared to depend on nothing more than the simple presentation of the appropriate stimulus.

(2) *The unconditioned response (UCR)*   This is in effect the measurable response to the UCS. Unconditioned responses (UCRs) are mainly reflexive in nature and are usually a response of the autonomic nervous system. [This may simply be because experiments in classical conditioning have, following Pavlov, traditionally utilised responses of the autonomic nervous system as the UCR. Recent evidence suggests that more complex skeletal responses can be conditioned using the classical conditioning procedure (see pp. 57–58).]

(3) *The conditioned stimulus (CS)*   This, at the outset of the experiment, is a biologically 'neutral' event which fails to elicit the *UCR*, and is paired with the *UCS* during the experiment. Typical examples are a bell, a metronome, flash of light, tactile stimulation, etc.

(4) *The conditioned response (CR)*   This occurs as a result of the pairing of the UCS and the CS, and is the learnt response to the CS. Although it resembles the UCR it is often not identical to it. For example, after the continual pairing of meat (UCS) with the sound of a tone (CS), salivation will eventually be produced during the presentation of the CS alone (CR). However, quantity of saliva is usually less than that to the original UCS and in some cases of a slightly different chemical constitution.

## Basic procedures

In a classical conditioning experiment, the experimenter presents two stimuli – the UCS and CS – in close temporal proximity. This is repeated a number of times. Usually, after a few pairings of UCS and CS, the CS itself will evoke a facsimile UCR (the CR). In schematic form the procedure can be represented as shown in figure 2.1.

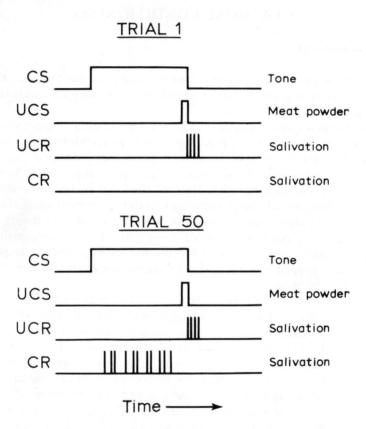

Figure 2.1   The classical conditioning paradigm. On early trials responses occur only to the unconditioned stimulus; however, after several pairings of the conditioned and unconditioned stimuli, responses are also elicited by the conditioned stimulus.

The variety of responses, organisms and stimuli successfully used in classical conditioning studies has been very great. Appetitive UCSs range from food used to condition salivation and general activity levels (Slivka and Bitterman, 1966; Zamble, 1967), to access to a member of the opposite sex in order to condition courtship behaviour (Rackham, 1971). The most widely used aversive UCS is electric shock, and this can be used to condition such behaviours as forepaw flexion in the dog (Soltysik and Jaworska, 1962; Wahlsten and Cole, 1972), eyelid closure in the rabbit (Gormezano, 1965; Smith, 1968) and galvanic skin response (GSR) and heartrate responses (Black, 1971; Fromer, 1963). Apart from the wide variety of responses that can be conditioned using this procedure, such learning apparently can occur in the most fundamental of organisms such as planaria (Thompson and McConnell, 1955), and even during the earliest stages of development; classical conditioning can occur prior to birth in human neonates (Spelt, 1948) and conditioning of physiological processes in birds has been demonstrated while the embryo is still in the egg (Hunt, 1949).

Apart from using a training procedure where the UCS immediately follows the CS, a number of different combinations of CS and UCS associations can be used. As we shall see later, a number of these provide control procedures which either help to identify responding produced by factors other than the simple pairing of CS and UCS or help to assess the conditions under which CS and UCS will successfully become associated. Most of these control procedures can be defined in terms of the relationship (usually temporal) between CS and UCS. These are presented schematically in figure 2.2.

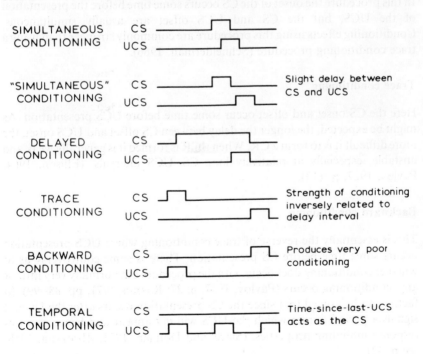

Figure 2.2  Various methods of relating CS and UCS in classical conditioning. See text for description and explanation.

## Simultaneous conditioning

In this procedure the UCS and CS are presented simultaneously for a brief period of time; or as in some experiments, with a slight delay between onset of the CS and onset of the UCS (see the example presented above). This is the normal experimental procedure used to produce classical conditioning and it can be seen as the CS signalling the imminent presentation of the UCS. The optimal interval between CS and UCS for 'maximum' conditioning is uncertain. A number of experiments have addressed themselves to this problem but a single appropriate answer seems unlikely to be found. Kimble (1961) concludes that an interval 'in the quarter second range on either side of 0.5 seconds' (p. 158) was optimum, although it is more likely that the optimum interstimulus interval will depend on other factors such as the type of response being conditioned or the nature of the CS. Pavlov in his studies of salivary conditioning in dogs often used inter-stimulus intervals of up to 30 s with success.

## Delayed conditioning

In this procedure the onset of the CS occurs some time before the presentation of the UCS, but the CS and UCS offset are usually simultaneous. Conditioning effects using this procedure are commonly stronger than using a trace conditioning procedure (Schneiderman, 1966).

## Trace conditioning

Here the CS onset and offset occur some time before UCS presentation. As might be expected, the longer the delay between CS offset and UCS onset, the more difficult it is to form a CR. When a CR is formed it is generally weak and unstable, especially at relatively long CS–UCS intervals, (Ellison, 1964; Pavlov, 1927, p. 113).

## Backward conditioning

This is essentially the reverse of trace conditioning where UCS presentation occurs some time before CS presentation. There is some controversy as to whether conditioning does occur with the procedure but normally very little, if any, conditioning occurs (Pavlov, 1927, p. 27; Razran, 1971, pp. 88–99). In fact, it can be argued that since the CS presentation occurs *after* the UCS, it signals a period during which the UCS will *not* appear and hence acquires response-inhibiting properties, (Siegel and Domjan, 1971; Zbrozyna, 1958, see p. 32).

### Temporal conditioning

In this procedure the UCS is presented periodically such that time-since-the-last-UCS acts as the CS. Pavlov has used this procedure to condition salivation in dogs using an inter-UCS interval of up to 30 min; when the UCS is omitted on a particular occasion when it is due, salivation still occurs at that time, indicating that time-since-the-last-UCS has come to function as a CS for salivation.

## Some common behavioural phenomena

### Acquisition

Figure 2.3 illustrates an acquisition curve which relates the number of CS-UCS pairings (each pairing is called a 'trial') to the probability or 'strength' of the CR. Although the number of trials needed to acquire an

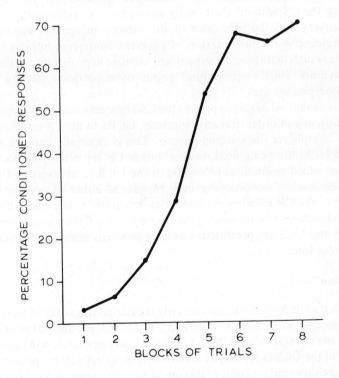

Figure 2.3   A typical acquisition curve in classical conditioning. In this experiment an airpuff acted as the UCS eliciting eyelid responses in rabbits; when a neutral CS was paired with the airpuff, acquisition of eyelid responses to the CS followed this S-shaped function. (After Schneiderman, Fuentes and Gormezano, 1962).

asymptotic level of conditioned responding varies with the different response systems used, the basic shape of the acquisition function is similar in the vast majority of classical conditioning experiments. A period of positively accelerated learning is followed by a levelling off to an asymptote, eventually producing an 'S'-like function. Why different response systems should condition at different rates is unclear; this may reflect differences in response measures or even differences in the CSs and UCSs used. Nevertheless, it is reassuring that even though these differences in learning rate occur, the overall learning function remains similar.

In most classical conditioning studies the UCS follows CS presentation during both training and testing (that is, as in the simultaneous or delayed conditioning procedures illustrated in figure 2.2). This contingency between CS and UCS is often called *reinforcement*, and witholding the UCS after the presentation of the CS is called *non-reinforcement*. The use of the term in this context must be distinguished from its use in operant studies where it has a functional rather than operational definition (see p. 35ff.). Thus: in classical conditioning 'reinforcement' = the contingent presentation of the UCS following the CS without there being any contingent relationship between UCS delivery and the behaviour of the subject; in operant conditioning 'reinforcement' = the presentation of an event contingent upon a class of behaviours such that there is a subsequent increase in probability of that class of behaviours. Further procedural points of importance relating to the acquisition process are

(1) It is usually advisable to present the CS alone on a number of trials prior to conditioning in order that any 'spurious' UCRs to the CS may habituate and not complicate the learning process. This is especially important when aversive UCSs are being used, since a 'novel' CS can often produce startle responses which resemble components of the UCR to an aversive UCS.

(2) The level of conditioning may be gauged either by measuring the strength of the CR on what are known as 'test trials' (trials where the CS is presented alone), or by measuring the strength of the CR on every trial where both CS and UCS are presented. Learning proceeds more rapidly using the latter procedure.

## Extinction

Even when a CR has been acquired, fairly regular pairings of the CS and UCS are still necessary to maintain responding. If the UCS is withheld for a number of trials the strength of the CR will diminish and responding will cease if the absence of the UCS is prolonged. Pavlov originally called this operation, and later the behavioural outcome of this operation, *extinction*. A great number of factors affect the rate of extinction of a CR and resistance to extinction is often used as an index of the strength of learning. It is not intended to deal here with the question of why animals should cease to respond when the UCS is withheld, although this problem has been a perplexing one for learning theorists. It is one thing to account for how responses might be acquired, it is quite another to explain the processes by which a response disappears from an organism's repertoire. The interested reader is referred to Mackintosh (1974, chapter 8) for a detailed discussion of theories of extinction.

One phenomenon associated with the process of extinction does deserve mention however, and this is the effect known as *spontaneous recovery*. If the CR is extinguished by giving a sequence of trials on which only the CS is presented, and then this is followed by a relatively long rest interval, further presentation of the CS alone after this interval will elicit a 'rejuvenated' CR (that is, a level of conditioned responding much higher than that exhibited on the last extinction trial before the delay interval). What spontaneous recovery seems to imply is that the extinction trials have not resulted in the animal gradually 'forgetting' the conditioned response, but that they have inhibited its emission in some way. The long delay interval alleviates this inhibition and allows the CS to again evoke the CR (Pavlov, 1927, pp. 52–53).

### The CS–UCS relationship

In most classical conditioning experiments the occurrence of the CS predicts the forthcoming UCS with a probability of 1. It is possible of course to manipulate this correlation so that it is less than perfect, and even, in some cases, negative. Reducing the 'predictiveness' of the CS can be done in two ways: (1) by failing to present the UCS after some of the CS presentations, or (2) by presenting 'surplus' UCSs during the inter-trial intervals (ITIs). Reduction of the correlation from 1.0 usually results in weaker conditioning in terms of amplitude of the response when this is measurable (Brogden, 1939c; Fitzgerald, 1963), and acquisition of asymptotic performance takes a greater number of trials (Leonard and Theios, 1967). The importance of this finding is that it is not the absolute number of occasions on which the CS and UCS are *paired* that is important to learning, but the *predictive significance* of the CS. It is not simply enough to continuously associate CS and UCS, it appears to be necessary that in the experimental situation the CS be the most reliable 'predictor' of the UCS. Thus the better the CS-UCS correlation, the more rapid will be CR acquisition.

### Stimulus generalisation and discrimination

If a dog is trained to salivate in the presence of a 1000-Hz tone and subsequently on test trials tones of different frequencies are substituted for this CS, the strength of the CR will diminish in proportion to the difference between the new test CS and the original training CS. From studies of this kind 'generalisation curves' can be plotted, an example of which is shown in figure 2.4. Although most generalisation curves are of the 'concave upward' variety as in figure 2.4, other types have been found (for example, variation of the stimulus along the dimension of 'intensity' produces a linear fucnction; see chapter 3, pp. 111–112).

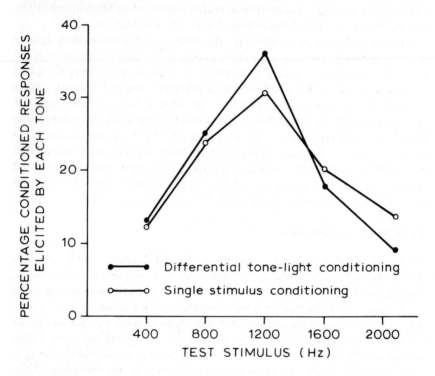

Figure 2.4    Examples of generalisation curves in classical conditioning. The original CS was a tone of 1200 Hz which through pairing with paraorbital shock in rabbits had come to elicit a reliable nictitating membrane response. Subsequent presentation of tones of varying frequency produced a bell-shaped function with the strength of the conditioned response diminishing with the distance of the test stimulus from the original CS. (After Liu, 1971).

Generalisation is an extremely basic and reliable conditioning phenomenon which can be found with the widest variety of responses and stimuli. Its adaptive significance is fairly obvious – unless there were such a process as stimulus generalisation new behaviours would rarely occur other than in the restricted conditions in which they were acquired. To illustrate this further, Bolles points out:

Even in a Pavlovian experiment in which extraneous stimulation and inadvertent variation among stimulus conditions are minimized, there still has to be some variation in the total pattern of stimulation from one trial to the next. It may make some difference whether the animal is inhaling or exhaling at the moment the CS comes on. But the important factor is not that the stimulus situation is slightly different from trial to trial, but that it is largely similar from trial to trial, so that the learning that occurs on one trial can be manifest on the subsequent trial.

(Bolles, 1975, pp. 32–33)

In the natural environment where stimulus conditions are likely to differ even more from 'trial to trial' generalisation provides a useful method for transferring adaptive behaviours from situation to situation.

Discrimination, on the other hand, can be considered as the opposite to generalisation. During discrimination training, one CS is 'reinforced' (that is, paired with a UCS) and a second different CS is not reinforced. The reinforced CS is known as a CS + (sometimes simply called S + ) and the nonreinforced CS as a CS − (sometimes simply called S − ); these two stimuli can be qualitatively different, for example, a tone and a light, or be qualitatively similar but differ merely among a single dimension such as frequency or intensity. Although the CS − will initially elicit a conditioned response, continued nonreinforcement will extinguish this until a CR is elicited only on presentation of the CS + . Differential responding to the two stimuli is taken as an indication that a discrimination has been formed. Like stimulus generalisation, this process is fundamental to learning. For instance, discrimination must occur in almost all conditioning experiments where the animal has to distinguish between stimuli important to learning and extraneous background stimuli. This can be illustrated by the fact that during the classical conditioning of salivation in the dog, salivation occurs often during inter-trial intervals in the early part of training (Sheffield, 1965). Subsequent 'non-reinforcement' of these background stimuli eventually extinguishes salivation during inter-trial intervals until only the experimental CS (which is the only stimulus reliably paired with the UCS) elicits a CR.

However, the processes of generalisation and discrimination are not in reality as straightforward as they have been painted here. Both processes produce behavioural phenomena interesting to the learning theorist and these will be discussed in more detail in chapter 3.

## The nature of the conditioned response

So far, we have talked about the CR in classical conditioning experiments as being a relatively faithful facsimile of the UCR. Although this is usually the case the CS often comes to elicit a wider range of behavioural changes. For example, when using an appetitive UCS such as food, the CS will generally elicit increased levels of activity as well as more discrete responses such as salivation or orientation (Holland, 1977; Konorski, 1967); aversive UCSs such as electric shock will similarly result in the acquisition of 'emotional' responses in the presence of the CS – such responses include changes in heartrate, freezing, defecating, etc. (Black, 1971; Borgealt, Donahoe & Weinstein, 1972). These responses can be considered as being 'preparatory' rather than 'consummatory' in nature, and they have led some theorists to suggest that the adaptive significance of classical conditioning is that it allows the animal to 'prepare' itself generally to receive the UCS (Zener, 1937). The range of responses elicited by a CS will depend on a number of factors related to both the conditioning procedure and the nature of the stimuli used. If the animal is restrained during conditioning, thus prohibiting the appearance of UCRs typical of the UCS (for example, UCS orientation and approach), then the range of CRs elicited by the CS is liable to be likewise limited. On the other hand, in some studies where the full range of UCRs to the UCS have been restricted, a 'substitution' of responses has been observed. For example, when a rat in a confined chamber is shocked to the feet (UCS), the typical UCR is jumping and prancing. However, the CR to a stimulus signalling this shock is usually freezing. Both jumping and freezing are known to be UCRs to electric shock but in the experimental situation one is elicited by the UCS and the other by the CS. Thus, the CS does not simply come to elicit only those responses which the UCS elicits during training, but can, under certain conditions, come to elicit CRs which did not appear during training and which appear to be part of the animal's normal repertoire in response to the UCS. The nature of the CR in classical conditioning has posed many questions for theorists, it is most often a simple facsimile of the UCR, but also can be of a more preparatory or anticipatory nature. Fuller account will be taken of these questions in chapter 7.

**Sensitisation and pseudo-conditioning**

Although presentation of the UCS prior to conditioning usually results in adaptation (see p. 31), the opposite effect can sometimes occur. Sensitisation refers to such occasions when a CS comes to elicit responses without having been paired with a UCS; prior exposure to the UCS can often bring about this effect. For example, Harris (1943) measured the activity levels of rats in response to electric shock when this shock was signalled by a buzzer. The buzzer produced increased activity levels when presented alone prior to conditioning, but produced even greater activity increases in a group of rats that had just previously been given a number of electric shocks. Prior presentation of the UCS caused the CS to elicit behavioural changes even prior to conditioning.

Pseudoconditioning is a phenomenon closely related to sensitisation but is characterised by the elicitation by the CS of the actual UCR to be conditioned. For example, Grether (1938) frightened monkeys on a number of occasions with a powder flash. Following this a bell was sounded, again on several occasions, and the bell reliably elicited fright reactions which it had not done prior to the presentation of the powder flashes. Pseudoconditioning is largely confined to situations involving aversive stimuli and appears to reflect the possibility that exposure to aversive stimuli eventually activates the defence reactions of the animal to most subsequent novel stimuli. If this is the case then it can be considered not as a malfunction or by-product of the conditioning process, but as an important defensive mechanism valuable to the animal's survival (Wells, 1968; also chapter 6, p. 214).

Nevertheless, the possibility that pseudoconditioning is occurring in classical conditioning experiments must be considered by the experimenter. Two main control procedures can be adopted to assess this possibility: (1) the CS can be presented alone on a number of trials prior to conditioning, and the animal's reaction to it assessed, and (2) a group of subjects can be given independent presentation of CS and UCS on different trials; elicitation of a CR by the CS in this group will suggest that responding in the experimental group (who receive CS and UCS on the same trial) is not necessarily a result of pairing the CS and UCS (Rescorla, 1967). This latter condition is now considered as a most valuable and necessary control procedure in the majority of classical conditioning studies and it is known as the truly random control.

## Inhibition of responding

So far we have talked almost exclusively about excitatory effects that occur during classical conditioning; that is, situations where new responses are formed or existing responses strengthened. However, there are a number of conditions which produce inhibition of responding, that is, either a reduction in response strength to the CS or a retardation of the learning process. (In this context inhibition is being considered merely as an observable behavioural effect, that is, the reduction of response strength or the blocking of the acquisition of a response as a result of some experimental operation.) It is probably wrong to consider all the following phenomena as reflecting properties of a single underlying inhibitory mechanism, but all illustrate ways in which the acquisition process can be 'blocked' or retarded.

### Adaptation

Presentation of the UCS alone prior to conditioning can lead to reduced response strength when the UCS is subsequently paired with a CS. For example, if human subjects are given 50 presentations of electric shock to a finger (UCS), subsequent classical conditioning of the finger retraction response is less effective than if no prior adaptation trials are given (MacDonald, 1946).

### External inhibition

When a novel stimulus is presented during the course of conditioning, the CR on that trial can often be disrupted or fail to appear. Pavlov himself considered this merely to be the result of orientation reflexes competing with the conditioned response (Pavlov, 1927, p. 44). Continued presentation of the novel stimulus results in it rapidly losing this inhibitory effect. However, novel stimuli can also *increase* response strength: when a novel stimulus is presented during the extinction of a CR, this can often produce a fleeting increase in response strength which is known as *disinhibition*.

### Inhibitory conditioning

When there is a negative correlation between a CS and UCS (that is, the CS predicts non-reinforcement and is therefore a CS −), the CS is often found to acquire inhibitory properties. It is unlikely that it is merely being ignored by the animal since it is extremely difficult to subsequently condition that CS as a stimulus that reliably elicits a CR (Konorski and Szwejkowska, 1952; Marchant, Mis and Moore, 1972). Similarly, presentation of such a CS −, either during the performance of a CR or in compound with a CS +, will usually reduce the strength of the CR, suggesting that the CS − has actually acquired response suppressing properties (Pavlov, 1927, p. 76; Rescorla, 1966). The importance of this is that animals will not only learn to respond when an event reliably predicts reinforcement, but will also learn to withold the response when an event reliably predicts non-reinforcement. They do not simply learn to ignore the stimulus but learn to inhibit responding in its presence.

### Latent inhibition

If an animal is pre-exposed to a CS for a number of trials before conditioning commences, the acquisition of a CR to that stimulus will be retarded. This could be due to (1) the animal learning to ignore a stimulus that in effect signals nothing of significance, or (2) the CS acquiring suppressive properties as a result of it signalling non-reinforcement prior to the conditioning trials. The latter is unlikely since presentation of such a CS in conjunction with an already established CS + fails to suppress the CR to the CS +. The most likely explanation of latent inhibition is that the animal has simply learnt to ignore, or ceased to attend to, a stimulus which has no important consequences. Subsequent failure to attend to it during conditioning trials will obviously retard response acquisition.

### Inhibition of the unconditioned response (UCR)

Although pairing of the CS and UCS usually produces conditioning, on some occasions it can also have the effect of suppressing the response to the UCS and sensitising the response to the CS. For example, Kimble and Ost (1961, in Kimble, 1961) found that when a light CS and a puff of air into the eye (UCS) were paired, there was a sudden decrease in the magnitude of the unconditioned response. When the CS was removed from the pairing, the UCS evoked a UCR of original magnitude.

*Overshadowing and blocking*

The strength of conditioning to a particular stimulus crucially depends on whether it is presented alone or in a compound. Pavlov found that when a compound CS, consisting of auditory and visual stimuli, was conditioned and subjects subsequently tested with each of the individual components in isolation, only one of the components elicited a CR (Pavlov, 1927, p. 142). The acquisition of control of the CR by only one component of a compound CS appears to depend on such factors as the relative strengths or intensities of the components, and even the predictive value of the components: if one component predicts reinforcement more accurately than the other, then that is the one which will acquire control of the CR and 'overshadow' all others (Wagner, 1969). Blocking is a similar phenomenon to overshadowing and was first observed by Kamin (1969). If a light is established as a CS + and this is then compounded with a tone for a further number of reinforced trials, subsequent presentation of the light and tone alone will reveal that CRs are occurring only in response to the light and not the tone. That is, prior conditioning with the light prevented the tone from being established as an effective CS + when the light and tone were presented together: prior conditioning to the light 'blocked' conditioning to the tone. The actual reasons why blocking occurs are difficult to assess at present. However, one possible explanation suggested by Kamin (1969) and others is that animals do not condition to new stimuli when these stimuli provide no new information about the learning situation (as is the case with the tone in the above example). However, if, when the light and tone are initially paired, the nature of the UCS is altered slightly (either by manipulating its intensity, or presenting some event simultaneously with it) the blocking does not occur (Gray and Appignanesi, 1973).

# OPERANT CONDITIONING

Before the publication of Skinner's influential book *The Behaviour of Organisms* (1938), it was generally believed that the majority of learned behaviour came about as a result of association of responses with prior events known as stimuli. The laws which governed the bonding of stimuli and responses were the principles of classical conditioning that we have just discussed, and stimuli were considered to elicit responses in a reflexive fashion. Skinner's book was responsible for a number of innovations in the way learning theorists viewed the learning process. He emphasised the distinction between classical and operant conditioning: that is, a distinction between behaviour which is controlled by its antecedents and that behaviour which is controlled by its consequences. Skinner writes:

> The kind of behaviour that is correlated with specific eliciting stimuli may be called *respondent* behaviour and a given correlation a *respondent*. The term is intended to carry the sense of a relation to a prior event. Such behaviour as is not under this kind of control I shall call *operant* and any specific example an *operant*. The term refers to a posterior event. . . . An operant is an identifiable part of behaviour, which it may be said, not that no stimulus can be found that will elicit it, (there may be a respondent the response of which has the same topography), but that no correlated stimulus can be detected upon occasions when it is observed to occur. It is studied as an event appearing spontaneously with a given frequency.
>
> (Skinner, 1938, pp. 20–21)

Thus respondent conditioning corresponded to traditional classical conditioning where a response was 'elicited' by its controlling stimulus, operant conditioning referred to the conditioning of more 'voluntary' behaviours which appeared to be 'emitted' by the animal, and were not obviously controlled by any one identifiable stimulus. Instead of being controlled by antecedent stimuli, the frequency of operant behaviour was controlled by its consequences; 'reinforcing' consequences increased their frequency, 'punishing' consequences reduced their frequency. Hence the name *operant conditioning* since the behaviour 'operates' on the environment to produce a change in that environment. (This type of conditioning is also known as 'instrumental' conditioning since the behaviour is instrumental in obtaining rewards and punishments.) Although it is perhaps difficult to maintain the operant-classical conditioning dichotomy on the basis of the distinction originally formulated by Skinner (see p. 52ff.), this original formulation did have a number of valuable results. (1) In the mid-1930s learning had become to all intents and purposes synonymous with classical conditioning and all the involuntary-reflexive connotations that that procedure entailed. It was becoming increasingly difficult to talk of complex patterns of learned behaviour in these terms and operant conditioning provided a more viable framework for analysis of these behaviours. (2) Operant conditioning allowed behaviour to be studied as it occurs in a free-flowing context. Local changes in rate and frequency of behaviour could be studied more readily than in the discrete trial method that was necessary with classical conditioning. (3) Skinner revived Thorndikean ideas concerning the importance of the consequences of behaviour. The notion of operant reinforcement was essentially a more hard-headed interpretation of Thorndike's law of effect.

## Basic concepts

The operant conditioning paradigm can be represented in what is known as a three-term contingency:

discriminative stimulus     :     Operant response $\longrightarrow$ reinforcing stimulus

$$S^D \qquad : \qquad R \longrightarrow S^R$$

The operant response is emitted and is reinforced in the presence of a discriminative stimulus. It is perhaps best to talk about these three notions individually.

### Reinforcement/punishment ($S^R$)

Reinforcers and punishers are the consequences of behaviour which come to affect the subsequent frequency of behaviour. In Skinnerian terms a reinforcer is defined entirely on the basis of its effect on behaviour such that a reinforcing stimulus is 'any stimulus which, when made contingent upon a class of behaviours, increases the future probability of the occurrence of a member of that class of behaviours'. So there are no classes of events which can be said to be intrinsically reinforcing, reinforcers are identified only on the basis of their observed effect on behaviour. Although this may seem an unnecessarily *post hoc* definition, it is useful in a number of ways. (1) It does not allow the theorist to slip into the quagmire of defining reinforcers *a priori*; for example, although animals will usually respond to obtain biologically important events (such as a hungry animal requiring food) there are still important conditions under which they will not do this (see chapter 5). So because of the inevitable exceptions to the rule, to label 'food for a hungry animal' as intrinsically reinforcing provides little or no insight into the workings of the learning process. (2) Labelling certain events as being intrinsically reinforcing leads to the question of how reinforcers reinforce. This question has again proved to be something of a sterile one for learning theorists who have tackled it. Skinner, however, in defining reinforcement empirically has avoided this debate – when it is defined functionally, reinforcement becomes merely a *procedure* for controlling behaviour. Once behaviour can be controlled it can be systematically studied and the principles which govern the frequency and patterning of behaviour more easily formulated. Skinner's emphasis on the use of schedules of reinforcement in studying behaviour rather than on study of the reinforcement process itself is an illustration of this approach and will be dealt with in more detail in the following chapter.

In contrast to reinforcers, punishers are those events which, when made contingent upon a class of behaviours, decrease the probability of an instance of that behaviour class reoccurring. Both reinforcers and punishers can be either positive or negative, these adjectives referring to the nature of the consequence. If the operant response *adds* something to the situation (for example, produces a food pellet) then this is a positive event; if the response *removes* something from the situation (for example, switches off an electric shock) then it is a negative event. Therefore, positive reinforcement would be a situation where a class of behaviours is increased in frequency (reinforcement) by an instance of that class of behaviours producing a stimulus event. Reinforcement/punishment refers to the direction of change in the frequency of behaviour, positive/negative refers to the nature of the stimulus change following behaviour. This distinction in terminology is illustrated more fully in figure 2.5. (The stimulus change resulting in an increase in frequency of the operant response is usually known as the 'reinforcer' or the 'reinforcing stimulus'. The term 'reinforcement', however, is usually reserved to describe the increase in frequency brought about by the contingency between response and consequence, that is, it describes a process rather than an event.)

|  | **REINFORCEMENT** | **PUNISHMENT** |
|---|---|---|
| **POSITIVE** Responding adds an environmental event | Receiving a food pellet for pressing a lever RESULT : Response rate increases | Receiving shock for pressing a lever RESULT : Response rate decreases |
| **NEGATIVE** Responding removes an existing part of the environment | Pressing a lever serves to avoid or escape shock RESULT : Response rate increases | Pressing a lever produces time-out from food RESULT : Response rate decreases |

Figure 2.5   Reinforcement and punishment as procedures in operant conditioning. Each cell contains a typical example of positive and negative reinforcement and positive and negative punishment. See the text for further explanation.

## The operant

The operant has again been defined empirically by Skinner as 'a class of behaviours on which a reinforcer is made contingent'. Although this avoids the problem of topographically defining a response class (for example, a rat can press a lever with a paw, with its tail, or with its nose), and also emphasises the functional nature of behaviour in operating on the environment, it does raise problems concerning the logical status of Skinner's definitions. That is, a reinforcer increases the future probability of an operant, but an operant is a class of behaviours on which a reinforcer is made contingent. The definitions are circular. Apart from this problem of circularity, questions also arise concerning the status of the 'operant' when considering such phenomena as superstition (see p. 42), extinction, and transfer of learning. If an operant is defined only in terms of its contingent relationship to a reinforcer, then how do we talk about classes of responses when there is no response-reinforcer contingency (as is the case with superstitious reinforcement) or when there is no reinforcer (as during extinction)? Clearly the problem is more of a logical than practical one. Even though we might have no water-tight taxonomy of operant behaviour, its relationship to its consequences, and the effect of these consequences on future instances of that class, study of conditioning processes can still carry on relatively unhindered. However, the interested reader is referred to Schick (1971) for a critical review of the notion of the operant.

## Discriminative stimulus

Although in his initial account of operant conditioning Skinner emphasises that it is not necessary to account for stimuli which precede the response, he does note that environmental stimuli can come to exert some control over the emission of operant responses. If, for example, a hungry rat is reinforced with food pellets for pressing a lever in the presence of a bright light, but never reinforced for pressing the lever in darkness, then the light eventually comes to control responding: lever-presses occur only when the light is on. Thus, the light has become a *discriminative stimulus ($S^D$)* for responding. In Skinner's terms a discriminative stimulus differs from the conditioned stimulus (CS) of classical conditioning experiments because the former does not reflexively 'elicit' responses but acts as a signal 'in the presence of which' responses are emitted. Although there are many similarities between the way in which conditioned and discriminative stimuli affect behaviour the difference being emphasised here is in the strictness with which the stimulus controls each individual occurrence of the response.

# Basic apparatus

The majority of the early work relating to operant or instrumental conditioning was carried out on either the laboratory rat (Skinner, 1938) or the pigeon

(Ferster and Skinner, 1957). This was mainly because these animals were easy to breed, easy to keep and, in the case of the pigeon, emitted a response (pecking) which was fairly easy to record and which the animal could emit for long periods without fatigue. The wisdom of using such a limited number of species in studying phenomena which were only assumed to be universal is equivocal (this issue is covered more fully in chapter 5). However, rats and pigeons are still convenient organisms for the study of many learning phenomena, and we shall come across them very many times in this book. It is therefore fairly important that at this early stage the reader should be familiar with the standard rat and pigeon chambers used in operant conditioning studies. Figures 2.6 and 2.7 show the interiors of a rat conditioning chamber and a pigeon conditioning chamber (the former is also called a 'Skinner-box' after its inventor). Both contain a means of delivering food to the animal: food is usually a small food pellet or a small amount of sucrose solution for the rat, and a limited period of access to mixed grain for the pigeon. Both chambers contain one or more response manipulanda: a lever conveniently placed for the rat to press, or a pecking-key (a small illuminated disc) usually placed at head-height for the pigeon to peck. A grid floor in the Skinner-box allows electric shock to be presented when required to the feet of the rat, and both rat

Figure 2.6  A rat conditioning chamber, normally called a Skinner-box after its inventor. A small lever acts as the operant manipulandum, and the rat receives food pellets in a small aperture next to this lever. The lamps on the front wall can be programmed as discriminative stimuli and mild electric shocks can be delivered through the grid floor to the rat's feet.

Figure 2.7   A pigeon conditioning chamber. In this particular chamber three pecking keys are mounted at head-height on the front wall; stimulus patterns can be projected onto each key, or, as is more commonly the case, they can be illuminated with coloured light. The aperture below the centre key is where the pigeon receives limited periods of access to grain.

and pigeon chambers provide a means for presenting both visual and auditory stimuli. Stimulus lamps are usually placed in strategic positions on the front wall of the Skinner-box, whereas visual stimuli for the pigeon are projected on to the pecking-key itself.

Another specialised piece of equipment used by operant conditioners is the *cumulative recorder*. This is a modified polygraph recorder which enables the experimenter to assess at a glance both local and overall rates of responding (see figure 2.8). Paper is continually fed out from the recorder at a regular speed; this axis represents time. Each response the animal makes on the manipulandum steps the pen one unit upward. As you can deduce from this, if the animal responds quite frequently the slope of the resulting graph will be much steeper than if it only responds occasionally. Thus, the experimenter can

Figure 2.8   A cumulative recorder which is used for plotting the local fluctuations in the rate and patterning of operant responding. See text for further explanation.

look closely at local rates of responding, perhaps in relation to events such as the delivery of food. Food delivery is indicated on a cumulative record by a small downward 'hatch' mark (see figure 2.9). Thus the cumulative record provides a detailed history of the animal's responding throughout an experimental session. Moreover, as we shall see in the next chapter, different schedules of reinforcement produce different patterns of responding which are characterised by distinctive cumulative records.

## Some common behavioural phenomena

### Response shaping

The basic tenet of operant conditioning states that if a reinforcer (such as food for a hungry animal) is presented following the occurrence of a particular behaviour, then instances of that class of behaviour will occur more frequently in the future. From this it is obvious that a particular behaviour must first occur for some unspecified reason before its frequency can be controlled by reinforcement. When a particular behaviour is commonplace in an animal's repertoire the problem of its initial occurrence is not an important one: the experimenter merely waits for it to occur and then reinforces it. However, when the experimenter wishes to reinforce behaviour patterns which are rare in an animal, or which are particularly complex, the problem of the first occurrence of such behaviours is a salient one. For example, the circus trainer

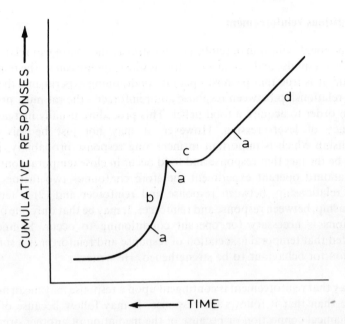

Figure 2.9   An example of a cumulative recording. The record moves from left to right with the subject's rate of responding gradually increasing until response rate is relatively high (point 'b'). At point 'c' the subject abruptly ceases responding for a period of time until responding is resumed at a steady intermediate rate (point 'd'). The three 'hatch-marks' designated 'a' represent the points at which a reinforcer was delivered.

would be waiting a lifetime before a horse would walk a dozen or so paces on its hind legs, or a dog would jump spontaneously through a hoop of fire. Instead of waiting indefinitely for the first occurrence of a particular behaviour, the way round this problem is to reinforce *successive approximations* to the desired act. For instance, to speed up the acquisition of lever-pressing in the rat, approximations to this behaviour can be successively reinforced. First, the animal is given a food pellet for going to the part of the chamber where the lever is situated – a behaviour which has a high probability of occurring. When the rat is consistently returning to this part of the box, the response requirement is made stricter by specifying that he should sniff the lever, and so on, step by step, until he is eventually touching and then depressing the lever regularly. Finally, reinforcement is provided only for successful lever presses. It can be envisaged that quite complex patterns of behaviour can be shaped relatively quickly using this method, which combines a knowledge of the principle of operant reinforcement with the fact that even the most strictly controlled behaviour will vary in topography from occurrence to occurrence.

### Superstitious reinforcement

The functional definition of reinforcement stresses the contingent relationship between a response and a reinforcer; that is, the response *causes* the reinforcer to occur. It is true that in most operant conditioning experiments there is a causal relationship between response and reinforcer – the rat must press the lever in order to acquire a food pellet. This procedure usually increases the frequency of lever-pressing. However, it may not just be this causal relationship which is important in increasing response probability, it may simply be the fact that response and food occur in close temporal contiguity. The standard operant experiment therefore confounds two factors, (1) a causal relationship, between response and reinforcer and (2) a temporal relationship, between response and reinforcer. It may be that only one of these conditions is necessary for operant conditioning to occur. Skinner has suggested that temporal association of response and reinforcer is a sufficient condition for behaviour to be strengthened. He writes:

> To say that reinforcement is contingent upon a response may mean nothing more than that it follows the response. It may follow because of some mechanical connection or because of the mediation of another organism; but conditioning takes place presumably because of the temporal relation only, expressed in terms of the order and proximity of response and reinforcement. Whenever we present a state of affairs which is known to be reinforcing at a given level of deprivation, we must suppose that conditioning takes place even though we have paid no attention to the behaviour of the organism in making the presentation.
>
> (Skinner, 1948)

Skinner (1948) attempts to support his arguments with a simple, but effective, experiment. Food was delivered to a pigeon periodically every 15 s, regardless of what the pigeon was doing at the time. After a number of food presentations, two independent observers were agreed that six out of eight subjects were exhibiting stereotyped behaviours throughout the interval between reinforcers. One bird would turn counter-clockwise about the chamber, another repeatedly thrust its head into one of the corners of the chamber, and a third developed a response which consisted of tossing its head back and forth beneath a seemingly 'invisible' bar. Skinner has suggested that these particular behaviours were strengthened as a result of their accidental correlation with food delivery. That is, even though there is no causal relationship between behaviour and food, the subject still has to be doing something at the time of food delivery and it is this behaviour that is more likely to occur in the future. Although this effect is known popularly as 'superstitious reinforcement', it is best to think of it not as a result of the animal perceiving a causal relationship where none exists, but as stressing the importance in conditioning of the contiguous temporal relationship between behaviour and consequences. Nevertheless, although superstitious reinforcement can be demonstrated quite reliably (Blackman, 1974, pp. 17–29), it appears to play only a transitory role in shaping adaptive behaviour. An extended replication of Skinner's original experiment has suggested that, although superstitious behaviours occur in the initial stages of non-contingent reinforcement, behaviour patterns soon drift towards behaviours characteristic of food-getting in the organism being conditioned (Staddon and Simmelhag, 1971). Although stereotyped behaviour patterns can still be observed at this later stage, their maintenance (in pigeons at any rate) cannot readily be explained in terms of accidental correlations between behaviour and food (see chapter 5, p. 163ff.).

## Intermittent reinforcement (schedules of reinforcement)

In operant conditioning it is certainly not necessary to reinforce every instance of a response class in order to maintain responding. In fact, stronger conditioning and more interesting data are obtained when only a selected few of the responses are reinforced. The responses one chooses to reinforce are selected on the basis of a set of rules known as *schedules of reinforcement*. It is intended to discuss schedules of reinforcement in detail in the following chapter, but it should be mentioned here that there are two basic methods of selecting responses to reinforce, (1) on the basis of time (interval schedules) and (2) on the basis of number of responses (ratio schedules). The construction of such schedules is schematised in figure 2.10 and discussed more fully on p. 62ff. of chapter 3. The importance of the intermittent reinforcement procedure rests on the facts that first, each schedule of reinforcement produces its own characteristic pattern of responding; this tells us something about the way the organism adapts to reinforcement contingencies, and secondly, response rates on intermittent reinforcement schedules are usually higher than with continuous reinforcement; this gives a sound baseline level of responding on which the effects of other variables (for example, drugs) can be assessed (see chapter 3, p. 88ff.).

## Extinction

Extinction in operant conditioning is the procedure of witholding the reinforcer after the emission of responses that had previously been selected for reinforcement (see figure 2.11). This produces a gradual decline in response frequency with the rate of this decline depending on the schedule of reinforcement previously in operation. The more predictable that the occurrence of the reinforcer was during training, the more rapidly responding will extinguish when reinforcement is withheld: that is, if reinforcement were given for every response during training then extinction would be quicker than

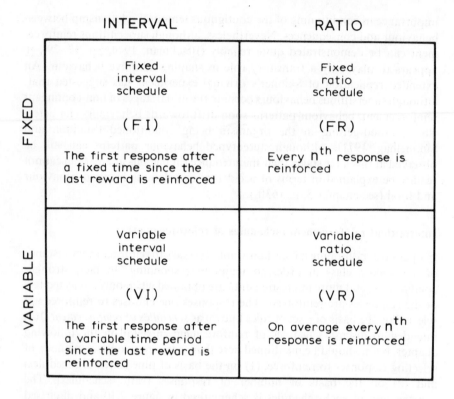

Figure 2.10  The four basic schedules of reinforcement. Reinforcers can be delivered either on the basis of the number of responses the subject has emitted, or on the basis of the amount of time that has elapsed since the last reinforcer. Note that each of these schedules is response-dependent, that is, the reinforcer is delivered only on the occurrence of a response.

if food occurred periodically for, say, the first response emitted every 60 s (an FI60-s schedule). Similarly, if the time between food deliveries were made even more unpredictable by varying the time between each food availability, extinction of the response would procede more slowly than in the fixed interval case. This relationship appears to reflect the certainty with which the animal can distinguish between the stimulus conditions during training and the stimulus conditions during extinction. When, during training, every response is reinforced, food occurs regularly, and at clearly discriminable times (that is, immediately after the subject has responded). When reinforcement has been discontinued, the change in conditions is fairly readily distinguishable. However, when training on an intermittent food schedule has preceded extinction, the change from reinforcement to extinction is less readily discriminable. During training, many responses may have had no consequences; therefore, when responding has no effect during extinction, the

# EXTINCTION OF AN OPERANT RESPONSE

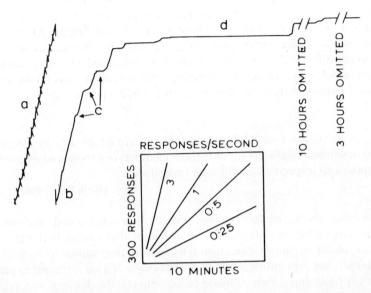

Figure 2.11  After being trained to key-peck for food on a FR60 schedule, responding in this pigeon was subsequently extinguished. Point 'a' represents the cumulative record for responding on the FR schedule and point 'b' designates the introduction of extinction. Points marked 'c' show the first signs of pausing and at point 'd' responding has almost ceased completely. (After Ferster and Skinner, 1957, p. 60).

conditions will seem little different from those earlier. So it appears that the speed with which a response will extinguish may depend crucially on whether the animal can discriminate the difference between conditions during training and conditions during extinction. If this difference is enhanced by providing external stimulus changes during extinction (for example, by switching off the chamber light, or sounding a continuous tone), then the learned response extinguishes much more rapidly than if no external stimulus change was made. Thus, contrary to the first hypothesis intuition might lead us to make about extinction, the opposite is in fact true! That is, it does not follow that the greater the probability of reinforcement for a response, the greater the resistance to extinction of that response. So perplexing was this finding that it was originally labelled 'Humphrey's paradox' after L. G. Humphrey, who was the first to demonstrate the effect. However, instead of being paradoxical, it appears that rather than simply reflecting strength of conditioning, the rate of extinction of an operant response more likely reflects the subject's ability to discriminate between conditions of reinforcement and conditions of extinction.

### Conditioned reinforcement

It had traditionally been popular to divide reinforcing stimuli into two broad categories: (1) primary reinforcers, and (2) conditioned or secondary reinforcers. Primary reinforcers are considered to be those events which are of immediate biological importance to the organism, for example, food, water, sex, or painful stimuli such as electric shock. Conditioned reinforcers are those events which come to act as reinforcers through their association with a primary reinforcer. Kelleher defines a conditioned reinforcer quite succinctly as:

> . . . a stimulus whose reinforcing properties are established by conditioning; it will be a reinforcer for only those members of a species who have been exposed to a specific conditioning procedure.
>
> (Kelleher, 1966, p. 162)

Although the distinction between primary and conditioned reinforcers is not a major issue nowadays, there is no doubt that stimuli that were once neutral to an organism can, through conditioning, come to control the behaviour they are contingent upon. For example, if a rat is trained to press a lever for food and a tone is made to accompany the delivery of food, the animal will subsequently press the lever to produce only the tone. The fact that the tone is acting as a reinforcer can be demonstrated in two ways: (1) by presenting the tone following lever presses during extinction. Animals that receive the contingent tone presentation during extinction continue responding for longer than the animals that do not receive the tone or have not had the tone paired with food during training; (2) by presenting the tone contingent upon the performance of a new response, for example, pressing a nose-key instead of a lever. However, the durability of conditioned reinforcers is poor without intermittent association with 'primary' reinforcers; they certainly never achieve the powerful control over behaviour that primary reinforcers do, but they can help maintain response patterns by bridging a temporal gap between behaviour and a primary reinforcer. This should become clearer as we discuss some of the procedures in which responding can be maintained by conditioned reinforcement, and also some early theories of conditioned reinforcement.

Perhaps the most impressive demonstration of the effect of conditioned reinforcement can be demonstrated with the use of chaining techniques. If a stimulus $(S_1)$ is a discriminative stimulus for responding $(R_1)$ to acquire food, then a second response $(R_2)$ can be developed and maintained if that new response merely produces $S_1$. This procedure can be replicated again so that a third response $(R_3)$ produces $S_2$, and so on as in figure 2.12.

Complex chains of behaviour such as this can be developed and maintained as long as the chain is terminated by a primary reinforcer. The length and

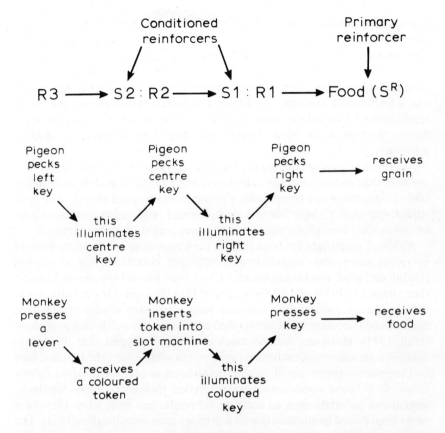

Figure 2.12   A schematic representation of a chain schedule of reinforcement with examples using first a pigeon and secondly a monkey as subjects. In the first example the changes in key-colour act as conditioned reinforcers for the pigeon and in the second example the coloured token and illuminated key act as conditioned reinforcers for the monkey.

complexity of such chains can be illustrated by a study carried out by Pierrel and Sherman (1963). They trained a rat, called Barnabus, to 'climb a spiral staircase, mount a platform, climb an open ladder, jump into a vehicle which he propelled by turning a paddle-wheel to the foot of an additional flight of steps, which he also ascended. At the top the rat ran through a tube, clambered into a model lift, the descent of which was started by raising a flag; when the lift reached the lowest level the rat jumped out, and approached and pressed a lever, thereby receiving a pellet of food; whereupon, he set off on this journey once more!' (Blackman, 1974, p. 97). Although there will obviously be a limit to the length of such chains, this example does clearly illustrate how animals will respond to 'switch on' discriminative stimuli which eventually lead to primary reinforcers. The Pierrel and Sherman example is known as a *heterogeneous* chain, since each $S^D$ maintains a different response. How-ever, chains of responding can be maintained where the response is the same in each of the different components. Such *homogeneous* chains can be found in what are known as *second-order schedules of reinforcement*. Here, food plus a brief stimulus may be presented for lever-pressing on the basis of one schedule (for example, an FI5-min schedule) while during the inter-reinforcement interval the brief stimulus may be presented following lever presses on the basis of a diferent schedule (for example, an FI30-s schedule).

At the risk of oversimplifying the results of this kind of schedule, suffice it to say here that overall response rate during the FI5-min schedule with added tone is higher than in a comparable FI5-min with no added stimuli. The brief stimulus appears to have become a conditioned reinforcer which maintains lever-pressing throughout the whole FI5-min period (Kelleher, 1966).

Although much light has been shed on the nature of conditioned reinforcers in recent years, the necessary and sufficient conditions for producing conditioned reinforcers is still unclear. Clark Hull, one of the eminent learning theorists of the 1930s and 1940s suggested that the important operation was the contiguous pairing of a stimulus with a primary reinforcer. That is, conditioned reinforcers were established by a classical conditioning procedure (Hull, 1943). However, there is much evidence to suggest that this is too simplistic an account. Conditioned reinforcers when established do not take on the essential properties of primary reinforcers as one might expect from a simplistic classical conditioning interpretation (Schuster, 1969). Similarly, stimuli can be established as conditioned reinforcers even when they have never been paired contiguously with a primary reinforcer (Stubbs, 1971). The alternative traditional view of conditioned reinforcers is that originally propounded by Skinner (1938) who suggested that only discriminative stimuli ($S^D$s) could be established as conditioned reinforcers. In the initial example of the tone being paired with food and the tone subsequently supporting lever-pressing during extinction, Skinner claims that the tone comes to act as a discriminative stimulus for leaving the response lever and approaching and eating food; that is, lever-pressing is maintained by the tone (a conditioned reinforcer) which is in turn a discriminative stimulus for a different pattern of behaviour – that of approaching the food tray. The fact that complex chains of behaviours can be established using conditioned reinforcers is evidence that discriminative stimuli do function as conditioned reinforcers. However, whether all conditioned reinforcers must also by necessity be $S^D$s for responding is open to question.

A more recent theory of conditioned reinforcement which has enjoyed popularity throughout the 1960s is that which alludes to conditioned reinforcers providing the animal with *information*. Egger and Miller (1962, 1963) carried out an experiment which suggests that the more informative a stimulus is with regard to predicting the delivery of primary reinforcers, the more likely it is to become a conditioned reinforcer. If two stimuli, $S_1$ (duration 2 s) and $S_2$ (duration 1.5 s), are presented immediately prior to food delivery, then subsequent tests reveal that $S_1$ becomes the more powerful conditioned reinforcer. On the basis of information content, $S_2$ is redundant, it tells the animal no more, in fact less, than the longer stimulus, $S_1$. However, if extra presentations of $S_1$ are made between food deliveries (that is, it becomes a less reliable predictor of food), then $S_2$ becomes the more powerful conditioned reinforcer and the conditioned reinforcing abilities of $S_1$ are considerably reduced. Subsequent experiments have demonstrated that animals will work to provide themselves with more information about the reinforcement contingencies in operation (Bower, McLean and Meacham, 1966; Steiner, 1967; cf. Hendry, 1969).

Although there is still much speculation concerning the nature of conditioned reinforcers, there is no doubt that they are an important method of maintaining behaviours in the immediate absence of primary reinforcement. This is especially so when one considers human behaviour. So little of our own behaviour is followed immediately by such events as food, water or sex, but generalised conditioned reinforcers such as money, cheques, affection, praise, attention, etc. occur quite frequently to act as conditioned reinforcers which bridge the temporal gap between the everyday behaviours we emit and those events which are biologically important to us. It may seem an impossible task to talk about the complexity of human behaviour as being maintained initially by conditioned reinforcers and eventually by primary reinforcers, but there is no doubt that conditioned reinforcers can exert considerable influence over human behaviour. This is a fact that has been utilised with the advance of behavioural technology. We shall discuss in the last section of this book both the theoretical value of considering human behaviour in these terms and also the uses to which generalised conditioned reinforcers have been put in manipulating human behaviour.

**Stimulus generalisation and discrimination**

Stimulus discrimination in operant conditioning is produced in much the same way as it is in classical conditioning: the organism is reinforced for responding in the presence of one stimulus (the $S^D$ or $S+$) and not reinforced for responding in the presence of a second stimulus (the $S^\Delta$ or $S-$). This can be carried out in two basic ways. First the $S^D$ and $S^\Delta$ can be presented *successively*. For example, a pigeon may be reinforced on a particular schedule of reinforcement for pecking the response key when it is illuminated by a green light; this will last for a period of minutes until the key light is changed to red, during which he is never reinforced for pecking. The two stimuli are alternated successively in this fashion. Eventually, key-pecking is only maintained to the green key and extinguishes to the red key. Alternatively, the two stimuli can be presented *simultaneously*: two pecking keys may be present in the chamber, one illuminated by the green light and the other by a red light. Only pecking at the green key is reinforced. The latter is usually the more efficient and effective way of establishing a discrimination since, (1) both $S^D$ and $S^\Delta$ are present at the same time and thus should be easier to tell apart (this is especially helpful when $S^D$ and $S^\Delta$ are quite similar in appearance), and (2) when the two stimuli are presented simultaneously, making the correct response precludes the animal from making the incorrect one and it appears to be the case that if during discrimination training errors can be kept to a minimum, then the eventual discrimination will be finer and less susceptible to disruption (cf. Terrace, 1963, 1966a).

There are two main methods by which the degree of discrimination can be assessed. The first, and most simple, is to calculate a *discrimination ratio* from response rates during the two stimuli. This is simply to divide the response rate during the $S^D$ by the response rate during the $S^D + S^\Delta$, the larger the resulting figure the greater the difference in responding between the two stimuli. However, as useful as this may seem, it does not really tell us very much about what is happening in the learning of the discrimination. A more widely used and more valuable method is to construct a *generalisation gradient* by presenting the organism, after training, with a number of different stimuli which approximate in appearance the $S^D$ or $S^\Delta$. Figure 2.13 shows an idealised generalisation curve obtained from pigeons by Guttman and Kalish (1956). In Guttman and Kalish's experiment, the pigeons were first of all trained on a variable-interval schedule of food reinforcement to respond when one particular wavelength of light illuminated the key. When responding to this stimulus was fairly stable, other wavelengths of light were introduced during a session and rate of responding to them was recorded. Note that responding to all of the wavelengths of light during this testing period was never reinforced: that is, they were tested *during extinction*. If one continues to reinforce the original $S^D$ but not the test stimuli during generalisation testing, then the discrimination between $S^D$s and $S^\Delta$s will be sharpened (Blough, 1967). Guttman and Kalish's particular experiment did not train the pigeons with an $S^\Delta$ as well as an $S^D$, they were merely interested in how their subjects would react when presented with a new stimulus which approximated the training $S^D$. It was not, therefore, particularly necessary for the pigeons to take any notice of the key colour at all during training since key-pecking was reinforced throughout each training session, the key merely being illuminated to draw the bird's attention to the manipulandum. However, the results do indicate that their subjects were associating key colour with reinforcement since it was the original $S^D$ that evoked most responding during generalisation testing. So generalisation gradients do tell us something about what the organism is learning in a discrimination experiment. Take the example where periods of $S^D$ and $S^\Delta$ alternate successively, and responding to the $S^D$ is acquired and responding to the $S^\Delta$ is eliminated. Is the animal learning merely to respond to the $S^D$, or is he learning *not* to respond to the $S^\Delta$, or both? We know that the organism will learn to associate the $S^D$ with food,

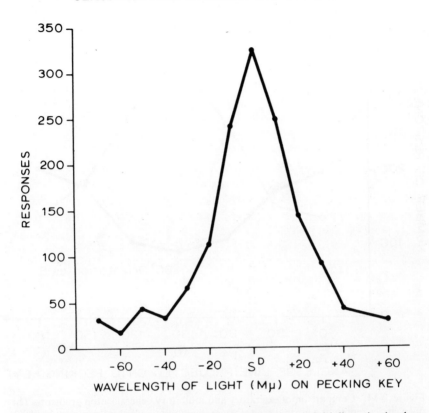

Figure 2.13   An operant generalisation curve. The pigeon is initially trained to key-peck for food with the pecking key illuminated by light of a particular wavelength. During testing the pecking-key is periodically illuminated with light of different wavelengths. The number of responses emitted in the presence of each test stimulus is inversely related to its distance from the original training stimulus on the wavelength continuum. This produces a typical bell-shaped generalisation gradient. (After Guttman and Kalish, 1956).

the Guttman and Kalish experiment suggests this much. Even so, rather than just learn to ignore the $S^\Delta$ as an irrelevant stimulus, it appears that animals will actively learn not to respond in its presence. For example, if $S^D$ and $S^\Delta$ are stimuli on different continua (for example, the $S^D$ is a colour and the $S^\Delta$ is the angle of orientation of a line projected on to the pecking key), then generalisation testing can be subsequently carried out along both continua. Sure enough, we find an inverted-U-shaped gradient of response rate for the $S^D$ but we also obtain what appears to be a mirror image of this for the $S^\Delta$. Figure 2.14 illustrates this difference. The $S^D$ gradient is known as an excitatory gradient of responding and the $S^\Delta$ function as an inhibitory gradient (see Hearst, Besley and Farthing, 1970; Jenkins, 1965 for detailed accounts).

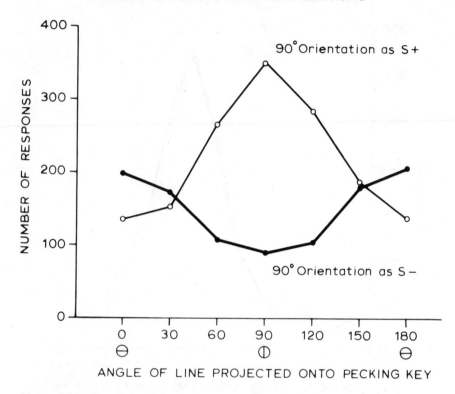

Figure 2.14 Comparison of excitatory and inhibitory generalisation gradients. The open circles represent a normal excitatory gradient obtained after the subject has been initially trained with a vertical line as an S$^D$ for food. The filled circles show the inhibitory gradient obtained if the vertical line has been used as a stimulus which signals the absence of food. (After Honig, Boneau, Burstein and Pennypacker, 1963).

These, then are some of the basic facts of operant discrimination learning, and you can no doubt see that stimulus control in operant conditioning shares many of the same characteristics as stimulus control in classical conditioning. There are many more important phenomena associated with stimulus control in operant conditioning, and we shall come across these in chapter 3. The phenomena we shall be discussing there not only tell us about what an animal learns during discrimination training but also gives us insights into the workings of the perceptual processes of animals.

## DISTINCTIONS BETWEEN OPERANT AND CLASSICAL CONDITIONING

We have talked so far in this chapter about what appear to be two seemingly different methods of modifying an organism's behaviour. Although it can be said without question that operant and classical conditioning are procedurally different – that is, the experimenter sets up different types of contingencies in the two approaches – can we also say that these procedural differences tap different underlying learning mechanisms or do they produce results which reflect different aspects of just one underlying adaptive mechanism? The history of this problem has been long and difficult with evidence available to support either viewpoint. However, although we can come to no definite conclusion on this matter yet, we can put the problem into some kind of perspective. This perspective can be subdivided by considering (1) the roots from which the traditional dichotomy stemmed, and (2) more recent evidence which bears more directly on the validity of such a dichotomy.

## Traditional distinctions between operant and classical conditioning

Although the two different conditioning procedures had been established long before any theoretical distinction had been postulated between operant and classical conditioning, it was probably the initial practical problems involved in studying the two types of conditioning that led to the suggestion that there might be two underlying learning mechanisms. For example, in order to be able to collect reliable data from classical conditioning experiments the experimenter had to choose responses which would be easily and consistently produced by eliciting stimuli (UCSs). The types of responses that fitted this need were almost all of a reflexive, involuntary or autonomic nature. Similarly, in order to be able to study adequately the effects of response consequences (operant conditioning) the experimenter needed a response which was less stimulus bound and which occurred with some frequency in a free-flowing manner. More complex skeletal responses such as lever-pressing or maze- and alleyway-running supplied these requirements. Therefore, since classical conditioning experiments used responses of the autonomic nervous system, and operant conditioning experiments used responses of the somatic nervous system, it was not long before this was being considered as a theoretical rather than a merely methodological distinction between the two (Skinner, 1938). The list in table 2.1 was considered until recently to contain the essential differences between operant and classical conditioning. Although the majority of these distinctions pertain to differences in methodology, they were often also considered as implying underlying theoretical differences (Konorski and Miller, 1937). Thus, questions related to the study of conditioning mechanisms were turned on their head: instead of attempting to find evidence for two separate conditioning mechanisms, two separate mechanisms were assumed to exist in the first place and experiments were conducted in an attempt to discover how far one mechanism could be reduced to the other. Early attempts at reducing the two processes to one were clearly

**Table 2.1   Traditional Distinctions between Operant and Classical Conditioning**

| Classical Conditioning | Operant Conditioning |
|---|---|
| 1. The experimenter has full control over the response | Experimenter has little direct control over the emission of responses |
| 2. A new response is formed (the conditioned response) | A response in the organism's repertoire is 'strengthened' |
| 3. The conditioned response does not manipulate or change the environment | The response produces environmental consequences |
| 4. Responses are almost always a response of the autonomic nervous system | Responses are usually made by the somatic nervous system using skeletal movements |
| 5. Responses are measured in magnitude and latency | Responses are measured in terms of response rate, frequency, probability and sometimes latency |
| 6. The conditioned response can have an initial strength of zero | The response has to occur at least once 'for other reasons' in order to be reinforced |
| 7. Partial reinforcement results in rapid loss of the conditioned response during extinction | Partial reinforcement produces a resistance to extinction |

(Mainly from Kimble, 1961, pp. 98–108)

unsatisfying. In order to explain operant conditioning in Pavlovian terms, appeal was very often made to hypothetical 'internal' CSs which elicited motivational states and approach behaviour (see chapter 7, p. 228ff.). The verification of the existence of such internal stimuli was difficult, if not empirically impossible. Similarly, attempts to encompass classical conditioning within an operant framework were made by suggesting that the UCS, since it often follows the elicited CR, reinforces the CR in an operant manner. Thus, presenting the dog with a meat powder as it salivates may well operantly reinforce salivating. Although this is a complication which has to be considered in many classical conditioning experiments, it is extremely unlikely that the formation and maintenance of a CR in such experiments can be wholly explained by operant principles. First, the initial occurrence of the CR still has to be explained (operant principles do not discuss how a response occurs in the first place so that it can be reinforced), and secondly many studies have shown that where the UCS is withheld on the occurrence of the CR (omission training, see below), the CR often fails to be significantly suppressed (Gormezano and Hiller, 1972; Sheffield 1965). That is, if the CR were sensitive to its consequences it should be eliminated or at least drastically reduced in probability.

Nevertheless, perhaps the greatest stumbling block for those who wished to attribute the two conditioning procedures to one conditioning mechanism was the fact that the two procedures appeared to apply to different types of responses. As we mentioned earlier, responses of the autonomic nervous system appeared to be susceptible to only classical conditioning procedures, and responses of the somatic nervous system, which involved skeletal movements, were amenable only to modification by operant methods. This subsequently led to the belief that classical conditioning provided a mechanism by which the organism could regulate its 'internal' environment in a preparatory fashion, while operant conditioning was the means by which the organism adapted to its 'external' environment. If there was basically only one underlying mechanism, then it had to be shown that autonomic responses could be operantly conditioned and skeletal responses could be classically conditioned. This problem was to become a focus of attention during the 1960s, and it is worth relating the results of this work in detail.

## Autonomic and skeletal response systems

### Operant conditioning of autonomic responses

After a number of years spent searching for appropriate and practical techniques, Neal Miller and his associates at the Rockerfeller University finally succeeded in demonstrating that autonomic or visceral responses could be modified by operant conditioning (see Miller, 1969, for a review). The problems involved in such a task are manifold, and it will probably be instructive to consider the experimental evidence in the light of these problems. First, one has to ensure that any increase in the frequency of visceral responses during operant conditioning is not the result of any hidden or implicit classical conditioning contingency. Secondly it has to be ensured that changes in the frequency of visceral responses are not mediated by skeletal responses. For example, we would easily operantly reinforce increases in heartrate by reinforcing increases in activity: increases in activity thus producing increased heartrate. It has to be shown that operant reinforcement directly affects the visceral response itself. Using operant conditioning procedures Miller *et al.* have been able to demonstrate changes in such visceral responses as salivation in dogs (Miller and Carmona, 1967), heartrate in rats (DiCara and Miller, 1968a), intestinal contractions in rats (Miller and Banauzizi, 1968), rate of urine formation in the kidneys of rats (Miller and DiCara, 1968), the amount of blood in the stomach wall in rats (Carmona, Miller and Demiere, quoted by Miller, 1969, p. 440); and in 'peripheral' visceral responses such as the amount of blood in the tails of rats (DiCara and Miller, 1968c), vasodilation in one of a rat's two ears (DiCara and Miller, 1968b), and systolic blood pressure in rats independent of heartrate (Dicara and Miller, 1968c). These changes have been brought about using both appetitive and aversive behavioural consequences.

Nevertheless, how have the problems of interpretation been solved? There are two methods of ruling out implicit classical contingencies. First, by using a 'yoked' control design. If a rat has to produce heartrate changes in order to avoid a signalled electric shock (a discriminated avoidance procedure, see p. 132), then the contingency between signal (CS) and electric shock (UCS) might be enough to produce classically conditioned changes in heartrate to the CS. However, if a second subject is 'yoked' to the first 'experimental' subject, that is, he receives shocks only when the experimental rat fails to respond appropriately, then he experiences the classical contingency but not the operant contingency. Changes in the yoked animal's heartrate do not determine whether shocks occur or not. If the yoked animal does produce heartrate changes similar to the experimental animal, then this would suggest that it is the classical rather than operant contingency that is affecting behaviour. Secondly, the problem of implicit classical contingencies can also be countered by reinforcing both increases and decreases in the same visceral response. For example, Miller has been able to demonstrate that it is not only possible to reinforce increases in the rate of such visceral responses as salivation, heartrate, intestinal contraction, renal blood flow, etc. but that it is also possible to reinforce rate *decreases* with these responses. It is highly unlikely that any implicit classical contingencies could in some situation produce an increase in these responses, and in other apparently identical stimulus situations produce a decrease.

The problem of mediation by skeletal responses is technically a more difficult one to overcome. Miller and his colleagues tackled this problem by reinforcing visceral responses when the skeletal system of the animal had been rendered paralysed by the drug curare. Keeping the animal alive by means of an artificial respirator and using intracranial stimulation (ICS) as the reinforcer, he was able to show reliable changes in visceral responses as a result of operant reinforcement. (Whenever the rat made the appropriate response a small electric current was passed into the 'pleasure centre' of the brain via an implanted electrode (Olds, 1969). Rats apparently find this stimulation reinforcing and will press levers at unusually high rates simply to achieve it.) Similarly, fairly specific and peripheral visceral responses can be changed in frequency or amplitude without any effect on more 'central' visceral responses. For example, changes in the vasomotor response in the left ear of a rat (that is, gorging the ear with blood) can be operantly reinforced without concomitant changes in overall heartrate or vasomotor responses in the 'non-reinforced ear' (Carmona, 1967). Although these experiments seem to demonstrate the direct effect of operant reinforcement on fairly specific visceral responses, a note of caution in interpreting these results should be introduced. As Black (1971) has pointed out, although the skeletal system of the animal is paralysed by curare, changes in visceral responses might still be brought about as a result of mediation by the neural 'correlates' of skeletal responses. To put it in layman's language, increases in heartrate might still be brought about by the rat 'thinking of being active'. Nevertheless it does not seem in doubt that at least *some* visceral responses can be operantly reinforced in *some* situations. This has gone a long way to restructuring our conception of the operant–classical distinction. Although we might tentatively suggest that the evidence points to the fact that both skeletal and visceral responses can be affected by operant conditioning, it is still not in doubt that the responses of the central nervous system are more readily influenced by operant contingencies than are responses of the autonomic nervous system. Miller offers a reason for this:

> The skeletal responses mediated by the cerebrospinal nervous system operate on the external environment, so that there is survival value in the ability to learn responses that bring rewards such as food, water, or escape from pain. The fact that the responses mediated by the autonomic nervous system do not have such direct action on the external environment was one of the reasons for believing that they are not subject to instrumental learning.
>
> (Miller, 1969, p. 443)

### Classical conditioning of skeletal responses

Even before experiments were specifically addressed to the question of classically conditioning skeletal responses there was some evidence that at least some very simple skeletal responses could be classically conditioned, and also that skeletal responses often occurred concomitantly with autonomic changes in traditional classical conditioning experiments. Behaviours such as the knee jerk, leg flexion to electric shock, and eye-blink to a puff of air on the cornea can all be readily classically conditioned even though they are primarily controlled by the skeletal response system. Nevertheless to be more convinced that responses of the cerebrospinal nervous system were indeed amenable to Pavlovian conditioning it needs to be demonstrated that, (1) more complex and integrated skeletal responses could be classically conditioned, and (2) that any effect on skeletal responses by a Pavlovian procedure could not be accounted for in terms of covert operant contingencies.

Evidence on the first point comes from studies of the phenomenon known as *autoshaping*. Brown and Jenkins (1968) discovered that if a brief illumination of a response key is paired with food, then hungry pigeons will eventually come to consistently peck the key when it is illuminated. Thus, simply by the pairing of food (UCS) with an illuminated key (CS), the normal UCS to food (pecking) can come to be elicited by the CS. (A more detailed account of autoshaping is given in chapter 5, p. 180ff.) This phenomenon has been further studied to show that quite integrated behaviour patterns can apparently come to be elicited by, and even directed at, previously neutral stimuli. If a stimulus is paired with food then the organism's species-typical repertoire of 'food-getting' behaviours will be elicited during the presence of that stimulus. This has been demonstrated with pigeons (Brown and Jenkins, 1968), rats (Peterson, 1975), quail (Gardner, 1969), and even certain kinds of fish (Squier, 1969). In the case of the rat, a number of relatively integrated response patterns comprising sniffing, licking, biting and pawing can be elicited by a CS which has been paired with food (Peterson, 1975). Even courting behaviours in male pigeons can be conditioned to a CS previously paired with access to a female conspecific (Rackham, 1971). Although there is still some speculation as to the precise principles which govern autoshaping (see p. 186ff.), there is no doubt that the procedure used to produce autoshaped responding is explicitly Pavlovian. The principal alternative to a Pavlovian interpretation has been to suggest that autoshaped responses such as key-pecking in the pigeon might be adventitiously reinforced by the chance occurrence of a peck in close temporal proximity to food delivery (that is, superstitious reinforcement). This introduces the second point we mentioned at the outset of this section – the possible reinforcement of skeletal responses by implicit operant contingencies. This possibility can be examined by using the omission training procedure mentioned earlier. Williams and Williams (1969) superimposed an omission contingency on to a normal autoshaping procedure with pigeons. That is, whenever the pigeon pecked the key prior to food delivery, food at the end of that trial was withheld. Thus pecking the key essentially had aversive consequences. Although this reduced the overall number of key-pecks, it only suppressed pecking by around 30–40 per cent. It appeared as though the contingency between key-light and food was enough to significantly maintain key-pecking. So, this again suggests that certain types of skeletal responses might be conditioned by Pavlovian means. However, it is important to note that a number of recent experiments which have used omission contingencies in the classical conditioning of skeletal responses have found the response in some degree to be sensitive to its consequences; that is, the frequency of the CR can be affected by its consequences when compared with appropriate control animals (Barrera, 1974; Gormezano and Hiller, 1972; Lucas, 1975; Schwartz and Williams, 1972; Wahlsten and Cole, 1972). One tentative conclusion that might be made from this evidence is that although skeletal responses do appear to be amenable to classical conditioning procedures, their frequency is still quite sensitive to operant contingencies.

## Characteristics of responses amenable to operant and classical conditioning

Instead of operant and classical conditioning being applicable to differing response systems as was originally believed, it might be that there are certain more subtle characteristics of responses which make them more amenable to one type of conditioning than the other. Two characteristics which have recently become popular in this respect are (1) the 'reflexiveness' of the response, and (2) the amount of feedback provided by the response. These characteristics do not involve well-defined response classes but describe continua rather than categories. There is some evidence that both factors play a part in determining the susceptibility of a response to control by operant or classical contingencies.

### Reflexiveness

In chapter 5 we discuss what are popularly known as 'constraints on learning'; that is, instances where the traditional principles of conditioning appear to be infringed. In the majority of these instances the relationship between the response and the reinforcer seems to be important; in an operant conditioning experiment the response will be acquired more readily if the reinforcer also *elicits* that response. For example, pecking in pigeons can be easily and quickly established using operant reinforcement, this ease may be a result of the food reinforcer also being a UCS for pecking. On the other side of the coin, Sheffield (1965) attempted to teach food-deprived dogs *not* to salivate in order to obtain food. In this experiment it was almost impossible to suppress salivation, since it appeared that the food reinforcer reflexively *elicited* salivation. The conclusion to be drawn from this and other similar experiments (see chapter 5, p. 189ff.) appears to be that if the response being operantly conditioned is to some extent 'reflexive' (that is, elicited by certain UCSs), the presence or absence of the UCSs for that response may either facilitate or retard conditioning. Mackintosh summarises this point:

> The reinforcer used in any instrumental experiment is also a potential classical UCS. A set of responses will therefore be reflexively elicited by the reinforcer and be classically conditioned to stimuli reliably accompanying its delivery. If these classically conditioned responses are incompatible with the performance of the instrumental response specified by the experimenter, then instrumental learning will be inefficient. In this sense, therefore, the potency of the Pavlovian principle of reinforcement dictates that responses completely unrelated to the reinforcer, and, in particular, responses incompatible with behaviour elicited by the reinforcer, will be instrumentally trained only with the greatest difficulty. Not only are nonreflexive responses unavailable for classical conditioning, but if sufficiently unrelated

> to the reinforcer they may be equally unavailable for instrumental training.
> (Mackintosh, 1974, p. 138)

Thus, the more dependent a response is on antecedent stimuli, the more likely it is that the presence or absence of those stimuli will affect operant conditioning of the response. This approach to the problem denies a distinct operant-classical conditioning dichotomy and instead suggests that the two types of conditioning *interact* – the extent of the interaction depending on the nature of the reinforcer and the dependency of the response on causal stimuli.

### Feedback

It was originally believed that responses of the automatic nervous system were not susceptible to operant conditioning because they provided little or no feedback to the organism. That is, the animal could not discriminate whether or not he had made the required response (Miller and Konorski, 1928; Mowrer, 1960). One can divorce the factor of feedback from its association with the visceral response system and suggest that response feedback itself will play an important role in determining the ease with which a response will be operantly conditioned. Providing exteroceptive stimulus feedback for responses which inherently give little or no feedback supports this supposition. First of all the vast literature that has accumulated on biofeedback techniques (see chapter 14, p. 384) demonstrates the success with which organisms can manipulate their own visceral responses given that they are provided with appropriate feedback aids (see Blanchard, Young and Jackson, 1974, for a review). Nearly all responses provide some kind of feedback – it is clearly a matter of degree. Thus, those responses which provide a substantial amount of feedback should operantly condition quite readily; those that provide only a minimum amount of feedback should operantly condition only with great difficulty. Experiments with animals that have in some way been able to manipulate the degree of feedback from a response have supported this viewpoint (Brener and Hothersall, 1966; Taub and Berman, 1968).

## Conclusion

Perhaps the best way of summarising such a complex issue is to emphasise a number of points of importance which can be extracted from the preceding discussion. First, operant and classical conditioning are two distinct conditioning *procedures*. However, although the actual explicit procedures cannot be easily reduced to one, this does not mean that classical conditioning might not be produced by *implicit* operant contingencies, or vice versa. Nevertheless, considering the state of our knowledge at the moment the conclusion that fits most of the facts is certainly an *interactionist* one – that is, there are two learning processes which will readily interact with one another. The behaviour of animals is sensitive not only to stimulus–reinforcer contingencies (classical conditioning) but also to response–reinforcer contingencies (operant conditioning). Both types of contingency are found in learning situations, and which of these two types of contingency comes to exert most influence on the organism's behaviour will depend on a number of factors. We have discussed some of these in detail previously, but to summarise them briefly, (1) in order to be classically conditioned, a response must have a natural eliciting stimulus (UCS); that is, it must be to some extent 'reflexive'. The more reliably a response is tied to an eliciting antecedent stimulus, the more readily it can be classically conditioned; (2) it appears that a response must produce some kind of feedback for the organism if it is to be operantly conditioned – the greater the amount of feedback, the easier it is to condition; (3) operant reinforcers will not only increase the frequency of contingent responses, but they may also act as UCSs eliciting unconditioned species-specific responses. These elicited responses may become classically conditioned to experimental cues and retard conditioning if they are incompatible with the operant response.

So finally then, the traditional distinction between operant and classical conditioning on the basis of response type now seems inappropriate, and similarly, it seems inappropriate to attempt to reduce the two types of conditioning to one. Most responses in an organism's repertoire appear to be amenable to both types of conditioning; the ease with which they can be conditioned appears to depend on certain specific characteristics of the response in question and not on its membership of a particular response system.

# 3 The Appetitive Control of Operant Responding

Having discussed some of the fundamental characteristics of classical and operant conditioning, we now move on to look in more detail at some of the behavioural phenomena that occur during appetitive operant conditioning. Here we begin to ask such questions as 'What are the important variables that regulate an animal's behaviour during periods of intermittent reinforcement?', 'In what way is rate of learning related to magnitude of reinforcement?', 'When there is more than one source of reinforcement, what laws govern the animal's choice of behaviours?' and so on. It is from asking more specific questions such as these that we may eventually come to form some impression of the learning mechanisms which underlie adaptive behaviour in animals. Remember that although one can consider operant reinforcement as a principle of adaptive behaviour which directly reflects some underlying reinforcement mechanism, the unassuming empiricist is perhaps more concerned with considering reinforcement merely as a method for controlling behaviour, and – without making assumptions about reinforcers being 'selective' with regard to behaviours or 'stamping in' particular behaviours – manipulation of stimulus, response and reinforcer parameters can yield an unbiased insight into the conditioning process.

## SCHEDULES OF REINFORCEMENT

Schedules of reinforcement are often mentioned in the same breath as operant conditioning, and it is true that they are probably the most widely adopted means of studying operant behaviour. In fact, so prevalent has been the use of schedules of reinforcement that rather than being considered as a methodological tool for the study of behaviour, they have on some occasions been elevated to the status of phenomena worthy of individual enquiry and theory (Ferster and Skinner, 1957; Schoenfeld, 1970; Schoenfeld and Cole, 1972). It is important before we start discussing schedules of reinforcement to put them into some kind of perspective. Many text books on learning discuss them as behavioural phenomena which are described independently of underlying adaptive mechanisms – it is not the intention to do that here. Although we shall discuss the important schedules of reinforcement individually, it is not with the object of formulating a theory which describes responding on each respectively; we shall discuss schedules in order to see what light the characteristic behaviour on each can shed on more basic underlying learning mechanisms. In the previous chapter (p. 43) we discussed the relationship between reinforcement and responding on the four basic schedules; fixed-interval, fixed-ratio, variable-interval, and variable-ratio. The following sections discuss in more detail the behaviour maintained by each.

## Fixed-interval

### Description of behaviour on FI

A fixed-interval (FI) schedule is one which reinforces the first response that occurs after a specified period of time. For example, an FI60-s schedule is one that would reinforce the first response that occurred one minute after the preceding reinforcer was delivered. As an alternative to this, fixed-interval can be timed 'by the clock' that is, each interval is timed from the termination of the last one regardless of whether a reinforced response has occurred or not. Although this latter method often produces more irregular inter-reinforcement times (since intervals are timed regardless of when a reinforcer is delivered), the patterning of behaviour eventually produced by the two methods is to all intents and purposes identical (Ferster and Skinner, 1957).

After a period of training the behaviour generated by this schedule consists of a fairly long pause in responding after reinforcement (the postreinforcement pause) followed by a gradually accelerating rate of responding which continues up to the delivery of the next reinforcer. Figure. 3.1 gives a number of examples of this behaviour in the form of cumulative records. This characteristic positively accelerated response pattern is known as the fixed-interval *scallop* and is a fairly robust phenomenon, occurring with a wide variety of responses and reinforcers, and in pigeons even with fixed-interval durations of up to 27 hours (Dews, 1965)! This scalloped pattern of responding on FI schedules has more theoretical importance however; it is said to reflect a temporal discrimination on the part of the animal. Certainly the point in an individual interval when the animal reaches its 'terminal rate' of responding (that is, the highest rate of responding which is maintained up to the next reinforcer delivery, see figure 3.1), varies with the value of the fixed-interval itself – the longer the fixed-interval value, the longer into the interval it is before the animal initiates its terminal rate. In fact, after prolonged training on an FI schedule the scallop takes on a *break-and-run* pattern where the postreinforcement pause (PRP) and terminal rate of responding can be clearly identified and separated. The point at which the pause is terminated and the terminal rate initiated has been found to be approximately two-thirds of the fixed-interval value (Schneider, 1969). That

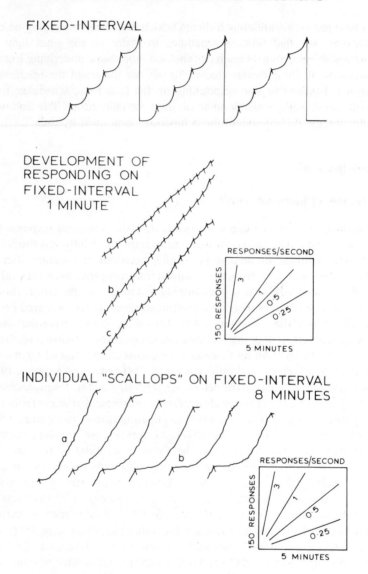

Figure 3.1   Cumulative records of fixed-interval responding. The uppermost record shows a stylised example of fixed-interval *scalloping* with the rate of responding being positively accelerated between each reinforcer. The middle record shows the development of the fixed-interval *scallop* over sessions 2, 3 and 4 of training (After Ferster and Skinner, 1957, p. 143). The bottom panel shows individual *scallops* from a pigeon that has been well trained on an FI8-min schedules. In scallop 'a' the subject reaches its terminal rate of responding quite early in the interval, whereas in scallop 'b' responding is preceded by a typically lengthy postreinforcement pause. (After Ferster and Skinner, 1957, p. 170).

is, if it is an FI-60-s schedule, the animal will on average terminate his post-reinforcement pause after about 40-s, on an FI-5 min after 200-s, and so on. This direct proportionality between PRP duration and FI-value apparently reinforces the view that the animal is making a temporal discrimination of some kind (although there are other possible explanations as we shall see later) and certainly it is a proportionality which extends over a very wide range of FI values. This relationship breaks down only at very short FIs of less than 10-s when the PRP is proportionally longer (Starr and Staddon, 1974).

## Explanations of FI behaviour

What are the important aspects of the FI schedule which control an animal's behaviour and generate the characteristic FI performance? There are two points to make before we embark on any theoretical analysis of this kind. First, it must be recognised that schedules of reinforcement are such that detailed analysis is not easy! For instance, the animal is constantly making responses in a free-flowing fashion and the emission of these responses can be related to or controlled by any one of a large number of factors, each of which cannot easily be isolated. To put it bluntly, a schedule of reinforcement allows too many variables to operate at once. This should become clear when we discuss the possible explanations of schedule responding. Secondly, we must be wary of what we are going to accept as an 'explanation' of schedule performance. It is easy to suggest that on an FI schedule the animal is making a temporal discrimination and that this is the explanation of FI scalloping. Quite so, but even if this were the case, temporal discrimination is a hypothetical process we have inferred from the relationship between time and response probability; a relationship between responding and the temporal parameters of a schedule can be brought about in many different ways. One must be careful not simply to redescribe what is observed, but to isolate and identify controlling variables as far as is possible.

### Response chaining

One of the earliest attempts to explain the FI scallop was to suggest that the responses in each interval constitute a chain.

> Since in FI regular responding occurs between the reinforced responses the suggestion is that in each interval the responses constitute a chain. A chain is a sequence of responses in which every response serves as a discriminative stimulus for the next response; it is maintained since the final component in the chain is reinforced.
>
> (Harzem, 1969, p. 316)

This approach implicitly assumes that the time between each response sets the occasion for a subsequent inter-response time of shorter duration; hence the positively accelerated response gradient. Thus, there appears on the face of it to be no need to invoke the notion of temporal discrimination. To give a hypothetical example to illustrate this point – a rat may turn a full circle then press the lever; this behaviour acts as an $S^D$ for turning half a circle and pressing the lever, and this is an $S^D$ for turning a quarter circle and pressing the lever and so on. This would produce progressively shorter inter-response times and a positively accelerated response curve, and the animal need make no discrimination of the temporal parameters of the schedule. However, there are a number of problems with this interpretation which make its viability doubtful. First, the postreinforcement pause duration still has to be accounted for. As we have noted, it is usually proportional to the FI value. This means that if an animal is transferred from FI30 s, through FI60 s to FI5 min and (as is most likely) the post-reinforcement pause duration changes accordingly, then the animal *must* be taking some account of the temporal parameters involved, even if it is only varying the duration of the components of its response chain. Secondly are those experiments which have shown that interrupting the sequence of responses on an FI schedule does not substantially disrupt the overall pattern of responding (Dews, 1962, 1965, 1966; Ferster and Skinner, 1957; Wall, 1965). Dews (1962) trained pigeons on an FI500-s schedule and, when FI performance had stabilised, introduced periods of time-out (extinguishing the houselight) every 50 s. Figure 3.2 shows that although the positively accelerated response pattern was disrupted, the subjects' rate of responding still increased during subsequent houselight-on periods through the interval. In a similar study, Wall (1965) trained rats on a discrete-trial FI60-s schedule. A retractable lever was presented every 60 s and a press by the rat would produce the reinforcer. After this training the lever was inserted at varying times during the 60-s interval and the latency with which the rat responded was recorded. Wall found that the latency of response was inversely related to the time in the interval that the lever was inserted – the longest latencies occurred early in the interval and the shortest latencies occurred late in the interval. Thus, there was certainly no opportunity for chaining of lever responses, yet probability of lever-pressing was directly related to time into the interval. To conclude from these studies, it seems that response chaining is not a necessary and essential determinant of the FI scallop. For the necessary factors we must look elsewhere.

*Delay of reinforcement*

Both Dews (1962; 1965; 1966) and Morse (1966) have suggested that the positively accelerated scallop is a result of differing delays of reinforcement. That is, responses early in the interval are reinforced only after a long delay, while responses late in the interval receive more immediate reward; thus response strength early in the interval is weakened and response strength later in the interval is increased, hence the positively accelerated response rate.

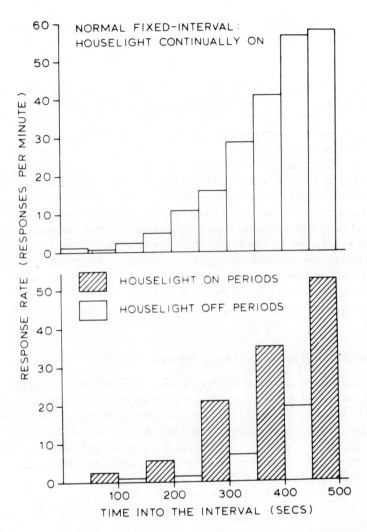

Figure 3.2    The effect of multiple S^ periods on responding on an FI schedule. The top panel shows the response rate of a pigeon in successive 50-s segments of a normal FI500-s schedule. The bottom panel shows the response rate when the houselight is extinguished during alternate 50-s segments. (After Dews, 1962).

Proponents of this hypothesis have also suggested that there is no need to resort to notions of temporal discrimination since delay of reinforcement should, in principle, be able to explain the FI scallop. This seems a difficult position to uphold as Mackintosh explains:

. . . different delays of reinforcement for responses emitted at different points in time could produce differences in rate of responding only if subjects were discriminating between those different points in time.

(Mackintosh, 1974, p. 171)

There is further evidence that delay of reinforcement alone cannot account for the scallop. Both Trapold, Carlson and Myers (1965) and Zamble (1969) have shown that FI scalloping can appear very rapidly if animals are given prior training on a fixed-time (FT) schedule of the same value. That is, if for a number of sessions the animals are given food every 60 s independently of their behaviour and then transferred to an FI60-s schedule, they will acquire FI scalloping relatively quickly. This implies that the scallop is a result of the animal learning something about the temporal parameter of the FT schedule and transferring this knowledge to its responding on FI; it was unlikely that delay of reinforcement could have produced the response scalloping in these studies because the scallop appeared after only a relatively small amount of exposure to the response-contingent FI schedule. So, whatever kind of explanation we seek it seems that in some way the animal is learning something about the temporal parameters of an FI schedule and distributing its responses accordingly. Even so, we still have to ask *how* it manages to do this.

### Temporal control by the reinforcing stimulus

When considering any schedule performance it is important to consider what events are present that might regulate responding or on which discriminations could be based. In the formal fixed-interval schedule there are essentially only two: the animal's own behaviour and the delivery of the reinforcer. The occurrence of the reinforcer is perhaps most important on FI because it provides the animal with an event which in essence not only predicts the occurrence of the next reinforcer but also signals a period of non-reinforcement. Ferster and Skinner considered that the pause after reinforcement was a product of the stimulus properties of reinforcement. They suggest that

. . . residual stimuli – from food in the mouth, swallowing, etc. – may extend past the moment of reinforcement. . . . other behaviours may also be set in motion (*e.g.* washing for the rat) which may also control a low-rate of responding because of its relationship to reinforcement.

Ferster and Skinner (1957, p. 135)

Thus the internal cues and behaviours generated by a food delivery effectively acted as $S^\Delta$s which controlled a low-rate of responding – the animal learned not to respond in the presence of, for example, crumbs of food which remained in the mouth. This hypothesis however has difficulty coping with the fact that different FI values produce different PRP durations. Presumably residual stimuli from eating food would last the same amount of time on FI60s as on FI5min. So how come the different pause durations? However, it is possible to retain the idea that the reinforcer acts as an $S^\Delta$ without having to postulate hypothetical 'residual' stimuli which survive after consummation of the reinforcer. Staddon (1972, 1974) has suggested that a distinction should be drawn between what he calls *situational* stimulus control and *temporal* stimulus control. Just as it can be shown that a stimulus may control responding in its presence (situational control), Staddon suggests that responses can also be controlled by stimuli that have occurred previously.

> . . . if Event A (a stimulus) occurs at a certain point in time and can be shown to determine the time of occurrence of Event B (a response), which occurs at a later point in time, the label *temporal control* is proposed for the relationship – no matter what the events A and B, no matter how long or short the time separating them, and no matter what other contextual dependencies may exist.
>
> (Staddon, 1972, p. 213)

Thus, instead of considering that residual stimuli control not responding, it is considered that the delivery of reinforcement controls the time of occurrence of the first response after reinforcement (thus determining the duration of the post-reinforcement pause). It is to all intents and purposes a more standardised way of saying that the animal is making a temporal discrimination, but by specifying a controlling event (the reinforcer) we are more able to test this assumption. What evidence is available lends support to this interpretation. For instance, if the reinforcer is a stimulus controlling the temporal placement of the first response in the interval, then manipulation of the properties of the reinforcer should affect that temporal discrimination. First, simple studies which have omitted scheduled reinforcers or presented a neutral stimulus such as a tone or light in lieu of reinforcement have found that the pause is either substantially reduced in duration or that the animal will continue to respond straight through to the next reinforcer with no pause at all (Staddon and Innis, 1966, 1969; Zeiler, 1972a). A more interesting study is by Kello (1972) who, using pigeons, manipulated the number of stimulus components present at the time of reinforcement. The stimulus complex at reinforcement consisted of food, magazine light and general blackout of the chamber. The pigeons were trained on an FI schedule with all three components present at reinforcement until performance was stable. Following this, different combinations of these components were presented in lieu of reinforcement. Kello found that the rate of responding was lowest immediately following a combination of all three, somewhat higher following just magazine light + blackout, higher still following blackout alone, and highest following unsignalled non-reinforcement. These results suggest that, during training, the stimulus complex at reinforcement had come to control a particular PRP duration, and that this temporal control breaks down in proportion to the difference between the event in lieu and reinforcement. Thus, it certainly seems that on an FI schedule the reinforcer acts as a stimulus controlling postreinforcement pause duration. The implicit assumption in the temporal control approach is that, just as varying the properties of a situational $S^D$ will affect the control it exerts over responding in its presence, then varying the properties of a temporal inhibitory stimulus will affect the accuracy of a temporal discrimination. (This is what Staddon (1972, p. 233) calls an event which exerts temporal control.)

So, to sum up fairly briefly, this hypothesis considers that the fixed-interval scallop reflects the temporal control of responding by the reinforcing stimulus as a temporal inhibitory stimulus. The gradual increase in response rate after the termination of the PRP being attributed to a gradual deterioration in temporal discrimination with passage of time since the controlling event.

*Summary*

We have looked in some detail at three hypotheses relating to fixed-interval responding. The evidence appears to support the view that the fixed-interval scallop reflects the temporal control of responding by the reinforcing stimulus rather than the chaining of responses or factors associated with delay of reinforcement. It must be pointed out though that we have only really assessed what factors control the *patterning* of responses, but we have hardly touched on those factors which control, for instance, terminal *rate* of responding on fixed-interval schedules.

# Fixed-ratio

### Description of behaviour on FR schedules

A fixed-ratio (FR) schedule is one which reinforces only every *n*th response the animal makes. For example, on an FR20 schedule only every 20th response would be reinforced. After a period of training on this schedule, stable performance is characterised by a *break-and-run* pattern of responding. That is, there is a short pause after reinforcement followed by an abrupt transition into a steady terminal rate of responding (see figure 3.3). By gradually increasing the FR requirement, behaviour can be maintained even on ratio schedules as high as FR900 (Ferster and Skinner, 1957). But high FR values will often produce what is known as ratio strain; this is characterised by lengthy pauses in responding which occur while the animal is in the middle of a ratio run (Ferster and Skinner, 1957, p. 63). In general, PRP duration is directly related to the ratio value: the higher the ratio, the longer the pause after reinforcement (Felton and Lyon, 1966). However, the terminal rate of responding appears to be constant over different ratio values for an individual animal (Felton and Lyon, 1966; Ferster and Skinner, 1957, pp. 49–57).

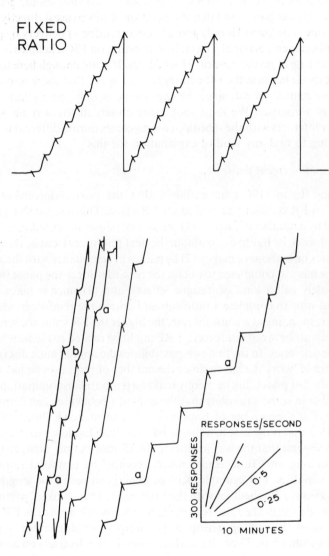

Figure 3.3  Cumulative records of fixed-ratio responding. The top record shows a stylised example of typical *break-and-run* behaviour developed on fixed-ratio schedules. The bottom records show responding by two different pigeons on FR200 and FR120 respectively. Points marked 'a' illustrate the *break point* between pausing and responding, and 'b' marks a ratio run which is uncharacteristically preceded by little or no post-reinforcement pause. (After Ferster and Skinner, 1957, p. 52).

**Explanations of fixed-ratio behaviour**

When considering fixed-ratio behaviour there are two main facts which require attention. First, it is interesting that such a schedule should generate a postreinforcement pause, and that the duration of this pause is directly related to ratio value. The animal is only going to delay reinforcer delivery by pausing after reinforcement. Secondly, overall responding on FR schedules is usually higher than rate of responding on FI schedules. It is not enough here to simply state that this is because the FR contingencies require that more responses be emitted for reinforcement, since the contingencies only specify number and not *rate* of responses. We must look more closely at the way an animal's behaviour interacts with the schedule contingencies during different phases of FR training to find any kind of explanation for this.

*The postreinforcement pause*

Felton and Lyon (1966) have shown that the postreinforcement pause duration on FR is directly related to the FR value. This relationship could be explained by a number of factors: (1) the animal pauses in accordance with the amount of work he has to do to obtain the next reinforcer (that is, the number of responses he is about to make); (2) he pauses in accordance with the amount of work he has just completed to obtain the reward prior to the pause (what we might loosely call a kind of 'fatigue' effect); and (3) since it takes a finite amount of time to complete a ratio run and obtain the reinforcer, and if the animal is responding at a constant rate, the higher the FR value the longer will be the intervals between reinforcers. PRP might be related to the time between reinforcer deliveries. In the first two possibilities the animal must discriminate the amount of 'work' he has to do or the number of responses he has to emit, whilst in the last possibility he merely makes a temporal discrimination. While it is possible in some situations that the animal's behaviour can come under the control of the number of responses it has to make (Rilling, 1967) it is unlikely that that is the case in a simple FR schedule. Evidence from a number of studies suggest that the way an animal distributes its time between pausing and responding on FR schedules reflects control by temporal rather than response variables. For example, Killeen (1969) compared the length of the PRP on various FR schedules with the PRP emitted by a 'yoked' partner who did not have to meet the ratio requirement. That is, whenever the FR animal obtained food this set up a response-contingent reinforcer for a partner animal in another box. Thus, this partner animal was effectively on an interval schedule whose inter-reinforcement times were the same as the FR animal. Killeen found that postreinforcement pause durations from the two animals were almost identical: being required to make $x$ responses during the inter-reinforcement interval had no effect on PRP duration, this appeared to be controlled solely by the duration of the inter-reinforcement interval itself. Further evidence to support this view comes from a study by Neuringer and Schneider

(1968). They trained pigeons on an FR15 schedule but the time between reinforcers was varied by switching off the key-light for various periods of time after individual responses: thus, number of responses per inter-reinforcement period remained constant but inter-reinforcement time could be systematically varied. Sure enough, postreinforcement pause duration was proportional to the length of the inter-reinforcement interval, even though the amount of 'work' the animal had to do remained constant. So, we might say with some confidence that the postreinforcement pause on FR schedules appears to reflect a temporal discrimination. If this is so, we might expect that the reinforcer also acts as a temporal inhibitory stimulus exerting temporal control as is the case with FI schedules. Certainly, manipulation of the reinforcing stimulus on FR, either by omitting it, presenting a stimulus in lieu of it, or by manipulating its magnitude, has exactly the same effects on the pause as it does on FI schedules (Lowe, Davey and Harzem, 1974; McMillan, 1971). Perhaps what is more interesting about this finding is that the animal comes under the control of factors which are apparently counterproductive: the longer he pauses the fewer reinforcers he will obtain in a fixed period of time. This does illustrate a couple of points. First, animals are not invariably able to 'maximise' reinforcer acquisition, even under what seem to be fairly simple contingencies. Secondly, it does illustrate the power of temporal variables in controlling and regulating behaviour; this is a phenomenon we will come across quite regularly in discussions of schedule performances.

*Rate of responding*

Why does a fixed-ratio schedule maintain a higher rate of responding than a comparable fixed-interval schedule? There are two important hypotheses put forward to explain this. The first is indirectly related to the fact that responding on relatively high FR schedules is sometimes difficult to establish. If an animal is trained up on an FR schedule one can be pretty sure that the subject is eventually either going to respond at a fairly high overall rate or in contrast hardly respond at all (more often the former). Herrnstein and Morse (1958) have suggested that this tendency to polarity on FR schedules is because rate of responding might be determined by rate of reinforcement (compare Herrnstein, 1970). If this were the case then FR schedules provide a direct relationship between response rate and reinforcement frequency – the faster the animal responds the greater the rate of reinforcement, the slower he responds the lower the rate of reinforcement. Thus, take the example where an animal – for whatever reason – increases its rate of responding; this will increase the rate of reinforcement which, according to the proposed principle which relates response rate to reinforcement rate, should further increase the response rate, and so on, until supposedly the animal's response rate reaches a 'ceiling' – it cannot respond any faster. It can be seen that the obverse would be true with a decrease in responding, the effect would theoretically 'snowball' until the animal ceased to respond altogether. Obviously this process has been blandly caricatured and local fluctuations in rate will disturb the process, but if this process can be considered in relation to FI schedules the point becomes clearer. Changes in rate of responding on FI schedules do not substantially affect rate of reinforcement since each reinforcer is scheduled to become available after a fixed period of time, regardless of how many responses have been made in the inter-reinforcement interval. Thus, on FI schedules response rate 'ceiling' effects would not be generated unless the rate of reinforcement were relatively high (that is, the FI value was small). The fact that the terminal rate of responding on different FR values is invariant also lends support to the possibility that ratio schedules in general may be generating a terminal rate of responding which the animal cannot exceed.

A second account of the differences in rate between FR and FI schedules is one originally proposed by Ferster and Skinner (1957, pp. 133–4). They suggest that an FI schedule differentially reinforces long inter-response times (IRTs) whereas an FR schedule does not (see p. 81 for a fuller description and discussion of inter-response times). This produces lower rates of responding on FI than FR. On, for example, an FI60-s schedule, the longer the animal waits before responding, the more likely it is that the next response will be reinforced, and it is true that reinforced interresponse times on FI are generally longer than those that precede them (*cf*. Dews, 1969, 1970). Thus, if one adopts the assumption that inter-response times can be reinforced and thus occur more frequently in the future (see p. 76) then FI selectively reinforces long IRTs as opposed to short ones.

## Variable-interval

### Description of behaviour on VI

A variable-interval (VI) schedule reinforces the first response that occurs after a specified period of time has elapsed since the previous reinforcer. However, unlike FI schedules, the specific period between reinforcers varies from interval to interval. Thus, if you like, the organism cannot predict when the next reinforcer is due. To say that a schedule is a VI1-min schedule is to say that the *average* inter-reinforcement time is 1 min: the minimum inter-reinforcement time the animal encounters may be as small as 5-s and the longest as long as 5 min, but over a representative sample of intervals the mean is 1 min. Because it is impossible to predict the availability of forthcoming reinforcers, animals usually emit a steady constant rate of responding on VI schedules (see figure 3.4). However, a closer look at local rates of responding reveals some interesting interactions between the parameters of VI schedules and behaviour. First, it is certainly the case that overall rate of responding is a function of the frequency of reward: the shorter the mean inter-reinforcement

VARIABLE-INTERVAL

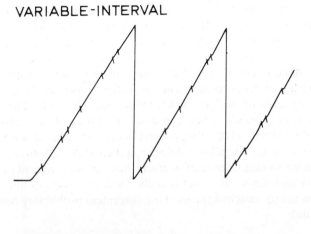

VARIABLE-INTERVAL
2 MINUTES

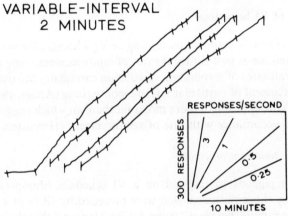

Figure 3.4 Cumulative records of variable-interval responding. The top record shows a stylised example of the steady constant rates of responding normally developed on VI schedules. The bottom record shows an actual example of the responding produced by a pigeon on a VI2-min schedule. (After Ferster and Skinner, 1957, p. 330).

interval the faster the overall rate of responding (Catania and Reynolds, 1968; Ferster and Skinner, 1957). However, it can also be shown that local rates of responding fluctuate in accordance with the local probability of reinforcement. Take, for example, a normal VI1-min schedule which has a maximum inter-reinforcement interval of 5 min. The more time that has elapsed since the previous reinforcer, the more probable it is that the next reinforcer is due – and when 5 min has passed since the previous reinforcer the probability of reinforcement for the next response is by necessity 1.00. So what is in fact found on normal VI schedules is that just as the probability of reinforcement increases with time since the previous reinforcer, so does the animal's rate of responding (Ferster and Skinner, 1957, pp. 331–332). If cumulative records of VI performance are scrutinised closely, instead of finding a truly constant rate of responding between reinforcers, one should find a very shallow *scallop*. The dependence of local rate of responding on local probability of reinforcement can be further illustrated by quoting an experiment by Catania and Reynolds (1968, pp. 339–354). They trained pigeons on a VI schedule which contained a surplus of very short inter-reinforcement intervals so that the probability of reinforcement was relatively high *immediately* after a reinforcer delivery; it then declined before increasing again as the maximum inter-reinforcement interval was reached (see figure 3.5). Catania and Reynolds found that the local response rate functions of their pigeons neatly matched the local reinforcement probability function of the VI schedule.

**Explanation of VI behaviour**

Since there is very little response patterning on VI schedules the main factor in need of explanation is how different rates of reinforcement come to produce different overall rates of responding. Two main candidates are usually cited: (1) the reinforcement of particular inter-response times (Anger, 1956; Shimp, 1967); and (2) the appeal to a more molar mechanism which regulates rate of responding in accordance with rate of reinforcement (Herrnstein, 1970).

*Inter-response time reinforcement*

After training pigeons to respond on a VI schedule, Shimp (1967) then reinforced only those responses that were preceded by IRTs of a particular duration (for example, between 0.9 and 1.2–s). He found that the emission of IRTs within this band subsequently increased in frequency. Similarly, Blough (1966) found that he was able to reinforce responses preceded by pauses which recently had occurred less frequently than would be expected by chance. These two experiments, plus the extensive literature on differential reinforcement of low rate schedules (see p. 80ff.) strongly suggest that inter-response times could possibly act as unitary operants which can be increased in frequency by operant reinforcement. But first, how does this relate to rate of responding on VI schedules, and second, what possible factors might mediate the reinforcement of particular IRTs? The answer to the first question has been discussed in some detail with regard to response rate differences between FR and FI schedules (pp. 73–74). That is, the longer the mean inter-reinforcement interval, the more likely it is that a long IRT will be reinforced. Therefore, if we assume that IRTs that immediately precede food are the ones that are reinforced – the longer the mean inter-reinforcement time, the longer will be the IRTs that are reinforced, and hence the lower the overall response rate. An alternative slant

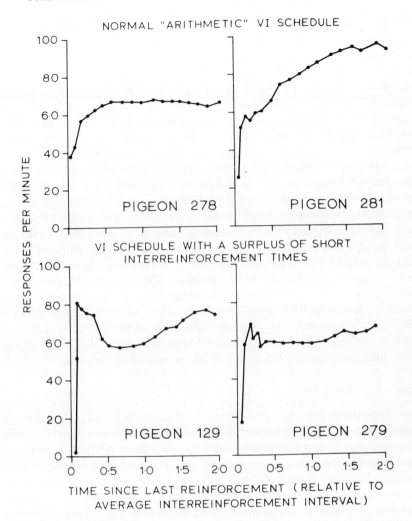

Figure 3.5   Local response rates on a normal and synthetic variable-interval schedule as a function of time since the preceding reinforcer. The top two functions represent typical local response rates on a normal 'arithmetic' VI schedule. The bottom two functions represent data from pigeons trained on a VI schedule which contained a surplus of very short inter-reinforcement intervals. (After Catania and Reynolds, 1968).

to this approach is to suggest that it is not particular IRTs *per se* that are reinforced but particular response topographies (Blough, 1966; Reynolds and McLoed, 1970). For example, the shorter the mean inter-reinforcement time, the more likely it is that a pigeon will be reinforced for standing directly in front of the key and emitting a 'burst' of responding. It may be that these 'bursts' of responding act as a unitary response that is reinforced on VI schedules. The smaller the VI value the more likely it is that these 'bursts' will continue to be reinforced, thus maintaining a relatively high rate of responding. Although IRT reinforcement theory appears very attractive, it also has a number of problems associated with it. First, VI schedules do not explicitly specify the reinforcement of any particular IRT, thus IRT reinforcement theory assumes that adventitious correlations between a particular IRT and a reinforcer will result in superstitious reinforcement of that IRT. Although it is quite possible that superstitious reinforcement operates during the early stages of operant learning it is currently not seen as a process which maintains behaviour in a stable fashion over long periods of time (Staddon and Simmelhag, 1971; see also pp. 163–165) and, since VI responding remains quite stable over long periods of training, its rôle in determing VI response rates should be treated with some caution. Secondly, it is not always the case that the IRT that immediately precedes the reinforcer is representative of the rest of the IRTs that an animal emits during an inter-reinforcement interval. We have noted this in relation to FI schedules (Dews, 1970; Williams, 1968), and in a similar vein Reynolds and McLeod (1970) report that they were able to reinforce particular PRP durations on a modified interval schedule without making any changes in the frequency of reinforced IRTs. Thus a theory of IRT reinforcement as an explanation of response rate on VI schedules may need some modification before it can adequately fit all the facts.

*Rate of reinforcement*

An alternative approach to explaining rate of responding on VI schedules is to appeal directly to the relationship between reinforcement probability and response rate. It is an empirical fact that there is a direct relationship between reinforcement probability and response rate and a number of writers have suggested that, at the present time, a 'descriptive' theory of this kind is as good as any (Herrnstein, 1970; Catania and Reynolds, 1968). There are two points to be made about this kind of theory of VI responding. First, it is a descriptive theory – it merely describes the data – and so it is empirically true; any alternative theories must take this fact into account. Secondly, it is a macro-theory: in its broadest form it simply states that 'overall rate of reinforcement will determine overall rate of responding'. Therefore, it does not mean that, for example, IRT reinforcement is necessarily an incompatible alternative theory – overall rate of reinforcement may determine overall rate of responding, but this may come about on a micro-level by selective IRT reinforcement. A problem with this approach is that it may be *too* descriptive, and thus its explanatory value is in doubt (Reynolds and McLeod, 1970, p. 91). One of the ways in which it could be elevated to a level beyond the purely descriptive, is to suggest that the animal is sensitive to either overall or momentary reinforcement probability, and this affects some hypothetical underlying variable such as 'response strength'. However, it has been argued that the data on rate of responding on VI schedules do not warrant the postulation of such hypothetical variables and can be explained more adequately at the micro-level (Mackintosh, 1974, p. 179).

## Variable-ratio

### Description of behaviour on VR

A variable-ratio (VR) schedule is one which reinforces the animal for emitting a particular number of responses between reinforcers. However, the number of responses specified varies for each successive reinforcer. Thus, if the animal is on a VR60 schedule, the average number of responses required to be emitted per reinforcer is 60, but on any one occasion this may be as low as 3 or 4, or as high as 200. This type of schedule generates very high steady rates of responding with only very short postreinforcement pauses (see figure 3.6). The actual patterning of responding is thus very similar to VI schedules, but the overall rate of responding on VR is usually much higher than on comparable VI schedules (Thomas and Switalski, 1966). In fact, rate of responding may, on many occasions, be so high that it has been calculated that the energy the animal expends in responding so vigorously is not compensated for by the calorific value of the food reinforcement he obtains (Skinner, 1957a)!

### Explanation of VR behaviour

Here again, the important fact in need of explanation is why VR schedules should maintain response rates higher than those on corresponding VI schedules. Thomas and Switalski (1966) trained pigeons on a VR40 schedule and yoked them to a second group of birds. That is, whenever the VR bird received food this set up a reinforcer for a yoked partner. Thus the yoked partner had the same distribution of reinforcers but did not have to complete the VR requirement (it was as though it were on a VI schedule). They found that the VR birds responded at a much higher rate than their yoked partners, even though reinforcement frequency and distribution were the same for the two groups. The arguments which relate to rate differences between FR and FI schedules are relevant here (see pp. 73–74).

## Differential reinforcement of low rate

There is another schedule of reinforcement which is worth discussing in this context since analysis of the characteristic behaviour patterns it generates provides interesting insights into some of the important variables that determine schedule behaviour. This schedule is known as a differential

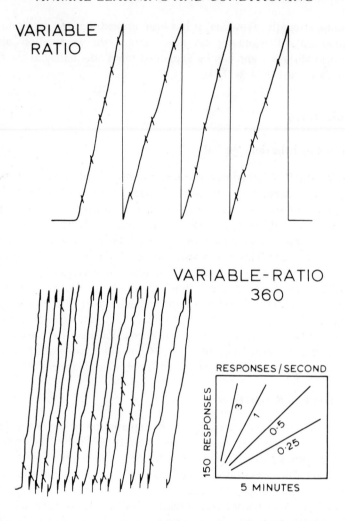

Figure 3.6 Cumulative records of variable-ratio responding. The top record shows a stylised example of the constant high rate of responding engendered by this schedule. The bottom record shows an actual example of the responding produced by a pigeon on a VR360 schedule of food reinforcement. (After Ferster and Skinner, 1957, p. 396).

reinforcement of low rate (DRL) schedule since it produces a characteristic low rate of responding in the animal.

## Description of behaviour on DRL

A DRL schedule is one which specifies that a minimum amount of time should have elapsed between two responses before the reinforcer is delivered. For example, on a DRL20-s schedule the animal has to wait *at least* 20 s since the last response before he can obtain a reinforcer; if he responds before this minimum period of time has elapsed, that response goes unreinforced and the timer is reset to zero. Typically, this schedule produces a low rate of responding with, after some training, the modal IRT corresponding to the DRL criterion (figure 3.7). There is a further way of analysing this behaviour to give us a more detailed insight into what is happening under DRL contingencies. Since response rate is so low, we really need to know in

BEHAVIOUR OF A RAT ON A
DRL 15 SECONDS SCHEDULE

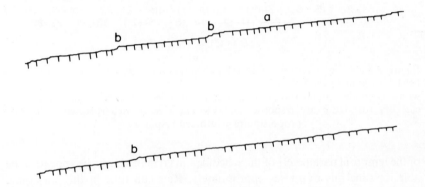

Figure 3.7   Cumulative record of a rat responding on a DRL15-s schedule. At point 'a' the subject is spacing his responses accurately enough to obtain a constant flow of reinforcers. At points marked 'b' the subject emits 'bursts' of responses which are not sufficiently separated in time to meet the DRL criterion.

more detail how long the animal is pausing between responses. This can be achieved by constructing what is known as an IRT frequency histogram. In figure 3.8 all the IRTs that occurred during an experimental session are sorted into a frequency histogram. In this particular histogram they are arranged into 2-s categories or 'bins', such that the number of IRTs of 2 s or less are represented by the first bar of the histogram, the number of IRTs between 2 and 4 s are represented by the second bar, and so on. The shaded area represents those IRTs which were successfully reinforced because they exceeded the DRL criterion. Figure 3.8 is typical of the general shape of IRT histograms constructed from behaviour on DRL schedules. They are characterised by two peaks or 'modes'. The first mode is of very short IRTs or 'bursts' of responding. These are very short and well below the criterion specified by the schedule. The second mode occurs around the DRL criterion value. This suggests that the animal is able to make a successful discrimination

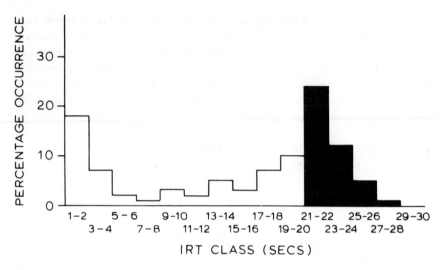

Figure 3.8   An inter-response time (IRT) histogram composed of IRTs taken from behaviour emitted on a DRL20-s schedule. Such histograms are normally bimodal with one mode around the DRL criterion and a second mode around the 1–2-s category (the latter are response 'bursts'). The shaded area represents the total proportion of reinforced responses.

of the temporal parameters of the schedule – in layman's terms he appears to be able to 'time' his responses appropriately. Rats can time in this way quite efficiently up to DRL values of 2–3 min (Ferraro, Schoenfeld and Snapper, 1965; Richardson and Loughead, 1974b), whilst pigeons are less efficient – efficient DRL behaviour starts to break down at DRL values above 20 s when they begin to emit too many short IRTs (Staddon, 1965; but see Richardson and Clark, 1976). (This appears to be related to the type of response used in the DRL situation rather than any phylogenetic differences in 'timing ability' (see pp. 190–191).) The experimenter can sharpen the animal's discrimination of the temporal parameters of the schedule by imposing what is known as a limitedhold (LH) contingency. On a DRL20 LH5 schedule, only a response which occurs between 20 and 25 s following the previous response will be reinforced. This tends to sharpen the mode around the criterion IRT and reduces the number of very long IRTs. Apart from the obvious characteristics of DRL behaviour there are more subtle sequential dependencies in the IRTs which an animal emits. For example, an IRT long enough to be reinforced is more likely to follow either a very short IRT (a 'burst') or a reinforced IRT than it is to follow an IRT that is just short of the DRL criterion (Ferraro *et al.*, 1965; Kelleher *et al.*, 1959; Malott and Cumming, 1964). Similarly, a 'burst' is more likely to follow an IRT just short of criterion, than it is to follow any other category of IRT (Harzem Lowe and Davey, 1975b). These factors have interesting theoretical implications as we shall see later.

**Explanation of DRL behaviour**

The two outstanding facts in need of explanation are: (a) how the organism comes to be able to make such fine temporal discriminations (that is, emit a large percentage of his IRTs around the DRL criterion), and (b) why such a large proportion of very short IRTs or 'bursts' should persist, even though they are never reinforced.

*Explanation of spaced responding*

*Collateral or mediating behaviour*   It has been observed on a number of occasions that animals being trained on a DRL schedule come to emit stereotyped patterns of behaviour (Laties, Weiss, Clark and Reynolds, 1965; Wilson and Keller, 1953). For instance, Wilson and Keller report that one of their rats:

> . . . went to the water bottle and climbed on it, climbed the ventilation holes in the rear of the cage, poked its nose at the glass cover over the cage, and returned to the bar.
>
> (Wilson and Keller, 1953)

It repeated this behaviour religiously after each bar press. The assumption here is that stereotyped chains of behaviour such as these come to be adventitiously reinforced and thus enable the animal to span the temporal interval specified by the DRL schedule. It is claimed that with this kind of interpretation the animal does not explicitly make a temporal discrimination, it merely repeats a sequence of behaviours which happens to fill the desired interval. There is some support for the proposition that mediating behaviour does play a role in efficient DRL behaviour. First, disruption of any ongoing stereotyped collateral responding does tend to disrupt the DRL performance (Laties, Weiss and Weiss, 1969); secondly, providing the animal with opportunities to perform collateral behaviours also increases DRL efficiency (Laties *et al.*, 1969; Schwartz and Williams, 1971; Zuriff, 1969); and thirdly, training the animal in restricted environments or in situations where it cannot produce overt chains of mediating behaviour reduces DRL efficiency (Frank and Staddon, 1974; Richardson and Loughead, 1974a). However, despite these positive points there are a number of factors which testify against mediating behaviour playing an important and long-term role in DRL performance. Firstly, there are many instances where efficient DRL behaviour has been established without any overt collateral behaviour being observed (Kelleher, Fry and Cook, 1959). This does not imply that no 'covert' mediating behaviour took place, but it does weaken a strong collateral responding theory. Secondly, the argument supporting the implication of mediating behaviour in efficient DRL performance stresses that chains of collateral behaviour are built up through their adventitious or superstitious reinforcement. We have already mentioned that current opinion does not see superstitious reinforcement as a major determinant of behaviour, and certainly chains of superstitiously reinforced behaviour are very labile in nature. In view of the extremely efficient and stable DRL performances that animals exhibit it seems highly unlikely that such a 'fragile' process as superstitious reinforcement could be solely responsible. Finally, it must be mentioned that mediating behaviour cannot really act as a substitute for temporal discrimination. As Nevin and Berryman (1963) point out, even if an organism is emitting a chain of behaviour which fills a specified time interval, to do this consistently the animal must make some discrimination of the duration of each component of the chain. So we still have the problem of how an organism discriminates temporally. Nevertheless, to put this argument into some kind of perspective, it does seem that collateral behaviour does aid the *acquisition* of efficient DRL behaviour, but to understand how efficient DRL behaviour is maintained really requires us to look elsewhere.

*Temporal control* An animal's responding on a DRL schedule shares many similarities with that on an FI schedule. Both schedules specify fixed temporal contingencies and both schedules produce pauses or latencies which indicate a temporal discrimination might be taking place. Thus, the temporal control hypothesis outlined by Staddon (1972) might be applicable to the spaced responding observed on DRL (see p. 69). There is one important difference between application of this hypothesis to FI and DRL schedules. On FI schedules the only temporally predictive stimulus is the occurrence of the reinforcer; on DRL we have also to postulate that the animal's own behaviour, that is, non-reinforced responses, can act as stimuli exerting temporal control. This is because the animal on DRL appears to be able to space his responses almost as efficiently after non-reinforced responses as after reinforced responses. A number of studies which have varied the nature of the reinforcer on DRL schedules (such as varying its magnitude or presenting a stimulus in lieu) suggest that this event does exert some control over the animal's ability to space his responses accurately (Caplan, 1970; Lowe, Davey and Harzem, 1976). However, Staddon (1972, p. 213–214) has suggested that typical DRL performance might result, not from temporal control, but simply from temporal regularity: that is, behaviour might be controlled by some hypothetical internal cyclical process. One experiment in particular has shown that this is unlikely, and that accurate spaced responding on DRL schedules can be controlled by reinforced and non-reinforced responses. Harzem *et al.*, (1975b) trained rats on a 'two-component' DRL schedule; that is, after reinforced responses the DRL criterion was one value (for example, 20-s) and after non-reinforced responses the DRL criterion was raised to a different value (for example, 40-s). They found that the rats came to emit differential pause durations after reinforced and non-reinforced responses (see figure 3.9). This indicates that non-reinforced responses can come to function as temporal inhibitory stimuli controlling pauses of a different duration to reinforced responses; this provides support for the argument that spaced responding on

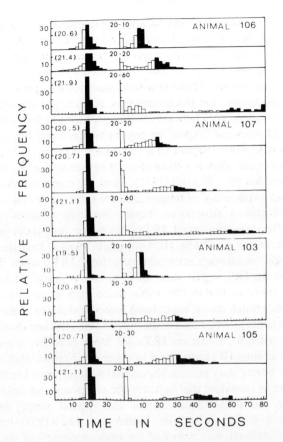

Figure 3.9   Relative frequencies of IRTs on a two-component DRL schedule. In each case the first component was held constant at 20 s and the second component varied between 10 and 60 s. The figure in brackets in the upper left-hand corner gives the duration in seconds of the median IRT in the first component. See text for further explanation. (From Harzem, Lowe and Davey, 1975b, copyright © 1975, by the Society for the Experimental Analysis of Behaviour. Reprinted by permission).

normal DRL schedules is controlled by responses acting as temporal inhibitory stimuli. However, the acid test of this hypothesis is a difficult one since it is not easy to manipulate the properties of a response in the same way that it is possible to manipulate the properties of a reinforcer.

*Explanation of response 'bursts'*

Response 'bursts' (IRTs of 2 s or less) are extremely persistent on DRL schedules, even though they are the IRT category most remote from the DRL criterion. Therefore, appeal to the schedule contingencies alone seems inadequate to explain this persistence.

*Frustration*   In the context in which we are to discuss it here, frustration is defined as a hypothetical motivational process which is considered to energise responding and is elicited when an 'expected' reward is omitted, (Amsel, 1958; 1962, see also pp. 94–95). It has been suggested that response 'bursts' may reflect frustration elicited by the non-reinforcing of a response. If this is the case, it must be shown that the animal was 'expecting' a reward at the time when the 'burst' occurred. There are obvious problems in discovering what the animal does or does not 'expect' at any particular time, but there are two points which lend some support to a frustration account. First, 'bursts' are more likely to occur after a non-reinforced response if a run of reinforced responses preceded this (Bradley, 1971). Thus, we might say that the animal has built up an 'expectancy' of reward, and the subsequent non-occurrence of that event elicited a frustrative 'burst'. Secondly, 'bursts' occur quite frequently after non-reinforced IRTs which are just short of the DRL criterion (Ferraro *et al.*, 1965). Here we might suggest that the animal has come to 'expect' reinforcers after relatively 'long' IRTs; when this does not occur because an IRT is just short of criterion, a 'burst' of responding is elicited. The evidence makes this explanation look quite attractive but its validity depends much on the logical and theoretical status of the concepts of frustration, a concept which has come in for some criticism (Staddon, 1970).

*Response feedback*   Criterion IRTs are more probable after very short IRTs than after long IRTs just short of criterion (Ferraro *et al.*, 1965). This suggests that 'bursts' may provide the animal with response feedback: that is, the animal can be confident that he has made a response and can time the next interval from this point. Thus 'bursts' may persist simply because they facilitate better timing behaviour. Kelleher *et al.*, (1959) found that if exteroceptive feedback was provided for each response – in the form of an audible 'click' – the number of 'bursts' declined: evidence which supports a feedback interpretation.

It is not inconceivable that both of these factors operate to maintain response 'bursts' on DRL schedules. However, it must be pointed out that runs of very short IRTs or 'bursts' of responding are found under all kinds of schedule contingencies, so it must also be considered that they might be a facet of behaviour which is insensitive to schedule control. This possibility has been mooted on a number of occasions, especially in relation to the pigeon's key-peck response (Blough, 1966; Schwartz, 1977; R. F. Smith, 1974).

## Summary

The DRL schedule is useful in the study of conditioning processes because it allows us to analyse closely the control of temporally based responding without excessive contamination by those excitatory factors which control response rate. It might be called a 'single-state' schedule in that it generates only states that we can call 'pausing'. Most other schedules could be labelled 'two-state' in that they generate a state of pausing followed by a state of responding. When this happens it is often difficult to isolate those factors controlling one state from the factors controlling the other. A DRL schedule is relatively free from this confusion.

## Response-independent schedules

There is perhaps another category of schedules which requires a mention because of the increased use it has recently enjoyed. These are schedules where reinforcers are presented without regard to the animal's ongoing behaviour; they are normally called response-independent schedules or schedules of non-contingent reward. The two important members of this category are called fixed-time (FT) and variable-time (VT) schedules. They are quite simply the non-contingent counterparts of fixed-interval and variable-interval schedules. They have so far had three main uses: (i) to assess the kinds of behaviour that might be *elicited* on schedules of reinforcement if there were no response contingency (Falk, 1969; Staddon and Simmelhag, 1971;), (ii) to assess if behaviours generated by contingent reinforcement will persist if the con- tingency is removed (Halliday and Boakes, 1972, 1974; Lowe and Harzem, 1977), and (iii) to assess the learning of the temporal parameters of schedules in the absence of a specified response (Trapold *et al.*, 1965; Zamble, 1967). Not all schedules need be response-contingent, and indeed it has often been argued that the response contingency only helps to obscure a clearer view of the variables which control responding in schedules of reinforcement (Staddon, 1977)!

## Schedule-controlled behaviour: a summary

At this point it is probably worth reiterating what was said at the beginning of this section – schedules of reinforcement do tend to allow too many variables to operate at once. This has invariably led those who wish to construct theories of responding on individual schedules of reinforcement to emphasise multiple causation in the determination of characteristic schedule perfor- mance (Ferster and Skinner, 1957; Reynolds and McLeod, 1970). Nevertheless, I think we can divide the controlling variables into two main groups. Those that control the *patterning* of responding and those that control the *rate* of responding. With regard to response patterning, temporal variables are by far the most important. We have seen that animals are extremely sensitive to the temporal parameters of reward even when these are not explicitly defined by the schedule (for example, on FR schedules). First, as a very general and loose principle we might state that animals will tend not to respond at times when reward is unavailable and will respond at times when reward is available (Staddon, 1974). At a more micro-level we have to discover what factors are important in enabling an animal to discriminate these different times. Secondly, the control of rate of responding is a more difficult topic to summarise. As we have mentioned, overall rate of responding does seem to be determined by overall frequency of reinforcement; how this comes about in practice is unclear – we have discussed one or two theories relating to this.

So, schedules of reinforcement do throw light on some of the basic principles governing adaptive behaviour in animals. However, they can also be used as important techniques in the study of other aspects of animal behaviour.

## Schedules as a basic technique in the study of behaviour processes

### Behavioural pharmacology

Schedules of reinforcement can be used not only to study processes of conditioning and learning, but also to generate stable patterns of behaviour on which the effects of other variables can be assessed. This is especially true of drug studies. Drugs not only have biochemical effects on the organism they also have behavioural effects as well, and it has now become clear that drugs often have behavioural effects which cannot be predicted simply by know-ledge of their chemical make-up or their effects on physiological states within the organism. So, in the study of the behavioural effects of drugs, schedules of reinforcement have a number of advantages. First, they maintain a steady, stable rate of responding (as in the case of VI schedules) or a stable pattern of responding (as in the case of the FI 'scallop') over very long periods of time. This has two benefits: (a) some drugs do not start to have their effects until many hours after administration; thus, animals can be placed into the experimental situation immediately after drug administration with little fear that the behavioural baseline will fluctuate irretrievably before the drug has had time to take effect; and (b) because schedule behaviour is stable over long periods, it can be used to assess any 'tolerance' effects the animal might acquire to the drug. (With repeated administrations of the same drug animals develop a 'tolerance' to it. That is, they show less and less intense reactions to the drug with each administration.) Since steady-state schedule performance varies little from day to day and experimental session to experimental session, the animal can be given repeated administrations of the drug and any tolerance to the drug effect that builds up can be fairly reliably detected. Secondly, schedules of reinforcement can be used to produce many different rates and patterns of behaviour. It has been discovered that the effect of a drug can depend crucially on the initial rate or patterning of responding (Sanger and Blackman, 1976). For example, (dl)-amphetamine, a central nervous system stimulant, will increase response rate on schedules which initially generate a low rate of responding, but will often *decrease* response rate on schedules which initially produce a high rate of responding (Dews, 1958). The reasons for these differential effects are complex, but this example does illustrate the way that schedules of reinforcement can be used to highlight the behavioural effects of drugs when these effects are not easily predicted from the physiological effects of the drug. The interested reader is referred to Blackman (1974, pp. 170–180), Thompson and Schuster (1968), and Iversen and Iversen (1975, pp. 138–165) for detailed discussion of the effects of drugs on schedule performance.

### The study of perceptual processes

Since animals will respond in the presence of stimuli which predict reinforcement but tend not to respond in the presence of stimuli which signal non-reinforcement (a *multiple* schedule) we can in effect use this to 'ask' the animal questions about its perceptual processes. For example, a rat is trained to press a lever for food in the presence of a 70dB 1000-Hz tone ($S^D$) and never reinforced for responding when there is no tone ($S^\Delta$) we can then vary either the intensity or the frequency of the tone. Varying the intensity of the tone will give us some idea of the auditory threshold of the rat, that is, what is the minimum intensity of sound the rat can hear? If the rat is unable to detect the tone, response rate during it should be similar to that during no tone. This can also be done with tone frequency to tell us something about the limits of the rat's sensitivity to auditory pitch. This example is an idealised and simplified one and the practical difficulties involved in assessing the perceptual limits of an animal are manifold. One of the first workers to try and come to terms with these problems was Blough (1958). Using operant discrimination procedures like those outlined above, Blough was able to plot the visual threshold of pigeons. His actual apparatus was extremely sophisticated, as figure 3.10 shows. Similarly the experimental design was quite an elegant one. Blough more or less asked the pigeon to peck key A if a light was on and key B if it was not. That is, it was not a simple GO–NOGO situation, as the first examples were, but an active discrimination. If there was no light on the bird was reinforced for pecking key B, if there was a light on he was reinforced for pecking key A; light–no light trials were randomly interspersed. By progressively reducing the intensity of the illuminated light a point would be reached where the pigeon would be more likely to peck key B because its perceptual system was not capable of detecting the light – that is, to all intents and purposes the pigeon responded as if there were no light. For simple perceptual problems, these techniques are usually quite successful, but there are many instances when the procedure needs to be so complex that adequate schedule control is never acquired in the first place.

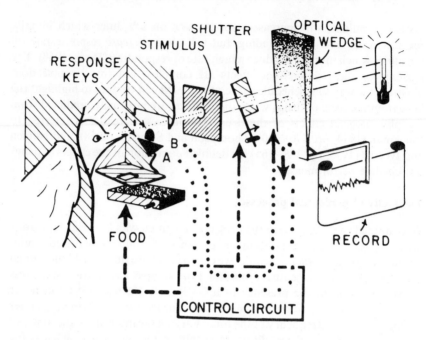

Figure 3.10 A schematic picture of the apparatus used by Blough to assess psychophysical thresholds in pigeons. (From Blough, 1958, copyright © 1958, by the Society for the Experimental Analysis of Behaviour. Reprinted by permission).

### Summary of the uses of schedules of reinforcement

The uses to which we can put reinforcement schedules are primarily threefold. First, they can be used to tell us more about the kinds of variables which influence learning in an operant conditioning situation. Secondly, they can be used to generate steady, stable baseline rates of responding on which we can assess the effects of other variables such as drugs. And thirdly, they can be used to 'ask' the animal questions about the limits of its perceptual and attentional processes.

## PARAMETERS OF THE REINFORCER

Before discussing some results of experiments which have manipulated aspects of the reinforcing stimulus we need to introduce a notion which, for many learning theorists is extremely important—the notion of *motivation*. In nearly all theories of conditioning two main factors are emphasised: (1) that the organism forms *associations* between events (for example, between stimuli in the environment or between external stimuli and behaviour), and (2) the organism needs to be motivated; this is usually conceptualised as a 'driving force' or 'energy' which in some way facilitates the learning of associations and may even 'direct' behaviour (see chapter 7 for elaboration of the theories relating to this). The distinction between these two factors is not a clear one: some theorists emphasise that they are relatively independent contributors to conditioned behaviour (Hull, 1943; Spence, 1956) while some see the two as inextricably inter-dependent or in some cases even as an unnecessary dichotomy (Rachlin, 1976; Skinner, 1953). Factors which are hypothesised as affecting the level of motivation in conditioning experiments are numerous, but such things as level of food deprivation, quality and quantity of the reinforcer and delay between response and reinforcer give the reader a flavour of the types of variables we are dealing with. Level of motivation is said to affect conditioned responding in two ways—it can affect either *learning* or *performance*. In the former, motivational level is said to influence the rate at which associations are formed. That is, it affects the rate of acquisition of asymptotic behaviour (the point at which the rate and patterning of behaviour ceases to change from session to session). In the latter, an asymptotic level may have been reached but motivational variables will affect the translation of the learned associations into appropriate behaviour. For example, a hungry pigeon may have learned to peck the one of two response keys which is illuminated with a red light in order to obtain food. Now, if the pigeon is given unlimited access to food prior to the experimental session the accuracy of his discrimination is likely to break down; he is also unlikely to peck so frequently at the response keys. In this situation, it is not that the conditioning has been 'unlearned', but that on this occasion the need to obtain food is less important than it usually is. This is commonly seen as an effect on performance rather than on learning. This distinction will crop up a number of times in the course of discussing the effects of manipulating the parameters of reinforcement.

## Magnitude and delay of reinforcement

### Magnitude

There are a number of different ways in which the magnitude of the reinforcer can be varied. One can increase the number of individual units or pellets of food given to the animal, the duration of access to food can be manipulated, or the concentration of the reinforcer can be varied where this is possible (for example, with a sucrose solution). Although results using these different manipulations are tolerably comparable, in some situations the method of manipulating quantity of reward can cause subtle variations in results (Daly, 1972; Traupmann, 1971).

In general, the greater the magnitude of reward the greater the efficiency of operant or instrumental performance. This has been noted with a number of different measures in many different conditioning experiments. For example, rate of responding on FI schedules varies directly with the magnitude of reinforcement (Guttman, 1953; Hutt, 1954; Stebbins, Mead and Martin, 1959) as does rate of responding on continuous reinforcement (crf) (Guttman, 1953); response latency in discrete-trial lever-press experiments decreases with increases in magnitude (Michels, 1957); speed of running in an alleyway increases with increases in magnitude of the goal-box reward (Daly, 1972; Goodrich, 1960; Hill and Wallace, 1967; Kraeling, 1961; Roberts, 1969). Similarly operant discriminations can be formed more quickly the greater the amount of reward given for correct discrimination (Waller, 1968; Weisinger, Parker and Bolles, 1973).

How should we interpret these apparently very neat results? We need to look at them in two different lights. First, in relation to the effect of magnitude on acquisition of response; and secondly, in relation to the effect of magnitude on asymptotic or steady-state responding. A number of studies have shown that the greater the magnitude of reward, the faster is the rate of acquisition, but the resulting level of asymptotic behaviour, when reached, is similar for different magnitudes (Black, 1969). However, practically just as many studies have shown that differential asymptotic levels of responding can be maintained over many sessions (Crespi, 1942; Guttman, 1953; Zeaman, 1949). It is difficult to tease out the crucial variables here since experimental design, type of response, method of varying magnitude etc., vary considerably between studies. However, one factor does appear to be important and this is the way in which different reinforcer values are contrasted. If an animal is trained for long periods on a single reinforcer value before being changed to a different value, the apparent difference in asymptotic response rate on each value is slight (Jensen and Fallon, 1973; Keesey and Kling, 1961). If different reinforcer values are contrasted within each session, then their differential effect on response rate is enhanced (Harzem, Lowe and Davey, 1975a). This type of phenomenon appears to fall into the general category known as *contrast effects*. That is, if two values of an independent variable are contrasted closely in time, any differential effect on behaviour they may have is enhanced (see also p. 112ff.). A further illustration of this type of effect in relation to magnitude of reward is supplied by a study by Crespi (1942). He trained a number of groups of rats to run up an alleyway for food pellets and found that if a subject was trained on 20 pellets per reinforcement and then changed to 1 pellet per reward, his running speed was considerably less than a subject who had only ever received 1 pellet per reward (negative contrast). He also found the opposite effect: subjects changed from 1 pellet to 20 pellets ran much faster than animals who had only ever received 20 pellets (positive contrast). This enhancement or suppression of running rate occurred on the first few trials after change from one reward amount to another. Since contrast effects appear to have such abrupt effects on behaviour we must seriously consider the possibility that magnitude of reward effects result simply from the effects of different motivational levels on performance and not to their effects on rate of learning. This has traditionally been, and still is, a difficult effect to unravel, even though the data appear on the face of it to be so clear cut. I am afraid we must leave it by saying that some theorists see the effects of magnitude of reward as reflecting simply performance differences (Kimble, 1961), while some see them as reflecting differences in the formation of underlying response strength (Black, 1969; Capaldi, 1967; Rescorla and Wagner, 1972). It looks as though we must await more knowledge of the processes which underlie operant learning before we can adequately pinpoint the locus of the magnitude of reward effect.

## Delay

Many theories of conditioning emphasise that for efficient learning the reinforcer must follow the response with the minimum of delay (Mowrer, 1960; Skinner, 1948; Spence, 1956); as the time interval between response and reinforcer is increased, so learning would be retarded. As we shall see, different theorists proposed different reasons for this. Many studies have shown that the efficiency of the learnt response is inversely related to the delay interval between response and reinforcer (Logan and Spanier, 1970; Perin, 1943; Perkins, 1947; Renner, 1963; Sgro, Dyal and Anastasio, 1967). Again, we must ask how should we interpret these results? Two hypotheses are the most prominent. First, decrements in performance with long delays may result from the fact that competing responses may have intervened between response and reinforcer and have the effect (a), of being superstitiously reinforced because they occurred immediately prior to food (Skinner, 1948), or (b), of causing interference in memory, thus disrupting the formation of an association between response and reinforcer (Revusky, 1971; Revusky and Garcia, 1970; Spence, 1956, pp. 153–163). Secondly, it might be suggested that, in delay of reinforcement experiments, the strength of the conditioned response may depend on the availability of cues which could act as conditioned reinforcers to bridge the gap between response and primary reinforcers. For example, if a rat is trained to run up an alleyway and then detained in a distinctive goal-box for say 30 s before being given food, the distinctive cues in the goal-box may come to act as conditioned reinforcers which maintain running in the alley. Consideration of these two hypotheses suggests that what happens during the delay interval, or what cues are available during the delay interval are extremely important factors. Taking this into consideration, it should be possible to design experiments to test which of the two hypotheses have the most validity. If an animal is trained on a two-choice task and both right and wrong responses are accompanied by the same stimulus cues during the delay period (for example, in a left–right discrimination, in a T-maze, the rat enters similarly decorated goal-boxes whether he makes the right or wrong response), then learning is more adversely affected than if differential right–

wrong cues are used (Grice, 1948; Perkins, 1947; Wolfe, 1934). This suggests that conditioned reinforcement may play some part in bridging the gap between response and reinforcer, but it does not mean that long delay learning is impossible without it (Mackintosh, 1974, p. 158). Studies which have restrained the animal during the delay interval so that it cannot make competing responses (Spence, 1956) or removed animals completely from the experimental apparatus during the delay interval, have found fairly successful and efficient learning with delay intervals of up to 8 min (Lett, 1973). Both the formation of conditioned reinforcers and interference by competing responses affect the efficiency of learning in a delay of reinforcement experiment. This, of course, is not so startling; what is surprising is the length of delay intervals over which animals will learn to associate two events. The most dramatic is the association of food with a subsequent toxic effect, the latter often occurring many hours after the food has been consumed (see chapter 6, p. 206ff.).

## Reinforcement omission

Another way in which a property of the reinforcer can be manipulated is to omit the event altogether. If a reinforcer is omitted on an occasion when it was scheduled to be delivered, a characteristic 'reinforcement omission effect' is obtained; the rate or vigour of responding is usually enhanced following the omission. This is true of omitting a scheduled reinforcer on FI schedules (Jensen and Fallon, 1973; Staddon and Innis, 1966, 1969), VI schedules (Dickinson and Scull, 1975), FR schedules (McMillan, 1971), VR schedules (Priddle-Higson, Lowe and Harzem, 1976), and DRL schedules (Caplan, 1970). Amsel and Roussel (1952) developed a special apparatus in which the effects of reward omission could be studied. This is called the double alleyway and is shown schematically in figure 3.11. It consists of a start-box followed by a first alley (A1) leading to the first goal-box (G1). The exit from G1 leads to a second alley (A2) terminated by the second goal-box (G2). The rat is always given food in G2 but receives food only intermittently in G1. Thus running speed in A2 can be compared following reward and non-reward. Running speed is found to be faster following non-reward than following reward (Amsel and Roussel, 1952). Omission of reward also appears to have a non-specific invigorating effect. It can increase the general level of goal-box activity (Dunlap, Hughes, Dachowski and O'Brien, 1974; Gallup and Hare, 1969; Tacher and Way, 1968); increase the force of lever-pressing when this is measured immediately after intermittent reinforcement of chain-pulling (Levine and Loesch, 1967); and increase the level of aggression when this is measured immediately after the intermittent reinforcement of running (Gallup, 1965). So, let us sum up these facts: the important effects of reinforcement omission appear to be (1) to increase the subsequent rate of responding on schedules of reinforcement; (2) to increase subsequent running

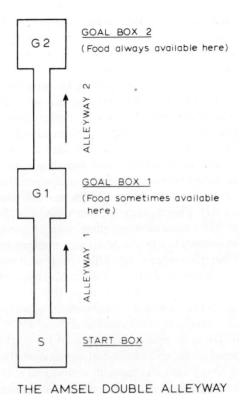

THE AMSEL DOUBLE ALLEYWAY

Figure 3.11   Diagrammatic representation of the Amsel Double Alleyway. See the text for a fuller explanation.

speed in the double alleyway; and (3) to have, on some occasions, a general and non-specific energising effect which influences behaviours other than those which are directly reinforced. Now how can this effect be interpreted? The traditional explanation has been one propounded by Abram Amsel (Amsel and Rousel, 1952; Amsel, 1958, 1962). He suggests that the absence of, or delay of, a reward event in a situation where it has been previously presented induces a motivational state which he calls 'frustration'. Frustration is 'an increase in the vigour of responding which immediately follows frustrating events' (Amsel, 1958, p. 103). (The reinforcement omission effect as we have called it here is perhaps more commonly known as the 'frustration effect' because of its traditional association with Amsel's theory.) Support for this interpretation has come from a study by Wagner (1959). One crucial factor in Amsel's theory is that frustration can be elicited only after the omission of an *expected* reward. Wagner trained two groups of rats to run the double alleyway. One group received food only intermittently in G1, the second group never received food in G1. Thus it is hypothesised, the first group would have a greater 'expectancy' of food in G1 than the second group who would have no 'expectancy' of food. Thus 'no food' in G1 should elicit frustration in Group 1 but not in Group 2: running rate in A2 after non-reward should be faster for Group 1 than Group 2. The results confirmed this hypothesis. The notion of frustration is an attractive one and has been used to explain a number of behavioural phenomena (Amsel, 1962). However, alternative interpretations of the reinforcement omission effect have been postulated and it is perhaps worth mentioning them here. Instead of non-reward eliciting frustration which invigorates subsequent responding it has been suggested that the reinforcement omission effect reflects the absence after non-reward of the demotivating effects of reinforcement (McHose, 1970; Seward, Pereboom, Butler and Jones, 1957). That is, response rate is low after food because this produces a kind of 'short-term satiation effect' which temporarily suppresses responding. This is absent when food is omitted. Some credibility is lent to this hypothesis by the fact that response rate following food is inversely related to the amount or concentration of food reward (Carlson, 1968; Jensen and Fallon, 1973; Lowe et al., 1974; Seward et al., 1957). However, the most convincing evidence in favour of this 'demotivation' hypothesis as it is known is provided in experiments by Dickinson and Scull (1975) and Platt and Senkowski (1970). Using an operant analogue of the double alleyway, rats were never reinforced at the end of the first component (the G1 analogue). After a period of training food was suddenly introduced at the end of the first component; this had the almost immediate effect of suppressing response rate during the second component (the A2 analogue). These results provide strong evidence that reward does have a suppressive effect on subsequent response rate. Whether this is due to 'demotivation' or 'short-term satiation', however, is another matter.

A third hypothesis is one that we have already come across earlier in this chapter. Staddon (1970, 1972) has similarly suggested that the omission effect might be due to the absence on non-reward occasions of the inhibitory or suppressive effects of reward. However, demotivation theory emphasises that these inhibitory effects are unconditioned while Staddon suggests that they are learnt during conditioning. For example, on an FI schedule, food predicts a period of non-reward and therefore acquires $S^\Delta$ or response suppressing properties (see p. 69). Omission of the reinforcer consequently removes its inhibitory effects thus apparently increasing response rate. This hypothesis has also been applied to the double alleyway omission effect: food in G1 is spatially (rather than temporally) some distance from the next reinforcer and hence acquires response suppressing properties. Omission of food in G1 removes this inhibitory effect and enhances running speed in A2. Two experiments lend support to this theory. Staddon (1970) trained pigeons on a schedule of food reward which developed a pattern of responding similar to an 'inverted' scallop: response rate was highest immediately after food and subsequently slowed down as the next reinforcer delivery became due. In this type of schedule it is claimed that food acquires excitatory rather than inhibitory properties; therefore, if food is a signal for responding rapidly, the presentation of a neutral stimulus in lieu of food should *reduce* rather than increase subsequent response rate. This is exactly what Staddon found. The result not only supports this 'generalisation decrement of inhibitory after-effects' hypothesis but is also contrary to both demotivation hypothesis and frustration theory. Similarly, the study of Kello (1972) that we have already mentioned (p. 69) provides further support for generalisation decrement. Without reiterating the details of the experiment the results have the following implications. First, frustration theory predicts that the more the stimulus events present at non-reinforcement approximate those present at reinforcement, the greater should be the elicited frustration and hence the higher the subsequent response rate; generalisation decrement hypothesis predicts the opposite, the greater the approximation the *lower* the subsequent response rate and if an event is physically similar to reinforcement it should have similar (inhibitory) effects as reinforcement. Kello found the latter case to be true in practice.

We have mentioned three important hypotheses which have attempted to account for the reinforcement omission effect. There are however a number of other interpretations which should be given credence, (Scull (1973); Mackintosh 1974, pp. 360–367). In summary, there are a number of points that should be made about interpretation of the reinforcement omission effect: (1) it is difficult to find evidence in favour of frustration theory which is not uncontaminated by other factors, that is, could not be interpreted in terms of generalisation decrement or demotivation; (2) although we can say with some confidence that food reward does have an unconditioned suppressive effect on subsequent behaviour, there are a number of situations where this fact cannot be implicated in an interpretation of the results of omission studies (Dickinson, 1972; Kello, 1972; McMillan, 1971; Staddon, 1970); (3) although generalisation decrement hypothesis can claim much experimental support, it has only really been tested on operant responses maintained by schedules of reinforcement, that is, its applicability to runway studies is untested; (4) the fact that reward omission does produce non-specific invigorating effects on behaviour (Dunlap *et al.*, 1974) lends more support to frustration theory than to any other.

Perhaps the most reasonable conclusion to be drawn from this might be to say that food *does* have an unconditioned suppressive effect, food *does* often acquire conditioned inhibitory properties, and omission of expected food *does* invigorate many aspects of an animal's subsequent behaviour, Which of these factors will exert more influence in an omission of reinforcement study will thus depend to a large extent on properties of the reinforcer, the nature of the response, and the structure of the reward schedule.

# CHOICE BEHAVIOUR

So far we have talked about behavioural phenomena which relate to a single source of reinforcement. However, there are many situations, especially in the wild, where the organism will have a choice of food sites, a choice of qualitatively different reinforcers, and a choice of different responses in order to obtain these different rewards. Given situations where the animal is confronted with two or more sources of reinforcement, what factors will govern his choice of response?

## Experimental evidence

### Discrete-trial probability learning

In discrete-trial studies of probability learning, an animal is placed in a two-choice situation on each trial (for example, he must choose to enter one arm of a T-maze or press one of two levers in a Skinner-box). The probability of any one of these choices being reinforced can be allotted a particular probability. For example, food can be placed in the left arm of a T-maze on 70 per cent of the trials and on the remaining 30 per cent of the trials it would be found in the right arm of the maze; both choices are intermittently reinforced, but turning right less regularly than turning left. The eventual pattern of choices that an organism settles on appears to depend less on the relative probability of reinforcement of the two alternatives than on what that organism is. Looking at a choice situation like this from our own viewpoint, there are two idealised strategies that the organism could adopt. He could *maximise*; that is, continually choose the option with the greater probability of reinforcement. Or he could *match*; that is, over a given number of trials he could match his choice between the alternatives in accordance with the relative probability of their reinforcement. Although no organism appears simply to adopt one or other of these two strategies certain animals do tend towards one or the other. For example rats will tend to maximise. In an experiment with a 70:30 distribution of rewards rats will usually choose the arm with the greater probability of reinforcement on over 90 per cent of the trials (Bitterman, Wodinsky and Candland, 1958; Johnson, 1970; Roberts, 1966). Mammals such as monkeys (Wilson and Rollin, 1959) and cats (Poland and Warren, 1967) also maximise whereas the only animals which consistently show a tendency to match in such choice situations are fish (Bitterman *et al.* 1958; Behrend and Bitterman, 1961, 1966; Mackintosh, Lord and Little, 1971). Studies of probability learning in pigeons tend to disagree as to whether they maximise or match. Some studies have reported matching (Bullock and Bitterman, 1962), but most have suggested a tendency to maximise (Mackintosh *et al.*, 1971; Mackintosh, 1970; Bitterman, 1971). The underlying reasons for these species differences in probability learning is unclear although there does seem to be some kind of phylogenetic progression which has led some theorists to propose that vertebrate classes differ qualitatively in the way they tackle probability learning: mammals tend to maximise, fish tend to match (Bitterman, 1965a). An alternative, however, is not to suggest qualitatively different methods of learning but simply to suggest that some animals are more *efficient* at learning the problem than others. That is, rats make fewer 'errors' than birds, that is, choose the lower-probability alternative less often) and birds make fewer 'errors' than fish (Mackintosh, 1969). Some support is lent to this latter view by the fact that even within the mammalian species there are variations in the rate at which different members will learn to maximise.

## Concurrent schedules of reinforcement

Another way of looking at choice behaviour is slightly different from the discrete-trials method. This is where the organism is confronted with two manipulanda and the frequency of reward on each manipulandum can be varied. For example, a pigeon may be confronted with two pecking keys: on the left-hand key a VI1-min schedule is in operation, on the right-hand key a VI5-min schedule is in operation. This is known as a concurrent VI:VI schedule and allows us to assess the proportion of responses emitted on each key in relation to the relative frequency of reinforcement available on each key. On concurrent schedules of this kind it is usually found that animals will match the proportion of responses on each key in accordance with the relative frequency of reward on the keys (Autor, 1960; Herrnstein, 1961, 1970; Pliskoff and Brown, 1976). On some occasions however matching of this kind does not occur, very often because the animal will develop a 'preference' for one or other of the manipulanda, or because it develops a superstitious pattern of alternating between choices. In order to eliminate bias of this kind a *change-over delay* (COD) is introduced. If the pigeon switches from one key to the other, then the COD arranges that the subject cannot obtain a reward for a brief period after the switch.

A further elaboration of the concurrent schedule procedure is known as the *concurrent chain schedule*. This is pictured schematically in figure 3.12. Initially both keys are white but a peck on one of them makes the other one inoperative; once the animal has completed the first link in the chain on the operative key, this changes the key colour and produces the terminal component. The initial and terminal components can be similar schedules or they can be completely different (Autor, 1960; Fantino, 1969). Concurrent chain schedules are useful in that they allow the experimenter to separate rate of responding between two schedules from the initial choice between them. Once the pigeon has made his initial choice between keys he is committed to obtaining reinforcement according to the schedule in operation in the terminal

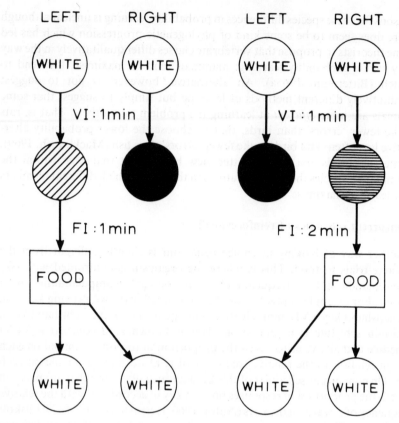

Figure 3.12   Schematic example of a concurrent chain schedule. At the beginning of each trial both keys are illuminated with white light. If the pigeon pecks only on the left key the colour of this key will eventually change in accordance with a VI1-min schedule that is operating on both keys. When this happens the right key is darkened and becomes inoperative; food can now be obtained according to an FI1-min schedule on the left key only. Once food is obtained the two keys are both re-illuminated with white light. The right hand side of this figure illustrates what would happen if the pigeon pecked the right key.

link of the chain. The initial choice of keys is usually independent of the schedule in operation in the initial link but is dependent on the frequency of reinforcement in the terminal link.

## The matching law

In choice experiments on concurrent schedules and concurrent chain schedules a fairly simple relationship holds between proportion of responses made on each key and relative rate of reinforcement on each key. This is

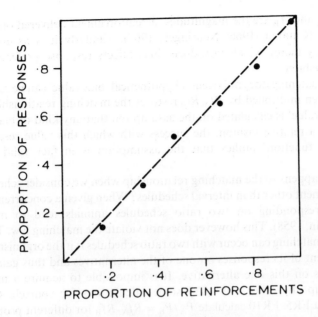

Figure 3.13   The relative frequency of responding to one alternative in a two-choice concurrent VI:VI situation as a function of the relative frequency of programmed reinforcement on the two VI schedules. Data are taken from Herrnstein (1961).

illustrated in figure 3.13 with data taken from concurrent VI : VI schedules run by Herrnstein (1961). This relationship can also be expressed in the form of an equation:

$$\frac{P_A}{P_A + P_B} = \frac{Rf_A}{Rf_A + Rf_B}$$

where $P_A$ and $P_B$ stand for the rate of responding on the two alternatives, and $Rf_A$ and $Rf_B$ are the rates of reinforcement scheduled on the two alternatives. This can be simplified to:

$$\frac{P_A}{P_B} = \frac{Rf_A}{Rf_B}$$

This relationship holds if all other things are equal. However, if the magnitude of reinforcement is greater for one of the two alternatives then matching, according to this simple equation, fails to occur. But the balance of the equation can be redressed because parameters of the reinforcer such as magnitude also affect choice in a relative fashion such that

$$\frac{P_A}{P_B} = \frac{Rf_A}{Rf_B} \times \frac{A_A}{A_B}$$

where $A_A$ and $A_B$ are the magnitudes of reinforcement delivered on the two schedules (Catania, 1966; Neuringer, 1967). Similarly it is assumed that almost any factor which introduces bias affects response preference in a relative fashion.

When matching does not occur a hypothetical 'bias value' can be calculated which when multiplied by $Rf_A/Rf_B$ restores the matching relationship. Since the 'bias value' is calculated on the assumption that any kind of bias affects choice in a relative fashion, the success with which this value restores the matching function implies that the assumption is in fact valid (Baum, 1974).

What happens to the matching relationship when we consider schedules of reinforcement other than interval schedules? When given a concurrent choice between responding on two ratio schedules animals tend to maximise (Herrnstein, 1958). This however does not violate the matching law. The only way that matching can occur with two ratio schedules is if the organism makes 100 per cent of its responses on one of the alternatives and thus gains all its reinforcers on this one alternative. It is impossible to acquire a matching relationship any other way. (Try some calculations for yourself. Given a concurrent FR5: FR10 calculate $P_A/P_B = Rf_A/Rf_B$ for different proportions of responses; the only proportion that matches is 100:0!)

While the matching law appears to have extensive applicability on concurrent schedules of reinforcement, there are a number of points to be made about its value to our understanding of what processes underlie choice behaviour. First, it is a simple empirical law that relates response rate to reinforcement rate. It has been closely argued that the apparent dependence of rate of responding on relative rate of reinforcement reflects the organism's ability to 'sample' density of reinforcement over a period of time and adjust his rate of responding accordingly (Baum, 1973; Hineline and Herrnstein, 1970). Much evidence which suggests that this is the manner in which the animals are able to adjust their rate and choice of responding comes from studies of avoidance learning and will be discussed in the next chapter (p. 140ff.). Suffice it to say here that proponents of this view believe that the matching law as a 'molar' law reflects a 'molar' mechanism. As we shall see, other theorists believe a more adequate analysis of choice behaviour can occur at the molecular level. Secondly, it may not have escaped the reader's notice that most studies on discrete-trial probability learning report maximising behaviour, while concurrent schedules generally produce matching behaviour. How are these two groups of findings to be reconciled? There are, of course, procedural differences between the two, and one talks of choice in terms of relative *rate of responding* while one talks of choice in terms of relative *proportion of choices*. However, it has been argued that all choice behaviour can be thought of as maximising, and that matching on concurrent schedules may be an artifact of using change-over delays (Mackintosh, 1974, pp. 193–194; Shimp, 1966).

There are a number of points to be made in relation to this suggestion: (1) matching only occurs on concurrent interval schedules, concurrent ratio schedules produce maximising (although we have stated that this latter fact is not inconsistent with the matching law, it is in reality only 'matching' in a trivial sense); (2) the majority of concurrent schedule studies use a change-over delay to discourage the animal from alternating between choices: the longer the COD the less the animal alternates (Shull and Pliskoff, 1967). This has the effect of tending the animal towards a maximising strategy, especially with CODs of 20 s or more. We should consider the possibility that with short CODs the simple response rate data might be contaminated by the adventitious reinforcement of alternation producing results more akin to matching than maximising. A ratio schedule is much less likely to reinforce alternation since, having switched from one key to the other, in most instances the subject would still have to make a number of responses to obtain food; an interval schedule – since it is solely time based – would be more likely to set up a reinforcer for the first few responses after an alternation; (3) a third way of considering behaviour on concurrent schedules is to think of the animal as not being sensitive to the relative rate of reinforcement but to the momentary probability of reinforcement. Mackintosh explains how this might account for matching on concurrent interval schedules and maximising on concurrent ratio schedules:

> On a concurrent (FR:FR) schedule . . . the alternative offering the highest probability of reinforcement . . . must *always* be the alternative correlated with the smaller ratio. The crucial point about concurrent (VI:VI) schedule is that both VI schedules continue to run while the subject is responding to only one of the alternatives. Since, on any interval schedule, the probability of reinforcement increases with time since the last reinforcement, it must necessarily be the case that the probability of reinforcement becoming available on the other alternative will eventually surpass the probability of reinforcement for continued responding on the same alternative. Hence, a shift to the other alternative is in accordance with a maximizing principle.

(Mackintosh, 1974, p. 194)

This is an attractive proposition since it would account for maximising behaviour in discrete-trial probability learning and concurrent ratio schedules, and also account for the apparent matching behaviour on concurrent interval schedules. Similarly, there is empirical evidence that pigeons can in fact track the momentary probability of reinforcement when it is manipulated experimentally (Fantino and Duncan, 1972; Shimp, 1966).

## Summary

Choice behaviour is that area of conditioning where molar and molecular approaches to the study of learning come into closest conflict. The matching law, as an example of the molar view, appears to have acceptable generality, but as an empirical law it does not necessarily tell us anything concrete about the mechanisms underlying choice. However, since it is an empirical law, explanation of choice behaviour on any other level must take the matching law into account. On the other hand the most acceptable example of the molecular view is that choice is determined by momentary probability reinforcement. The arguments supporting the account are compelling but experimental support for the claim that this is actually what is happening on concurrent schedules is not all that substantial at present.

# DISCRIMINATION AND STIMULUS CONTROL

We talked briefly in chapter 2 about establishing operant discriminations and also about the use of methods to measure stimulus discrimination and generalisation. This section intends to look more closely at factors affecting the formation of discriminations and also some of the behavioural pheno-mena which characterise discrimination learning procedures. Before continu-ing with this section the reader is urged to familiarise himself with the discussion of generalisation and discrimination in chapter 2 (pp. 49–52).

## Excitatory and inhibitory generalisation gradients

### Factors affecting the slope of excitatory and inhibitory generalisation gradients

Figure 2.14 (p. 52) illustrates examples of excitatory and inhibitory gradients produced by presenting various approximations to $S^D$ and $S^\Delta$ respectively. A flat function implies poor discrimination and good generalisation, a function with steep gradients implies good discrimination and poor generalisation. The slope of these functions can be influenced by a number of important factors.

*Differential conditioning*

We noted in chapter 2 that a good discrimination is best obtained when the animal is given differential training with both an $S^D$ and $S^\Delta$. If the animal is trained only with an $S^D$ (that is, is continually reinforced in the presence of a particular stimulus with no intervening $S^\Delta$ periods) and subsequently tested with various values of $S^D$, a completely flat generalisation curve is often found (Jenkins and Harrison, 1960). We noted, however, (p. 50) that Guttman and

Kalish (1956), using an identical training procedure, did find some evidence of discrimination. Whether any kind of discrimination is found after non-differential training seems to depend in part on what aspect of the environment the animal is attending to. Intuitively it is not hard to imagine that a pigeon will attend closely to visual stimuli at the end of its beak (for example, the colour of the pecking key) and associate this with food – Guttman and Kalish varied key colour and obtained a generalisation gradient which suggested some discrimination of the training value. Jenkins and Harrison, however, varied the frequency of an auditory tone. Again, intuitively it is difficult to see how a pigeon could fail to attend to a continuously presented auditory stimulus. However, there is evidence suggesting that pigeons will more readily associate visual stimuli with food than they will associate auditory stimuli with food (Foree and Lolordo, 1973, 1975, see chapter 6, p. 205). Thus, in this situation the pigeon may come to associate some visual aspect of the environment with food and not the auditory tone.

*Schedule of reinforcement*

The lower the rate of reinforcement during $S^D$ periods the flatter will be the generalisation gradient (Hearst, Koresko and Poppen, 1964). Thus, a flatter gradient is obtained on a VI4-min compared to a VI15-s schedule (Haber and Kalish, 1963), and also on a DRL schedule compared to a VI1-min schedule (Hearst *et al.*, 1964). Also, a flatter relative gradient of generalisation is found on a VR schedule than on a comparable VI schedule (Thomas and Switalski, 1966). A number of possible explanations present themselves to account for this relationship between level of discrimination and rate of reinforcement during training. First, lower rates of reinforcement usually produce lower rates of responding and thus less opportunity for reinforced responses to be associated with the training $S^D$. Secondly, the lower the rate of reinforcement during $S^D$ the more difficult it is for the organism to distinguish between conditions of reinforcement and extinction, thus he may take longer to distinguish the relevance of the $S^D$ and $S^{\Delta}$ as the events that signal reinforcement and non-reinforcement. Thirdly, the flatter generalisation gradient might simply be an artifact of 'floor' effects. That is, the organism responds less the lower the rate of reinforcement, so if response rate is very low (for example, as on a DRL schedule) it is fairly obvious that different rates of responding to different values of the $S^D$ will be less likely to show up than if rate of responding to the $S^D$ is initially high. A final possibility has been put forward by Hearst *et al.*, (1964) and this is that on some schedules successful responding needs to come under the control of other stimuli as well as the exteroceptive $S^D$. When this is the case, these 'other' stimuli are still present during generalisation testing and thus should produce behaviour quite similar to that produced during training. For example, DRL schedules require the animal to attend to temporal cues as well as the $S^D$, and ratio schedules 'make proprioceptive feedback from rapid responding a positive discriminative cue for additional responding' (Thomas and Switalski, 1966, p. 236). Thus generalisation gradients with these two types of schedule should be relatively flat compared with VI schedules. This has been shown to be the case:

### Errorless learning

Initially, during differential discrimination training, responses will occur to $S^\Delta$; intuitively this might be considered important since it enables the animal to learn that no rewards are delivered for responding during $S^\Delta$. So what happens if the organism makes no errors? Terrace (1963) developed a technique for producing discrimination without errors. The animal is first trained to respond solely in the presence of $S^D$, and then the $S^\Delta$ is 'faded in'. That is, the duration of its first presentation is very brief and its intensity very low; these parameters are very gradually increased until the $S^\Delta$ stimulus is the same duration and the same intensity as the $S^D$. Very few errors, if any, occur using this 'fading in' technique. However, the interesting point about errorless learning is that if generalisation testing is given for $S^D$ and $S^\Delta$, the generalisation gradient for $S^D$ is usually as sharp, and discrimination sometimes better than if training occurs with errors. However, the generalisation gradient for $S^\Delta$ is essentially flat and does not show a minimum around the $S^\Delta$ value, (Terrace, 1966a, b). The flat generalisation gradient for $S^\Delta$ presumably reflects the fact that since the animal never responded to $S^D$ he was never able to learn that $S^\Delta$ signalled non-reinforcement.

### Interactions between gradients

#### Summation

If a pigeon is trained to respond for food in the presence of a red key and also reinforced for responding in the presence of a yellow key, what will be the bird's reaction to being confronted with an orange key? Figure 3.14 shows the hypothetical excitatory gradients we might obtain for the two $S^D$s and the broken line illustrates the gradient we might obtain for the orange key if response strength from the two gradients was additive. Some experiments using key colour have demonstrated summation of this kind – but although response rate to the intermediate colour is higher than to the two training $S^D$s, total summation is rarely achieved (Blough, 1969; Kalish and Guttman, 1957, 1959). One point we must make in relation to summation is that although overall response rate to the intermediate stimulus might be higher than to the two training $S^D$s, it does not mean that the animal is responding at an overall higher 'tempo'. Rate of responding on most schedules of reinforcement can be characterised as 'bursts' of responding interspersed with pauses of different duration; rate of responding within each 'burst' is fairly constant but overall

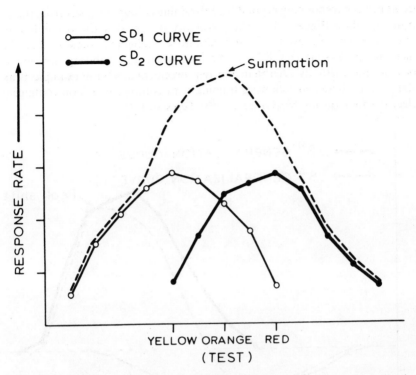

Figure 3.14  A stylised example of the summation of two excitatory generalisation gradients. Responding is developed separately in the presence of yellow and red S^Ds. When subsequently the subject is trained with an intermediate colour stimulus (in this case orange), the two generalisation gradients summate to form the curve represented by the broken line.

rate of responding is most often varied by variations in the length of the pauses which intervene between 'bursts'. The shorter the pauses between 'bursts' the greater the *perseveration* of responding – what appears to be happening during summation is greater perseveration of responding. Variation in the perseveration of responding rather than changes in overall tempo most often account for the differences in response rate observed to different values of $S^D$ and $S^\Delta$ during generalisation testing (Gray, 1976; Migler, 1964).

*Peak shift*

Inhibitory gradients are usually shallower and flatter than their excitatory counterparts. So what happens for example when a pigeon is presented with an orange key after being trained with a red $S^D$ and a yellow $S^\Delta$? The result is generally known as *peak shift* (Hanson, 1959). If we add together the positive and negative values of excitatory and inhibitory gradients this produces a shift in the peak of responding beyond the S$^D$ training colour and away from the S$^\Delta$ training colour (figure 3.15). The possibility that peak shift does reflect an interaction between excitatory and inhibitory gradients is supported by the fact that peak shift can still be obtained when an inhibitory gradient is produced by methods other than non-reinforcement in S$^\Delta$. For example, peak shift is even obtained following training with contingent or non-contingent electric shock during S$^\Delta$ (Grusec, 1968; Terrace, 1968).

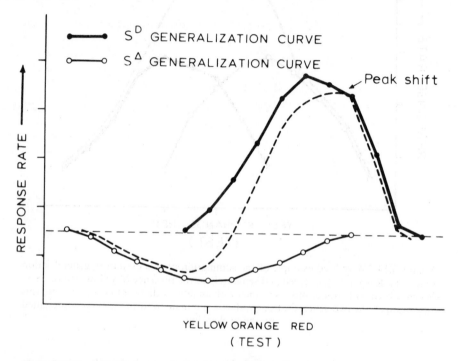

Figure 3.15  A stylised example of peak shift. In separate training procedures a red stimulus is established as an S$^D$ for responding, thus producing an excitatory generalisation gradient; and a yellow stimulus is established as an S$^\Delta$ with its associated shallow U-shaped generalisation function denoting inhibitory control. When the subject is subsequently trained with an intermediate stimulus (orange) the two gradients interact to produce an excitatory curve in which the peak is shifted away from the S$^\Delta$ end of the continuum.

## Parameters of the stimulus

### Stimulus compounds

During the course of learning a discrimination it must be remembered that there is much more happening in the animal's environment than just discrete presentation of $S^D$ and $S^\Delta$. In some instances responses may become conditioned as much to background stimuli as to the experimentally defined events. This being so, it is important to know more about the way in which conditioned stimuli interact in a learning situation. We have already noted that in classical conditioning the interaction between background and experimental stimuli and even between different experimenter-defined stimuli is quite complex (see p. 33). For this reason it is necessary that we look more closely at the characteristics of stimulus compounds and their components in operant conditioning. Since stimulus compounds in operant conditioning often behave in a similar manner to those in classical conditioning it is suggested that the reader acquaints himself with the latter before continuing.

*Selective attention*

Reynolds (1961b) conducted a discrimination experiment with two pigeons. A white triangle on a red background was the compound stimulus that signalled reinforcement, and a white circle on a green background signalled non-reinforcement. After training he presented each bird with the separate individual components of the discrimination: that is, red, green, triangle and circle. He found that one bird had learned to respond only to red, while the other had learned to respond only to a triangle. This selective attention experiment has been repeated many times with similar results (Eckerman, 1967; Keehn, 1969) and suggests that, instead of associating the whole compound $S^D$ with responding, the animal will 'select out' one relevant aspect of the compound and associate responding with this one element. The fact that animals do tend only to associate one element of a compound with responding is perhaps not surprising if we make the reasonable assumption that animals have a limited capacity for processing concurrently available information. But apart from the perceptual limitations of the animal, motivational variables also play a large part in selective attention. For example, an animal is more likely to attend solely to a single element of a compound stimulus if the magnitude of reward is increased or the level of food deprivation increased (Cohen, Stettner and Michael, 1969; Telegdy and Cohen, 1971). Similarly, the greater the response strength to one element, the less will be the response strength to the remaining elements (Sutherland and Holgate, 1966). It seems therefore that the animal not only has a limited capacity for processing information but also a limited store of 'associative strength', that is, the more it associates element $x$ with responding, the less it can associate element $y$ with responding (Rescorla and Wagner, 1972; also pp. 223–225).

*Stimulus compounding*

An animal is trained to respond for food on a VI schedule when either a tone or a light is present, with the absence of both signalling non-reinforcement.

Subsequent presentation of the tone and light *concurrently* will produce a response rate roughly equal to that if response rate during tone alone and light alone are added together (Weiss, 1964, 1971; Wolf, 1963). This is known as *additive summation* (Weiss, 1972) and seems to suggest that the response strength maintained by two individual $S^D$s will summate when the two are compounded. However, it must be pointed out that this only seems to occur, (1) when $S^\Delta$ periods are interspersed between $S^D$ presentation (Lawson, Mathis and Pear, 1968; Weiss, 1971) – this seems to reflect the importance of differential training in establishing excitatory stimulus control, and (2) when response rates maintained by $S^D1$ (tone) and $S^D2$ (light) are similar. If tone and light maintain behaviour on VI and DRL schedules, respectively, then response *averaging* occurs. That is, response rate to the compound is intermediate to that maintained by the two components (Weiss, 1967, 1969). If IRT distributions are analysed it can be shown that this averaging does not reflect an overall rate of responding intermediate to that maintained by VI and DRL, but to periods of responding appropriate to DRL interspersed with periods of responding appropriate to VI (Weiss, 1969, 1972). This again stresses the point that simply considering overall response rate during discrimination and generalisation studies tells the experimenter very little about what is actually happening to the behaviour. In fact response rate, when it is used as a dependent variable in these situations, is very often a statistical artifact of combining mixtures of behaviour over time (Migler, 1964).

Response averaging not only occurs when $S^D1$ and $S^D2$ signal VI and DRL, but seems to occur simply when they signal schedules which maintain vastly different rates of responding (Adams and Allen, 1971). There are a number of other interesting interactions which occur with the compounding of $S^D$s. For example, if $S^D1$ and $S^D2$ are separate stimuli which predict unavoidable electric shock superimposed on a baseline of VI behaviour (Conditioned Suppression procedure, see p. 142ff.), then VI responding is suppressed during their presence; if they are compounded, the suppression of responding in the presence of the compound is additive (Miller, 1969; Reberg and Black, 1969; Van Houten, O'Leary and Weiss, 1970). Similarly, when $S^D1$ and $S^D2$ maintain avoidance responding, the $S^D1\,S^D2$ compound produces a rate of avoidance responding equal to the rates $S^D1 + S^D2$ (Emurian and Weiss, 1972). Thus, whether additive summation or response averaging occurs with stimulus compounding depends less directly on the schedule of reinforcement, or the nature of the controlling events (appetitive or aversive) but more on whether the $S^D$s come to acquire excitatory or inhibitory control over responding. Even so, the nature of 'excitatory' and 'inhibitory' control is only poorly understood, but Weiss sees the stimulus compounding paradigm as one which 'might further help in clarifying and placing in contextual perspective, what are currently thought of as 'excitatory' and 'inhibitory' mechanisms in learning – thereby even stimulating the development of more advanced unifying principles of behaviour control' (Weiss, 1972, p. 206).

## Stimulus intensity

If a stimulus is established as an $S^D$ for responding, subsequent manipulation of the intensity of that stimulus has a direct effect on its control over behaviour. For example, when responding has stabilised on a mult VI:EXT schedule in the presence of a tone $S^D$, subsequent increases in the intensity of the tone result in corresponding increases in response rate during $S^D$ periods (Blue, Sherman and Pierrel, 1971; Hearst, 1969; Miller, 1971; Pierrel, Sherman, Blue and Hegge, 1970). This direct relationship between intensity of the stimulus and the degree of control it exerts over behaviour can be illustrated in a number of ways. Acquisition of discriminated avoidance learning is facilitated as a function of the intensity of the warning stimulus (Kessen, 1953; Bower, Starr and Lazarowitz, 1965); stable-state avoidance performance is more efficient the greater the intensity of the warning stimulus (Myers, 1962); in simple food reinforced discrimination procedures, probability of responding within a specified period after the onset of a stimulus increases as a function of stimulus intensity (Nevin, 1970); and percentage of correct choice responses in both forced-choice and GO–NOGO discrimination procedures increases as a function of stimulus intensity (Mentzer, 1966; Terman, 1970). These *stimulus intensity dynamism* effects, as they are known, are found not only in operant conditioning studies, but also in classical conditioning (Frey, 1969; Gray, 1965b), human vigilance tasks (Thurmond, Binford and Loeb, 1970), and also human reaction time experiments (Grice and Hunter, 1964; Murray, 1970).

Now how is this relationship between stimulus intensity and stimulus control to be explained? There have been two influential theories which have addressed themselves specifically to the animal studies; these can be called, (1) the stimulus trace hypothesis (Hull, 1949, 1952), and (2) the generalisation of inhibition hypothesis (Gray, 1965a; Logan, 1954; Perkins, 1953). In the first theory, Hull claimed that stimulus intensity had a direct dynamogenic effect; that is, increases in intensity resulted in increases in the intensity of a neural trace which in turn 'energised' behaviour. This produces increases in response rate, and response magnitude, and decreases in response latency. The second theory claims that stimulus intensity dynamism is merely a special case of generalisation and discrimination phenomena. For example, if on a mult VI:EXT schedule a tone is scheduled as the $S^D$ and no-tone as the $S^\Delta$, then the tone acquires excitatory control of responding while no-tone acquires inhibitory effects. Now, if the intensity of the tone is relatively low, one would expect poorer discrimination between $S^D$ and $S^\Delta$ such that the inhibitory effects of no-tone would generalise to the low intensity tone $S^D$. Thus, this generalisation of inhibition to tones of low intensity would tend to suppress response rate in their presence. Conversely, the greater the difference in intensity between $S^D$ and $S^\Delta$ the less will be the generalisation of inhibition from $S^\Delta$ to $S^D$ and hence the higher the response rate in the presence of the $S^D$.

This latter theory is supported by the fact that if the *highest* intensity stimulus is scheduled as the $S^\Delta$ and the *lowest* intensity as the $S^D$, response rate becomes an inverse function of stimulus intensity (Gray, 1965a; James and Mostoway, 1968). This suggests that stimulus intensity does not inevitably have an energising effect on behaviour but that its effect is dependent on the behaviour controlled by that stimulus. If a stimulus controls not-responding ($S^\Delta$) then increases in its intensity will reduce response rate. Results of this kind make the generalisation of inhibition hypothesis an extremely attractive one, but there are still data which suggest that stimulus intensity in some cases may have a direct dynamogenic effect. For example Sadowsky (1966) found that when the intensity difference between $S^D$ and $S^\Delta$ was held constant at 10-decibels, rate of extinction to $S^\Delta$ was significantly more retarded when $S^\Delta$ was the more intense stimulus and the discriminanda were located at the high intensity end of the continuum. Similarly, Spence (reported in Hull, 1947) found that rats made fewer errors in choice of runway when a white runway was the positive stimulus and a black runway the negative stimulus than when the signal value of the black and white runways were reversed. A number of other studies (Birkimer and Drane, 1968; Hearst, 1969) have shown that the acquisition and efficient performance of responding are facilitated when high intensity stimuli are used as the S + and low intensity stimuli as the S − rather than vice versa.

So, in conclusion, most of the data on stimulus intensity effects align well with the generalisation of inhibition hypothesis; there are, however, a small number of studies which also point to higher intensity stimuli having a greater potency for controlling behaviour.

## Contrast effects in discrimination learning

The literature on animal conditioning is peppered with phenomena which can be called contrast effects. We have covered a number of candidates already in this chapter when discussing contrasting conditions of reinforcement (Crespi, 1942; Dunham, 1968; Mackintosh, 1974, chapter 7; also p. 92), but perhaps those contrast effects which have aroused most research in recent years have been related to discriminative stimulus control. Contrast effects are characterised by the fact that when two or more differing conditions of reinforcement are presented either concurrently or in close temporal proximity, the difference in performance in the two conditions is exaggerated. If responding is relatively low in one condition and relatively high in the other when the two conditions are presented in isolation, then subsequent presentation of the two conditions in close temporal proximity will tend to increase responding in the high condition (positive contrast) and decrease responding even further in the low condition (negative contrast). Of course, this does not imply that all behavioural phenomena that are labelled 'contrast effects' are labelled so because they derive from the same or similar underlying processes. An explanation of contrast effects resulting from different magnitudes of reinforcement (see p. 92) is probably quite different from an explanation of the contrast effects observed in multiple schedules of reinforcement.

## Transient contrast in multiple schedules of reinforcement

### *Description of the phenomenon*

Figure 3.16 illustrates data from a study by Nevin and Shettleworth (1966). Using pigeons, they alternated 3-min presentations of a VI2-min schedule

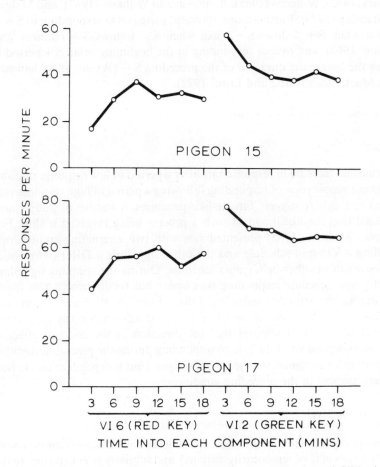

Figure 3.16   Rate of responding in successive 30-s segments of a mult VI6-min VI2-min. Early segments on the component with the lower reinforcement frequency (VI6-min) have a lower than average response rate (negative contrast), while early segments on the component with the higher reinforcement frequency (VI2-min) have a higher than average response rate (positive contrast). (After Nevin and Shettleworth, 1966).

(green key-light $-$ S $+$) with 3-min presentation of VI6-min schedule (red key-light–S $-$). (S $+$ is used to refer to the stimulus signalling the schedule with the higher rate of reinforcement and S $-$ to the stimulus signalling the lower rate of reinforcement). It can be seen from their results that at the beginning of the 3-min period of VI2-min (S $+$) response rate was temporarily enhanced (positive contrast) while at the outset of the 3-min period of VI6-min (red key) it was temporarily depressed (negative contrast). In nearly all cases, after the initial enhancement or depression of responding during the first 30 s of the new stimulus, response rate returned to a fairly stable intermediate level. Similar results have been obtained with both rats and pigeons by Boneau and Axelrod (1962), Williams (1965), Bernheim and Williams (1967), and Malone and Staddon (1973). Furthermore, transient increases in responding to S $+$ is greatest when S $+$ follows S $-$, than when S $+$ follows S $+$ (Boneau and Axelrod, 1962), and rate of responding at the beginning of an S $+$ period is greater the longer the duration of the preceding S $-$ (Wilton and Clements, 1971; Mackintosh, Little and Lord, 1972).

*Explanation of transient contrast*

Fatigue

The transient increase in response rate after a period of low response rate and a transient suppression of responding following a period of high response rate seem at first sight to suggest 'fatigue-like' processes. A number of studies have suggested that the likelihood of such a process being involved is slim. For example, Williams (1965) presented rats with two alternating stimuli, one signalling a VI5-min schedule and the other signalling a DRO (differential reinforcement of other-behaviour) schedule. During the stimulus signalling the VI-5 min schedule responding was *higher* but reinforcement rate *lower* than during the stimulus signalling DRO. Even so, Williams found that response rate at the outset of the VI5-min period was temporarily suppressed (negative contrast). It appears that the direction of the contrast effect is relatively independent of the rate of responding during the preceding stimulus (contrary to an explanation in terms of 'fatigue') but is dependent on the rate of reinforcement in the preceding condition.

Motivational level

Following exposure to the stimulus signalling the higher rate of reinforcement (S $+$) animals may be temporarily 'satiated' and similarly after exposure to the stimulus signalling the lower rate of reinforcement (S $-$) animals may be temporarily 'more hungry'. These motivational factors may be reflected in the transient increase and decreases in response rate following the change in reinforcement conditions. However, there is one overriding criticism of this interpretation of transient contrast, and that is that transient contrast effects do not appear immediately at the beginning of training as a motivational account would imply; similarly the effect tends to disappear with extended training (Williams, 1965).

## Relative 'value' of the $S^D$

The evidence we have presented so far suggests that transient contrast effects are more related to the rate of reinforcement in $S+$ and $S-$ than to the rate of responding in the two conditions (we shall come across this difference in more detail during the discussion of behavioural contrast). Both Nevin and Shettleworth (1966) and Williams (1965) have noted this point and suggested that the contrast effects observed may be an indirect result of $S+$ and $S-$ acquiring relatively excitatory or inhibitory properties depending on their association with a high or low rate of reinforcement. Thus, according to the principle of Pavlovian induction positive and negative transient contrast occur as the result of enhanced excitation immediately following an inhibitory stimulus, and greater inhibition following an excitatory stimulus. The problem with this approach is that it is difficult in some cases to ascribe one stimulus with inhibitory properties and another with the qualitatively different property of excitation. Presumably in Nevin and Shettleworth's experiment both stimuli should have been to some degree excitatory since they both maintained reinforced responding. Malone and Staddon (1973) have also addressed this problem, suggesting that the stimuli in transient contrast studies do not acquire qualitatively different properties of excitation and inhibition, but that they acquire properties (whatever they may be) which simply differ quantitatively along a single dimension. One suggestion is that the stimulus acquires a 'value' which is related to the relative rate or magnitude of reinforcement in its presence. In this case positive contrast becomes an 'elation' or overestimation of the value of reward in $S+$ following the relatively 'bad' conditions of reward in $S-$, and negative contrast is a 'deflation' or underestimation of the value of reward in $S-$ following the comparatively rich reward conditions of $S+$. As Bloomfield writes:

It appears that the most tenable hypothesis about contrast is that it is the result of elation at, or exaggeration of the value of reward in a discrimination situation which is caused by worse conditions (absence of reward) in $S-$. This exaggeration is a misinterpretation of the situation by the pigeon, which although it has parallels in human behaviour, is not justified by the reward contingencies.

(Bloomfield, 1969, p. 239)

### Behavioural contrast

*Description of the phenomenon*

Transient contrast effects are characterised, quite logically, by their transient nature; positive and negative contrast occur immediately after S+ has followed S − or S − has followed S +, with response rate eventually returning to an intermediate level. However, under some conditions of discrimination training these contrast effects do become relatively permanent. For example, S1 and S2 are key colours which alternate periodically; the pigeon however is reinforced during *both* on a VI1-min schedule (mult VI:VI). When responding during this non-differential period has stabilised, the VI schedule during S2 is replaced by extinction (mult VI:EXT). What usually happens is that as response rate during S2 predictably extinguishes, response rate during S1 actually increases above the pre-discrimination baseline level. Figure 3.17 illustrates this contrast effect, which lasts over a number of sessions and is quite marked in some animals (Reynolds, 1961a). How does this effect differ from transient contrast? The main difference is that the sudden introduction of the contrasting conditions of reinforcement in Reynold's study produced an elevation of responding to S1 which was maintained throughout the S1 stimulus and over very many sessions, until the original training conditions are reintroduced (Mackintosh, 1974, pp. 372–375; Terrace, 1966a).

*Explanations of behavioural contrast*

The first task that confronted investigations of the behavioural contrast phenomenon was to discover if the enhanced response rate to S1 was a result of (1), the reduction of responding in S2 or (2), the reduction in reinforcement in S2 or (3), a combination of both. The resultant evidence suggested that reduction in response rate in S2 was not a necessary nor a sufficient condition for positive behavioural contrast to occur; an actual reduction in reinforcement *rate* during S2 was, however, an important factor. For instance, positive behavioural contrast fails to occur reliably when reinforcement rate during S2 is equated with S1 but response rate during S2 is reduced in various ways. Such methods include (1), the punishment of a percentage of the responses occurring during S2 (Brethower and Reynolds, 1962; Terrace, 1968) (2), scheduling reinforcers on a DRL schedule during S2 (Terrace, 1968; Weisman, 1969) or (3), presenting free-food on a VI schedule during S2 (Halliday and Boakes, 1972, 1974). Similarly, if response rate during S2 is artificially increased by scheduling food on an FR schedule or a DRH (differential reinforcement of high-rate) schedule, enhancement of response rate in S1 occurs only if this schedule change produces a reduction in reinforcement rate during S2 regardless of an increase or decrease in response rate to S2 (Bloomfield, 1967a; Hemmes and Eckerman, 1972). However, having clarified the conditions under which positive behavioural contrast

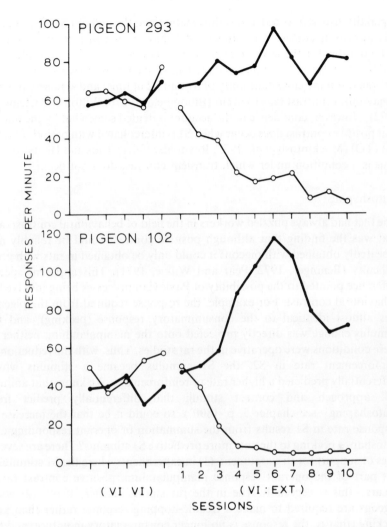

Figure 3.17 Response rates of pigeons during the final five sessions of non-differential training (VI:VI) and the first ten sessions of discrimination training between VI (●) and Extinction (O). (After Halliday and Boakes, 1971).

occurs, this does not point unequivocally to the processes which underlie the phenomenon. So how has theory attempted to integrate these facts?

Conditioning of transient contrast

The starting point is to suggest that positive behavioural contrast is merely a transient contrast effect that has become adventitiously reinforced. That is, the higher rate of responding occurring at the beginning of S1 may have been superstitiously reinforced and is thus maintained throughout S1. Support for this approach comes from studies of behavioural contrast in conditions unfavourable for the development of transient contrast. For example, if S1 and S2 are presented on different days, transient contrast does not occur (since it depends on the close continuity of S1 and S2 in time) and similarly positive behavioural contrast fails to occur (Bloomfield, 1967b; Wilton and Clements, 1972). However, confidence in the account is dented somewhat by the finding that positive contrast does occur when S1 is intercollated with periods of time-out (TO) (Mackintosh *et al.*, 1972; Reynolds, 1961a; Taus and Hearst, 1970). This is a condition under which transient contrast does not occur.

Autoshaping

One fact had always puzzled workers in the field of behavioural contrast, and that was the finding that although positive contrast could be reliably and repeatedly obtained with pigeons it could only be obtained in rats with great difficulty (Beninger, 1972; Pear and Wilkie, 1971). This apparent species difference pointed to the possibility of Pavlovian processes being involved in behavioural contrast. For example, the response required from the pigeons was almost identical to the consummatory response (pecking), and the stimulus change was directly projected onto the manipulandum; neither of these conditions were operative in the rat studies. Thus, with the reduction of reinforcement rate in S2, the S1 stimulus became a stimulus which differentially predicted a higher rate of reinforcement. We know that animals will approach and contact stimuli that differentially predict food (autoshaping – see chapter 5, p. 180ff.), so could it be that the increase in response rate in S1 results from the summation of operant responding and autoshaped pecking to the now more predictive S1 stimulus? There are several lines of support for this argument: (1) in experiments where the S1 stimulus is not part of the operant response manipulandum, positive contrast rarely occurs – this is especially true in the rat studies (Keller, 1974); (2) when pigeons are required to make a treadle-hopping response rather than key-pecking (that is, the response is no longer consummatory in nature, and the manipulandum is independent of S1), positive contrast fails to occur (Hemmes, 1973; Westbrook, 1973); (3) Gamzu and Williams (1973) have demonstrated that even when there is no response contingency, key-pecking will occur to S1 if it differentially predicts a higher rate of reinforcement; that is, no pecking occurred if S1 and S2 predicted the same density of free-food, but when food during S2 was withheld, pecking to S1 was elicited; (4) positive contrast only occurs if S1 contains a positive feature at which the pigeon can peck (Morris, 1976); if S1 is distinguished by the absence of a feature on the key (for example, S2 is a black line on a white key and S1 is merely a white key) then positive contrast fails to occur.

This evidence provides compelling support for an interpretation of positive behavioural contrast in terms of autoshaping. However, we must also note those occasions on which contrast has been found in conditions unfavourable for autoshaping to develop – this is especially relevant when we consider that although a majority of rats fail to demonstrate reliable positive contrast, a number still do. Perhaps the point to stress is that both the transient contrast and the behavioural contrast paradigm are experimentally untidy – they allow too many variables to operate. What we observe in these conditions clearly seems to be a complex interaction between both Pavlovian and operant discriminative processes.

# 4 The Aversive Control of Behaviour

Those events which are biologically important to an animal can generally be characterised in one of two ways. First, there are those things which an animal requires for normal physiological and behavioural functioning (appetitive events), and secondly, there are those things which damage or disrupt this normal healthy state (aversive events). Manipulation of either of these two classes of events in relation to an animal's behaviour is the basis of most animal learning. However, it is not necessarily wise to consider control by aversive events as merely the obverse or mirror image of appetitive control. Behaviours controlled by the two methods have important differences – differences which have been reflected in theories of appetitive and aversive conditioning. We shall come across behavioural phenomena which are peculiar to aversive procedures in this chapter, but before proceeding it is important to clarify the difference between an aversive event and a punishing stimulus. I have defined aversive events as those which are generally damaging, painful or simply uncomfortable for the animal – such things as electric shock, toxic substances, loud noises, etc. However, you will remember from chapter 2 that a punishing event is one which 'when made contingent upon a class of responses reduces the future probability of an instance of that class recurring'. Thus, there are *a priori* no intrinsically punishing events. This distinction has been reiterated for one main reason: so-called aversive events can on some occasions act like positive reinforcers. For instance, after certain training conditions both rats and monkeys will respond to produce response-contingent electric shock (McKearney, 1969, 1972). Similarly, the example of cigarette smoking is an all too familiar one: even though nicotine is harmful to health, human beings will continue to light up and smoke cigarettes!

For the sake of clarity this chapter has been arbitrarily divided into two sections. Response-correlated procedures involve those procedures where the aversive event is scheduled in a contingent relationship to responding. That is, responding either produces the aversive event (for example, punishment), removes the aversive event (for example, escape) or delays the aversive event (for example, avoidance). Response-independent procedures include those paradigms which involve non-contingently superimposing aversive events on a baseline of appetitively maintained responding (for example, conditioned suppression) or look at the effect of unavoidable shock on the subsequent acquisition of responding (for example, learned helplessness).

# RESPONSE-CORRELATED PROCEDURES

## Punishment

### Procedural aspects

It was mentioned earlier that punishment should not be considered simply as the obverse of reinforcement. It has, through various stages in history, been considered an important tool for controlling behaviour, notably because when scheduled efficiently its effects are immediate and very often long lasting. However, punishment is a fickle process which from the point of view of controlling behaviour can produce many undesirable side-effects – we will come to discuss these eventually. But first we need to consider what are punishing stimuli and what are the best ways to study the effects of punishment. Thorndike, it may be remembered, talked of events called 'annoyers', which when made to follow a behaviour reduced the strength of that behaviour (see chapter 2, p. 20). Azrin and Holz have formalised Thorndike's 'law' into a functional definition such that a punishing event is

> A consequence of behaviour that reduces the future probability of that behaviour. Stated more fully, punishment is a reduction of the future probability of a specific response as a result of the immediate delivery of a stimulus for that response. The stimulus is designated as a punishing stimulus; the entire process is designated as punishment.
>
> (Azrin and Holz, 1966, p. 381)

Stimuli which have usually been found to be effective punishers in experimental conditions include foot-shock delivered through a grid floor for rats (Azrin and Holz, 1966), electric shock to the tail region of pigeons via implanted electrodes (Azrin and Holz, 1966), a blast of air with cats (Masserman, 1946), and time-out from positive reinforcement in terms of a black-out of the conditioning chamber (Coughlin, 1972; Herrnstein, 1955). One important factor in punishment studies however is that there needs to be a pre-punishment baseline level of behaviour so that the suppressive effects of punishment can be assessed. The non-reinforced or naturally occurring 'operant rate' of some behaviours have been used in punishment studies (Baron and Antonitis, 1961), but more usually a steady baseline rate of responding (such as lever-pressing) is developed through appetitive conditioning on a schedule of reinforcement (usually a variable-interval schedule). A second procedural requirement that needs to be taken into account is that the punishing stimulus needs to be one which the subject cannot easily escape or avoid. Foot-shock is usually adequate for this purpose, but rats have been known to adopt elaborate strategies for avoiding the punishing event. Such examples include a rat learning to press the response lever for food while lying on its back, its hair thus insulating it from the grid floor shock (Azrin and Holz, 1966, p. 385). Others include the rat learning to stand on the earthed bars of the grid floor when grid floor shock is not scrambled, or even to hang from the ceiling of the conditioning chamber by its teeth! Although we might have great admiration for a rat's ingenuity, and in most cases empathy with his motives, such situations do not make for good experiments in punishment! So, you can see that even procedurally punishment is not merely the obverse of reinforcement. The experimenter has many more factors to take into account when designing an experiment on punishment.

### Behavioural aspects

Figure 4.1 is an idealised representation of some of the effects of punishment on rate of appetitively maintained responding. This diagram illustrates a

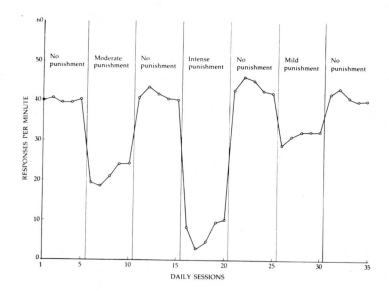

Figure 4.1   A hypothetical punishment experiment. This figure illustrates the effect on response rate of variations in punishment intensity during alternating periods of punishment and no punishment (From *A Primer of Operant Conditioning* by G. S. Reynolds, copyright © 1968, by Scott, Foresman and Company. Reprinted by permission).

number of facts about punishment. First, the degree of response suppression is directly related to the intensity of the punishing stimulus (Appel, 1963; Azrin, 1959b, 1960; Azrin, Holz and Hake, 1963; Boe and Church, 1967; Cohen, 1968), with very intense punishment usually abolishing responding totally (Azrin, 1960). It is important to note that, with most intensities of punishment, behaviour is only partially suppressed and the removal of the punishment contingency usually leads to either a partial or total recovery of responding. However, if responding becomes totally suppressed (that is, response rate becomes zero) then responding does not recover for a long period of time after the removal of the punishment contingency (Azrin, 1960; Masserman, 1946; Storms, Boraczi and Broen, 1962). Secondly, the suppression of responding by a punishment contingency is usually greatest when punishment is first introduced; a gradual recovery occurs in subsequent punishment sessions. It must be noted, however, that the method of introduction of the punishing stimulus is important. If a high intensity punisher is used on the first occasion, responding can be irreversibly suppressed (what is known as 'one trial learning' often occurs with intense punishment), but if the intensity of the punishing stimulus is gradually increased from mild to intense over successive days, suppression of responding is attenuated (Miller, 1960). Thirdly, although intense punishment has effects which in many cases are irreversible, mild and moderate punishment do allow the eventual recovery of responding. When responding is partially suppressed using either mild or moderate continuous punishment, what is known as a 'punishment contrast effect' is often observed following the removal of the punishment contingency. That is, after punishment has been discontinued, the rate of responding often increases to a level *greater* than the original pre-punishment baseline (Azrin, 1960). This increase lasts for a short period before returning to the original baseline level. The magnitude of this transitory increase in rate is usually directly related to the intensity of the punisher. Reasons for the appearance of this contrast effect are unclear except that it might be related to the behavioural contrast phenomenon we discussed in the previous chapter (p. 112ff.). That is, when a period of reduced reinforcement rate is introduced into a session, subsequent reintroduction of the original reinforcement rate produces a transitory increase in response rate (Ferster and Skinner, 1957; Nevin and Shettleworth, 1966). Since suppressing responding with punishment also reduces the number of reinforcers obtained, the formalistic similarity with transient local contrast effects in purely appetitive conditioning argues for a common interpretation. This interpretation is supported by the fact that when explicit stimuli signal periods of punishment and non-punishment, although responding during the $S^D$ for shock is suppressed, responding during the $S^D$ for no-shock recovers, often to a level exceeding the level present before punishment (Brethower and Reynolds, 1962; Honig and Slivka, 1964). Thus punishment contrast effects occur, as do appetitive contrast effects, in multiple schedule designs.

These then are the main behavioural effects of punishment. There are other variables which are important such as the schedule of appetitive reinforcement maintaining responding, the availability of alternative responses or escape responses, and the schedule of punishment itself. However, the interested reader is referred to two reviews by Azrin and Holz (1966, pp. 380–447) and Church (1969) for further details of these effects.

## Theories of punishment

Although it is quite clear that certain events, when made contingent upon a behaviour, do reduce the future probability of that behaviour, it is not at all clear how this comes about. A number of very different theories have been postulated to account for the suppression of responding in this way and we will discuss a number of them here.

### The negative law of effect

One possibility is to assume that the processes of reinforcement and punishment are symmetrical opposites (Estes, 1944) and, just as reinforcement might enable the learning of S–R connections, punishment might effectively unlearn them. Few theorists have supported this viewpoint for a number of reasons. First it was felt that the actual contingency between the response and the punishing stimulus was unnecessary for the suppression of responding and both the negative and positive law of effect rest firmly on the belief that the contingency *is* necessary for learning or 'unlearning'. However, subsequent experiments have demonstrated that although response suppression does occur with non-contingent punishment, greater suppression results from the administration of response contingent punishment (Azrin, 1956; Boe and Church, 1967; Camp, Raymond and Church, 1967; Schuster and Rachlin, 1968). So the contingency does play an important part in the suppressive effects of punishment. More damaging to the negative law of effect, however, is an analysis of what is meant by 'unlearning' of S–R connections. Presumably, if punishment results in the unlearning of the conditioned response then that conditioned response should never occur again when both punishment and reinforcement are discontinued. However, if a period of punishment is followed by a period of extinction (neither punishment nor reinforcement), then response rate does increase during the initial periods of extinction (Estes, 1944; Skinner, 1938). This cannot be easily explained by the negative law of effect. Skinner (1938), rather than maintain that responding was 'unlearnt' by punishment, suggested that animals merely learnt to anticipate punishment; this anticipation suppressed responding. Thus, when the punishment contingency was removed so was the anticipation of punishment, and hence responding would be 'disinhibited' and show at least a transient increase in rate.

*Punishers as discriminative stimuli*

There is a great deal of evidence that some stimuli which normally act as punishers can also acquire discriminative properties. For example, Holz and Azrin (1962) found that a response contingent mild electric shock was as equally effective as a response contingent change in key colour in acting as a discriminative stimulus for reinforcement. Similarly, if in a two-choice discrimination a mild electric shock accompanies the *correct* response, then learning proceeds more rapidly than if shock is not delivered (the 'shock-right' effect, Fowler, 1971; Muenzinger, 1934). Azrin and Holz (1966) have argued that some of the effects observed during punishment procedures may be due to punishing stimuli also acquiring discriminative properties. For example, they suggest that the recovery of responding in extinction observed by Skinner (1938) was due to the fact that the punishing stimulus merely became a discriminative stimulus for non-reinforcement; when the punisher was removed, stimulus conditions were similar to those during reinforced training and so it is reasonable to expect that responding would recur. Although punishers acting as discriminative stimuli can putatively explain many perplexing punishment phenomena, it still appears that punishing stimuli have effects on behaviour over and above those possessed by discriminative stimuli. First, although mild electric shocks can act as $S^D$s, intense shock has primarily suppressive effects. For example, in the 'shock-right' phenomenon as shock intensity is increased, so shock for correct responding becomes less likely to facilitate learning (Fowler, Goldman and Wischner, 1968). Secondly, if punishers were merely discriminative stimuli signalling non-reinforcement one would expect that non-contingent and contingent shock should suppress responding equally. As we have already noted, suppression is greatest when there is a contingency between response and punisher.

*Competing responses*

We shall see in the next chapter that both operantly reinforcing and punishing stimuli can act as UCSs which elicit certain species typical behaviours. For example, food will often elicit food-related consummatory responses in rats or pecking in pigeons. Similarly, aversive stimuli such as electric shock can elicit not only emotional responses but also a variety of what are commonly called species-specific defence reactions (see chapter 5, p. 192ff.). It might simply be the case that punished responding is suppressed because it is replaced by competing responses which are elicited by the punishing stimulus. These elicited competing responses then become associated through Pavlovian conditioning with stimuli in the experimental chamber. Support for this interpretation comes from studies which have attempted to punish these elicited behaviours and failed. For example, Morse, Mead and Kelleher (1967) found that monkeys who were punished for struggling on a leash showed an increase in struggling as the time for the next punisher became due. Thus electric shock did not suppress struggling but appeared to elicit it. Similarly, Walters and Glazer (1971) found that gerbils who were punished for adopting an alert upright posture showed a subsequent increase in this behaviour. The implications from these results is that punishers have their effect by eliciting incompatible behaviours which over-ride the operant response; if the operant response is similar to or identical with the elicited behaviour then it will fail to be suppressed. However, although punishing stimuli do in most situations elicit competing behaviours, there are a number of factors which suggest it cannot be a sole explanation of the effects of punishment. First, if intensity of punishment is high enough, even elicited behaviours can be suppressed (Azrin, 1970; Fantino, Weigele and Lancy, 1972; Melvin and Anson, 1969). Secondly, when a record of the behaviours emitted by an animal in a punishment situation is kept, it is sometimes observed that the suppression of the punished response often precedes by some time the development of an alternative or competing response (Dunham, 1971, 1972). Thus, when alternative or competing responses are observed in a punishment experiment, two distinct behaviour stages are observed: (1) the suppression of the punished operant, and (2) the subsequent development (for whatever reason) of the alternative behaviour. Thirdly, we continually harp back to the fact that contingent punishment produces greater suppression than non-contingent punishment. If punishers merely elicited competing responses without having a direct effect on the punished response one would expect suppression to be equal in the two procedures.

*Suppression of responding by events eliciting a mediating state of 'fear'*

None of the theories we have discussed so far provides a full account of the process of punishment, although this is not to say that in particular situations such factors as punishers being discriminative stimuli or eliciting competing responses do not suppress responding. The final theory we shall discuss is one which does have a more general appeal than those already discussed, but whose premises are more difficult to test. It has been suggested that an internal motivational state, which for convenience sake we will call 'fear', becomes conditioned via Pavlovian means to stimuli which precede the aversive event; this state of 'fear' either elicits competing responses or inhibits the appetitive motivational state and thus suppresses responding by inhibiting appetitive motivation. Evidence that stimuli signalling aversive events do suppress responding is found from studies of conditioned suppression (see p. 142ff.), so that part of this hypothesis is not in doubt. However, there are no obvious external stimuli which signal the aversive events in the punishment paradigm, so what elicits this state of 'fear'? One way round this difficulty is to suggest that the execution of the response itself is the event which signals the aversive stimulus; thus, either full or partial attempts to make the response will elicit 'fear' which will suppress either subsequent responding, or if the response is only partially executed, suppress the completion of the response. We shall see later in this chapter (p. 135ff.) that there is some evidence for a generalised hypothetical motivational state which one can call 'fear'.

## Summary of theories of punishment

At least two of these theories claim that the effects of punishment can be explained in terms of processes other than a specific punishment process, and as Mackintosh points out it should be noted that

> A general theory of punishment must allow that effective punishers have effects on behaviour over and above those attributable to other stimuli. But this is not to deny that shocks are stimuli, or that the result of some experiments are explicable in these terms.
>
> (Mackintosh, 1974, p. 283)

Although there is no definitive evidence favouring any one of the above theories, what the evidence does indicate is that a specific punishment process must be postulated to account for certain punishment phenomena. What the bare bones of this process look like is unclear, although it is almost certainly not a simple symmetrical opposite to the reinforcement process.

## Evaluation of punishment in the control of behaviour

### Procedures for maximum effectiveness

When one talks of punishment being maximally effective it is assumed that this means complete elimination of responding for an indeterminate period of time. We have already noted some of the factors which will achieve this end. The most obvious is to use intense punishment rather than mild or moderate punishment, and to instigate intense punishment at the outset of the punishment procedure. This usually results in the total suppression of responding with this suppression lasting for long periods of time even after the removal of the punishment contingency. Other factors which will maximise the efficiency of punishment are, first that punishment should be response contingent rather than non-contingent, and the punishing stimulus should be delivered as soon after the undesired response as possible; secondly, in order to avoid punishment contrast effects following the removal of the punishment contingency, intermittent rather than continuous punishment should be scheduled. One must point out, though, in the control of behaviour, punishment does not merely imply the elimination of behaviour, it might also be necessary simply to reduce in frequency a behaviour that is occurring at an undesirably high rate, or to suppress the occurrence of responding for only a specified period of time or in the presence only of specific stimuli. Appropriate scheduling of the punishment can usefully achieve these aims.

*Punishment as a method for eliminating behaviour*

Punishment is not the only way to eliminate behaviour, there are a number of others. So how does punishment compare with those other methods? Azrin and Holz (1966, p. 433) compared punishment with four other methods of eliminating behaviour: stimulus change, extinction, satiation and physical restraint. Their results are summarised in table 4.1.

**Table 4.1     Comparison of different methods of eliminating behaviour**

| Procedure | Immediate effect | Enduring effect | Complete suppression | Irreversible effect |
|---|---|---|---|---|
| Stimulus change | Yes | No | No | No |
| Extinction | No | Yes | No | No |
| Satiation | Yes | Yes | No | No |
| Physical Restraint | Yes | Yes | Yes | No |
| Punishment | Yes | Yes | Yes | Yes |

From this it can be seen that, under the conditions utilised by Holz and Azrin, punishment is by far the most effective. Remember though, that this list does not constitute the whole truth because changes in the parameters in any one of these procedures can make them either more, or less, effective. For instance, extinction is more effective after continuous reinforcement than after intermittent reinforcement and punishment only has irreversible effects when it is intense. Although punishment is now considered to be a most effective procedure in eliminating behaviour, it is interesting to note that there were times in its history when it was believed that a punishment contingency *per se* did not suppress behaviour at all (Thorndike, 1931; Skinner, 1953)!

*Comparison of reinforcement and punishment*

It is rather difficult to compare learning under contingencies of reinforcement and punishment: punishment is technically a procedure for eliminating behaviour whereas reinforcement is a procedure for developing new behaviours. It must be pointed out, though, that you cannot have a behavioural 'vacuum', so when a behaviour is eliminated by punishment some other behaviour has to fill its place. This implies that punishment procedures will be most effective if an alternative response is available, or if a desirable response is being concurrently reinforced as the undesirable response is being punished. Procedures which make alternative responses available do show that elimination of the punished response proceeds more rapidly under these circumstances, (Herman and Azrin, 1964). However, the point to be made here is that, as we have defined a punishing stimulus, you cannot teach a new response using only punishment. Punishment is merely a procedure for eliminating existing responses.

## Undesirable effects of aversive stimulation

The use of aversive procedures in the control of behaviour has recently fallen into disfavour, although in many periods of history it was considered a necessary and important part not only of behavioural control but also of the personal and moral development of the recipient. It is not our duty here to decide on the ethics of aversive control, although we will do that briefly elsewhere (chapter 15). What does concern us here is whether punishment might have disruptive effects on behaviour over and above the simple elimination of the punished responses. One of the main reasons why punishment has fallen into disfavour is because it is considered to engender emotional responses which can disrupt the efficient performance of un-punished behaviours. Although on some occasions it is fairly obvious that animals in punishment situations are displaying signs of emotional disturb-ance, the experimental evidence that is available suggests that this is more likely to be the exception than the rule. Studies by Hearst (1965) and Hunt and Brady (1955) have shown that punishment using electric shock produces few behavioural changes over and above the elimination of the punished response: rats showed no increase in defecation and an ongoing discrimination was only temporarily disrupted by shock presentation. So the claim that punishment produces disruptive emotional reactions does not appear to be founded in fact. From an adaptive viewpoint this is extremely important since if emotional disruption of normal day-to-day behaviours were produced every time a response was punished then the animal's survival would be severely jeopardised.

Another common complaint made against punishment procedures is that they elicit aggressive responses. Animals, when presented with a painful stimulus, will attack an available conspecific even if that animal played no part in delivering the painful stimulus (Azrin, Hutchinson and Hake, 1963; Ulrich and Azrin, 1962; Ulrich, Wolf and Azrin, 1964). This appears to reflect species-specific defence reactions when confronted with aversive situations (see also chapter 5, p. 192ff.). Given the availability of appropriate stimuli and the opportunity to make the response, most animals will either run away (flight), freeze (fright), or aggress towards certain aspects of the environment (fight). Punishment elicited aggression appears to be an example of the latter. However, it is difficult to say whether elicited aggression is a phenomenon specific to punishment, and thus it seems unfair to use this criticism specifically to condemn punishment as a method of behavioural control. Elicited aggression has been observed during a number of experimental procedures including even the appetitive reinforcement of responding (Azrin, Hutchinson and Hake, 1966; Delwesse, 1977; Flory, 1969; Thompson and Bloom, 1966; see also chapter 5, pp. 170–171). However, as far as the punishment procedure is concerned Azrin and Holz point out that

> The principal disadvantages of using punishment seem to be that when the punishment is administered by an individual, (1) the punished individual is driven away from the punishing agent, thereby destroying the social relationship; (2) the punished individual may engage in operant aggression directed toward the punishing agent; (3) and even when the punishment is delivered by physical means rather than by another organism, elicited aggression can be expected against nearby individuals who were not responsible for the punishment. These three disadvantages seem to be especially critical for human behaviour since survival of the human organism appears to be so completely dependent upon the maintenance of harmonious social relations.
>
> (Azrin and Holz, 1966, p. 441)

We shall return to these problems in Section Three.

## Escape and avoidance learning

In figure 2.5 we noted that there were theoretically two ways in which new behaviours could be operantly reinforced. The first of these was positive reinforcement and consisted of the presentation of an appetitive event contingent upon the new behaviour; the second was negative reinforcement where the defined behaviour either terminates or delays an aversive event. Termination of an aversive event is usually called escape learning, while the postponing or cancelling of a forthcoming aversive event is known as avoidance learning.

### Escape learning

If a rat in a Skinner-box is subjected to an electric shock and a bar press is programmed to terminate the shock, then rate of bar pressing will usually increase. It should not be difficult to envisage that in this paradigm schedules of negative reinforcement can be constructed almost as easily as schedules of positive reinforcement. For example, when a mild electric shock is given to the animal the experimenter may specify that only the 20th response will switch it off: thus, an FR20 schedule of escape responding is defined. Given that the intensity of the aversive event is not too great, a large diversity of escape schedules can be constructed and the patterning of behaviour on these is tolerably comparable with that on schedules of positive reinforcement, (Azrin, Holz, Hake and Ayllon, 1963; Sidley, 1963), that is fixed-ratio escape will produce break-and-run responding, fixed-interval escape will produce a 'scalloped' pattern of responding and so on. Dinsmoor and Winograd (1958) for example trained rats on a VI30–s schedule to escape from mild electric shock. This generated steady rates of responding which closely resembled the steady state behaviour found on VI schedules of positive reinforcement. Figure 4.2 illustrates some patterns of behaviour developed on different FR schedules of negative reinforcement; compare these patterns with those shown in chapter 3 for positive reinforcement. What does this formalistic similarity suggest? It does not imply that the actual processes of positive and negative reinforcement are identical, that is, that responses learned by the two methods are learnt in a similar fashion. A comparison of theories of avoidance learning and positive reinforcement will illustrate what is meant here. What it does suggest is that those factors which determine the *patterning* of behaviour in schedules of positive reinforcement are also operating in schedules of negative reinforcement. The point to be made here is that just because the patterning of behaviour is similar does not mean that the mechanisms for the *acquisition* of responding are similar.

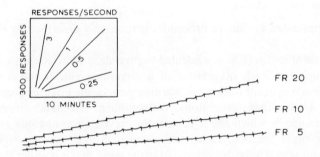

Figure 4.2 Cumulative recordings of fixed-ratio escape behaviour. Note that the 'break-and-run' pattern of responding found on appetitive FR schedules is still identifiable. (After Winograd, 1965).

Before we proceed, one other aspect of escape learning should be mentioned. This paradigm provides an excellent method for assessing an animal's threshold for different aversive events. For example, a rat may be subjected to a very mild electric shock, the intensity of this shock being stepped up by small amounts at regular intervals. If the rat makes a lever-press this will step the intensity down by an increment. This method is known as *titration* and what normally happens is that the rat will allow the intensity of the aversive event to reach a level where it becomes painful or uncomfortable and then responds to reduce the intensity. A dynamic relationship develops between shock intensity and the rat's lever-pressing such that the intensity of shock is maintained roughly around the value at which the rat begins to find it aversive. This procedure is especially valuable in assessing the effectiveness of certain drugs on behaviour, especially those drugs which are supposed to have analgesic or pain-reducing properties (Weiss and Laties, 1961).

## Avoidance learning

Once an animal has learned to escape from aversive events, the next beneficial strategy is to try and avoid those aversive events totally. The avoidance procedure differs from that of escape in that the occurrence of the aversive event is made predictable in order that the animal can response to postpone or avoid it. We have discussed escape responding first because avoidance learning almost inevitably involves learning to first of all escape from the aversive event. Indeed, avoidance learning proceeds more rapidly when an animal is initially allowed to escape from the aversive event (Carlson and Black, 1960).

*Avoidance learning procedures*

Discriminated avoidance

The discriminated avoidance procedure is succinctly described by Hoffman:

> A neutral stimulus (CS) is scheduled to precede, and in this sense serves as a warning for, each occurrence of a noxious event. If the S emits an appropriate operant during the warning period, the noxious event fails to occur. Under these conditions, discriminated avoidance represents a performance in which the S constantly prevents the noxious event, but seldom emits the operant in the absence of the warning signal. The behaviour is said to be discriminated in the sense that it is under the control of the extroceptive stimulus.
>
> (Hoffman, 1966, p. 499)

The response adopted for discriminated avoidance studies depends very much on the type of apparatus used. Although lever-pressing in rats is a commonly-used example, equally common is the study of avoidance learning in what is known as a *shuttlebox* (Miller, 1948). In this apparatus a rat is placed in one of two adjacent compartments and a warning stimulus is presented for a short period of time. Immediately after this an electric shock is turned on, and remains on, until the rat jumps into the adjoining compartment. After a brief inter-trial interval the warning stimulus (either a light or a tone) is again presented and, unless the rat jumps back into the original compartment, shock is subsequently presented. Eventually, after a number of trials the rat will learn to avoid shock by 'shuttling' into the adjacent compartment during the warning stimulus. During avoidance learning, attention is focused more on those variables which affect the rate of acquisition of avoidance learning than on the characteristics of steady-state avoidance responding (in free-operant schedules of positive reinforcement the latter commands more attention). The reasons for this should become apparent as we discuss theories of avoidance learning.

### Non-discriminated or 'Sidman' avoidance

In the non-discriminated avoidance procedure, no exteroceptive warning stimulus is scheduled; however, temporal variables can become adequate predictors of shock (Sidman, 1953). In a normal experiment brief electric shocks will be presented at regular intervals (the S–S interval); if, however, the animal makes an avoidance response at any time this will delay the next shock delivery for another specified period of time (the R–S interval). For example, shocks may be programmed to occur every 5 s (an S–S interval of 5 s) but if the avoidance response occurs (for example, a lever-press) this will delay the next shock presentation for, say, 20 s (an R–S interval of 20 s). Thus, two temporal parameters are experimentally defined, the S–S interval and R–S interval. In this situation each response will delay the next shock for 20 s from the occurrence of the response; if, however, a shock does occur, the shock–shock interval of 5 s is reinstated. Thus, if the rat presses the lever at least once in every 20 s he should be able to avoid ever receiving an electric shock. Under avoidance contingencies of this kind rats can learn to avoid shock efficiently. Since the contingencies specify fixed temporal relationships between shocks and between responses and shocks, animals normally come to distribute their avoidance responses in a way which suggests they are making a temporal discrimination. In the example we have quoted above, an animal that had successfully learned the avoidance response would not respond randomly, or in what would look like a 'panicky' fashion; but, if IRT distributions were analysed, it would show that the animals were normally pausing for just under 20 s between responses. This maximises avoidance of shock with the minimum of effort (Sidman, 1954, 1966). As we shall see, this fact has interesting implications for theories of avoidance learning.

*Variables affecting avoidance learning*

### Type of response

Without doubt the rate of acquisition of avoidance learning is dependent on the type of response. Whereas running up an alleyway to avoid shock is very readily acquired (Theios, 1963), it has frequently been noted that there are large numbers of rats who entirely fail to learn to avoid if lever-pressing is the response (Weissman, 1962). These 'failure' rats often adopt inefficient strategies such as lever-holding (Campbell, 1962), or develop competing behaviours such as freezing (Hoffman, and Fleshler, 1962; Hoffman, Fleshler and Chormy, 1961). Although elaborate procedures have been developed which reduce the interfering affects of lever-holding and freezing (Feldman and Bremner, 1963; Myers, 1959) the important question here is why lever-press avoidance should be so inefficient when compared with alleyway or shuttlebox avoidance. Why, for example, does freezing only rarely interfere with shuttlebox avoidance when it frequently retards or obviates lever-press avoidance? Bolles (1970) has suggested one explanation for this which is related to species-specific defence reactions (SSDRs). Since this theory is outlined in more detail in chapter 5 (p. 192ff.) we shall only discuss it briefly here. Bolles suggests that an avoidance response will only be efficiently learnt: (1) if it is a species-specific defence reaction, and (2) if that response produces stimulus changes which indicate 'safety'. That is, aversive events such as electric shock elicit defence reactions (such as running, freezing or aggression in the rat) and these defence reactions can be rapidly utilised as avoidance responses. Conversely, these elicited defence reactions will compete with and suppress the avoidance response if the latter is not a defence reaction. Thus alleyway running fulfills both of Bolles' criteria and is therefore readily acquired; lever-pressing fulfills neither of these criteria and is therefore acquired only with great difficulty. However, as a point of relevance, it is interesting to note that rats will often acquire bar-press avoidance by attacking the lever (the aggressive SSDR – Moore, 1973; Pear, Moody and Persinger, 1972) or by freezing onto it (Bolles, 1971). For a fuller account of the importance of response types in avoidance learning the readers is referred to chapter 5.

Intensity of the aversive event

The role of shock intensity in avoidance learning is rather complex. Although we might intuitively expect learning to progress more rapidly the more aversive the event to avoid, we also have to remember that the more intense the event to be avoided the more likely it is to elicit species-specific defence reactions. Thus, the effect of shock intensity will depend on the nature of the avoidance response; Lever-press avoidance is retarded by intense shock (Biederman, D'Amato and Keller, 1964; D'Amato and Fazzaro, 1966) presumably because competing SSDRs are more likely to be elicited, whereas one-way avoidance is facilitated (Moyer and Korn, 1966), again presumably because the running response necessary in one-way avoidance is more probable. This type of analysis appears to apply primarily to the *acquisition* of avoidance responses, because once the response is acquired, increases in shock intensity will in fact facilitate the efficiency of bar-press avoidance responding (D'Amato, Fazzaro and Etkin, 1967).

Temporal parameters

In discriminated avoidance, the optimum intertrial interval (ITI) for acquisition of avoidance learning is around 5 min (Brush, 1962). It is difficult to know exactly why ITIs shorter or longer than this produce poorer learning until we have a fuller understanding of avoidance mechanisms. However, Brush suggests that 5 min may be an optimum because it produces a balance between the acquisition of a state of 'fear' to the CS and extinction of 'fear' to background apparatus cues. As we shall see later, the conditioning to the CS of a motivational state labelled 'fear' is important to the acquisition of responding in two-factor theory of avoidance. Temporal variables are also important in the acquisition of non-discriminated avoidance. Although animals will learn to avoid if S–S and R –S intervals are equal, they generally learn more quickly when the R–S interval is longer. In fact, the shorter the absolute S–S interval (around 5 s or just less is an optimum) the more avoidance learning is facilitated. It has been suggested that this facilitation results from the reinforcing effects of interrupting a steady stream of frequent shocks (Sidman, 1966); that is, the response is reinforced because it produces a perceivable gap in the constant delivery of shocks (in this sense it is almost like escape learning).

## Theories of avoidance learning

Historically, avoidance learning has travelled an interesting course with theories of avoidance learning developing somewhat in isolation from the rest of conditioning. The reason for this is that avoidance responding raises two apparently perplexing questions, questions which are not directly encountered in other forms of conditioning. First, in what way can the *absence* of an event reinforce responding? That is '. . . how can a shock which is *not* experienced, that is, which is avoided, be said to provide either a source of motivation or of satisfaction?' (Mowrer, 1947). Secondly, avoidance responding is very resistant to extinction; that is, a high rate of responding is maintained for long periods after the removal of the avoidance contingency. Why should an avoidance contingency generate such persistence? So, any theory of avoidance learning must address itself primarily to these two questions and there are two theories which claim some degree of success: two-factor theory (Miller, 1948; Mowrer, 1947), and the shock frequency reduction hypothesis (Herrnstein, 1969; Sidman, 1962b).

Two-factor theory

In the discriminated avoidance procedure the warning signal is initially followed by shock on nearly all trials. This theory therefore postulates two stages to the acquisition of avoidance responding. First by a process of classical conditioning a central motivational state of fear is conditioned to the CS. (It then becomes a 'conditioned aversive stimulus' – a CAS.) Secondly, the avoidance response normally switches off the CS, consequently removing the stimulus which elicits fear. It is the reduction of fear which in this way reinforces the avoidance response. Mackintosh summarises this process in the three following propositions

(a) Pavlovian principles govern the conditioning of fear to stimuli correlated with aversive events such as electric shock. The strength of fear conditioned to a set of stimuli depends upon their correlation with shock. If shock occurs in the presence of A and not in its absence, then A will elicit fear and Ā will not. If shock occurs in the presence of A, but not in the presence of A + B, then B will become a conditioned inhibitor of fear.

(b) A response will be reinforced by a reduction in fear, i.e. if it changes the stimulus situations in the above examples from A to Ā, or from A to A + B.

(c) The magnitude of reinforcement for a response depends upon the degree of fear reduction. If A is correlated with a high probability of shock, and Ā or A + B with no shock, then reinforcement will be large; if Ā or A + B were simply correlated with a lower probability of shock, a response that changed A to Ā or A + B would receive some (but less) reinforcement. A second way of varying the effectiveness of reinforcement will be to vary the discrimability of A and Ā or of A and A + B for this will increase the generalization of fear from one stimulus situation to the other and thereby decrease the amount by which a change from one to the other will reduce fear.

(Mackintosh, 1974, pp. 307–308)

So, how successful is this theory in explaining the facts about avoidance learning? Perhaps the first task in this assessment is to find justification for the hypothetical motivational state called 'fear'. Postulating a concept such as 'fear' has been criticised for a number of reasons. First, and most obviously, using the term 'fear' often allows in by the back door all those connotations which are held by that word in the vernacular English. Thus, when it comes to predicting or describing the behavioural effects of an aversive procedure, we may ascribe to certain behaviours the characteristics of fear when this is unjustified. Secondly, 'fear' is normally defined as a 'classically conditioned, internal, unobservable response, accessible only through indirect measurement'. This is fine as long as 'fear' bears a predictable relationship to our indirect measurements. However, there is some doubt as to whether this is the case since it has been admitted that 'fear' is not linearly related to its measures and even simple relationships between 'fear' and behaviour can be obscured by 'other variables' (McAllister and McAllister, 1971). If this is the case, Hineline has argued that 'if a fear-eliciting stimulus is required to produce fear, and removal of that stimulus is required to remove fear, why not simply deal with directly measurable properties of that stimulus's effect on behaviour?' (Hineline, 1973, p. 537). That is, simply relate behaviour to the parameters of the CS (warning signal) and the aversive event. There are generally two arguments against this proposition. First, when a previously neutral stimulus becomes a conditioned aversive stimulus by being paired with an aversive event, it is not the stimulus that changes but some aspect of the animal; either its behaviour directly, or indirectly through the mediation of some central state (Fowler, 1971; Mowrer, 1939). As Anger argues

> . . . when a noise stimulus preceding an electric shock is said to have acquired an aversive property, this certainly does not refer to any change in the acoustic events in the environment of the animal. Instead, the 'acquisition of the aversive property' refers to a change in the reaction of the animal to the noise. Then why not call this new reaction a 'fear reaction'? Because 'aversive property' refers to just one part of a complex reaction, a precisely defined part. The whole reaction to a conditioned aversive stimulus is an extraordinarily complex reaction with many different components . . . many of which are difficult to measure.
>
> (Anger, 1963, p. 480)

This leads on to the second argument. This claims that conditioned aversive stimuli (CASs) acquire properties which are not easily explained by simply describing the operations by which the neutral stimulus became a CAS. A number of studies have shown that if a stimulus is established as a CAS and this is then presented to the animal during a second different learning task, the effects of the CAS will depend upon the motivational system maintaining responding during this second task. For example, if a CAS is presented during appetitively maintained operant responding, then response rate during its presence is significantly suppressed (Estes and Skinner, 1941; see p. 142ff.); similarly, if a CAS is presented during a session of non-discriminated avoidance, then response rate during the stimulus is enhanced (Herrnstein and Sidman, 1958; Scobie, 1972). It is certainly easy to conceive of these effects as the result of interaction between different or similar motivational states. A further experiment has relevance here. Solomon and Turner (1962) trained dogs to press a panel during a light to avoid shock; following this they were curarised, so that no skeletal responses could be made, and subjected to pairings of a tone with shock. Subsequently, when the effects of curare had worn off, five out of six subjects pressed the panel to switch off the tone on the first trial; this is interesting because there had been no previous opportunity for the panel-press to be reinforced by the termination of the tone. The implication here is that during training panel-pressing was reinforced by the removal of elicited fear, not the removal of the light CS. In effect, therefore, what the animal had learned to do was to terminate 'fear' and not any particular external CS. Thus the tone acquired fear-eliciting properties through its pairing with shock and on the first avoidance trial with the tone CS the dog merely responded to terminate fear – a response that it had already learnt. These, then, are the arguments for and against postulating a central mediating state called 'fear'. There are obviously some experimental results which are difficult to explain without its help, but because it is often difficult to relate the quantitative aspects of 'fear' to the measurements we do make in an aversive conditioning experiment, there are serious doubts about its predictive value at anything other than the very gross level. Although we have hardly been conclusive in discussing these arguments, it is instructive in the sense that this particular argument provides a good example of the debate between those learning theorists who see the necessity of postulating intervening variables in the explanation of behaviour and those who adopt a more radical behaviourist view which sees the explanation of behaviour in terms of an adequate description of the variables controlling behaviour (Hineline, 1973; Schoenfeld, 1969).

Now, given that for the moment we accept the usefulness of the concept of 'fear', how does the two-factor theory cope with the practicalities of avoidance learning? As we have seen, since responding terminates a stimulus which elicits fear, and in this way presumably reduces fear, we have now postulated a consequence of behaviour which could possibly be reinforcing. One implication of this, however, is that there must be some aspect of the environment which: (1) predicts shock, (2) then becomes a conditioned aversive stimulus eliciting 'fear', and (3) is then removed or changed by the avoidance response in order for fear reduction to occur and the avoidance response to be reinforced. Two-factor theory seems to be supported by the fact that acquisition of avoidance responding proceeds most rapidly when the response terminates a warning signal, less rapidly when the response merely produces a 'safety signal' and slowest when there is no warning signal at all (Bolles and Grossen, 1969; Bower, Starr and Lazarovitz, 1965; D'Amato, Fazzaro and Etkin, 1968; Kamin, 1956, 1957a). There are, however, certain avoidance procedures where it is difficult to identify any conditioned aversive stimulus whose termination will reduce fear. For example, in discriminated avoidance with a trace CS (that is, a brief stimulus which predates shock by a short but constant period of time) and also in non-discriminated avoidance, there are no exteroceptive stimuli which reliably precede shock. Neither of these examples has proved to be a stumbling block for two-process theories. In trace conditioning studies two-factor theorists have made much of the fact that avoidance learning is extremely inefficient (Kamin, 1954; Warner, 1932), presumably because there is no exteroceptive CAS to terminate and thus little fear-reduction to reinforce responding. Secondly, it has been postulated that non-discriminated avoidance learning occurs because temporal variables come to function as CASs. Non-discriminated avoidance specifies fixed S–S and R–S intervals: when the animal makes a response this terminates shock for a fixed period of time. Anger (1963) has suggested that an anticipation of shock builds up with time-since-the-last-response and that this anticipation is aversive. To put it more simply, the animal develops a discrimination of the temporal parameter of the schedule, and learns that periods of certain duration without a response result in shock: thus, time-since-the-last-response comes to act as the CAS predicting shock and thus elicits 'fear'. A response will set the probability of shock to zero and so result in fear-reduction which reinforces responding. On the face of it this seems to be a difficult hypothesis to test, but we will return to it again during the discussion of shock-frequency reduction hypothesis.

Having been able to bypass the problems of what reinforces avoidance learning, how does two-factor theory interpret the persistence of responding during extinction? If 'fear' is conditioned to the warning signal by Pavlovian means, then surely during extinction, even though responding still occurs – the continued absence of shock following the CS should eliminate any classically conditioned response to the CS. That is, since shock never occurs, 'fear' elicited by the CS should be extinguished and since no fear-reduction can occur, the avoidance responding should also be extinguished. There are two points to be made here. In the first place, how true is it that avoidance learning is resistant to extinction? Although many studies have shown the difficulty in extinguishing the response (Seligman and Campbell, 1965; Solomon, Kamin and Wynne, 1953) others have had no difficulty in eliminating it (Bolles, Moot and Grossen, 1971; Kamin, 1959). There are obvious differences in procedure, shock intensity and types of response between studies, all of which are of crucial importance in avoidance learning. However, given that some studies do show that avoidance responding exhibits resistance to extinction, how does two-factor theory account for this? The most obvious target for two-factor theory is to try and show that the CS still retains aversive properties even though its pairing with shock has ceased. Some success has been achieved in demonstrating this. For example, Kamin, Brimer and Black (1963) attempted to assess the aversive nature of the CS after 3, 9 or 27 consecutive avoidance responses. They achieved this by appetitively reinforcing an operant response and, at various times during a session, presenting the CS (Conditioned Suppression procedure, p. 142ff.). Suppression of responding during the CS was greatest in the group who had only made 3 consecutive avoidance responses and least in the 27 consecutive response group. So, with increasing efficiency on avoidance schedules, the CS does tend to lose its aversive properties; but the important fact is that the CS in the 27 consecutive response group was *still* aversive – it still suppressed appetitively maintained operant responding. Thus, if it was still aversive (albeit less so than originally) it would presumably still elicit 'fear' whose reduction would continue to reinforce avoidance responding. This still raises the question, however, of how a CS manages to retain its classically conditioned fear-eliciting properties over so many consecutive extinction trials – it is certainly not in line with the relatively rapid extinction of conditioned responses in traditional aversive classical conditioning experiments. However, one factor has been shown to be important in maintaining the resistance to extinction of avoidance responses. In some ways the continued occurrence of the avoidance response during extinction seems to protect the association between warning stimulus and shock from extinction (Soltysik, 1963). For instance, if the occurrence of the avoidance response is physically prevented during extinction (for example, by locking the door of a shuttlebox or immobilising the animal with curare), then subsequent extinction training results in relatively rapid elimination of the avoidance response (Baum, 1970; Black, 1958; Page, 1955). How the avoidance response helps to protect the CS–shock association from extinction is difficult to comprehend, but one suggestion has been made by Mackintosh (1974, p. 335). Since the avoidance response during extinction usually terminates the warning stimulus, the warning stimulus only very rarely runs its full duration. Now, it is the full duration of the CS that is paired with shock, so if the animal is sensitive to the temporal parameter of the CS – which is quite a reasonable assumption – we might suggest that it has only learnt that a 'long' CS predicts shock, whilst a 'short' CS predicts omission of shock. During extinction, if responses still occur, the animal never experiences the 'long' CS and so its aversive properties have no opportunity to extinguish. If the animal is physically prevented from making the avoidance response during extinction he does experience the 'long' CS and also the fact that it is no longer paired with shock; thus, the aversive properties of the CS do extinguish.

So, two-factor theory does appear to have some success in explaining avoidance responding, but one criticism often levelled at it is that the two-factor theorist often stretches the theory beyond the bounds of either its credibility or its testability in order to achieve these ends. Many of the criticisms have come from those who feel that avoidance responding can be explained in terms of a fairly simple dynamic interaction between the animal's responding and the parameters of the aversive schedule.

Shock-frequency reduction hypothesis

CS elicited 'fear' was postulated by two-factor theorists at least in part so that some kind of consequence of avoidance responding could be pointed to. Other theorists have suggested that this is unnecessary if one takes a more molar view of avoidance learning. Their view is that animals can perceive reductions or increases in the *frequency* of aversive events and that, if a response results in the reduction of shock frequency, this is sufficient to reinforce that response (Herrnstein, 1969; Sidman, 1962b). This is certainly a more molar analysis of avoidance learning and is based on similar principles to those by which response rate on appetitive schedules is presumed to be determined by reinforcement frequency (see p. 78–79). So, by this hypothesis, on discriminated avoidance procedures the CS does not become a fear-eliciting stimulus but a discriminative stimulus in the presence of which a response will reduce shock frequency. Similarly, on non-discriminated avoidance schedules responding simply results in reduced shock frequency and, as a result of this, is reinforced. Support for the possibility that there might be a dynamic relationship between avoidance responding and shock frequency derives from the fact that on non-discriminated avoidance schedules overall response rate does vary in accordance with R–S and S–S intervals in a fashion predicted by the hypothesis (Sidman, 1966). Indeed it has been empirically demonstrated that rats will learn an avoidance response when that response does not immediately terminate or delay an electric shock but produces a period where overall shock frequency is reduced (Herrnstein and Hineline, 1966). Certainly, on the evidence we have reviewed so far, the shock-frequency reduction hypothesis looks to be a more parsimonious and possibly less contrived explanation of avoidance learning than two-factor theory.

First, as we have already noted, responding on non-discriminated avoidance schedules is not random; response patterning does indicate some kind of temporal discrimination. Now, in its rawest form, shock-frequency reduction hypothesis only makes predictions about the rate of responding and not the temporal patterning of responding. Two-factor theory is reasonably capable of dealing with this fact (Anger, 1963) but shock-frequency reduction hypothesis requires further assumptions which are not necessarily derived from the hypothesis itself. However, it must be pointed out that although temporal patterning does occur on non-discriminated avoidance schedules it is not as regular or as efficient as two-factor theory would like. The accuracy of the temporal patterning of responding does tend to get overstated in such avoidance schedules and this has been emphasised by proponents of the shock-frequency reduction hypothesis (Hineline and Herrnstein, 1970). Secondly, if two-factor theory has its explicit assumptions then shock-frequency reduction hypothesis has its implicit ones. For example, ' . . . it must be assumed that animals are calculating variations in the overall rate of shock per session, and correlating these with variations in their own pattern of responding' (Mackintosh, 1974, p. 328). What evidence is available suggests that animals can do this only with great difficulty. For example, Bolles and Popp (1964) found that rats could not learn to respond when a response did not avoid the next programmed shock but did avoid subsequent shocks. Similarly, Hineline (1970) found that rats will learn to press a lever merely to postpone shock. That is, on any particular trial a lever-press did not avoid shock, it merely delayed it to a later time. Thus the schedule was arranged so that lever-pressing produced no change in overall shock frequency – yet avoidance responding still developed. So what evidence is available does suggest that animals often respond more in accordance with the immediate consequences of their behaviour than in accordance with the more molar aspects of behavioural consequences.

So, to summarise, argument still surrounds the relative abilities of two-factor theory and shock-frequency reduction hypothesis to explain avoidance responding. Both claim, in differing ways, to explain avoidance learning phenomena but both make either explicit or implicit assumptions about the avoidance learning process which their opponents claim to be either untrue or untestable. However, perhaps the greatest source of discord is the fact that the two theories really attempt to explain avoidance learning on different levels;

one by postulating intervening variables to mediate contingency behaviour interactions, the other by merely relating behaviour directly to its controlling variables.

# RESPONSE-INDEPENDENT PROCEDURES

Both punishment and avoidance procedures stipulate a contingency of some kind between responding and an aversive event; that is, responding in these procedures influences the delivery or withdrawal of the aversive events. As stated in the previous chapter, it is not logically necessary that one should particularly concentrate one's attention on response contingencies in learning or that response-consequence contingencies are the most crucial variables in a learning situation. The non-contingent delivery of aversive or appetitive events also has important and often striking effects on behaviour, and these phenomena can often illuminate aspects of underlying adaptive mechanisms. We have touched on the effect of non-contingent food presentation in various places in this book without collating them into an integrated discussion (see p. 87). This is mainly because data are sparse and theoretical interpretations consequently speculative. However, the study of response-independent aversive procedures has a longer history and is at a more developed theoretical stage. For this reason they have been subsumed into a single section.

## Conditioned Suppression

### Description and measurement

Responding that is maintained by intermittent appetitive reinforcement (for example, on a VI schedule of food reinforcement) will usually be suppressed during the presentation of a tone or light (CS) if that event reliably precedes a brief, unavoidable electric shock (UCS). Figure 4.3 illustrates this effect on response rate: responding is only suppressed during presentation of the CS, and it returns to its pre-CS rate immediately after shock presentation. Procedurally this is the superimposing of a Pavlovian contingency (the CS-shock pairing) onto behaviour maintained by an operant contingency (the schedule of food reinforcement). This phenomenon was first studied in 1941 by Estes and Skinner and was originally called conditioned anxiety because, at the formal level, it bore a striking resemblance to those behavioural effects we call 'anxiety' in our everyday life. They further suggested that the conditioned suppression paradigm provided a useful means for quantifying anxiety; that is, the amount of suppression of responding during the CS could be taken as an index of underlying anxiety. Whether this is a valid supposition or not we shall discuss later.

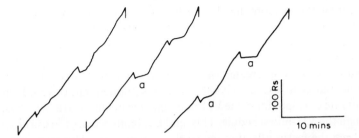

Figure 4.3   Cumulative recordings taken during a conditioned suppression experiment. Responding is maintained on a VI schedule of food reinforcement and the signal for unavoidable shock is indicated by a downward movement of the pen. Points marked a show a suppression of response rates in the presence of the signal for shock.

However, before proceeding to look at some of the variables that affect conditioned suppression, we must first discuss a method for calculating the degree of response suppression to the CS. Typically, suppression is measured in terms of a *suppression ratio* which can be calculated by using one of two simple equations

$$\frac{B}{A} \quad \text{or} \quad \frac{B}{(A + B)}$$

where $B$ is response rate during the CS, and $A$ is response rate in a corresponding period immediately prior to the CS. The second equation is the more widely adopted since it will provide an index of suppression on a scale from 0 to 1.0, where 0 is maximum suppression (no responding during the CS), 0.5 implies identical response rates prior to CS and during CS, and any ratio over 0.5 implies an enhanced response rate during the CS. Although the suppression ratio is a useful guide to what is happening to response rate during conditioned suppression procedures, it must be treated with some caution because: (1) there is no real theoretical justification for using this measure (Hoffman, 1969), and (2) the degree of suppression will often depend on the baseline rate of responding and there is no reason to believe that the CS will produce a similar relative decrement in responding with different baseline rates. Nevertheless, the suppression ratio does provide a useful shorthand for quantifying suppression effects, but its arbitrary nature must be borne in mind when it comes to theorising about the mechanisms underlying conditioned suppression.

### Variables affecting conditioned suppression

The severity of conditioned suppression appears to depend on both the baseline rate of responding and also the frequency of the reinforcement maintaining this responding. It can be envisaged that isolating these two variables is not an easy task since, for example, manipulation of reinforcement frequency will also influence rate of responding. Nevertheless, a number of studies have investigated these problems. For instance, Lyon (1963) found that suppression was greater on a VI4-min schedule than on a VI1-min schedule, but as Blackman, (1968b, 1972) has pointed out, this study did not control rate of responding: the VI4-min schedule maintained a lower response rate than the VI1-min schedule. This, and not reinforcement frequency, could have been the factor affecting the level of suppression. However, Blackman (1968b) conducted a number of experiments which were able to isolate the effects of response rate and reinforcement frequency. To generate different rates of responding with similar reward frequencies he used what is known as a response 'pacing' method. That is, overall reinforcement was scheduled according to a VI schedule of particular value, except that it was further stipulated that the reinforced response should be preceded by an IRT of a specified minimum duration. Different IRT criteria produced different overall rates of responding even though overall reinforcement rates were kept constant. By using a latin square design and manipulating the IRT criterion and the value of the VI schedule, Blackman was able to assess independently the effect of response rate and reinforcement frequency on conditioned suppression. The results showed that (1), conditioned suppression is affected by rate of operant responding, high rates being most suppressed and (2), the frequency of reinforcements obtained also has an effect, most suppression occurring when frequency is low. This finding has been used in differing ways to support different theories of conditioned suppression and so the possible locus of the effects of response rate and reinforcement frequency is best discussed in that context. However, perhaps more startling than the effect of rate of responding and rate of reinforcement is the effect on conditioned suppression produced by certain schedules of reinforcement. Variable-interval is the most widely adopted baseline schedule in conditioned suppression studies, simply because it develops a steady, constant rate of responding, upon which the effects of the aversive CS can be most reliably assessed. However, when responding is maintained by a differential reinforcement of low rate schedule, presentation of the pre-shock CS actually results in a *facilitation* of responding during the CS (Blackman, 1968a). This fact again has important theoretical implications which we will discuss shortly. Finally, if the baseline behaviour is maintained by negative reinforcement (that is, the avoidance of electric shock), presentation of a pre-shock CS also enhances rather than suppresses response rate during the CS (Rescorla, 1967; Sidman, Herrnstein and Conrad, 1957). Conditioned suppression of avoidance responding can be obtained if the CS duration is relatively short (Pomerleau, 1970). It appears then that conditioned suppression is something of a misnomer since suppression of responding during a pre-shock CS appears to be found only when responding is maintained by specified conditions of appetitive reinforcement.

## Theoretical interpretations of conditioned suppression

So, the procedure that initially appeared to Estes and Skinner to be a fairly simple measure of 'anxiety' turns out, on further investigation, to be quite a complex phenomenon. In the first place it has not been easy to find a comprehensive account of how a pre-shock CS disrupts operant responding, let alone also account for those circumstances in which responding is facilitated! Nevertheless, let us look at those theories which have addressed themselves to this problem.

### Competing responses hypothesis

Perhaps the most obvious way of interpreting conditioned suppression is to suggest that disruptive responses such as freezing, defecating, urinating and their covert autonomic correlates become classically conditioned to the pre-shock stimulus: when the CS is presented, the elicited CRs interfere with operant responding. On first sight this seems quite a reasonable and attractive hypothesis, but unfortunately it has a number of flaws: (1) what emotional CRs there are that are observable (such as freezing, defecating etc.) appear to be confined to the acquisiton stages and are not a feature of steady-state conditioned suppression; (2) there is little evidence that those classically conditioned responses that do occur in conditioned suppression experiments correlate highly with the suppression of responding, or develop at the same time as response suppression (Black, 1971; Brady, Kelly and Plumlee, 1969); (3) perhaps most damaging to the competing response hypothesis is the fact that a pre-shock CS on DRL and non-discriminated avoidance schedules actually facilitates responding. However, one might suggest that on DRL competing responses might disrupt the collateral behaviours which may have developed as necessary for efficient timing (Blackman, 1970). For instance, in studies where observable mediating behaviours have been experimentally disrupted, an increase in response rate is often observed (Laties, Weiss, Clark and Reynolds, 1966). Nevertheless, the fact that enhancement occurs with avoidance responding is still difficult to explain in terms of the interfering effects of conditioned respondents. It really seems likely that we need to look more closely at the interaction between appetitive reinforcers, aversive events, conditioned aversive stimuli and ongoing behaviour for a full account of conditioned suppression.

### Punishment hypothesis

Although shocks are delivered independently of the animal's behaviour it may be that the accidental occurrence of a response immediately prior to a shock delivery adventitiously punishes the animal for responding in the presence of the CS. This interpretation is supported by the fact that the higher the baseline response rate (and thus presumably the more likely a chance coincidence of response and shock) the greater the degree of conditioned suppression.

Similarly, Gottwald (1967) has found that conditioned suppression on trial $N$ is inversely related to the interval between the last response and shock on trial $N-1$. That is, when shock and response occur close together on a particular trial, suppression is enhanced on the following trial. Again, this hypothesis looks quite appealing but there are a couple of problems which are important. First, this hypothesis again has difficulty in accounting for response facilitation on avoidance schedules without making added assumptions about the effect on responding of the delay between response and non-contingent shock. Secondly, which is the phenomenon and which is the explanation? We have seen earlier in this chapter that some explanations of punishment regard contingencies such as those explicitly programmed in conditioned suppression studies as responsible for punishment effects (see pp. 126–127). It seems that a more realistic evaluation of the punishment hypothesis must wait on clarification of the processes which underlie punishment itself!

*Conflicting motivation hypothesis*

Instead of looking directly at the behavioural effects of the conditioned suppression procedure, it has been suggested that conditioned suppression (and conditioned enhancement) can be better understood at a motivational level (Estes, 1969; Millenson and de Villiers, 1972). That is, the pairing of CS with shock induces a 'negative' motivational state (let us call it 'fear') which is antagonistic to the appetitive motivational state maintaining operant responding. The assumption here is that the negative motivation elicited by the CS subtracts from the appetitive motivation maintaining responding, thus reducing the 'desire' to respond. Since these motivational states are not directly observable, their existence and mode of interaction must be inferred from the experimental data. Certainly, this hypothesis does deal quite well with the fact that a pre-shock CS *facilitates* avoidance responding. Just as negative motivation would subtract from appetitive motivation to produce conditioned suppression, so one would expect 'fear' motivation to summate with the motivational state maintaining avoidance responding, and thus enhance response rate during the pre-shock stimulus. Similarly, if it is granted that the higher the frequency of appetitive reinforcement, the greater the amount of appetitive motivation, then this hypothesis also explains why suppression is conversely related to rate of positive reinforcement (see p. 144). Difficulties do arise, however, when we consider the data from DRL studies. This is also an appetitive schedule so why does the aversive CS enhance responding? It could, of course, be argued that appetitive motivation on DRL schedules not only maintains lever-presses or key-pecks but also maintains collateral behaviour essential to efficient spacing of responses. The negative motivation generated by the CS-shock pairings might reduce the occurrence of collateral behaviours, thus leading to the more frequent occurrence of lever-pressing or key-pecking. So, this hypothesis, given certain assumptions about the relationship between behaviour and underlying motivational states, seems to deal reasonably adequately with most of the facts of conditioned suppression and conditioned enhancement. However, it does suffer from the drawbacks inherent in the two-factor theory of avoidance learning – no adequate empirical measure of these motivational states is available. If postulation of such motivational states is to be useful in anything other than a trivial sense we need behavioural assays which can more accurately predict their strength and mode of interaction (see Millenson and de Villiers, 1972, for a discussion of this).

*Discriminative control hypothesis*

Blackman has suggested that, rather than talk in terms of motivational states, an adequate account of conditioned suppression can be constructed in terms of the disruption of discriminative control by the pre-shock CS. He suggests that

> The results of the Estes–Skinner procedure may depend on the degree of discriminative control exerted by a schedule, rather than on the nature of the motivation maintaining the baseline performance. Acceleration of some types of avoidance responding may depend on poor discriminative control of that behaviour, as does acceleration of food-reinforced behaviour.
> (Blackman, 1972, p. 43)

Different schedules obviously exert different types of discriminative control over behaviour, and Blackman suggests that it is the way in which the CS–shock presentation interacts with this discriminative control that determines whether response suppression or enhancement will occur. For example, Blackman points out that conditioned enhancement effects appear to be confined to schedules which specify a fairly rigorous timing contingency (that is, DRL schedules, and schedules of non-discriminated avoidance), whereas conditioned suppression occurs on schedules which require the animal to position itself close to the response manipulandum and emit a constant flow of responses (notably variable-interval schedules). The supposition is then that the pre-shock stimulus merely disrupts this discriminative control, producing inefficient timing behaviour on temporally based schedules (resulting in response rate increases), and a failure to 'attend' to the response manipulandum and emit responses on schedules which generate relatively high response rates. Blackman (1970) has provided some support for this hypothesis by showing that conditioned enhancement occurs when CS-shock pairings are superimposed on responding maintained by a *non-discriminated* avoidance schedule, but that conditioned suppression occurs when CS-shock pairings are presented during responding maintained by a *discriminated* avoidance schedule. The assumption here is that non-discriminated avoidance responding is at least partially controlled by shock occurrence since shock occurrence predicts the time of the next shock delivery (the S–S interval); when a non-contingent shock is delivered this may provide false discriminative cues which evoke avoidance responding, thus producing conditioned enhancement. In discriminated avoidance, however, shock has less control over responding because the CS preceding avoidable shock should control avoidance responding; thus, superimposed non-contingent shock should not evoke responding. This 'discriminative' interpretation fits the facts of Blackman's experiment quite well, and certainly the fact that both suppression and enhancement were found with responding maintained by aversive means provides difficulties for the conflicting motivation hypothesis: this hypothesis should predict enhancement on both discriminated and non-discriminated avoidance schedules. Obviously the discriminative control hypothesis is not an easy one to apply – schedules of appetitive and aversive reinforcement control responding in complex and different ways. However, where we do know something about the way in which schedule behaviour is controlled it is possible to predict the effects a pre-shock CS might have on responding. Where this has been done the results are encouraging for the 'discriminative' account.

### The concept of 'anxiety'

We started off this section on Conditioned Suppression by mentioning Estes and Skinner's original paper (1941) called 'Some Quantitative Properties of Anxiety'. How does this discussion reflect on the concept of 'Anxiety'? It is a word we used quite freely in everyday speech and one which has at least some kind of meaning for most of us. However, when asked to define exactly what we mean by anxiety the concept becomes a little more difficult to pin down. Replies will range from the subjective kind such as 'I become gripped by a feeling of terror', 'my palms start perspiring', or 'I just can't seem to do anything' to more objective observations such as 'an increase in heartrate and blood-pressure' or simply as 'recordable physiological or behavioural changes in the presence of a pre-aversive stimulus'. Estes and Skinner believed that the conditioned suppression paradigm provided a method for studying not only the effects of anxiety on behaviour but also the potency of supposed anxiety-reducing drugs. However, the reader may have noticed that during our discussion of conditioned suppression, and also in theories of conditioned suppression, there is very little mention of the word 'anxiety'; so how useful is the term in a behavioural analysis of this kind? If we wish to retain the concept there are two ways in which we have to look at the problem, (1) Is anxiety quantifiable in behavioural or physiological terms? and (2) Do we have any evidence that anxiety is an internal state which *causes* the behavioural changes observed in conditioned suppression experiments? Taking this second problem first, there is very little evidence for a correlation between suppression of responding and changes in internal states. Brady, Kelly and Plumlee, using rhesus monkeys as subjects, monitored changes in heartrate and systolic and diastolic blood-pressure during a conditioned suppression experiment. They could find no evidence for any covariation of operant and autonomic activity and suppression of operant responding invariably developed before any detectable changes in the autonomic measures. They concluded that

> ... the weight of available evidence would still seem to support the view that the significant behavioural and autonomic effects are not causally dependent. Rather, it would appear that the cardiovascular and skeletal changes are more accurately represented as independently conditioned effects of the same environmental contingencies. This characterization of the emotional conditioning process, if correct, reflects unfavourably upon theoretical formulations that emphasize either the causal interdependence of behaviours and physiological events or the primacy of either one.
>
> (Brady, Kelly and Plumlee, 1969)

So there seems little experimental evidence to support the characterisation of 'anxiety' as an internal state which *causes* suppression of responding. Can we better think of anxiety merely as observed behavioural changes? Estes and Skinner compromised by suggesting that anxiety was a particular state of the animal but that it was only detectable through its effects on behaviour.

> Anxiety is here defined as an emotional state arising in response to some current stimulus which in the past has been followed by a disturbing stimulus. The magnitude of the state is measured by its effect upon the strength of hunger-motivated behaviour, in this case the rate with which rats pressed a lever under periodic reinforcement with food.
>
> (Estes and Skinner, 1941, p. 400)

The staunch behaviourist would here argue that if anxiety is an internal state serving no explanatory purpose and measured only in terms of its effects on behaviour, why not just label those particular observed behavioural changes as anxiety and dispense with thinking of anxiety as an internal state? Blackman is one who has made this point

> Anxiety is only manifested in these experiments through operant behaviour .... It might therefore be better to talk of anxiety as these disruptions of emitted behaviour, rather than as a cause of them. Anxiety would then become a description of observable interactions between certain specified operations and various patterns of emitted behaviour.
>
> (Blackman, 1972)

If we adopt this approach then 'anxiety' has merely become a label for a set of operations and their observable effects on behaviour; so why retain the concept at all? From a theoretical viewpoint there seems little use for the term, but there do appear to be benefits in retaining it when putting conditioned suppression into a broader framework. As Sidman (1960a) has pointed out, the conditioned suppression procedure is a normal learning situation producing predictable changes in behaviour in a variety of different species. If we grant that the behavioural and physiological effects of the conditioned suppression procedure are similar to those effects we call anxiety, then anxiety is a perfectly *normal* response to certain environmental contingencies and not a *pathological* condition which implies some kind of physiological or psychological dysfunction. This not only helps to remove any stigma that those who suffer from 'anxiety' might feel, but also, if we can understand the processes that cause 'anxiety' – point the way to developing successful therapeutic techniques.

## Learned helplessness

There is a long history to the study of non-contingent shock presentation, especially the relationship of unavoidable aversive conditioning to the development of what was known as 'experimental neuroses' in animals (Masserman, 1943; Stroebel, 1969). However, more recently the study of the 'emotional' effects of unavoidable aversive events has been focussed on a phenomenon known as *learned helplessness* (Maier and Seligman, 1976; Seligman, 1976). Quite simply, if an animal is presented with a series of unavoidable shocks prior to avoidance learning, his subsequent ability to learn the avoidance response is retarded – sometimes permanently impaired – when compared with animals who have not received the pre-exposure to unavoidable shock. The study of this seemingly minor experimental preparation has become prominent for two reasons: (1) it has led to a 'cognitive' interpretation of a conditioning phenomenon (learned helplessness theory) – something which has hitherto been avoided where possible by learning theorists, and (2) the phenomenon itself bears a striking formalistic resemblance to certain types of pathological depression found in man – a resemblance which certain theorists have not been slow to recognise (see specifically, Seligman, 1976). However, perhaps the most interesting aspect of the learned helplessness effect is that it has pitted cognitive theories of behaviour face on against traditional S–R accounts. But before we discuss these theories let us look more closely at the phenomenon itself.

The fact that prior exposure to unavoidable shock appeared to retard subsequent avoidance learning was first reported in dogs being taught to jump a small barrier to avoid shock (Overmier and Seligman, 1967; Seligman and Maier, 1967). Seligman reports the typical behaviour of a dog in the avoidance situation who had been previously presented with unavoidable shocks:

> (The) dog's first reactions to shock in the shuttlebox were much the same as those of a naive dog: it ran around frantically for about thirty seconds. But then it stopped moving; to our surprise, it lay down and quietly whined. After one minute of this we turned the shock off; the dog had failed to cross the barrier and had not escaped from shock. On the next trial, the dog did it again; at first it struggled a bit, and then, after a few seconds, it seemed to give up and to accept the shock passively. On all succeeding trials, the dog failed to escape.
>
> (Seligman, 1976, p. 22)

Impairment of avoidance learning by pre-exposure to unavoidable shock has been demonstrated in other animals including rats (Maier, Albin and Testa, 1973; Seligman and Beagley, 1975), fish (Padilla, Padilla, Ketterer and Giacolone, 1970), and man (Hiroto, 1974; Hiroto and Seligman, 1975). The 'learned helplessness' procedure also appears to retard the initiation of aggressive as well as defensive responses (Maier, Anderson and Lieberman, 1972), and in general produces an emotional syndrome resembling experimental neurosis (Hearst, 1965; Mowrer and Viek, 1948).

**Theories of learned helplessness**

Although the phenomenon itself has come to be labelled 'learned helplessness' this tag must be approached with caution since it implies only one of a number of possible explanations of the effect. Let us critically assess these accounts.

*Adaptation*

One possible interpretation is that animals who become 'helpless' because of prior exposure to unavoidable shock do so because they become adapted to trauma and no longer 'care' enough to respond. This suggestion is inadequate on a number of scores, but the most valid criticism is that if animals failed to learn avoidance responses because they had adapted to the trauma of shock, then animals pre-exposed to escapable shock should equally adapt to shock. This is patently not the case (Seligman, 1976, p. 66).

*Competing responses*

A further suggestion is that 'helpless' animals may have become so because during unavoidable shock presentations certain responses may (1) occur adventitiously at the termination of shock and thus be superstitiously reinforced, or (2) reduce the intensity of the shock and thus become a 'partial' escape response. Transfer of such competing responses to the avoidance learning situation may retard the acquisition of the new avoidance response. Alternatively, unavoidable shock may elicit species-specific defence reaction such as freezing: these 'passive' reactions to shock could be elicited by shock during the later avoidance conditioning phase and retard the acquisition of the 'active' avoidance response (see Levis, 1976 for a fuller account of competing response, or S–R interpretations of learned helplessness).

Intuitively this account seems quite appealing, but there are a number of points we must bear in mind here: (1) on logical grounds this explanation is somewhat weak since if a competing response is superstitiously reinforced it is more likely to occur in the future, therefore, it is just as likely to get punished as it is to get reinforced. Therefore one can argue as much for the elimination of active competing responses as for their maintenance; (2) competing response accounts of conditioning phenomena abound in the literature – we have covered a number of them in this chapter. This appears to be because it is so easy with many learning phenomena to evoke competing responses as an explanation *post hoc*; and as Welker, Hansen, Engberg and Thomas (1973) have pointed out 'as such they (competing responses) can account for facilitation or impairment in transfer tasks and still have no predictive value'. This, of course, does not rule out competing responses as an explanation of 'learned helplessness' but it does post a warning with regard to the usefulness of postulating, willy-nilly, the existence of competing responses; (3) there is some empirical evidence which challenges the validity of part of the competing response hypothesis. Maier (1970) trained a number of dogs to remain totally passive or inert in order to avoid shock (passive avoidance). He later found that they learnt an active avoidance response (barrier jumping) no slower than a group of dogs with no previous avoidance experience. The conclusion from this is that learning to 'freeze' or be totally inactive (one of the competing responses postulated by advocates of this account) does not impair subsequent learning of an active avoidance response. The impairment in learned helplessness studies, it appears, stems from some other source. (One procedural problem with Maier's experiment is that his animals were reinforced for every response, and Mackintosh (1974, p. 218) has pointed out that responses reinforced on *crf* extinguish more quickly than those which are intermittently reinforced. Presumably, the adventitious reinforcement of freezing would only be intermittent during unavoidable shock presentation.)

*The learned helplessness hypothesis*

As a somewhat radical alternative to S–R interpretations of the learned helplessness phenomenon, Seligman (1976) and Maier and Seligman (1976) have outlined a theory which has a more 'cognitive' flavour to it. They postulated that animals can learn that no contingency exists between their own behaviour and its consequences. Having learnt that responding is futile they transfer this 'cognitive set' to subsequent learning conditions where their behaviour does have scheduled consequences. Hence they fail to learn to avoid shock after pre-exposure to unavoidable shock. Although this theory is consistent with a majority of the experimental literature (see Maier and Seligman, 1976, for a review) it has implications which are not necessarily proven by these studies. For example, it implies that (1) animals over a sample period of time estimate the relationship of their behaviour to shock, (2) on the basis of this sampling they learn that there is no contingency between their behaviour and shock, and (3) they develop the expectation that responding and outcome will remain independent on future trials.

An appetitive counterpart to 'learned helplessness', called 'learned laziness', has similarly been postulated (Mellgren and Ost, 1971; Welker *et al.*, 1973). Pre-exposure to uncorrelated food presentation and key-illumination in pigeons retards subsequent acquisition of autoshaped responding when a positive contingency is arranged between key-light and food. Nevertheless, it must be emphasised here that all that appears to have been proven in these and learned helplessness experiments is that certain kinds of pre-training (for example, a nil correlation between behaviour and its outcomes) retards conditioning on tasks where a contingency is subsequently introduced. Whether this occurs as a result of the three stages implied by learned helplessness hypothesis is a different matter, and some investigators have suggested that a thorough examination of the role of more fundamental conditioning processes in learned helplessness might obviate the need to take the theoretically radical steps that learned helplessness theory has done (Gamzu, Williams and Schwartz, 1973).

# 5 Constraints on Conditioning: Reinforcer and Response Factors

In the preceding chapters dealing with conditioning, the reader will have noted that a number of general principles have been hypothesised to explain a diversity of conditioning phenomena. However, the reader may also have noted that the bulk of the data on which these principles are founded is derived from studies of rats, pigeons, and to a lesser degree monkeys. In the 1940s and 1950s there was a sustained belief that principles of learning derived from the study of a limited number of species would have limitless applicability; this belief was held by nearly all of the great theorists of the time including Hull, Tolman and Skinner. To these ends experimenters chose arbitrary responses and arbitrary stimuli in the hope that principles discovered with these types of events would be readily generalisable to real-life learning situations. However, even during the height of this endeavour, many biologists who were familiar with the vast diversity of the animal kingdom were sceptical of such an enterprise. This was so much so that Skinner's pioneering book *The Behaviour of Organisms* (1938) was often light-heartedly referred to as 'The Behaviour of the Rat' – since this was the only animal on which Skinner carried out his experiments! The desire to discover general laws of behaviour fostered a number of false assumptions about the nature of learning mechanisms and of learning situations. First, it was assumed that using arbitrary artificial learning situations such as the T-maze or the Skinner-box would provide principles uncontaminated by evolutionary determined species-specific response tendencies. This, as we shall see, has turned out to be wrong. Many of the simple learning situations studied inevitably implicated species-specific responding. For instance, the key-pecking response of the pigeon, which was first selected by Skinner because it 'can be easily executed, and because (it) can be repeated quickly and over long periods of time without fatigue' (Ferster and Skinner, 1957), is also the consummatory response of that organism when it is reinforced with grain. This factor inevitably complicates the analysis of some appetitive learning studies. Similarly, studies of learning which involve the aversive control of responding by arbitrary aversive stimuli such as electric shock (for example, avoidance learning, punishment, escape learning) inevitably induce defence and flight reactions which again contaminate the 'arbitrariness' of the learning situation with species-specific factors. For instance, in an aversive, fear-inducing situation, some animals will flee and some will freeze – such tendencies will obviously interact with the conditioned response and associated contingencies. A further point, which relates to this first basic assumption, is the question of how useful laws of behaviour which are derived from arbitrary learning situations would be to an understanding of the animal in the wild. Organisms, through evolutionary selection, are usually well adapted to their ecological niche – evolution providing them with a variety of species-specific behaviours favourable for survival. It is inevitable that the majority of organisms in the wild will be continually encountering situations in their 'niche' which will involve the operation of species-specific behaviours. If this is the case, general laws of learning may be inadequate to explain this adaptability without, at the very least, knowledge of the way they interact with species-specific response tendencies.

The second assumption fostered by the 'arbitrary' approach relates to the selection of elements in the learning situation. It has been an assumption of workers in the field of conditioning that the elements of a learning situation are arbitrary and interchangeable. Early evidence, especially from the work of Pavlov, tended to support this assumption. Teitelbaum sums up the attitude engendered by this approach

> We arbitrarily choose almost any act from the animal's repertoire and reinforce it with food, water or whatever else the animal will work to obtain. Although typically we teach a rat to press a bar or a pigeon to peck a key to obtain a pellet of food, we can readily train either to dance around the cage if we so choose. We usually use a light to signal the delivery of a pellet but we can use a tone or a buzzer or any other stimulus the animal can detect. . . . The same act can be used for any reinforcement . . . In effect, in any operant situation, the stimulus, the response, and the reinforcement are completely arbitrary and interchangeable. No one of them bears any biologically built-in fixed connection to the others.
>
> (Teitelbaum, 1966, p. 566)

Apart from the fact that it is now known that species-specific response tendencies do influence the course of conditioning, there are many situations in which particular stimuli are more readily conditionable to particular responses (see chapter 6). One instance which we will come across in more detail later, is the phenomenon of conditioned taste aversion: when a rat consumes a meal which later has toxic consequences, producing illness, it is only the *taste* of the food that acquires aversive properties – the rat will subsequently continue to sample foods which look the same and have similar external stimuli paired with them, but will consistently avoid foods with a similar taste. Finally, the use of a molar dependent variable such as response rate as the basic datum in a large number of instrumental conditioning studies also has its problems. It generated faith in the belief that some general mechanism of learning operates in all species. Members of a variety of different species appear to perform similarly on schedules of reinforcement: they generally produce positively accelerated responses gradients on fixed-interval schedules (the FI scallop), a break-and-run pattern on fixed-ratio schedules, and a steady, constant rate on variable-interval schedules. However, this similarity of patterning may hide important molecular differences between species in the way they adapt to contingencies. Formalistic similarity in response topography, which at the molar level has been constantly emphasised by operant conditioners (Skinner, 1969, p. 190), does not necessarily imply that the underlying mechanisms which generate these behaviour patterns are the same. For instance, White, Juhasz and Wilson have argued that formalistic similarity between the behaviour patterns of different species in an arbitrary learning situation tells us very little about the developmental history of learning mechanisms in different species

> Since the similarity between the different species could not possibly be due to phylogenetic relatedness, it must result from the fact that the contingency of reinforcement in the bar-pressing situation remains constant. Just as the wings of birds and bats evolved independently as parallel evolutionary adaptations to a similar environment, and yet rest on quite different causal bases (Hinde, 1966), so must the characteristic fixed-interval behaviour of different species have developed as a function of the constraints of the Skinner-box. It is not an underlying learning process that is basic to the behaviour but a similar environmental situation.
>
> (White, Juhasz and Wilson, 1973)

The possibility that such formalistically related behaviour may be the result of parallel but independent evolutionary adaptations also implies that the mechanisms underlying such topographically similar behaviours may be different in different species.

In recent years a slow accumulation of evidence has tended to erode the bases of many of the accepted principles of animal conditioning. This evidence has also tended to emanate from studies using species other than the traditional rat, pigeon and monkey, reinforcers other than the traditional food and water, aversive stimuli other than electric shock, and responses other than bar-pressing and key-pecking.

Some of the earliest studies of constraints on conditioning were carried out by Breland and Breland (1961, 1966). These workers were primarily struck by the difficulty they found in conditioning a number of animals to carry out relatively simple tasks in commercial advertising stunts. Although, initially, behaviours could be increased in frequency by food reinforcement, the topography of the response often drifted until it both delayed the availability of reward and necessitated expenditure of a great deal of energy. In one experiment a pig was reinforced with food for dropping a coin from its mouth into a piggy-bank. However, after a period of conditioning, the pig began to root the coin around and toss it into the air rather than place it directly into the slot. This not only increased the delay of reward but also increased the effort expended per reward. Something in the situation was obviously acting to counteract the conditioning process. In fact, increasing the degree of food deprivation, instead of increasing the likelihood of conditioning, only increased the tendency for the animal to display counterproductive behaviour. In another example the Brelands attempted to condition a racoon to pick up coins and deposit them in a piggy-bank. When food reinforcement was made contingent upon dropping two coins into the piggy-bank, the racoon often dipped them briefly into the piggy-bank without letting go and spent long periods rubbing the two coins together. These two examples suggested that, instead of taking the form most appropriate for obtaining reward, the behaviour of these animals often 'drifted' towards another behaviour more or less appropriate to some aspect of the conditioning situation. In both of these cases the 'rooting and tossing' behaviour of the pig and the 'dipping and rubbing' behaviour of the racoon are both species-typical food-related behaviours. Breland and Breland concluded that what they observed in these situations was a kind of *instinctive drift* where food related behaviour became directed towards aspects of the learning situation which were highly associated with food – in this case the coins to be dropped into the piggy-bank.

These early studies of Breland and Breland illustrate one particular aspect of constraints on learning: that stimuli previously established as reinforcers can be ineffective under certain conditions. There are many more instances in the literature of both inter- and intra-species anomalies in conditioning. The following list gives some idea of both the qualitative and quantitative extent of these anomalies:

(1) What reinforces one species need not reinforce another – and indeed, there may be consistent differences between sexes of the same species, (Denti and Epstein, 1972).

(2) Some kinds of conditioning or learning, such as imprinting or song-learning, tend to occur only during one period of an animal's life.

(3) There may be differences in the efficiency with which different species adapt to a particular learning task – for example, gerbils perform better in an active avoidance situation than do rats, but rats learn a passive avoidance response more quickly than gerbils (Galvani, Riddell and Foster, 1975).

(4) There may be variables which affect efficiency only within a specific species or a limited range of species – for example, seasonal variables affect learning performance in the goldfish (Shasoua, 1973); squirrels that have been

able to hibernate during an antecedent cold spell perform better at a conditioning task than squirrels denied the privilege (Alloway, Riedesel and McNamara, 1973); certain species of bees show similar learning curves when food is placed on a feeding tray in the open, but different learning curves when the tray is near to a prominent land-mark.

(5) There may be intra-individual anomalies in learning related to the characteristics of the learning situation. For instance, although pigeons normally exhibit the phenomenon of behavioural contrast when key-pecking is the response (Reynolds, 1961a; see chapter 3, p. 116) this does not occur when treadle-hopping is the response (Hemmes, 1973); rats rarely learn to avoid auditory stimuli when followed by an internal aversive event (poisoning), but readily associate auditory stimuli with external aversive events such as shock (Revusky and Garcia, 1970).

(6) Some species adapt differently to particular learning contingencies; for instance, whereas goldfish 'match' in a probability reinforcement situation, that is, distribute responding according to the probability of reinforcement of the choices, rats and monkeys 'maximise', that is, tend to continually opt for the choice with the higher reinforcement frequency (Bitterman, 1965a, b).

(7) Some species *never* adapt to a particular task whereas others do so readily; for example, monkeys and rats quickly adapt to a reversal learning situation (see chapter 11, p. 310) whereas certain species of fish never acquire such 'learning sets'.

This list at first appears quite daunting and capricious. Are we just simply to go ahead and catalogue the exceptions to the traditionally accepted principles of learning or should we be looking for new principles which fit the new data? As Shettleworth (1975) has pointed out, there is a tendency for investigators to put forward a multiplicity of *post hoc* explanations which are inadequate for predicting which of an animal's behaviours will be conditionable and for which reinforcers. It is probably too soon to say whether the traditional principles of conditioning will have to be abandoned in favour of a multiplicity of principles related to specific species or specific learning situations, or whether an adequate account of animal learning can be couched in terms of interactions between some general mechanisms of learning and species-specific adaptive mechanisms. At this stage, however, there are probably enough examples of species-specific constraints on conditioning to be able to fit them into categories according to the nature of the constraint. Although this approach does not necessarily give insight into the mechanisms involved in these constraints, this kind of initial classification does indicate some of the causal factors which play a role in the disruption of the conditioning process.

The initial classification divides constraints into three main packages which are related respectively to the nature of the response, the reinforcer and the discriminative or conditioned stimulus in the conditioning situation.

Nevertheless, some instances do not fall readily into any one of these categories but may equally well be included in more than one.

## 1. The Reinforcer

(a) Certain reinforcers will only increase the frequency of particular behaviours, often only in particular situations.

(b) Certain reinforcers act as UCSs for particular unconditioned activities. That is, they 'elicit' or 'induce' behavioural tendencies which in many learning situations may be incompatible with the response to be modified in frequency. For example, reinforcing a pigeon with grain will often elicit pecking which is often directed at the response key; this invariably makes it difficult to train pigeons to space their responses on an appetitive low rate schedule such as DRL (Hemmes, 1975; Reynolds, 1966).

## 2. The Response

Certain responses are so linked to a particular stimulus or motivational state that the response cannot be modified in frequency (and often will not occur at all) unless that specific 'causal' factor or state is present. Simple examples here are ear scratching in the cat and yawning in the dog (Konorski, 1967).

## 3. The Stimulus

(a) The control of a behaviour by a particular conditioned stimulus may be disrupted because that stimulus is a UCS or a 'releaser' for a competing response. For example, food reinforcement of immobility in a male stickleback only in the presence of another male stickleback would be extremely difficult to achieve since the male stickleback used as the $S^D$ would probably elicit aggressive attack behaviour on the part of the subject – behaviour which is, of course, incompatible with immobility.

(b) For various reasons the subject may not be able to distinguish between the relevant or irrelevant aspects of the learning situation. Either because it lacks the perceptual apparatus to make such a discrimination or because there appears to be a 'belongingness' or natural tendency for the organism to associate particular aspects of the environment with particular reinforcers or particular aspects of behaviour (Seligman, 1970).

Having now outlined the basis for this classification, a closer look at some of the phenomena contained in each is required. The latter category, dealing with 'The Stimulus' is dealt with in chapter 6. The remainder of this chapter will deal with reinforcer and response factors.

# FACTORS ASSOCIATED WITH THE REINFORCER

## Certain reinforcers may only reinforce certain behaviours

This category is a difficult one in that it contains a seemingly endless number of instances where reinforcers only reinforce under certain conditions. The nature of these instances can best be illustrated by a quote from an article by Stephen Glickman

> Male rats have been shown to run mazes for access to female rats in oestrus (Kagan, 1955), and female rats in oestrus will press levers for access to male rats with whom they can copulate (Bermant, 1961; Pierce and Nutall, 1961). Female rats (Wilsoncroft, 1969) and mice (Van Hemel, 1970) will bar press for the privilege of retrieving pups; and rats (Oley and Slotnick, 1970) and hamsters (Jansen *et al.*, 1969) will emit appropriate instrumental responses to gain access to strips of paper which are used for nest-building. Sand-digging, potentially functional in burrow construction, is an effective reinforcer in the deermouse (*Peromyscus leucopus*) (King and Weisman, 1964), and Fantino and Cole (1968) have noted that sand-digging requires no 'extra' reinforcers for its maintenance in laboratory mice.
>
> In addition to the 'constructive' acts which function as reinforcers in the situations noted above, some laboratory rats will run through a maze (Myer and White, 1965) or press a lever (Van Hemel, 1972) to obtain access to a mouse which is then killed with an appropriately directed bite.
>
> (Glickman, 1973, p. 226)

This quote raises a number of issues, not the least being that we seem to be dealing here with a 'rag-bag' category into which we can conveniently deposit all of those instances which for one reason or another do not fit comfortably into other categories! However there are a number of points to note: (1) Unlike food and water, many of the reinforcers we are dealing with here are species-specific and often specific to a particular sex; (2) Many of these reinforcers rely on the presence of a very discrete physiological state in the organism for their effectiveness; (3) A combination of these two points emphasises the futility in this context of asking such questions as 'Why is a reinforcer reinforcing?' There certainly does not appear to be any common factor amongst these examples which might give us any insight into the way in which reinforcers have their effects; (4) The final point here is related to the functional definition of operant reinforcement. Many of these instances of reinforcement, and similarly many we are to meet later in this chapter, deny the transituational nature of reinforcers. Unlike contemporary theorists such as Hull, Skinner (1953) by-passed the question 'Why are reinforcers reinforcing?' by placing more importance on merely identifying reinforcers and using them to predict and control behaviour. Many of his critics, however, pointed out that his definition of a reinforcer was circular since it was dependent on the 'increase in frequency of an operant' but an operant was 'that class of behaviours upon which the reinforcer was contingent'. Nevertheless, Meehl (1950) argued that this circularity could be broken by demonstrating the 'transituational' nature of reinforcers; that is, once a reinforcer was identified with the use of one operant it could be used in other situations to reinforce other behaviours. However, as we have just noted, the reinforcing nature of a stimulus can depend on factors other than the mere contingency between response and reinforcer; this obviously provides difficulties for the Skinnerian notion of reinforcement. None the less, rather than eliminating the notion of reinforcers from psychological study, these anomalies primarily stress the necessity of considering other contextual factors when studying learning phenomena. There appears to be no class of consequential events which invariably increase the probability of a response regardless of what that response is and regardless of the context in which the conditioning occurs.

In addition to those reinforcers which will only reinforce certain species, and then very often only during certain physiological states, there are those reinforcers which will only reinforce certain responses. For example, some reinforcers will reinforce 'choice' behaviour but will not reinforce other more straightforward operants. Chiang and Wilson (1963) found that rats in a choice situation will imbibe more of a saline solution than water, and will also approach the saline bottle more frequently. However, Pfaffman (1969) has shown that it is extremely difficult to train rats to run in a maze for saline solution.

Similarly, Shettleworth (1973) has shown that when hungry hamsters are reinforced with food for engaging in various responses, then digging, rearing, scrabbling and bar-pressing soon increase in frequency to relatively high levels of occurrence. However, the frequency of face-washing, scratching with a hind leg and scent marking is affected little, if at all, by the food contingency. Shettleworth (1975) has later attempted to interpret these results in terms of the conflicting or facilitative nature of behaviours which are elicited by anticipation of food. Activity in general, and especially action patterns such as digging, rearing and scrabbling are exhibited by hungry hamsters in the absence of any reinforcement contingency, whilst action patterns such as face-washing, scratching and scent-marking are more frequent in satiated and 'contented' animals in the home cage. Thus, she suggests, reinforcing a behaviour does not free it of control by anticipation of feeding, and in the conditioning situation operants which are compatible with these anticipation behaviours will readily increase in frequency whilst those which are incompatible will be suppressed by the conflicting action patterns. An alternative account of these results is provided by Vanderwolf (1971). He suggests that it is not the competing nature of food anticipatory behaviours which disrupts the conditionability of grooming, scratching and scent-marking, but that these behaviours are just not conditionable by instrumental means in any situation; that is, they are instances of 'autonomic' rather than 'voluntary' behaviours in the traditional respondent–operant sense. He found that rearing, digging, and climbing in rodents such as rats, guinea pigs and gerbils are all accompanied by rhythmical slow electrical activity of the hippocampus (theta waves), while grooming is not. The presence and absence of theta waves has been taken to accompany the performance of 'voluntary' and 'autonomic' behaviours respectively (Vanderwolf, 1969; see also pp. 119–200). A further possible explanation of this phenomenon can be derived from the account of instrumental learning put forward by Bolles (1972) (see chapter 7). In this situation what is being reinforced might not be the experimentally defined operant but an 'expectancy' of food which is correlated with the reinforced operant. Thus, what will increase in frequency is not necessarily the operant but behaviours indicative of food expectancy. However, whatever the mechanism which underlies this phenomenon, it produces an adaptive selectivity in the behaviours readily performed to get food, since those action patterns which involve active interaction with the environment are readily performed and those which do not directly function to get food in nature are not.

Another phenomenon which falls into this category is that some re-inforcers, although they will increase the frequency of a response, behave differently in the context of certain manipulations. For example, manipulation of their magnitude or frequency may have differing effects on the behaviour they are maintaining. Hogan (1967) and Hogan, Kleist and Hutchings (1970) reinforced male fighting fish (*Betta splendens*) with food, or with the opportunity to display in a mirror – the response in both cases was swimming through an alley. This experiment had a number of results: (1) Rate of acquisition of alley-swimming was the same for the two reinforcers; (2) Swimming for display extinguished much quicker than for food; (3) Increasing the fixed-ratio value of the swimming response did not alter the rate of responding for display but increased the rate of responding for food (see figure 5.1); (4) Varying the duration of display time between 5 and 40s had no effect on responding for display whilst research with other fish suggests that manipulating the magnitude of food reward will affect rate of responding (Rozin and Mayer, 1961); (5) Various aspects of the fishes' behaviour suggest that display was not a weaker reinforcer but that it had its effect on behaviour in a totally different way to food (for example, on some occasions the fish would swim much faster for display than it ever did for food).

Although there are problems involved in the comparison of the effects of different reinforcers which it is not intended to elaborate here (the interested reader is referred to Bitterman, 1965a, or Steiner, 1968), these results tend to suggest that display and food might have had their reinforcing effects via different mechanisms. Shettleworth comments

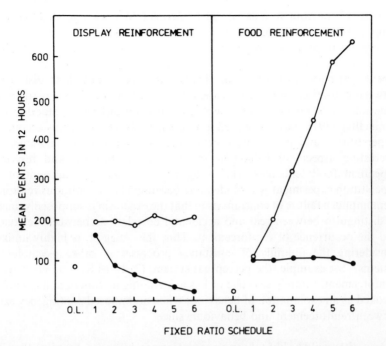

Figure 5.1   Mean response rates of groups of male *Betta splendens* reinforced on various fixed-ratio schedules for swimming through a hoop. For some of the subjects opportunity to display at their images in a mirror was the reinforcer (left hand panel), and for others food was the reinforcer (right hand panel). Open circles represent responses and filled circles represent reinforcements; O.L. = operant level. (After Hogan, Kleist and Hutchings, 1970).

. . . it is tempting to conclude that the differences between food and display with ratio reinforcement represent basic differences between regulatory systems designed to maintain homeostasis by increasing effort to obtain a constant amount of some variables such as food, and systems that do not maintain some internal state necessary for survival. . . . Clearly there is a need for further detailed comparisons between such reinforcers as food, water, and thermal stimuli, and those like the opportunity to perform aggressive or sexual behaviour which are not apparently homeostatic.

(Shettleworth, 1972a, p. 13)

## Effects on responding of behaviours elicited or induced by reinforcement

In chapter 2 we discussed briefly the 'superstition' experiment conducted by Skinner (1948) in which periodic presentation of food to a hungry pigeon regardless of what the bird was doing at the time led to a predominance of one particular behaviour such as turning half circles. Skinner suggested that what happened in this situation was a 'miscarriage of the process of conditioning'; the organism happens to be executing some response as the food is presented and as a result of this contiguity between behaviour and food he tends to repeat that pattern of behaviour. That is, the organism fails to distinguish between causation and mere accidental correlation, and hence it treats the situation as a response-dependent conditioning situation. Staddon and Simmelhag (1971) have proposed a number of objections to this account of 'superstition', and in a replication of the experiment noted a number of interesting aspects of behaviour in response-independent and response-dependent food schedules. They suggested that Skinner's account of the superstition experiment was inadequate because: (1) Adventitious reinforcement implies a failure of constancies in that the organism is supposedly unable to distinguish between real and accidental correlations between behaviour and the occurrence of reinforcement. This, they suggest, is highly unlikely considering the ubiquity of constancy processes in other psychological systems – for example, the perceptual system; (2) If, as Skinner considered, reinforcement is purely selective or has a 'stamping in' function, it cannot be invoked as an explanation of behaviour when no imposed contingency exists between reinforcement and behaviour, since

. . . . (it) would be like taking a population of white mice, breeding them for 20 generations without further selection for colour, and then attributing the resulting white population to the results of 'accidental selection'. In this case, as in the case of response-independent reinforcement, the outcome reflects a characteristic of the initial population (i.e. the mice gene pool, the nature of the organism), and not a non-existent selection process.

(Staddon and Simmelhag, 1971, p. 21)

(3) Following on from this second point, Staddon and Simmelhag point out the necessity in conditioning to consider principles of behavioural variation. If, in the 'superstition' experiment, certain behaviours predominate before the first food delivery, it is more likely that one of those particular behaviours will occur contiguously with food. It is highly unlikely that all behaviours occur with equal frequency before 'reinforcement' and so principles are needed which will predict such factors as the 'operant level' of behaviour in a particular situation prior to conditioning.

In their replication of the 'superstition' experiment Staddon and Simmelhag observed the behaviour of pigeons on schedules of both response-dependent and response-independent food delivery. They found that although initially the birds did tend to repeat the behaviour they were engaging in at the time of food presentation, this behaviour tended to drift slowly into stereotyped patterns of behaviour, some kinds of behaviour occurring only in the period immediately *after* food delivery and some occurring only towards the time when the next food delivery was imminent. The first class of behaviours they labelled *interim activities* and the latter class of behaviours *terminal activities*. In the majority of their subjects the response which tended to predominate at the end of the inter-food intervals was pecking of one kind or another; pecking at the magazine wall, at the hopper, or just at the air. Figure 5.2 shows the probability of behaviours exhibited by two of the birds as a function of post-food time. This shows quite markedly that certain behaviours increased in frequency as a function of post-food time, but similarly, other behaviours increased and then decreased in frequency early in the inter-food interval. Staddon and Simmelhag found that making the food delivery contingent upon pecking the response key had no effect on the frequency of pecking but merely located pecking *at* the key rather than at the air or the magazine wall. The results of this experiment suggest then that, even in the absence of response-reinforcer contingencies, certain types of behaviour seem to predominate in relation to food availability. Terminal activities, or behaviours

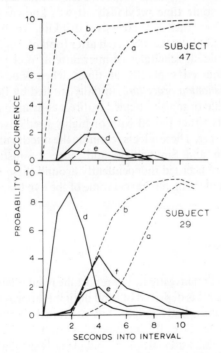

Figure 5.2 Probability of behaviours as a function of post-food time for two pigeons on a response-independent fixed-time 12-s schedule. a = pecking directed at the magazine wall, b = orientation towards the magazine wall, c = side-stepping along the magazine wall, d = turning 1/4-circles, e = pecking at the floor, f = holding head in the food magazine. The broken lines are examples of what Staddon and Simmelhag label 'terminal' behaviours, and the solid lines are examples of 'interim' activities.
(After Staddon and Simmelhag, 1971).

which in most instances are related to consummatory activities associated with the reinforcer, appear to predominate in stimulus conditions which are highly predictive of the occurrence of that reinforcer (in the Staddon and Simmelhag experiment the stimulus is a temporal one: time-since-last-food-delivery); interim activities appear to predominate in stimulus conditions which indicate a low probability of reinforcer occurrence. However, it is not obvious what kinds of behaviours will predominate as interim activities, or exactly why they occur, or if they serve any adaptive purpose. In order to account for the phenomenon observed in their 'superstition' experiment, Staddon and Simmelhag suggested that conditioning should be viewed as the outcome of two processes: a process that generates behaviour (that is, that embodies principles of behavioural variation) and a process that selects (that is, that selectively *eliminates* maladaptive or unnecessary behaviours).

Although this experiment was probably the first to integrate interim and terminal behaviour into the same context, such phenomena had been studied independently for some time previously. It was known, for instance, that *adjunctive behaviours* such as excessive drinking (polydipsia), running, aggression, wood-nibbling, etc. occurred after food on response-independent and response-dependent schedules of intermittent food reinforcement, and also that they occurred generally at times when food was unavailable. Similarly, the *autoshaping procedure*, initially studied by Brown and Jenkins (1968), suggested that organisms tended to direct consummatory activities at a stimulus which was either paired with or highly predictive of the imminent delivery of a reinforcer. These effects resemble the interim and terminal effects found by Staddon and Simmelhag, and since a substantial amount of knowledge has since accrued independently around these two phenomena a detailed discussion of them may reveal some of the dynamics which determine their occurrence.

### Adjunctive behaviours

Falk (1961) was the first investigator to report the phenomenon now known as polydipsia. When a rat is deprived of food until it is approximately 70–80 per cent of its free-feeding body-weight and is trained to earn food by lever-pressing in a Skinner-box, the post-pellet period is characterised by the excessive drinking of water from a concurrently available source – even though the animal is not water deprived! (Figure 5.3 illustrates the rate and pattern of this drinking on a VI1-min food schedule.) Such post-reinforcement drinking patterns have been repeatedly confirmed, not just with rats but also with rhesus monkeys (Schuster and Woods, 1966), chimpanzees (Kelleher, cited by Falk, 1971), and pigeons (Shanab and Peterson, 1969). The fact that this phenomenon has been observed in a number of different species

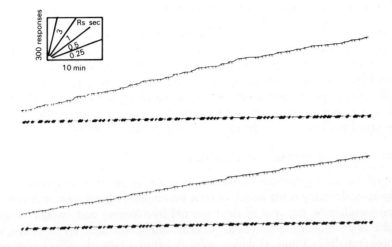

Figure 5.3   An example of the occurrence of schedule-induced polydipsia on a VI1-min schedule of food reinforcement. The upper channel is a conventional cumulative record of lever-presses and food occurrences. The lower channel shows every 12th lick on the water tube as a deflection. (From Falk, 1971, copyright 1971, by Pergamon Press Ltd. Reprinted by permission).

tends to suggest it is a fairly general behavioural phenomenon rather than a characteristic of a limited number of species. This, plus the fact that it was not easily categorised with any known behavioural effects, caused a considerable amount of attention to be focussed on *schedule-induced polydipsia*. As well as isolating some of the causal factors in polydipsia, this recent research has also shown that a number of other behaviours possess similar dynamic properties; so much so that they have been grouped into a relatively new category known as *adjunctive behaviours*.

## Schedule-induced polydipsia

As we have already outlined, polydipsia is excessive drinking in non-water deprived animals which is induced by operant schedules of food re-inforcement; since it was the first example of an adjunctive behaviour to be extensively studied it is perhaps instructive to discuss it in some detail.

This phenomenon arouses interest because under schedule conditions animals often drink around one-half their weight in water in a few hours, whereas states of water deprivation or heat stress do not stimulate nearly such a degree of water intake. Similarly, merely giving the animal a 'sessions-worth' of food in one meal will evoke nowhere near a comparable degree of drinking. The pattern and amount of water-intake is a function of a number of variables.

Relationship of drinking to food occurrence

When a water-bottle is freely available, polydipsic drinking usually occurs in the period immediately *following* consumption of the food pellet, and rarely in the period immediately preceding imminent food delivery. However, when the availability of the water-bottle is limited to periods later in the inter-reinforcement interval or when post-pellet drinking is obstructed, drinking will occur later in the interval with little if any diminution in the amount consumed per inter-food period (Flory and O'Boyle, 1972).

Intermittency of reinforcement schedule

One of the most important determinants of the occurrence and intensity of polydipsic drinking is the length of time between eating episodes. For rats at approximately 80 per cent of their normal free-feeding body-weight, water intake is an increasing function of mean inter-food time up to values of approximately 2–3 min, at which point polydipsia falls off to progressively lower values (see figure 5.4 and Falk, 1966a). At inter-food times of less than 4–5 s, and on continuous reinforcement schedules (FR1) polydipsia does not occur. However, under suitable conditions it occurs regardless of whether food is presented independently of ongoing behaviour or whether it is dependent upon the occurrence of a specific behaviour (for example, bar-pressing). Similarly, polydipsia has been found to occur on a variety of different schedules including fixed-interval (Falk, 1966a), variable-interval (Falk, 1967), and differential reinforcement of low rate (Segal and Holloway, 1963) as well as a variety of second-order and multiple schedules (Jacquet, 1972; Rosenblith, 1970).

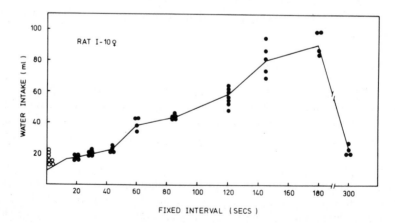

Figure 5.4   Schedule-induced polydipsia as a function of the value of a fixed-interval food reinforcement schedule. Open circles represent the pre-experimental 24-h water intake of the subject whilst on free-feeding in the home cage. (After Falk, 1966a).

## Food deprivation

As the food-deprivation regime is progressively eased so that the subject's weight increases to 95–100 per cent of his free-feeding weight, polydipsia progressively decreases even though rate of bar-pressing for food does not substantially decrease (Falk, 1969). This implies that the maintenance of operant responding and also eating are not sufficient conditions for maintaining schedule-induced polydipsia but that it is in some way causally linked to underlying motivational states.

## Amount of food per meal

Studies have generally disagreed over the effect on polydipsia of amount of food per meal. Some studies have shown that an increase in the magnitude of reinforcement leads to an increase in the degree of polydipsia (Bond, 1973; Hawkins, Schrot, Githens and Everett, 1972) whereas others claim to have demonstrated that polydipsia is a decreasing function of magnitude of reinforcement (Falk, 1967; Lotter, Woods and Vasselli, 1973). However, Couch (1974) has suggested that these latter two studies may have failed to control for the number of reinforcement periods occurring during a daily session. In a study which controlled for this variable, Couch found results concomitant with those of Bond (1973) and Hawkins et al. (1972).

## Type and palatibility of fluid

When the water available for polydipsic drinking is laced with a strong NaCl solution polydipsia is generally decreased below the level obtained with untampered water (Falk, 1964, 1966b). Similarly, studies which have substituted an ethanol solution for water have witnessed a decrease in polydipsic ethanol intake as the ethanol concentrations are increased to unacceptably noxious levels (Gilbert, 1974a).

## Physiological accounts of polydipsia

The failure of traditional behavioural principles to explain all of the facets of polydipsic drinking will be discussed later in association with other 'adjunctive' phenomena. However, there are a number of considerations which are particularly pertinent to polydipsia and so should be discussed here. Attempts to explain polydipsic drinking in terms of currently held physiological tenets have generally failed.

It might be argued at the physiological level that the maintenance of polydipsia is attributable to decreased renal concentrating ability. For instance, it has been noted that compulsive water-drinkers exhibit a decreased maximum urinary concentration (Barlow and DeWardener, 1959). However, it is unlikely that impaired renal function plays any role in schedule-induced polydipsia since, (1) schedule performance is not disrupted by the removal of the water reservoir as might be expected with renal malfunction, and (2) drinking in the home cage between sessions is completely normal in schedule-induced polydipsic rats (Falk, 1969). Alternatively, Teitelbaum (1966) has interpreted polydipsia as the result of dehydration caused by feeding pulling water into the stomach from the tissues; this, it is suggested, produces a state of thirst accompanied by a dry mouth. Variations of this account have come to be grouped under the general title of 'dry-mouth theories' of polydipsia. The fact that polydipsic drinking normally occurs immediately after pellet consumption appears to favour the tenability of such a hypothesis. However, there are a number of factors which suggest that the dry-mouth explanation is inadequate at least in part, if not *in toto*, as an account of schedule-induced polydipsia. First, when subjects are pre-loaded with water before the experimental session by means of a rubber catheter passed down the oesophagus, the major polydipsic effect is still present (Falk, 1969), and this initial hydration has no discernible effect on drinking during the first portion of the session, a finding contrary to the expectation of a 'psychogenic hyperdipsia' account. Secondly, polydipsia is still found when the reinforcer is at least one-third water by weight (Falk, 1969; Stein, 1964; Stricher and Adair, 1966) – meals which are unlikely to produce dry mouths after consumption. Thirdly, and most importantly from the behavioural viewpoint, is the fact that polydipsia depends to a large degree for its occurrence on the parameters of the food-schedule. As we have already noted, polydipsia does not occur on schedules with very short inter-food times (for example, FI < 5 s and FR1), factors which should be irrelevant in terms of a dry-mouth interpretation. Similarly, polydipsia has been observed following brief stimuli on second-order schedules (Corfield-Sumner, Blackman and Stainer, 1977; Rosenblith, 1970); that is, polydipsic drinking still occurs (albeit at a lower rate) following a neutral stimulus (for example, a tone) presented in lieu of a proportion of the scheduled food deliveries. Although the brief stimulus may possess the same temporal predictive significance as food it does not *directly* possess the qualities needed to produce a dry mouth. The final factor which appears to discredit the dry-mouth hypothesis is the excessive nature of schedule-induced polydipsia; in most conditions the amount of water consumed is patently in excess of that required to either 'rinse' the mouth or redress the tissue fluid levels following eating.

*Schedule-induced aggression*

Aggression, like drinking, can be induced in the period just after food delivery on a variety of schedules, including fixed-ratio (Gentry, 1968; Hutchinson, Azrin and Hunt, 1968), fixed-interval (Richards and Rilling, 1972), differential reinforcement of low rate (Knutson and Kleinknecht, 1970), and response-independent fixed-time schedules (Flory, 1969). Aggression and attack are usually induced by the presence of an appropriate 'target' for the aggression. For instance, pigeons will attack another bird, a stuffed model, a mirror or a colour slide when these are present in the experimental chamber (Azrin, Hutchinson and Hake, 1966; Cohen and Looney, 1973; Flory and Ellis, 1973). Squirrel monkeys will bite on a rubber hose in the immediate post-food period (Hutchinson *et al.*, 1968) and rats on either water or food schedules will attack another animal (Gentry and Schaeffer, 1969; Thompson and Bloom, 1966).

## Schedule-induced wheel-running

Wheel-running has been found to occur during inter-food periods in a number of studies using both water and food reinforcers (King, 1974; Levitsky and Collier, 1968; Skinner and Morse, 1958; Smith and Clark, 1974; Staddon and Ayres, 1975). Levitsky and Collier (1968) found that rats held to 80 per cent of their free-feeding body-weights by food deprivation did more wheel-running following food on a VI 1-min schedule than on a FR1 schedule. Similarly, Staddon and Ayres (1975) report that rate of wheel-running is an inverse function of rate of food presentation. That is, the smaller the inter-food interval the less likely it is that running will be induced. However, unlike aggression and drinking, running tends to occur during the middle of the inter-food interval, a finding which bears an interesting resemblence to results obtained with pigeons by Killeen (1975). Killeen found that when food was presented on a fixed-time basis to hungry pigeons, general activity levels (as measured by a stabilometer under the chamber floor) increased in the immediate post-food period and then decreased as the time for food occurrence neared. Thus, regardless of the duration of the inter-food interval, activity levels followed an inverted U-shaped function with the peak always around the middle of the interval. This suggests that induced wheel-running may reflect properties exhibited by the more complex relationship between food and activity rather than merely being a post-food phenomenon in the same way as drinking or aggression. A final interesting point with regard to wheel-running is its peculiarities as an operant. Unlike more commonly adopted operants such as bar-pressing and key-pecking, the overall frequency of contingent wheel-running fails to be affected by the frequency of reinforcement (Skinner and Morse, 1958). This may reflect the fact that induced wheel-running increases in frequency with decreases in the frequency of independently presented food ( Staddon and Ayres, 1975). Thus when food is presented contingently on wheel-running, the decrease in *reinforced* running with a decrease in reward frequency (as might be expected with more arbitrary operants) may be nullified by an increase in the rate of *induced* running under these conditions.

## Schedule-induced pica

When rhesus monkeys are exposed to a response-independent fixed-time

15-min schedule of food reinforcement, the post-pellet food period has been found to be characterised by a bout of ingesting wood-shavings which cover the floor of the experimental chamber (schedule-induced pica; Villarreal, 1967). Villarreal found that animals often selected the larger pieces and, although on some occasions they only stored them in their cheek pouches, they usually ingested them. The subjects were never observed to ingest wood-shavings in the absence of a food schedule, even in conditions where they were food-deprived and wood-shavings were freely available.

### Schedule-induced air-licking

When a continuous air-steam is delivered from a drinking tube, post-pellet licking of the air-stream developed in rats on a fixed-time 1-min food schedule (Mendelson and Chillag, 1970). When the deprivation level of the subjects is increased (that is, by reducing body-weight levels) schedule-induced air-licking increases (Chillag and Mendelson, 1971).

### Similarities in the characteristics of adjunctive behaviours

The schedule-induced behaviours we have just discussed are often classified together under the heading of 'adjunctive' or 'interim' behaviours. However, what is the evidence that these topographically very different behaviours constitute a 'class'? Do they serve similar adaptive functions? Are they controlled by similar factors, and are they modified in frequency or intensity by similar variables? These are all questions to be asked when considering the validity of grouping such behaviours together. As Falk has pointed out

> A new classification should be more than just suggestive of new ways of viewing known facts. It must prove itself useful not only in recognizing current data, but in making certain fresh lines of investigation compellingly obvious. Such studies either bear out the generality of the suggested scheme or render it unconfirmed.
>
> (Falk, 1971, p. 578)

With this in mind, what similarities of function and causation can be extracted from these schedule-induced phenomena?

### Parameters of the reinforcement schedule

One extremely important factor relates all of these induced behaviours: their frequency or intensity is primarily linked to the parameters of the reinforcement schedule. They do not occur under similar motivational conditions in the *absence* of a reinforcement schedule (Falk, 1969, 1971), and they are not excessive behaviours brought about as a result of the animal being moved from his home cage to a novel experimental chamber (Reynierse, 1966). In examining the effects of inter-reinforcement time on schedule-induced behaviours two studies on the effect of food rate on induced attack in pigeons have reported relatively similar results (Cherek, Thompson and Heistad, 1973; Cohen and Looney, 1973). Rate of aggressive pecking peaks at an inter-food interval of around 2–3 min and declines sharply at lower rates. In these studies it must be emphasised that periodic food is essential to induce attack, since it declines to a negligible level in extinction (Cherek *et al.*, 1973). When comparing this with studies of polydipsic drinking (Falk, 1971; Segal, Oden and Deadwyler, 1965) both drinking rate and attack rate fall off drastically at food rates less than one delivery every 2 min, but attack responding also shows a decline at *higher* food rates (food rates greater than one every 120 s). This difference does suggest a slight difference in the time courses of induced drinking and induced aggression; the reason for this is unclear, although it has been suggested that there may be a limit to the speed with which a tendency to attack can build up – no matter how strong the inducing factors (Staddon, 1977). The wheel-running response similarly shows an increase in frequency with a decrease in the inter-food interval (Staddon and Ayres, 1975), although this relationship is not bitonic as exhibited by schedule-induced drinking and aggression. Wheel-running also differs in its similarity to drinking and attack by the fact that it tends to occur in the middle of the inter-food interval rather than in the immediate post-food period. This could be due to the possibility that it is a measure of general activity, and thus it could be affected not only by those variables which influence schedule-induced phenomena but also by a larger variety of motivational factors which affect activity levels in general. Alternatively, Staddon (1977) has suggested that the occurrence of induced running later in the inter-food period may reflect the hierarchical structure of induced behaviours. Some behaviours will predominate over others when the presence of inducing factors and opportunity for their respective occurrences are equal. Thus drinking may override running in the immediate post-food period (Staddon and Ayres, 1975). Similarly, schedule-induced drinking has been shown to override schedule-induced aggression in the immediate post-food period when opportunity for either response is available (Knutson and Schrader, 1975). However, when, for example, only the opportunity to run is allowed (by enclosing the rat in the running wheel between food presentations), running will develop in the immediate post-food period (Levitsky and Collier, 1968).

Deprivation level

Since the frequency of these induced behaviours is related to frequency of food occurrence it might legitimately be expected that level of food deprivation would also affect the frequency of such behaviours. This is in fact the case. Decreases in deprivation level decrease the rate of induced polydipsia, even though rate of operant bar-pressing is unaffected (Falk, 1969). Rate of schedule-induced air-licking is an inverse function of percentage free-feeding body-weight (Chillag and Mendelson, 1971), and the duration of induced attack decreases to low levels in food satiated pigeons (Azrin et al., 1966).

Temporal locus of adjunctive behaviour

Schedule-induced behaviours occur with greater intensity (with only a few minor exceptions), in the immediate post-reinforcement period regardless of the type of schedule (fixed-interval, fixed-ratio, variable-interval, differential reinforcement of low rate) or the operation of a response contingency to produce food. This is true of induced drinking (Falk, 1961, 1966), induced aggression (Azrin et al., 1966; Knutson, 1970), induced air-licking (Mendelson and Chillag, 1970), induced pica (Villarreal, 1967), and induced running when the animal is confined in the running wheel (Levitsky and Collier, 1968). This temporal pattern exhibited by induced behaviour is not substantially affected by manipulation of the inter-food interval duration. However, there are a number of exceptions to this general rule. First, with fixed-interval values in the region of 3-min, drinking has been known to become distributed throughout the inter-reinforcement period in a series of short drinking bursts (Falk, 1971). Secondly, wheel-running, when the rat is not confined in the running wheel, tends to occur during the middle of the inter-reinforcement interval rather than in the immediate post-pellet period (Staddon and Ayres, 1975). Thirdly, on large fixed-ratio schedules, attack often occurs with highest frequency on termination of the post-reinforcement pause and initiation of the fixed ratio run (Thompson, 1965). This anomaly, it has been suggested, is the result of the aversive properties of long fixed-ratios (Falk, 1971). Fourthly, when an organism can obtain a second pellet quite quickly after a first pellet (for example, on a tandem DRL 30; fixed-ratio 5 schedule), polydipsic drinking is only observed following the second pellet (Falk, 1969). Fifthly, when drinking is prevented in the immediate post-food period but is allowed later in the interval, the rat will drink almost as much at this later stage as it would have done in the post-food period had water been available (Flory and O'Boyle, 1972; Gilbert, 1974b). Sixthly, and possibly most theoretically important (as we shall see later), is the fact that both induced drinking and induced attack occur in the period following non-reinforced responses on DRL schedules (Knutson and Kleinknecht, 1970; Segal and Holloway, 1963), and polydipsic drinking has been observed in periods following a brief stimulus presented in lieu of food – whether that stimulus has been paired with food (Rosenblith, 1970), or not (Corfield-Sumner et al., 1977).

Reinforcing properties of adjunctive behaviours

Polydipsia can be established as a reinforcing activity capable of sustaining schedule behaviour. If the availability of water is made contingent upon the completion of a fixed-ratio schedule rats will respond, even on quite large ratio values, in order to obtain it (Falk, 1966). Similarly, Cherek *et al.*, (1973) found that pigeons would respond on a fixed-ratio 2 schedule in order to produce a mirror to attack. Extinction-induced aggression will also support fixed-ratio responding. Pigeons will respond on a collateral response key at the outset of extinction in order to produce a target bird which is then attacked (Azrin, 1964). As Falk (1971) has suggested, the reinforcing nature of opportunity to indulge in an induced behaviour implies that the animal is not simply 'time-filling' during inter-food periods, or that they are reflexively elicited by reinforcement.

## Excessive nature of adjunctive behaviours

Falk (1967) has noted that the amount of water drunk during VI1-min sessions which last for around 3h is almost 10 times the amount drunk when the same number of pellets is given all at once and water intake over the ensuing $3\frac{1}{2}$ h is noted. As we have already discussed, there is no physiological basis for this mammoth consumption, it is just plainly excessive. In fact, so excessive is schedule-induced polydipsic drinking that a number of investigators have suggested that schedule-induced phenomena may throw some light on the factors which control excessive drinking behaviours – for example, alcoholism in humans (Gilbert, 1974a). Similarly, the aggressive responses induced on food schedules are also extremely violent in their nature, pigeons will often badly bruise and pull out the feathers of live target birds (Azrin *et al.*, 1966), as well as quickly destroy taxidermically prepared model targets (Flory, 1969).

## *Theoretical interpretations of adjunctive behaviours*

The preceding account of the functional and dynamic similarities between topographically different schedule-induced behaviours tends to suggest that they constitute a unitary behavioural phenomenon. This now leads to more specific theoretical considerations related to identification of the important controlling factors, and indeed, to the possible adaptive significance of such behaviours. Initially attempts were made to integrate these schedule-induced phenomena under existing behavioural principles. This, however, has generally proved unfruitful and has subsequently led to more radical theoretical approaches.

## Adventitious reinforcement

Particularly on time-dependent response-independent schedules of food presentation (such as fixed-time or variable-time) a behaviour such as drinking, attack, or running might occur towards the end of the inter-food period and immediately prior to the delivery of a pellet. Skinner (1948) has suggested that such accidental correlations between behaviour and reinforcement will increase the frequency of the contiguously occurring behaviour. Schedule-induced polydipsia, as an example of one schedule-induced behaviour, has been interpreted in this way (Clark, 1962; Segal, 1965).

Apart from the arguments against the permanence of adventitiously reinforced behaviour that we discussed earlier in this chapter (see pp. 164–165; Staddon and Simmelhag, 1971), there are several other reasons why this interpretation is difficult to uphold. First, as Falk (1969) has pointed out, the properties of schedule-induced behaviours (particularly polydipsia) have neither the 'static' nor 'dynamic' attributes of superstitious behaviour. The rapid development, prolonged stability and predictable post-pellet locus of polydipsia are all contrary to the relatively unstable nature of so-called 'superstitious' behaviours. Secondly, some schedules on which induced behaviours develop rarely allow for a correlation between the induced behaviour and food occurrence. For example, fixed-ratio schedules usually develop stable rapid terminal rates of lever-pressing in rats or key-pecking in pigeons. This means that it is highly unlikely that the subject will be indulging in the induced behaviour for a reasonable period of time prior to reinforcement (especially on long FR values). Thirdly, it has been pointed out that if induced behaviours had been adventitiously reinforced they should occur prior to food delivery and not just after it (Stein, 1964). Similarly, on schedules with long inter-food intervals, induced behaviours are usually observed to occur only after food delivery, leaving a long temporal gap between the last occurrence of the behaviour and reinforcement (Schuster and Woods, 1966; Stein, 1964). Fourthly, and perhaps most convincingly, studies which have superimposed a minimum delay between the occurrence of an induced behaviour and food have failed to eliminate the induced behaviour. Falk (1964) added to a VII-min schedule the additional contingency that a bar-press could not produce food if there was a lick at the water tube within a preceding 15 s period. This effectively eliminated the possibility of any contiguity between licking and food occurrence, but failed to reduce the level of polydipsic drinking in all subjects. Similarly, both Cherek et al., (1973) and Flory (1969) have added a 15-s reinforcement delay contingency to attack behaviours which had been induced in pigeons. This added contingency failed to disrupt the rate or temporal locus of schedule-induced attack. Finally, even if there were not this bulk of evidence against an adventitious reinforcement account of schedule-induced phenomena, it is difficult to see how adventitious reinforcement on its own could adequately account for the development of such behaviours. Even if accidental reinforcement did act to maintain behaviours, it does not explain why behaviours such as drinking, attack, running, wood-nibbling, or air-licking, which normally occur with a relatively low frequency in other situations (even with the availability of the appropriate initiating stimuli such as water bottle, or attack 'target') should predominate over behaviours which are much more likely to be adventitiously reinforced (for example, holding head in the food hopper, turning, rearing in the rat, wing-flapping or pecking in the pigeon, grooming, etc.).

Reinforcers as elicitors of adjunctive behaviours

Although adjunctive behaviours usually occur immediately following the reinforcer, it is unlikely that these behaviours are the result of unconditioned elicitation by the reinforcer. The fact that there are a number of very different behaviours which occur in the postreinforcement period, and also a variety of very different reinforcers which are equally likely to induce them, suggest that the relationship is far more complex. Although the reliability and excessive nature of adjunctive behaviours are characteristic of unconditioned reflexes, unlike elicited responses, polydipsia for one takes at least a number of sessions to develop (Falk, 1971). This is unlike the spontaneous, unlearned relationship between a UCR and UCS. However, unconditioned factors may play some role in the *initiation* of these induced behaviours and the reinforcement schedule may merely serve to exaggerate this base rate. For example, rats tend to drink just before and just after meals (Kissileff, 1969). It is quite possible that this normal post-prandial activity may initiate drinking but become exaggerated by factors associated with the reinforcement schedule. This suggestion is supported by the fact that polydipsia rarely develops after electrical brain stimulation reinforcement, a reinforcer which presumably does not normally elicit drinking (Cohen and Mendelson, 1974; Ramer and Wilkie, 1977). Further support for this conception accrues from the facts that, (1) animals whose stomachs are preloaded with water before an experimental session can be prevented from acquiring polydipsia (Chapman, 1969), whereas a similar manipulation on animals that have already acquired polydipsia does not suppress it (Falk, 1969); (2) the initial rates of attack and wheel-running on food schedules are similar to the base rate of attack and wheel-running in other situations (Azrin et al., 1966; Levitsky and Collier, 1968); and (3) under conditions in which water drinking is probable, air-licking is also probable (Hendry and Rasche, 1961). However, the stimuli which produce these base rates are largely unknown, but they are almost certainly different factors to those which determine the final patterning and intensity of adjunctive behaviours. Factors associated with the schedule of reinforcement appear to be responsible for this.

Further evidence against a simplistic account of adjunctive behaviour in terms of reflexes or elicited behaviour is available from those studies which have demonstrated the appearance of adjunctive behaviours in loci other than the immediate post-reinforcement period. For example, when drinking is prevented in the immediate post-pellet period, it occurs later in the interval (Flory and O'Boyle, 1972); attack and drinking occur after responses on *DRL* schedules (Knutson and Kleinknecht, 1970; Segal and Holloway, 1963); and polydipsia occurs after brief stimuli on second-order food schedules (Corfield-Sumner et al., 1977; Rosenblith, 1970).

Adjunctive behaviours as temporal mediators

We already know that on schedules which have reliably defined temporal parameters (for example, DRL schedules, FI schedules, etc.) stereotyped chains of behaviour have been observed which, it has been suggested, 'mediate' the temporal interval between responses or reinforcers (see chapter 3, p. 83; Wilson and Keller, 1953; Laties, Weiss and Weiss, 1969). Similarly, it has been suggested that adjunctive behaviours may serve a similar function by helping to mediate accurate timing of inter-food intervals where this is necessary (Segal and Deadwyler, 1965). However, it seems unlikely that this is the case for two reasons. First, adjunctive behaviours exhibit a robustness which has rarely been demonstrated with 'mediating' behaviours on temporally-based schedules; adjunctive behaviours persist in a relatively unmodified form over a large number of sessions, whereas 'mediating' or 'timing' behaviours are usually quite labile, drifting quite regularly in topography and usually disappearing in any recognisable form when efficient timing behaviour has been established. Secondly, adjunctive behaviours occur quite readily on variable-interval and fixed-ratio schedules where there is no regular scheduled duration that any 'mediating' behaviour is required to time.

Induction by stimuli signalling non-reinforcement

One fact appears to be relatively consistent across the evidence we have reviewed: like the interim behaviours observed in Staddon and Simmelhag's experiment, adjunctive behaviours nearly always occur at times when reinforcement is unavailable, and rarely at times when food is imminent. On the majority of schedules of reinforcement, food is a fairly reliable predictor of a period of non-reinforcement (this is especially so on fixed-interval, fixed-time, DRL, and in some cases, fixed-ratio schedules; see chapter 3, p. 68ff.), and, as noted, adjunctive behaviours tend to occur in the period immediately following food on these schedules. The possibility that it might be the $S^\Delta$ signalling property of reinforcement which is important in the development of adjunctive behaviours is supported by the finding that adjunctive behaviours also occur after unreinforced responses on DRL, and after brief stimuli on second-order schedules. Both of these events have similar $S^\Delta$ properties to food (Corfield-Sumner et al., 1977; Knutson and Kleinknecht, 1970; Rosenblith, 1970; Segal and Holloway, 1963). Further evidence in support of this account comes from studies of behaviours induced in the presence of stimuli which signal periods of extinction. For instance, Azrin et al. (1966) found that pigeons would readily attack a restrained target pigeon during the extinction period of a multiple schedule. Similarly, biting attack by squirrel monkeys has been found to occur on the transition into periods of signalled extinction (Hutchinson et al., 1968) and polydipsic drinking occurs during stimuli signalling extinction on multiple schedules of reinforcement (Panksepp, Toates and Oatley, 1972).

The general conclusion here appears to be that adjunctive behaviours are induced in the presence of, or immediately following, stimuli which are strong predictors of non-reinforcement. Although this tells us something about the occurrence of this class of behaviours it does not tell us much about the mechanisms underlying its appearance. For example, what adaptive significance, if any, do these behaviours serve and what factors determine which particular behaviours will occur in such 'extinction' periods? There have been a number of attempts to answer such questions, although at the present stage these accounts are rather sketchy. For instance, it has been suggested that periods of non-reinforcement are aversive to the organism (since the organism will readily respond to escape $S^\Delta$ periods) and this aversiveness induces motivational states which are antagonistic to the motivational state appropriate to the scheduled reinforcer. Hence on food schedules, periods of non-reinforcement will induce motivational states other than hunger; for example, thirst resulting in polydipsia or aggression resulting in attack etc. depending on the appropriate initiating stimuli (for example, drinking tube, target animal). However, it is not obvious which is the cause and which the effect – the aversiveness or the antagonistic motivational state. As Staddon (1977) has pointed out, it might merely be the case that stimulus conditions highly predictive of non-reinforcement just *do* induce antagonistic motivational states; the apparent aversiveness of these conditions might be explained for example by the fact that a 'thirsty' (polydipsic) animal, or an 'aggressive' animal (one in which a state of attack has been induced) might well try to escape from a food situation making that situation seem apparently aversive.

Falk (1971) has attempted to interpret adjunctive behaviours in a slightly different way, by relating them to concepts initially conceived by ethologists. He suggests that adjunctive behaviours might be instances of displacement activities – that is, 'an activity belonging to the executive motor pattern of an instinct other than the instinct(s) activated' (Tinbergen, 1952). Displacement activities as they are defined by the ethologists include a vast variety of behaviours which seem to be out of context with the stimulus situation and the behaviour immediately preceding or following. One such situation in which displacement activities occur is following the thwarting of consummatory behaviour. As Falk continues

> . . . displacement activities are described as occurring in situations where an animal under high drive conditions is engaged in a phase of the consummatory behaviour and for some reason is prevented from continuing this behaviour. These are also the conditions producing adjunctive behaviours: a lean animal engaged in eating is prevented from continuing this behaviour by the intermittence imposed by the feeding schedule. . . . In both adjunctive behaviour and displacement activity situations the interruption of a consummatory behaviour in an intensely motivated animal

induces the occurrence of another behaviour immediately following the interruption which is facilitated by environmental stimuli.

(Falk, 1971, p. 585)

Nevertheless, merely placing adjunctive behaviours into another category does not help explain them unless the functional and dynamic characteristics of the new category are known. However, the mechanisms which underlie displacement are themselves not fully understood, but the analogy with displacement activities may perhaps throw some light on the adaptive significance of adjunctive behaviours. For instance, Armstrong (1950) has pointed out that a species which can modify its behaviour via a general 'displacement' mechanism when circumstances change has a distinct advantage over species which have developed specific mechanisms to cope with a specific limited number of situations. Thus the apparent eccentric behaviours induced in such a way by schedules of reinforcement may indeed reflect characteristics of a basic and important adaptive mechanism.

### Autoshaping

In the response-independent food schedules studied by Staddon and Simmelhag (1971) it was found that particular types of behaviour tended to predominate as the time approached for the next food occurrence. In general these 'terminal' responses resembled behaviours related to the consummatory act – in the case of Staddon and Simmelhag, who used pigeons as their subjects, the most predominant terminal response was pecking of some form or other (see figure 5.2). It has been suggested (Hearst and Jenkins, 1974; Staddon, 1977) that this reflects the induction of consummatory related activities in stimulus conditions which are highly predictive of reinforcement. In the Staddon and Simmelhag experiment the stimulus was a temporal one: time-since-the-last-reinforcement. As the time since reinforcement increased, so the probability of imminent food delivery increased. Results which support this kind of interpretation have also been found in situations where the stimulus which predicts food occurrence has been made explicit, for example, by presenting a brief exteroceptive stimulus prior to food delivery.

In 1968 Brown and Jenkins developed a procedure which has since acquired the name of *autoshaping*. If a pecking key (the CS) is illuminated for several seconds prior to the operation of the grain hopper (the UCS), experimentally naïve pigeons will soon begin to peck the key, even though there is no contingency relationship between a response on the key and the occurrence of food.

Direct observation and a study of motion pictures made of pigeons . . . showed the following gross stages in the emergence of the key-peck: first, a general increase of activity, particularly during the trial-on period; second, a progressive centering of movements around the area of the key when lighted; and, finally, pecking movements in the direction of the key.

(Brown and Jenkins, 1968)

The procedure for developing autoshaped pecking is elaborated schematically in figure 5.5. However, perhaps even more startling to the traditional learning theorist were the observations made by Williams and Williams (1969) who studied the effect on autoshaped pecking of a negative reinforcement contingency. Trials in which no peck occurred terminated with reinforcement, as in the original Brown and Jenkins procedure; however, trials in which pecks were directed onto the illuminated key immediately turned it off and terminated the trial without reinforcement. Despite this negative response–reinforcer contingency pecking was maintained at quite high levels for a substantial number of trials. This phenomenon has since been tagged 'negative automaintenance'. Apart from the initial difficulty which reinforcement theory would have in coping with the simple fact of autoshaping, 'negative automaintenance' poses further, more searching questions.

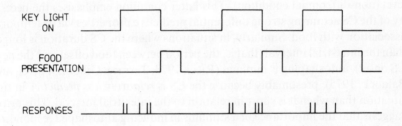

THE  AUTOSHAPING  PROCEDURE

Figure 5.5  Schematic representation of the original autoshaping procedure developed by Brown and Jenkins (1968). See text for explanation.

Although studies of autoshaping have largely been confined to pigeons, the phenomenon has been demonstrated in a number of species. These include the rhesus monkey (Likely, 1974; Sidman and Fletcher, 1968), squirrel monkey (Gamzu and Schwam, 1974), the rat (Myer and Hull, 1974; Peterson, 1975), bobwhite quail (Gardner, 1969), various species of fish (Squier, 1969), and even humans (Wilcove and Miller, 1974; Zeiler, 1972b).

*The stimulus-reinforcer relationship*

There are a number of important features of the stimulus–reinforcer (CS–UCS) relationship which are necessary for the development and maintenance of autoshaped responding. Figure 5.5 shows that in the original Brown and Jenkins study the offset of the CS immediately preceded the onset of the UCS (food), and that the inter-trial intervals are in general longer than the duration of the CS. Both of these factors are important for the support of autoshaped responding. In a study which varied the temporal relationship between CS and UCS presentation, Gamzu and Williams (1973) found that autoshaped key-pecking only developed if illumination of the key was positively associated with a difference in the average frequency of reinforcement. Figure 5.6 illustrates the different CS–UCS schedules that Gamzu and Williams studied, and the degree of autoshaped responding induced on each. The important facts to be gleaned from these results are, (1) that specific pairings of key and food are not necessary for autoshaping to occur, since autoshaping still occurs when some of the key presentations are not followed by food (the differential condition), (2) a differential positive association between key and food (defined in terms of relative densities of reinforcement) is necessary and sufficient to produce autoshaping. As long as food occurs during or immediately after CS occurrence (differential condition) autoshaping occurs. However, when food sometimes occurs during the inter-trial period, in the absence of the CS, autoshaped responding declines to a minimal level (non-differential condition). This latter condition emphasises the necessity of the CS acquiring strong differential predictive properties via its temporal association with food. Similarly, in situations where the CS duration is longer than the inter-trial interval (that is, the period between food offset and the next CS onset), autoshaping is minimal (Ricci, 1973; Terrace, Gibbon, Farrell and Baldock, 1975), presumably because the CS is *relatively less predictive* in this situation than when it is short in relation to the inter-trial interval. This again suggests that the importance of a stimulus in inducing autoshaped responding revolves not around the fact that it is merely *paired* with food, but that it *differentially predicts* food, that is, it precedes or signals food more reliably than any other stimulus.

*Aspects of the autoshaped response*

The type of behaviour induced in autoshaping situations depends primarily on two important factors: (1) the nature of the signalled UCS, and, (2) the nature of the CS.

The nature of the UCS

One of the most widely quoted experiments on autoshaping, and one which has important implications for a Pavlovian interpretation of autoshaping (see below) was carried out by Jenkins and Moore (1973) who studied the form of the autoshaped key-peck when either water or food was the UCS. They found that food-deprived birds presented with grain as a reinforcer responded on the key with a grain-pecking movement, that is, by opening the beak very slightly

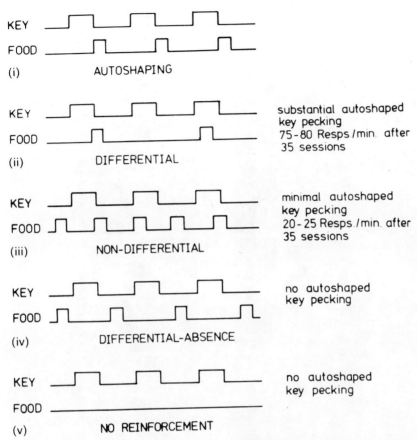

Figure 5.6   Schematic representation of the CS-UCS schedules used by Gamzu and Williams (1973); (i) the basic *autoshaping* procedure in which food delivery always follows the offset of the key light; (ii) the *differential* procedure in which food was presented randomly in time, but only during periods when the key was illuminated; (iii) the *non-differential* procedure in which food was presented randomly in time without regard to the illumination of the key-light; (iv) the *differential absence* procedure in which food was presented only during key-light off periods (the inter-trial interval), and (v) the *no reinforcement* procedure in which food was never presented. (After Gamzu and Williams, 1973).

on impact – a topography similar to that required for picking up and swallowing food. However, water-deprived birds who were presented with water as a reinforcer responded with drinking-like movements, that is, with a lower more 'deliberate' action and the beak closed on impact (see figure 5.7). In a further experiment where subjects were simultaneously deprived of food and water, they received one stimulus signalling food and another signalling water in a random series. The response to each stimulus resembled the

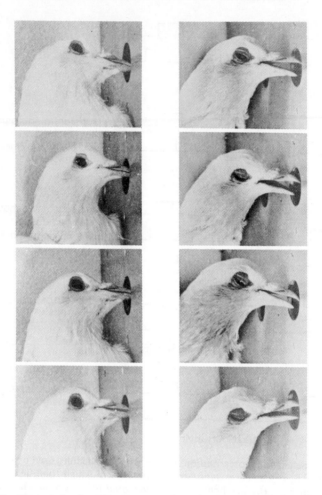

Figure 5.7 Typical food and water autoshaped responses of the pigeon at the moment of key contact. Photographs on the left show responses to a stimulus signalling water reinforcement; those on the right show responses to a stimulus signalling food reinforcement. (From Jenkins and Moore, 1973, copyright 1973, by the Society for the Experimental Analysis of Behaviour. Reprinted by permission).

consummatory response to the particular reinforcer signalled by the stimulus. The possibility that the CS comes to 'elicit' consummatory behaviours related to the UCS has received some support in other experiments. For instance, Gardner (1969) has reported that bobwhite quail peck at stimuli that have been paired with grain; and Squier (1969) has found that autoshaped *Tilapia* direct species-specific feeding movements towards the response key, and an autoshaped mullet reacted to the key with species-typical 'gobbling' movements. Similarly, Rackham (1971) exposed pigeons to repeated pairings of a stimulus light and a sexual UCS. The male and female of a pair were housed in adjacent halves of a large chamber separated by a sliding door. A stimulus light was turned on each day immediately prior to the opening of the sliding doors. Within 5–10 trials the male subjects approached the CS nodding and bowing, followed by cooing, strutting and pirouetting. These courtship reactions, like autoshaped eating and drinking responses, were directed towards the CS itself.

However, although these experiments all support the contention that, under optimal autoshaping conditions, behaviour appropriate to the UCS will be directed towards the CS, there are a number of studies whose results do not fit easily into this framework. For instance, chicks can be induced to *peck* a CS which is predictive of heat reinforcement (Wasserman, 1973), and similarly, rats will approach a CS predictive of reinforcing brain stimulation, a situation in which there is no appropriate overt consummatory activity (Peterson, Ackil, Frommer and Hearst, 1972). Furthermore, Rachlin (1969) has obtained key-pecking in pigeons when the CS predicted electric shock; this key-pecking resembling that which occurs when the subject is autoshaped with food as the UCS. However, in this particular experiment it has been argued (Moore, 1973) that shock may elicit aggressive behaviour, and pecking is an important part of the pigeon's attack repertoire. Indeed, photographs of Rachlin's subjects did suggest that other aspects of attack behaviour, such as feather-ruffling and wing-flapping accompanied the key-pecks. Finally, autoshaping procedures with organisms relatively high on the phylogenetic scale (and whose consummatory responses to, for instance food, are not as rigidly defined as that of the pigeon) have induced key-pressing responses which appear to be exploratory rather than consummatory in nature. These include rhesus monkeys (Sidman and Fletcher, 1968), squirrel monkeys (Gamzu and Schwam, 1974), and humans (Wilcove and Miller, 1974).

The nature of the CS

In general, the nature of the CS is relatively unimportant in influencing the type of behaviour that is autoshaped: for example, the colour and form of the CS have no effect on the rate of autoshaped pecking in the pigeon (Perkins, Beavers, Hancock, Hemmendinger, Hemmendinger and Ricci, 1975) unless the key colour is similar to the light illuminating the food hopper, in which case autoshaped responding is acquired sooner (Sperling, Perkins and Duncan, 1977). However, it is important that the CS be localisable in space. When a tone is used as the CS instead of a light located on the response key, autoshaped responding does not readily develop. However, in the instances that it has been found, the pigeon appears to be pecking at the air, or, when the speaker emitting the tone CS is visible, at the speaker itself (Jenkins, 1973). This tends to suggest that autoshaped behaviour is *directed at* rather than merely *elicited by* the CS; when the CS is localisable, autoshaped behaviour is directed towards it, when it is unlocalisable, such as with a tone CS, autoshaped responding is not easily developed.

*Theoretical interpretations*

We have already encountered autoshaping briefly in chapter 2. As a behavioural phenomenon it initially appeared to be something of a misfit because it did not fall comfortably into either the traditional operant or classical conditioning paradigms. Although the important condition for the development of autoshaped responses appeared to be the stimulus-reinforcer relationship, suggesting the operation of Pavlovian conditioning principles, the behaviour was of an integrated skeletal nature which suggested that operant principles might also be involved (since traditionally only involuntary or autonomic responses were considered to be amenable to the Pavlovian conditioning mechanism). However, perhaps the importance of the discovery of the autoshaping phenomenon is that it has helped to fragment the traditionally monolithic classical and operant conditioning paradigms into relatively simple and testable principles of behaviour modification. Thus, rather than attempting to incorporate autoshaping into the operant or classical conditioning paradigms *per se*, individual principles – in this case primarily the principles of adventitious reinforcement and stimulus substitution – have been extracted from these conditioning paradigms and applied *in puris naturalibus* to the autoshaping phenomenon.

The role of adventitious reinforcement in autoshaping

Although the development of autoshaped responding appears to be critically dependent on stimulus – reinforcer (CS–UCS) contingencies rather than response–reinforcer contingencies (Gamzu and Williams, 1973), attempts have still been made to explain the phenomenon in operant reinforcement terms. The assumption here is that at some point during the CS–UCS sequence pecking immediately precedes food occurrence and through this contiguity is thus adventitiously reinforced. On future trials the lighted key acts as a discriminative stimulus ($S^D$) for key-pecking. Apart from the problems of why, in the case of the pigeon, the bird's autoshaped response should always be key-pecking and not, for instance, turning, preening, inserting its head into the food hopper etc. there is an accumulation of evidence against this simple adventitious reinforcement account. For inst- ance, (1) The 'negative automaintenance' experiment of Williams and Williams (1969) has already been mentioned. If key-pecking is adventitiously reinforced, introducing a negative contingency between pecking and food delivery should immediately suppress key-pecking. This is not the case. Although some studies of 'negative automaintenance' have shown that this contingency does reduce the rate of autoshaped responding (Barrera, 1974; Lucas, 1975; Schwartz and Williams, 1972) it still remains at a substantial level, suggesting that factors other than the consequences of key-pecking are responsible at least for its maintenance if not for its development; (2) In situations in which the subject is physically restrained from pecking the key by placing a transparent barrier in front of it (Browne, 1974; Kirby, 1968), key-pecking occurs almost immediately on removal of the barrier. In this case actual key-pecks could not have been adventitiously reinforced prior to the removal of the barrier; (3) Jenkins (1973) reports that pecking will shift from a less predictive stimulus to a more predictive stimulus (that is, predictive of food), even when the reinforcer is produced only by a peck at the less predictive stimulus. If pecking in these circumstances were critically affected by its consequences, then pecking should occur at a higher rate to the less predictive stimulus – since that is the one which immediately produces food! (4) One of the difficult facts for the reinforcement account to incorporate is the finding that the form of the autoshaped key-peck depends critically on the nature of the reinforcer (Jenkins and Moore, 1973). In the situation where different stimuli precede either food or water reinforcers (see pp. 182–183) the form of the peck to each stimulus resembles the consummatory action pattern appropriate to the reinforcer signalled by the stimulus. Thus, although it might be argued that the frequency of key-pecking in this situation might be determined by its consequences, the *form* of the behaviour obviously depends on other factors, presumably factors related to the stimulus–reinforcer relationship.

The bulk of this evidence, then, suggests that simplistic notions of adventitious reinforcement are inadequate for explaining the autoshaping phenomenon. Perhaps the most widely touted alternatives are principles derived from Pavlovian conditioning, and notably the principle of stimulus substitution.

## Pavlovian interpretations of autoshaping

Many of the experiments discussed above illustrate the importance of stimulus–reinforcer relationships in the development of autoshaped respond-ing. Unless the CS differentially predicts the occurrence of the UCS, then autoshaped responding to the CS rarely occurs. The fact that the importance of the CS–UCS relationship in autoshaping situations parallels the similar importance of such contingencies in traditional Pavlovian experiments, using more 'reflexive' types of behaviour, has not gone unnoticed by learning theorists. This point, plus the fact that operant conditioning principles alone are inadequate to account for autoshaping, have led to a consideration of autoshaping in terms of some traditional Pavlovian principles. Two impor-tant facts support a broad Pavlovian interpretation. First, the importance of the stimulus–reinforcer contingency rather than response–reinforcer con-tingencies in the development of autoshaped responding is characteristic of classical conditioning processes; secondly, that the form of the autoshaped response is dependent largely upon the nature of the reinforcer suggests that UCRs elicited by the UCS (for example, pecking for food) become, as a result of the CS–UCS pairing operation, elicited by the CS (hence pecking at the key). More specifically, the Pavlovian principle of *stimulus substitution* fits these facts quite well (see pp. 219–220). When conducting their prototypical classical conditioning experiments with dogs, Pavlov and his associates often noticed that if the metronome, which was acting as the CS, was within reach of the restrained subject, the dog would often direct its salivating towards it – quite often licking or biting the object. What appeared to be happening here was that the metronome, through its pairing with food, had become a 'substitute' for food, and hence behaviours appropriate to food became directed towards it in quite a full-blown manner. The experiments conducted by Breland and Breland (1961, 1966) and cited earlier in this chapter, appear to provide further examples of stimulus substitution. The pig and the racoon, both of whom needed to drop coins into a piggy bank to acquire food, began to direct food-related behaviours at the coin – presumably because the coin was a stimulus highly associated with imminent food delivery. The principle of stimulus substitution appears to cope very well with the phenomenon of autoshaping. It accounts not only for (1), the importance of the CS–UCS relationship, and (2), the nature of the response depending on the nature of the reinforcer, but also for the fact that (3), autoshaped responding is *directed at* the CS and not towards any arbitrary aspect of the environment.

However, despite the appeal of a 'stimulus substitution' account of autoshaping, there are still a number of facts which do not fit well into this framework. For instance, in situations where the autoshaped response is not related to consummatory action patterns associated with the reinforcer, the notion of stimulus substitution appears inapplicable. For example, studies with heat reinforcement (Wasserman, 1973) and reinforcing brain stimulation (Peterson *et al.*, 1972) induced autoshaped responding which was not obviously similar to the consummatory response; similarly, autoshaping experiments with primates suggest that, rather than directing consummatory behaviours towards the CS, subjects merely orient towards it in an 'exploratory' fashion (Gamzu and Schwam, 1974; Sidman and Fletcher, 1968; Wilcove and Miller, 1974).

Certainly, Pavlovian notions such as stimulus substitution give some insight into the kinds of mechanism responsible for autoshaping, and the ability of such notions to encompass a greater bulk of the facts than principles derived from operant conditioning, suggests that the importance respectively of stimulus–reinforcer contingencies and 'elicited' behaviour has been greatly underestimated in operant conditioning situations. Indeed, Moore makes this point in relation to Skinner's notions of adaptive behaviour

Skinner's system describes organisms which are distinguished by their ability to transcend the limitations of elicited responses and to utilize operants of arbitrary form. The pigeon cannot conceivably be described in such terms. It is distinctly inferior to, and often appears to be the antithesis of the theoretical organism of Skinner's system.

(Moore, 1973, p. 174)

## Interactions between operants and behaviours induced by reinforcement

### Appetitive conditioning

Perhaps the most important fact to note from the preceding two sections on adjunctive behaviours and autoshaping is that under conditions where there is no explicit response–reinforcer contingency, certain types of behaviour still come to predominate. In a large number of cases, the factors which determine the particular behaviour or behaviours which will come to predominate are unclear (this is especially true of adjunctive behaviours). However, it is clear that one aspect of the conditioning situation – the reinforcer – often plays a major role in determining both the locus of such behaviours and their form. For instance, the occurrence of the reinforcer determines the temporal locus of adjunctive behaviour and even the actual form of the autoshaped response. These points have important implications for conditioning procedures. Since we know that a reinforcer may elicit certain species-typical behaviours, then the frequency and pattern of behaviours reinforced according to a particular schedule of reinforcement may depend crucially on the compatibility of the experimental operant with behaviours induced or elicited by the reinforcer. If the behaviours elicited by the reinforcer are similar in form to the operant being reinforced, then we might expect rapid conditioning and a relatively high rate of responding. If the two types of behaviour are highly incompatible, in that they cannot both be performed simultaneously, then we might expect conditioning to be relatively slow, and the typical patterning of behaviour (for example, the fixed-interval scallop) might be disrupted.

An interesting example which illustrates this interaction well is contained in an experiment carried out by Sevenster (1968). Using sticklebacks as subjects he reinforced either rod-biting or swimming through a ring with the opportunity to court a female. Figure 5.8 shows that Sevenster had much more success reinforcing swimming through a ring than reinforcing rod-biting with this reinforcer. As soon as the reinforcer was removed and a period of extinction was instigated, the frequency of ring swimming decreased slightly as might be expected. However, the frequency of rod-biting actually *increased* during the early extinction periods. This suggests that during conditioning some property of the reinforcer was acting to suppress rod-biting but not ring-swimming. There are two possible explanations for this diversity and both can be couched in terms of the interaction between reinforcer-induced behavioural states and the experimental operant. First, one might presume that the presentation of the female fish would elicit a state of high sexual

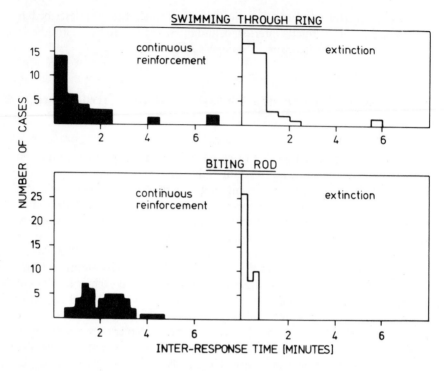

Figure 5.8   Frequency histograms of inter-response times (IRTs) during continuous reinforcement and extinction for male sticklebacks that were swimming through a hoop or biting a hose for the opportunity to court a female. (After Sevenster, 1968).

motivation; however, biting is a consequence of high aggressive motivation, a state which is unlikely to occur simultaneously with sexual motivation. Thus the prevention of this state by incompatible states elicited by the reinforcer may have removed some of the causal factors necessary for biting to occur. Secondly, in the stickleback courting involves swimming in a zig-zag fashion; thus, instead of actually biting the rod, Sevenster found that many of his subjects would merely zig-zag round it. Although this reinforcer-induced behaviour is incompatible with rod-biting, it is compatible with swimming through the ring, and thus might even be expected to facilitate conditioning of the latter behaviour.

This example illustrates failure in conditioning when the operant and reinforcer-induced behaviours are incompatible. Just as equally, failures in conditioning can occur when the operant and reinforcer-induced behaviours are somewhat similar. An example of this is provided by the performance of pigeons on differential reinforcement of low rate (DRL) schedules. Whereas rats are able to space apart their lever-presses with some accuracy on DRL schedules over 60 s, pigeons are notoriously inaccurate – efficient DRL behaviour breaking down at DRL values of beyond 15–20 s (Staddon, 1965). This, it was initially believed, merely reflected the superior ability of the rat at timing intervals; it was, after all, an animal 'higher' on the phylogenetic scale than the pigeon. Nevertheless, an experiment by Hemmes (1975) suggested that this account was incorrect and that the poor DRL performance of the pigeon was related more to the type of operant typically used in these experiments than any phylogenetic differences in timing ability. The operant commonly adopted with pigeons is, of course, key-pecking. When Hemmes studied the behaviour of pigeons on various DRL schedules with treadle-hopping as the response, she found that their DRL efficiency, even on relatively long DRL schedules, was favourably comparable with that of the rat. A further experiment by Kramer and Rodriguez (1971) suggested that the poor performance of pigeons with key-pecking as the operant might be due to the similarity of the operant to the consummatory activity in these situations. Kramer and Rodriguez, using rats as subjects and water reinforcement, found that DRL efficiency decreased markedly when licking and not lever-touching was the experimental operant; their subjects (as do pigeons with the key-pecking operant) were emitting many more responses than the schedule required, and were thus exhibiting what appeared to be an inability to 'wait' for the required interval between responses. Of course, in these experiments both licking in the rat and key-pecking in the pigeon are behaviours which are very similar in form to the consummatory response, and we already know from experiments discussed earlier in this chapter (Staddon and Simmelhag, 1971) that behaviours related to the consummatory activity are induced in conditions predictive of reinforcement. Thus, what appears to be suggested by the experiments of Hemmes, and Kramer and Rodriguez is that when the experimental operant is similar in form to the consummatory action pattern, DRL performance will be disrupted by reinforcer-induced additional responses which fail to be suppressed by the DRL requirement.

These examples of the way in which reinforcer-induced behaviours and experimental operants interact are theoretically important for two reasons. First, they suggest that the formulation of general laws of conditioning using only 'arbitrary' operants such as bar-pressing may give only a partial insight into the adaptive behaviour of organisms since they will exclude the effects of interaction with reinforcer-induced species-typical behaviours. These are interactions which could occur with great frequency in the wild. Secondly, they suggest that the inability of a particular behaviour to come under schedule control need not necessarily imply that that organism lacks the underlying mechanisms or the refinements in those mechanisms to cope with the conditioning contingencies. Failures in schedule control or conditioning in general may merely reflect competition between incompatible response tendencies.

*Aversive conditioning*

So far we have dealt primarily with the way in which behaviours induced by positive or appetitive reinforcers can interfere with conditioning. There is no reason to believe that similar effects do not operate in aversive learning situations. It has already been noted in chapter 4 that the efficiency of escape and avoidance learning is crucially dependent not just on response consequences but also on the *type* of response adopted as the escape or avoidance behaviour. For instance, avoidance learning in a shuttle-box may take around 100 trials for rats (Brush, 1966); with wheel-running as the avoidance response, subjects acquire relatively efficient avoidance behaviour in around 40 trials (Bolles, Stokes and Younger, 1966); with simply running down an alleyway as the avoidance response, the rat may learn in as little as 6–7 trials (Theios, 1963); merely jumping out of a box to avoid shock may need only one trial to establish robust avoidance behaviour (Maatsch, 1959). However, at the other extreme there are numerous reports of failures to train rats to avoid shock by, for example, bar-pressing in a Skinner-box (D'Amato and Schiff, 1964; Meyer, Cho and Weseman, 1960; Smith, McFarland and Taylor, 1961; Weissman, 1962). Bolles has made some attempt to systematise this list of failures and successes by relating the type of response used to the species-specific defence reactions (SSDRs) that the organism brings to the aversive situation. Broadly speaking, he suggests that if a particular avoidance response is rapidly acquired, then that response must necessarily be an SSDR. An animal in the wild, he continues, does not learn to avoid predators in the same way that an animal in the Skinner-box or shuttle-box learns to avoid signalled shock.

> The mouse does not scamper away from the owl because it has learned to escape the painful claws of the enemy; it scampers away from anything happening in its environment, and it does so merely because it is a mouse. The gazelle does not flee from an approaching lion because it has been bitten by lions; it runs away from any large object that approaches it, and it does so because this is one of its species-specific defense reactions. Neither the mouse nor the gazelle can afford to *learn* to avoid; survival is too urgent, the opportunity to learn is limited, and the parameters of the situation make the necessary learning impossible. The animal which survives is one which comes into its environment with defensive reactions already a prominent part of its repertoire.
>
> (Bolles, 1970, p. 33)

When the tame and domesticated laboratory rat is placed into a situation where it receives aversive stimulation a variety of its species-specific defence reactions are elicited. These generally include fleeing, or adopting some type of threat or pseudo-aggressive stance ('flight-fight-fright' reactions). Bearing this in mind, Bolles suggests that there are two major factors which are required for rapid avoidance learning to occur: (1) that the avoidance or escape response is compatible with one of the organism's species-specific defence reactions, and (2) that the response exert some 'change' in the stimulus conditions in the environment. Hence, responses which fit both of these conditions, for example jumping out of a box (Maatsch, 1959), or running up an alleyway (Theios, 1963) are rapidly learnt. They permit the animal to both 'flee' and change its environmental surroundings. Wheel-running and shuttle-box responding permit the animal to 'run' or 'flee', thus the response might be expected to be acquired relatively rapidly, but it does not substantially change the environmental stimuli: in the wheel-running situation, the surrounds are still the same no matter how fast the rat runs, and in the shuttle-box the rat has to jump on alternate trials back into the compartment where he has sometimes received shock. (In this sense it is not a 'safe' place; Bolles suggests that in order to be effective in reinforcing the avoidance response, the stimulus change effected by the avoidance response should never be associated with shock.) Finally, bar-pressing as an avoidance response fulfils neither of these criteria; it does not permit the animal to 'flee', and freezing (an alternative SSDR behaviour often observed in bar-pressing avoidance situations) is similarly not compatible with the avoidance response. However, rats will learn relatively quickly to press a lever to avoid shock if the lever opens a door through which they can run (Masterson, 1970). This again suggests that actually getting out of the place where the aversive events occur may be more reinforcing than just postponing shocks.

The dependency of good avoidance behaviour on the nature of the response has not only been observed in rats. For instance, it has proved extremely difficult to train pigeons to key-peck to avoid or escape shock (Azrin, 1959a; Hineline and Rachlin, 1969; Hoffman and Fleshler, 1959; Rachlin 1969; Rachlin and Hineline, 1967). However, certain other behaviours, related mainly to the pigeon's natural defence responses, have proved relatively easy to condition as avoidance responses. Such behaviours include locomotion (MacPhail, 1968), wing-flapping (Rachlin, 1969) and flying (Bedford and Anger, 1968). When studies have been successful in conditioning key-pecking as an avoidance response, it is often accompanied by wing-flapping and feather-ruffling, suggesting that it may be fostered or even maintained by an aggressive reaction elicited by the aversive stimulation (Rachlin, 1969). An interesting study carried out by Schwartz and Coulter (1973) gives some insight into the interaction between species-specific defence reactions and avoidance contingencies. They initially trained pigeons to key-peck on a VI food schedule before switching their subjects to a shock-avoidance schedule where key-pecking was the only effective means of avoiding or terminating shock. Although the VI schedule had developed a high rate of key-pecking for food in all subjects, they found that the newly introduced shock avoidance contingency could not maintain key-pecking. Subjects emitted at least 30 or more avoidance/escape responses in the first session of the avoidance procedure, but this subsequently declined to a negligible number. Schwartz and Coulter rightly conclude that 'the difficulty of key-peck avoidance procedures extends beyond the shaping of key-pecks to the maintenance of key-pecks which have already been shaped and are well-established'. That is, bar-pressing and key-pecking are not difficult to establish as avoidance responses merely because species-specific defence reactions stop them from occurring in the first place (and thus being reinforced); even when, in the Schwartz and Coulter case, key-pecking is readily occurring at a high rate, the introduction of aversive stimulation disrupts this behaviour (even though it avoids or terminates shock). Thus the 'SSDR model' outlined by Bolles appears not to be just a discussion of the variables which affect operant levels, but a discussion of the active interaction between species-typical defence behaviours elicited by aversive stimulation and the avoidance or escape contingencies set up in the conditioning situation.

Nevertheless, as appealing as the SSDR hypothesis seems the 'loose' way in which it is couched does tend to avoid important problems. For instance, what are the factors that determine which of an organism's defence reactions occur in particular situations? Is there a dominance hierarchy in which certain defence reactions suppress others? Implicit in the SSDR account of avoidance learning is the assumption that fleeing is the dominant SSDR of the rat (and of the pigeon for that matter), with freezing and fighting being suppressed when flight is possible. In this way avoidance responses which enable running and a 'change of scenery' are more rapidly acquired than those which are incompatible with fleeing. The second assumption is that running is the form in which the flight response will be executed. As Riess (1971) has pointed out, if pseudo-aggressive responses were the predominant defence reaction in the rat then it might be expected that bar-pressing would be one of the most efficient forms of avoidance response; the classic pseudo-aggressive threat response in the rat involves rearing with forepaws extended, a posture almost tailor-made for the acquisition of bar-pressing. However, the evidence available on the hierarchical structure of SSDRs is sparse and it appears that a more detailed exposition of the 'SSDR avoidance model' must wait upon the collection of this evidence.

*Modes of interaction*

This section has provided a variety of examples of the ways in which behaviours elicited by reinforcement (appetitive or aversive) interact with experimentally defined operants. Although, as we stressed at the beginning of this chapter, there is no general consensus as to the way in which constraints on learning should be related to traditional laws of conditioning, the examples in this section raise a number of interesting points. We know that reinforcer-induced behaviours can disrupt the conditioning process, but what are the specific dynamics of this process? Is there more than one process? Some hypothetical possibilities are outlined below and related to some of the experiments quoted in this section.

First, it may be that there are certain 'channels' in the learning mechanism which allow only certain responses and certain reinforcers to become associated. This notion has come to be known as the concept of 'belongingness' (Rozin and Kalat, 1971); certain types of responses can be linked to certain types of consequences with ease because there are predetermined links in the learning mechanism between that response and reinforcer; responses which do not have this predetermined link can be conditioned only with great difficulty. Although the majority of the examples which relate to this hypothesis are concerned with discrimination learning, it might be considered, for instance, that there is a predetermined link (and thus a 'belongingness') between flight avoidance responses and aversive events. This allows them to be conditioned more easily than those responses which do not have predetermined links. Examples of the latter would include bar-pressing.

Secondly, it might be considered that reinforcers simply elicit behaviours which are incompatible with the experimental operant, and thus the opportunity for the occurrence of the operant is limited. There are two slants to this approach. It might be that elicited behaviours actively override experimentally acquired operants and thus suppress their occurrence even when that operant has been well established under other conditions. Alternatively, it might be stressed that the induced behaviours merely affect the acquisition of an operant, in which case the induced behaviours would reduce the probability of the operant occurring *in the first place*, and thus being reinforced. Although this latter effect is certainly found in many conditioning situations, the former may also be true as the experiment by Schwartz and Coulter (1973) demonstrated; even when key-pecking has been established through operant conditioning, the transfer of this response to an aversive conditioning situation failed. This might suggest (although there are other alternatives as we shall see later) that shock elicited behaviours which were incompatible with key-pecking and thus they *actively suppressed* key-pecking.

A third approach might be to suggest that reinforcers need not directly elicit overt behaviours which interfere with the occurrence of operants, but that they induce motivational states contrary to the motivational states which normally control the occurrence of the operant. The validity of this type of approach is suggested by interpretations of the experiment conducted by Sevenster (1968) on conditioning in sticklebacks (see pp. 189–190). The use of a female stickleback as a reinforcer for male sticklebacks may have induced a state of sexual motivation in the subjects; however, one of the operants Sevenster attempted to reinforce was biting – a behaviour normally under the control of an aggressive motivational state. The elicitation by the reinforcer of a motivational state contrary to that needed to produce the operant may have led to the failure to condition rod-biting in his subjects. This type of explanation could equally well fit the results of the Schwartz and Coulter experiment. Pecking, a food-related behaviour in the pigeon, might be easily conditioned in an appetitive situation because the motivational state induced by the reinforcer is conducive to the appearance of pecking. However, in the aversive situation the motivational state elicited by electric shock (presumably a state which we might loosely call 'fear' or 'fright') is incompatible with the state necessary for the occurrence of pecking, and thus 'avoidance pecking', is rarely found in pigeons.

Finally, the fourth approach reflects more general phenomena related to failures in the transfer of learning. It may be that when a reinforcer is changed (as in the Schwartz and Coulter case), the topography of the behaviour is also significantly altered since the reinforcer might directly determine the nature of the operant response. The section on autoshaping suggests that the type of reinforcer used can significantly affect the nature of the response (see pp. 182–183), and it may be that the potency of the reinforcer in this respect overrides the response–reinforcer contingencies set up by the experimenter. The implicit assumptions of this approach are that (1), there are predetermined links between reinforcers and responses: when a particular reinforcer occurs only one particular class of responses will predominate and (2), response–reinforcer contingency plays only a minor and indirect role in learning; arbitrary responses, even though reinforcer presentation is contingent upon their occurrence, are suppressed by the predetermined behaviours elicited by the reinforcer. This approach is based not on defining the types of interactions between species-specific behaviours and traditional conditioning processes but on a radical reappraisal of the conditioning process itself. It is considered that constraints on conditioning do not reflect interactions between induced behaviours and traditionally conceived conditioning processes, but that they reflect our misconceptions of the conditioning process. Since this will be dealt with in more detail in chapter 7 it is not intended to elaborate such viewpoints here. However, the interested reader is referred to Moore (1973) for a more detailed exposition.

Thus, although there are now many recorded examples of the way in which induced behaviours apparently interact with the experimentally defined operants of a learning situation, the processes which underlie these interactions are still unclear. One or more of the possibilities we have discussed above may be the appropriate one.


## FACTORS ASSOCIATED WITH THE RESPONSE

In outlining his account of the theory of reinforcement, Skinner emphasises that any operant response must first occur 'for other reasons' before it can be reinforced. He writes

. . . contingencies (of reinforcement) remain ineffective until a response has occurred. The rat must press the lever at least once 'for other reasons' before it presses it 'for food'.

(Skinner, 1969)

These 'other reasons' were initially considered to be unimportant, or at least unnecessary for an account of operant reinforcement. As we have already seen in this chapter, factors which influence the operant level of a response, or the initial 'degree of behavioural variability' may well be a crucial factor in conditions where there is no strictly defined reinforcement contingency (Staddon and Simmelhag, 1971). Similarly, as we shall see in this section, the 'other reasons' which initiate a behaviour may play an important and necessary role in the operant conditioning of that behaviour.

We are then in essence talking in this section about the specific causal factors which initiate or control the occurrence of particular behaviours. There are a number of ways in which these may affect conditioning. They can be summarised as follows: (1) The operant being reinforced may be so linked to its causal factors that those causal factors need to be constantly present if the behaviour is to be effectively reinforced; (2) If only a discrete element of some fuller action pattern is being reinforced, but the causal agent for the full action pattern is constantly present, then the full-blown action pattern may persist; (3) Some behaviours may need their specific causal agents present during the early stages of conditioning, but may come under the influence of reinforcement to the extent that these causal factors can later be removed.

An example of the first category of constraint is provided by an experiment carried out by Black and Young (1972). They trained rats to bar-press or to drink to obtain food or to avoid electric shock. First, their rats were trained to bar-press for food in the presence of one stimulus and to drink for food in the presence of another. After this a similar discrimination training was attempted using shock-avoidance as the reinforcer. Figure 5.9 show that although the rats could be trained to drink to avoid shock, their responding was still under the fairly strict control of its causal factors (in this case a state of 'thirst' or water deprivation); the subjects only drank at a high rate if they were water-deprived. When avoidance drinking did occur, the rats drank mainly in the presence of the appropriate stimulus signalling that drinking would avoid shock. It has been suggested from a number of quarters that behaviours which require their causal factors to be constantly present during the conditioning are of the 'reflexive' or 'involuntary' kind (Segal 1972; Solomon, 1964; Teitelbaum, 1966; Vanderwolf, 1969, 1971). For instance, Solomon (1964) has suggested that 'reflexive, short-latency' responses are less susceptible to reinforcement or punishment contingencies than are 'non-reflexive, longer-latency' responses. Teitelbaum (1966) and Vanderwolf (1969, 1971) have attempted to pin down in more detail what is actually embodied in the term 'involuntary'. Vanderwolf suggests that those behaviours which can be

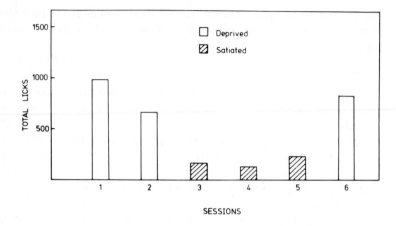

Figure 5.9   Drinking (recorded as licks per session) to avoid electric shock in a Sidman avoidance procedure during six consecutive sessions. On deprived days (open bars) the rat was required to drink tap water to avoid shock after a 24-h water-deprivation period. On satiated days (hatched bars) the rat was required to drink tap water to avoid shock after having had water freely available in the home cage during the previous 24-h. (After Black and Young, 1972).

identified by the *absence* of hippocampal theta rhythms during their performance are 'involuntary' in the sense that they rely totally on the presence of specific causal factors for their occurrence. They are not susceptible to modification by their consequences unless these causal factors are present. Behaviours which have been identified as 'involuntary' in this sense are grooming and drinking in the rat. We have already noted the difficulty in reinforcing drinking unless its causal 'state' is present; similar difficulties in the conditioning of grooming responses have been reported in a number of different animals including hamsters (Shettleworth, 1975), and pigeons (Hogan, 1964), although, since grooming is such a complex action pattern, an alternative explanation for these failures is possible. Reinforcement may have been presented contingently on different aspects of the grooming response, thus producing 'sham' grooming and even adventitiously reinforcing movements which are unrelated to the grooming response (Hogan, 1964). This notion of 'involuntary' must be distinguished from 'autonomic' in the very general sense that this latter term is used. In chapter 2 it has already been noted that certain autonomic responses are fairly amenable to modification via operant reinforcement (Dicara and Miller, 1968 a, b, c; Miller and Carmona, 1967; Miller and Dicara, 1968; Shapiro, Crider and Tursky, 1964). However, these types of operantly conditionable 'involuntary' responses are what Segal (1972) calls 'nonspecific', that is, they occur with a certain periodicity or rhythm 'even in the absence of well-defined, discrete, eliciting stimuli' (heart-rate and blood pressure are convenient examples).

Involuntary responses which have proved difficult to condition are those whose occurrence waits upon the presentation or induction of discrete, fairly well-defined causal stimuli or 'states'. Nevertheless, the notion of 'involuntary' as it is used here is still very sketchy. There does appear to be some link between the absence of hippocampal theta rhythms and insusceptibility to reinforcement, but to label these behaviours as 'involuntary' may be misleading. A wider variety of behaviours needs to be studied before any concrete conclusions can be drawn.

The second category of constraint is less of a constraint and more of an 'over-reaction' on the part of the organism. If the causal agent for a particular behaviour pattern is present during conditioning, or is induced during the conditioning process (for example, the induction of a particular motivational state), then that behaviour pattern may tend to over-ride the experimentally defined operant. For example, Azrin and Hutchinson (1967) reinforced a pigeon for merely pecking a target bird with a particular predetermined force. Although this response could be developed as a good operant and could be conditioned fairly well, it often developed from its initial 'polite' or 'cold' pecking to full-blown aggression, with the subject exhibiting wing-flapping, feather-erection and the like. This became so predominant that it even continued during periods of food presentation. What appears to be happening here is that causal states related to pecking a conspecific ('aggression' towards the target bird) have been induced, and the behaviour repertoire related to that causal state has thus been invoked, even to the extent of diverting the animal from its original task of obtaining food! Similar results have been obtained with pigeons by Reynolds, Catania and Skinner (1963) and rats by Ulrich, Johnston, Richardson and Wolff (1963). So here, although the purpose is to try and reinforce only a small aspect of a wider set of behaviours, the induction of the causal state for the whole set of behaviours necessitates their occurrence *in toto*.

So far we have dealt with behaviours which apparently depend indefinitely on the presence of their causal factors in order to be maintained as reinforced operants. (No long term studies have been carried out in order to substantiate this claim, although it is implicit in the theorising of a number of workers (Black and Young, 1972; Vanderwolf, 1969, 1971).) Other experiments tentatively suggest that in some cases the causal agents may only be needed during acquisition. Konorski (1967) reports a number of experiments related to this point. He reinforced cats for a variety of responses including face-rubbing, ear-scratching and anus-licking. Initially these behaviours had to be elicited by the appropriate unconditioned stimuli such as gum arabic on the face, cotton wool in the ear, or soap on the anus. However, these stimuli were not required throughout training and could be disposed of as the particular operant was acquired. Nevertheless, two factors suggest that this point should be treated with some caution. First, Konorski did report that the eliciting stimulus for ear-scratching (a wad of cotton wool in the ear) quite often had to be reinstated at the beginning of the first few sessions, even though reinforcement had substantially increased the frequency of ear-scratching in previous sessions. Secondly, on the removal of the causal factors for the originally reinforced behaviour, reinforcement may come to maintain only 'sham' behaviours or abbreviated forms of the original behaviour. For instance, Konorski (1967) found that a dog which was reinforced with food for yawning continued merely to open its mouth as if in a yawn, but the frequency of full-blown yawning was affected very little (see figure 5.10). Abbreviated forms of grooming also frequently appear in lieu of the full-blown grooming behaviours which are originally reinforced. Thorndike (1911) reported that the behaviour of both cats and chicks trained to preen to escape from a puzzle-box deteriorated appreciably during training until preening was only a fraction of its original form. Shettleworth (1975) has also noted strikingly short bouts of face-washing in hamsters when this response was reinforced with food. Abbreviated forms of grooming responses have been reported in other situations where this behaviour has been reinforced (Beninger, Kendall and Vanderwolf, 1974; Hogan, 1964; Lorge, 1936).

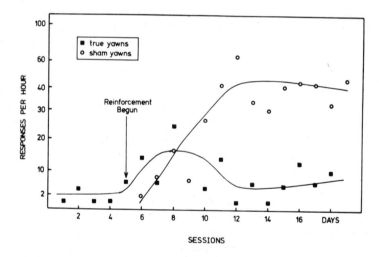

Figure 5.10   Rate of yawning (filled squares) and 'sham' yawning (open circles) in a dog reinforced for yawning with food. (After Konorski, 1967).

The fact that behaviours in this latter category often deteriorate into abbreviated topography or 'shams' of their original form following the removal of their causal factors suggests a permanent inability to be affected by reinforcement unless these natural causal factors are present. Thus, they are rather more important than the 'other reasons' which Skinner suggests *merely initiate* operants in order that they may be initially reinforced. These natural causal factors actually appear to be a *necessary* condition both for the acquisition and for the maintenance of many behaviours. However, it is worth posting a warning on accepting such an interpretation without further consideration. Most of the behaviours we have discussed in this section are of a complex nature topographically and thus present problems when it comes to actually deciding when they have been completed and the reinforcer should be administered. On this score reinforcers have usually to be presented on the arbitrary decision of an experimenter and not on the instant decision of a microswitch which is closed by the organism itself. Thus it is quite possible that abbreviated or 'sham' behaviours might be artifacts of inappropriate reinforcer presentation and not a result of the absence of causal factors as we have suggested. However, the fact that the reinforcement of relatively simple behaviours such as drinking and 'aggressive pecking' (both of which can be easily recorded, and thus reinforcers can be fairly accurately meted out on their occurrence) is affected by the presence or absence of their natural causal agents suggests that we are dealing with a fairly real constraint on conditioning. The extent to which this type of constraint occurs (as opposed to artifactual behaviour forms fostered by 'loose' reinforcement presentation) must, however, await more extensive evidence and possibly more refined techniques of contingency management.

# 6 Constraints on Discrimination Learning

In the previous chapter we dealt with those constraints on conditioning which primarily involve the relationship between response and reinforcer (UCS). For certain reasons reinforcers may only selectively reinforce certain types of responses – either because they elicit competing responses under some conditions, or because the response is so tied to its antecedents that the response–reinforcer contingency can have little effect in the absence of these antecedents. Since in chapter 5 we were primarily concerned with the nature of response–reinforcer relationships, much of the discussion centred around constraints on operant conditioning. However, because the contents of this chapter involve discussion of stimulus factors, they will relate a little more to anomalies that occur during certain types of classical conditioning experiments. For example, can any neutral stimulus which is perceptible be utilised as a Pavlovian CS? Are there predispositions for an animal to associate certain kinds of CSs with certain kinds of UCSs? Or even certain CSs with certain types of responses? Certainly the interrelationships between CSs, UCSs and responses has not been found to be as simple as Pavlov had originally envisaged them.

First, let us consider the kinds of stimulus constraints that might disrupt either discrimination learning in operant conditioning or the acquisition of a Pavlovian conditioned response to a CS. Most obviously, of course, the animal must be able to perceive the stimulus; that is, it must have the physiological apparatus necessary for detecting it. For example, diurnal primates such as the squirrel monkey and marmoset, whose retinae are rich in both rods and cones, can make much finer visual discriminations than nocturnal primates such as the owl monkey or the bush baby whose retinae consist mainly of rods (Ordy and Samarajski, 1968). Similarly, using a tone of 40 kHz as a Pavlovian CS for a rat will be much more successful than using the same CS for a human being since a rat can 'hear' ultrasonic tones up to a frequency of around 60 kHz whilst for man the upper limit is only 15 kHz (Miller, 1970). Many constraints such as this may seem trivial in their obviousness, and they tend to reflect peripheral characteristics of the perceptual systems. But what is obvious should not be ignored. It must not be forgotten that nature has given animals specific adaptations to fit them for survival in their ecological niche, and the fact that certain animals have more

202

flexible basic senses than others appears to be only the thin end of the adaptive wedge. For instance, some animals may have the ability to sense a wide range of values along a particular sensory dimension (for example, colour), but evolution has equipped them with a *predisposition* to detect certain values in that continuum more readily than others. For example, the octopus can discriminate horizontal–vertical differences far more readily than the differences between opposite obliques (Sutherland, 1957), and the frog has a predisposition to respond more to 'blue' than other colours (Muntz, 1964). In the case of the frog, not only does this animal readily detect a blue area in its visual field it also has a tendency to jump towards it. It is reasonable to suppose that the adaptive significance of such an inbuilt mechanism is to direct a frightened frog to jump into water and away from optimistic predators. It is perhaps instructive to remain with the frog while we discuss perceptual predispositions and their adaptive significance as this amphibian is a rich source of built-in visual analysers. The peripheral visual system, primarily the retina and the optic fibres, can be considered more as a filter than as a photographic plate – it sends only the most useful information to the brain. Perhaps the most efficient way of doing this is to have channels especially designed to seek out and transport to the visual centre specific information. This seems to be exactly what evolution has given the frog. Lettvin, Maturana, McCulloch and Pitts (1959) in a revealing series of studies, found that the frog had optic nerve fibres whose functions were to transmit only specific information from the retina to the optic tectum (the chief visual centre in the frog). For instance, one fibre only fired when sharp edges entered the visual field; a second fibre fired only when small moving objects appeared in the visual field (the 'bug detector'); a third and fourth channel only fired when the retina was confronted with onset and offset of illumination respectively. The 'bug detector' comes to mind as being the most valuable of filters: a frog's main source of food is, of course, flies, and the sooner he can detect them against the visually chaotic background of his habitat the better. In fact, a very large number of those specialised 'detectors' that have been identified can be seen to have definite appetitive (food getting) or defensive (predator avoiding) functions. Indeed, the moth's auditory system is even equipped with what can only be called a 'bat detector' (Roeder, 1963)!

This preceding discussion has illustrated two points about the perceptual system of animals. First is the obvious fact that animals differ in their sensitivity to the *range* of stimuli they can detect; and secondly, the animals differ in their *predispositions* to detect quite specific features of their environment. Both of these facts reflect physiological adaptations which have in general benefited the animal, and ensured the survival of the species. But how do these adaptations affect us as learning theorists? The first implication is fairly straightforward: if an animal has a tendency to select certain specific aspects of its environment for attention, then it seems quite likely that the animal will be more likely to use these aspects as cues during learning. That is, they are more likely to become conditioned stimuli or discriminative stimuli for responding than more arbitrary aspects of the environment. This, if it is true, contravenes Pavlov's law of equipotentiality of stimuli. The second implication is one which we might justifiably extrapolate from the physiological evidence. If animals have predispositions to detect certain stimuli, why should they not also have predispositions to *associate* certain stimuli? For example, the learning mechanisms of an animal may be pre-wired so that specific UCSs can only be associated with a small number of relevant CSs. This idea has been expounded elsewhere (Rozin and Kalat, 1972; Seligman, 1970) and in particular learning situations its adaptive significance has not gone unnoticed (Garcia, Hawkins and Rusiniak, 1974).

# CONSTRAINTS ON CUE UTILISATION

## The nature of CS and UCS

### The nature of the CS in appetitive and aversive associations

A number of recent studies have demonstrated that the particular aspects of a compound CS or $S^D$ that come to control responding will depend on the nature of the UCS or UCSs in the learning situation. Two particular groups of studies serve to illustrate this point.

The first concerns a pair of experiments by Shettleworth (1972a, b) on chicks. She trained young domestic chicks to discriminate between plain, palatable water and water made 'unpalatable' with quinine or with a mild electric shock which the chicks received through their beaks on touching the water. The two types of water were distinguished on the basis of the presence or absence of visual, auditory or compound visual–auditory cues. When either the colour of the water or the flashing of the chamber houselight was the $S^D$, chicks learned within a few trials to avoid drinking the 'unpalatable' water. But when a clicking sound or a loud interrupted tone was the $S^D$, the chicks only appeared to learn the discrimination when the beak-shocks were made particularly intense. In all cases, drinking came under the control of the visual element and never under the control of the auditory element. These differences were not due to the chicks being unable to hear the sounds since on early conditioning trials both visual and auditory stimuli produced an increase in the latency of drinking. The results of this experiment have a number of possible implications, the most obvious of these being that, (1) chicks may have a predisposition to associate aversive stimuli with visual cues, or, more specifically, (2) they have a tendency to associate the immediate aversive consequences of drinking with visual, but not auditory stimuli. A second experiment by Shettleworth ruled out the first possibility. In a classical conditioning procedure with chicks she paired either a flashing light or clicks with unavoidable foot-shock. The chicks rapidly showed signs of associating the clicks with shock: after only a few trials they began to run around, jump and shrill call when the auditory cue came on. The flashing light CS, however, produced much less dramatic changes in behaviour. Thus in this situation the chicks had readily associated auditory stimuli but not visual stimuli with shock. Although there were differences in procedure, these results suggest that chicks do tend to associate visual stimuli with the immediate consequences of drinking.

A series of experiments by Foree and Lolordo (1973, 1975) further clarify the nature of the relationship between aversive/appetitive UCSs and the salient components of compound CSs. In an early study (Foree and Lolordo, 1973), they found that if one group of pigeons was trained in the presence of a tone/light $S^D$ to press a treadle to obtain food, and a second group was trained to make the same response to avoid shock, then the first (appetitive) group learned only to respond to the light while the latter (aversive) group learned only to respond to the tone aspect of the compound $S^D$. From these results they hypothesised that the auditory features of a compound $S^D$ might acquire control of responding either (1) when there is an avoidance contingency in operation, or (2) when a painful stimulus like shock is presented. In a later series of experiments (Foree and Lolordo, 1975) they found that neither of these factors was a sufficient condition for auditory features to acquire control over responding. If a pigeon is concurrently responding (key-pecking) to obtain food and also responding (treadle-hopping) to avoid shock in the presence of a compound tone–light $S^D$, then the light and not the tone component acquires control over avoidance responding. What these results seem to suggest is that the presence of an appetitive reinforcer in the experimental task is a sufficient condition for the visual element of a compound stimulus to acquire control of responding – whether that respond-ing be to obtain food or avoid shock. This would explain why Shettleworth observed control by visual features in the avoidance study, but in the Pavlovian procedure, where no appetitive reinforcer was present, she observed control by auditory features. At this stage one can only speculate as to why pigeons and chicks should have a predisposition to associate visual cues with responding in appetitive situations and auditory cues with responding in an aversive situation. I am sure the reader can conceive of numerous benefits that such a predisposition could have for the animal in nature, but as Shettleworth (1972a, p. 41) points out ' . . . saying that animals can only learn things that would be relevant in nature is as much an oversimplification as saying that they can learn anything'! The full extent of these predispositions, and the precise conditions under which they occur have yet to be fully assessed; not to mention the fact that the above studies have only used birds – other species, who do not rely so heavily on vision in their search for food might display quite different associative predispositions.

### Conditioned taste aversion

*The characteristics of conditioned taste aversion*

One of the most interesting and well-known constraints on cue utilisation is that known as conditioned taste aversion. As everyone who has attempted to eradicate wild rats has found, they eventually develop a 'shyness' for a toxic bait if that bait is used often enough. First, how do they come to recognise the bait as poisonous and how, in the first place, are they able to associate that particular bait with toxic effects that presumably occur long after the poison has been consumed? An interesting experiment by Garcia and Koelling (1966) did much to highlight the individual nature of this phenomenon. They trained rats to drink 'bright, noisy water'. They did this by attaching an electrode to the drinking tube so that whenever the rat touched the spout with its tongue it produced a flashing light and a loud noise. For one group of rats drinking 'bring, noisy water' was followed by shock, for a second group the consequence of drinking was illness which was induced some time later by the rats being X-irradiated or injected with a toxic substance (lithium chloride). A third group of rats was offered water without noise or light but which was laced with saccharin – this was followed by immediate foot-shock. A fourth group of rats was given saccharin-laced water followed by induced toxicosis, as with group 2. They found that, of the first two groups, when offered 'bright, noisy water' the following day, only the electric shock group avoided drinking it. Of the saccharin groups, the subsequent test showed that only the group that had been given consequential toxicosis avoided drinking saccharin-laced water (see figure 6.1). The implications of these results are that whereas rats will learn to associate exteroceptive stimuli, such as lights and sounds with immediate electric shock, they will only learn to associate *taste* with subsequent toxic consequences (Revusky and Garcia, 1970). This effect seems to violate two of Pavlov's most fundamental principles of classical conditioning, (1) the equipotentiality of conditioned stimuli, which specifies that any perceptible stimulus can come to function as a Pavlovian CS regardless of the nature of the UCS; and (2) the principle of contiguity: to become well-established as a Pavlovian CS, the UCS must follow the CS only with the briefest of delays. In conditioned taste aversion studies only taste becomes associated with the toxic consequences (and hence only similar *tasting* foods are avoided in future), there do not seem to be enough trials to establish learning (only one association is usually necessary), and the delay between CS and UCS seems much too long (the rat will still avoid the food even if illness occurs hours later). Quite an impressive list of anomalies! However, this predisposition to associate taste with subsequent poisoning is not universal to all animals. Squirrel monkeys for example, appear to be able to associate both taste and colour of the food with poisoning (Gorry and Ober, 1970). Birds, on the other hand, tend to associate visual cues with eventual toxicosis (Capretta,

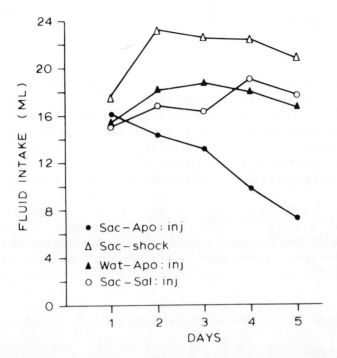

Figure 6.1   Fluid intake for four groups of rats subjected to pairings of water or saccharin with either shock or apomorphine injection. The only group to exhibit a substantial suppression of drinking was the group which had received pairings of saccharin with a toxic apomorphine injection. (After Garcia and Koelling, 1966).

1961; Wilcoxon, Dragoin and Kral, 1969). Wilcoxon *et al.*, (1969) trained quails to consume water which was both sour and blue and then subjected them to toxicosis. A subsequent test showed that these birds exhibited a pronounced aversion to unflavoured blue water but only a small aversion to uncoloured sour water. The visual cue appeared to be dominant over the flavour cue in this case. This difference in associative predispositions between species can tentatively be explained in terms of their different eating habits. Rats are natural foragers and grind their food in their mouth – thus taste is an important and salient cue. Birds, however, are mainly visual eaters, they more readily select food on the basis of its visual appearance, and grinding of the food generally occurs after ingestion. What is important to us as learning theorists, however, is not so much what cue is selected for association as, (1) why only one aspect of food is utilised as a cue, and (2) how such long-delay learning can occur and whether it reflects a preparedness of the underlying learning mechanism to accept only associations between specific CSs and specific UCSs.

Before we continue to look at theories of conditioned taste aversion it is perhaps interesting to note that we are not dealing here with an anomaly which is specific to the traumatic experience of poisoning. Conditioned taste aversion does appear to have a positive appetitive counterpart. For example, if thiamine-deficient rats are fed a flavoured solution and are then injected with thiamine, the rats subsequently exhibit an increased preference for that flavour (Campbell, 1969; Zahorik and Maier, 1969). If rats are intragastrically injected with food after consuming a non-discriminative but flavoured substance they eventually develop an increased preference for that food substance (Holman, 1969). So what we appear to be dealing with here is a phenomenon which reflects what the animal is able to learn about gustation in general and not something which is specifically linked to toxicosis.

## Explanations of conditioned taste aversion

On the theoretical front, conditioned taste aversion has been tackled in one of two ways. First, the violation of basic Pavlovian principles has been acknowledged and attempts made to incorporate them into a specific model of gustatory learning. Or secondly, these violations have been denied as only apparent and evidence has been sought to show that conditioned taste aversion can be assimilated under traditional principles of conditioning.

### Learned safety

An hypothesis which strikes one as being qualitatively different from most of the others we will discuss is one put forward by Kalat and Rozin (1973). They suggest that rats initially show a reluctance to consume novel foods (neophobia), and that this reluctance is dispelled over time as the animal learns that the new food is safe. This is contrary to the usual assumption that the animal forgets the taste over time. They tested their hypothesis with a neat experiment. One group of rats was fed a novel food substance and poisoned 4 h later; a second group was given the food and poisoned $\frac{1}{2}$ h later; a third group was given the food twice – 4 h and $\frac{1}{2}$ h before being poisoned. If forgetting of the taste was an important factor in the association between taste and toxicosis, then the second and third groups, who both experienced the food $\frac{1}{2}$ h prior to poisoning, should show similar conditioned aversions to it. However, this was not the case, in fact groups one and three, who both had food 4 h prior to poisoning, showed similar levels of aversion. This strongly suggests that the rat was learning something about the safety of the food over time rather than associating the toxicosis with a decaying memory of the food. This hypothesis is also consistent with the findings that conditioned aversion to novel foods is greater than to familiar foods (Revusky and Bedarf, 1967). However, all this being so, it still leaves a fundamental question about conditioned taste aversion unanswered.

Why, for the rat, is only taste and not other aspects of the food such as colour, shape and associated audio-visual stimulation associated with poisoning?

## Operant punishment of eating

It can be considered that eating or drinking in the presence of certain $S^D$s such as flavour or colour of food is operantly punished by poisoning and thus eating in the presence of these $S^D$s is less likely to occur in the future. An experiment by Domjan and Wilson (1972) suggests that this is unlikely. They curarised rats so that they could neither drink nor swallow, and flowed flavoured water across their tongues. They then poisoned these animals and later tests showed that the rats exhibited a significant tendency to avoid foods of that flavour. If no response of eating was executed then it logically could not have been punished. A more feasible approach to this problem is to consider conditioned taste aversion simply as an example of classical conditioning with the animal associating CS (taste) and UCS (poison) – the CS coming to elicit the avoidance responses that poison normally would.

## 'Preparedness'

What is more important than whether the conditioned taste aversion paradigm is operant or classical is how rats come to associate taste with poisoning after such long delays, and why they cannot do this with exteroceptive visual or auditory stimuli. One possibility is to suggest that the rat has two learning systems: one to learn about events in its external environment, and a second to learn about events in its internal environment. The two systems operate according to different principles of association and utilise different cues (Garcia et al., 1974; Seligman, 1970). As Garcia, McGowan and Green suggest

Natural selection has designed the rat with distal and proximal defence perimeters to handle threats from the external environment. The animal localizes a distant event by sight and sound mediated by the head receptors and if that distant event is followed by an insult to the cutaneous surface it quickly learns defensive reaction. The distal and proximal afferent categories converge to a somatic centre where potential pathways favour their integration. Other afferent categories are not so favoured.

Natural selection has designed the rat with another distal-proximal system to cope with the internal environment. Foodstuffs are chemically analysed by gustatory and olfactory receptors when sniffed and eaten. Later as the food is absorbed, internal receptors report on the ultimate effects with the internal environment. These two afferent categories converage upon a visceral centre which is relatively insulated from stimuli arising from the external environment. Since food absorption takes time, this system has become specialized to handle longer interstimulus intervals.

(Garcia, McGowan and Green, 1972, p. 39)

This theory is appealing in its simplicity and logic from an adaptive viewpoint, but on closer scrutiny it can be considered as something of a *post hoc* explanation. It might not be thought an expedient step in behavioural science to postulate the existence of new associative principles, especially when the theory seems conveniently designed to fit the facts that have recently become available. The rule of parsimony should encourage us first to search more thoroughly for an explanation in terms of well-tried existing principles and laws than to suggest, for example, that long-delay learning is now possible if we postulate a new learning mechanism within which it can operate. This might appear to be an unwarranted attack on a perfectly reasonable theory, but evidence we will discuss later suggests that such theoretical convervatism should be borne in mind. Still, to recap, this theory in its strongest form suggests that, because of evolutionary determined neural pathways, (1) taste becomes associated with poisoning only after long delays, and (2) exteroceptive stimuli become associated with external aversive events only when short delays intervene.

Having outlined this more radical explanation of the conditioned taste aversion phenomenon, let us look at accounts which stress the role of accepted learning principles and see how these relate to 'preparedness'.

Interference effects

Revusky (1971) has proposed that conditioned taste aversion does not necessarily provide an exception to the rule that animals can only learn CS–UCS associations when the two are temporally contiguous. He suggests that animals can in fact learn associations between events which are separated by quite long delays, but that this normally fails to occur because other stimuli and events intervene during the delay interval to interfere with the formation of the association. The greater the similarity between the CS and the intervening stimulation, the less likely it is that an association will be formed. If this is the case, long CS–UCS delays with audio-visual CSs are likely to produce little conditioning because most of the intervening stimulation will be audio-visual. With taste as the CS, however, very little stimulation of a similar nature (that is, different tastes, flavours, etc.) intervene to disrupt the formation of an association. There are a couple of factors which support this view. First, contrary to what we might be led to believe by contiguity theorists, it seems that animals can learn associations when the reference events are separated by a relatively long delay (see chapter 3, pp. 93–94) – but only when interference effects are minimised. Secondly, Revusky himself has carried out a number of experiments (1971, pp. 187–209) in which differently flavoured substances have been administered to the rat between CS presentation (saccharin flavoured water) and poisoning. In each case the subsequent avoidance of the saccharin flavoured water was inversely related to the amount of interference from intervening flavoured substances. Nevertheless, even though this account explains the long-delay learning of conditioned taste aversion without postulating radically new principles, it still has difficulty in explaining why taste only associates with toxic consequences and audio-visual stimuli with external consequences such as electric shock: why, for example, has it been frequently found that taste does not become associated with subsequent shock at short CS–UCS intervals (Braveman and Capretta, 1965; Dietz and Capretta, 1967; Garcia and Koelling, 1966)? There should be even less interference in this situation than in the long-delay taste aversion paradigm. This predisposition on cue utilisation is explained by what Revusky calls 'relevance'

> . . .the associative strength of a flavour is high when E-post (the UCS) is toxicosis and low when E-post is an electrical shock; the associative strength of a telereceptive or proprioceptive stimulus is low when E-post is toxicosis and is high when E-post is electrical shock. 'Relevance' is simply a term used to refer to such relationships.
>
> (Revusky, 1971, p. 168)

In explanatory terms this really helps little. It merely redescribes the observed fact that taste is associated with toxicosis and auditory-visual stimuli with shock! The first step in clarifying the reasons for these associative predispositions is to try and establish whether they reflect innate factors such as predetermined neural specificity or whether they reflect artifacts or biases which are inherent in the learning task. The next hypothesis throws some light on this.

Persistence of stimulus traces

Krane and Wagner (1975) conducted an experiment to assess the aversive properties of either saccharin flavoured water or a light–tone compound when the two were paired with electric shocks after different delay intervals. First they trained rats to drink from a tube, each lick producing either an audio-visual signal or contact with saccharin flavour, or no explicit cue. At the end of the drinking period the tube was made inaccessible and the subject was presented with a brief intense footshock either 5, 30 or 210 s after the termination of the CS. When later given access to the drinking tube they found that animals given the light–tone compound showed significant suppression of drinking with a CS–UCS interval of 5 s, but none at 210 s. However, animals given the saccharin CS showed no suppression of drinking at the short interval but substantial suppression at the longer, 210 s, CS–UCS interval. Although the failure to find an association between taste and shock at short delay is consistent with data we have already reviewed, the fact that taste could become associated with shock at *long* delays is something which hitherto had not been investigated. Krane and Wagner assert that these results 'could be generated by conventional principles of Pavlovian conditioning plus the assumption that the stimulus trace occasioned by saccharin consumption is more persistent that that occasioned by exposure to a light–tone compound' (Krane and Wagner, 1975, p. 887). The assumption here is that memories of audio-visual stimuli decay more rapidly than memories of taste stimulation. Now, this being the case, taste does not become associated with shock at brief CS–UCS intervals because the 'trace' of the flavour CS extends beyond the offset of the UCS: that is, memory of the taste is experienced both *before and after* the UCS presentation. It is well known that CSs that occur *immediately after* the presentation of an aversive UCS (Backward conditioning, see p. 24) often acquire the properties of a 'safety' signal (Denny, 1971) so they fail to become conditioned aversive stimuli. This is possibly why taste fails to acquire aversive properties when paired with shock after short delay intervals.

Summary

Theories such as learned safety or preparedness view the phenomenon of conditioned taste aversion as reflecting in-built mechanisms especially developed to cope with the problems of gustatory learning. In this sense they accept the apparent violation of fundamental associative principles. Other attempts have been made, however, to integrate the anomalies of conditioned taste aversion under existing principles of conditioning and thus preserve the 'generality of the laws of learning'. Preliminary attempts to do this have had some success at explaining long-delay learning (interference hypothesis – Revusky, 1971) and the predisposition of rats to utilise taste as the associative cue for poisoning (Krane and Wagner, 1975). But, as Krane and Wagner point out, so little is known about the quantitative effects of the variables involved in conditioned taste aversion that it is perhaps unwise at this stage to draw any positive conclusions either about the possibility of a specific gustatory learning mechanism, or about the generality of existing associative principles.

## The nature of the CS and the nature of the response

The previous section has been concerned with associative predisposition between CSs and UCSs. Some evidence has also been found that certain kinds of CSs or $S^D$s are more effective with certain kinds of responses. Although an animal may be responsive to a stimulus (that is, he can perceive the stimulus and react to it in other ways), in some instances he just cannot learn to emit a particular piece of behaviour in response to it. One such example which illustrates this point is the octopus' reaction to stimuli of differing weight. This animal can compensate appropriately for the weight of objects when it picks them up, but cannot learn to distinguish between objects that differ only according to weight (Wells, 1961). However, perhaps the only extensive study of this type of phenomena is one reported by Dobrzecka, Szwejkowska and

Konorski (1966). Dogs were trained in two different discrimination tasks. In one task they were trained to place the left foreleg on the feeder when a metronome sounded from in front of them and to place the right foreleg on the feeder when a buzzer sounded from behind. When their performance on this task had stabilised they were tested with the buzzer in front of them and the metronome behind. They found that most dogs responded mainly on the basis of the location of the sound and not its quality (for example, they placed their right foreleg on the feeder if a noise sounded from behind them, regardless of whether it was a buzzer or metronome). However, when dogs were trained to move a foreleg in response to a metronome in front and *not* to move in response to a buzzer behind (a GO–NOGO discrimination), tests later revealed that they were responding mainly on the basis of the quality of the sounds and ignoring position. So, to summarise, dogs would readily come under the control of the location of the sound in a left–right discrimination, but only come under the control of the quality of the sound in a GO–NOGO discrimination.

There is so little documented evidence concerning constraints on stimulus-response relationships that it is difficult even to make a guess at the reasons underlying such predispositions. Similarly, the constraints that have been demonstrated–such as the above example – are not simple ones, but relate characteristics of the stimuli to quite sophisticated characteristics of the response. However, for the interested reader discussions of this and similar phenomena can be found in Shettleworth (1972a, pp. 32–34).

## CONDITIONED STIMULI AS ELICITING STIMULI

When selecting a stimulus to act as a classical CS or an operant S$^D$ one has to be aware of the possible unconditioned eliciting effects of that stimulus. The CS might also be a UCS which elicits competing UCRs, or it might be a stimulus to which the animal is particularly sensitive; (for example, in pseudoconditioning effects, the CS can elicit fear reactions even when it has not been paired with an aversive UCS, if the animal has been exposed to that UCS just previously). Some CS–response interactions are fairly predictable given that one understands the significance of the CS for the animal. For example, Razran (1971) reports a case where the running response of a hare to an object moving in its field of vision plus a tactile stimulus applied to its neck (UCS) was conditioned to a sound imitating the animal's lip smacking in only 3 trials, and could not be extinguished 1½ years later after 300 extinction trials. The same response could only be conditioned to a metronome in hundreds of trials and then was readily extinguished. This type of effect is known to Russian theorists as a 'phyletic readying of learning' which they call 'natural unconditioned reflexes' (Razran, 1971, p. 20–21). the lip-smacking stimulus presented to the hare is surely one which has some behavioural significance to the animal prior to conditioning: it may even be a species-specific warning stimulus which unconditionally elicits flight reactions, hence the rapid conditioning and resistance to extinction when compared with the more neutral metronome CS.

Further processes which should be considered in this context are those of sensitisation and pseudoconditioning (see chapter 2, pp. 30–31). Sensitisation is the process by which repeated presentation of a single stimulus in isolation (that is, not in association with a UCS) eventually comes to produce greater and greater reactions of some kind on the part of the animal. It is not necessary to go into the adaptive significance of this non-associative mechanism here (Razran, 1971, chapter 4; Rachlin, 1976, pp. 102–107) but its operation is found particularly in prevertebrate chordates and early vertebrate animals. Certainly if the animal has been pre-exposed prior to conditioning to a CS whose repeated presentation produces diffuse behavioural effects (such as increased activity levels, autonomic changes, etc.) then – depending on the nature of the to-be-conditioned response – conditioning will either be facilitated or retarded. A more specific example of this type of effect can be illustrated by describing experiments carried out by Myers (1962, 1964). He found that rats learned two different avoidance responses more rapidly when a buzzer was the warning signal than when either a light or a tone was the CS. Furthermore, when the rats were given shocks unpaired with the presentation of a signal, he still found that response rate during the signal was higher if that signal was a buzzer than if it was a light or tone. These results seem to be very similar to those reported by Grether (1938, p. 30) on pseudoconditioning, and reflect the more general phenomenon that many animals, when in an aversive situation, will respond to raucous sounds as though they were aversive, even though they have never been paired with aversive events. In Grether's case the noise elicited startle and fright responses but in Myer's experiment the noise elicited responses which, under other conditions, avoided electric shock.

Although there has recently been a readiness to acknowledge that the CS might elicit specific CRs which either interfere with or facilitate conditioning, there has been – barring one or two exceptions (Bolles, 1970; Garcia *et al.*, 1972) – a neglect of the role of pseudoconditioning and sensitisation in adaptive responding. It seems quite reasonable to assume that such processes are not simply 'gremlins' in the conditioning process but truely significant mechanisms whose adaptive value is of some importance. When one considers sensitisation and pseudoconditioning as behavioural processes in their own right it should become an easier task to analyse some of the reasons why CSs can have differential effects on different types of responses.

# 7 Theoretical Analyses of Associative Learning

The earlier chapters in this first section deal with a variety of procedures that produce conditioning in animals. These procedures normally have predictable effects on behaviour but under some conditions they fail to work in the way we expect them to (chapters 5 and 6). There are two points to emphasise here. First, do the underlying mechanisms of conditioning bear any resemblance to the conditioning procedures we use (for instance, in classical conditioning the experimenter sets up a relationship between two stimuli, the CS and UCS; but is this the association that the animal himself learns?)? Secondly, what are the implications of the discovery that these procedures do not seem to work on some occasions (see particularly chapter 5)? For instance, because operant reinforcement does not work under certain conditions may mean that we are wrong to assume that there is an underlying operant reinforcement process at all (Bolles, 1972). It is questions like these that pose the most important theoretical question of all: what does the animal learn during conditioning? More precisely, what are the associations that the animal learns and how are these associations translated into the behaviour that we observe in the conditioning situation? This has been a problem that has perplexed learning theorists for the past 60 years. Of the traditional theorists Guthrie and Hull asserted that the animal learnt primarily S–R (stimulus–response) associations, Pavlov and Tolman claimed that the animal learnt only S–S (stimulus–stimulus) associations, while implicit in the writings of Skinner is the assumption that animals can learn directly about the consequences of their behaviour (that is, they form R–S (response–stimulus) associations). (It is not intended to enter into the details of these traditional theories of learning unless they concern the ensuing discussions. However, for comprehensive historical reviews of the theories of Guthrie, Pavlov, Hull, Tolman and Skinner, the reader is referred to Bolles (1975) and Hilgard and Bower (1966).)

In the remainder of this chapter we will look at the evidence for the formation of different kinds of associations during classical and operant conditioning, and the ways in which these associations might account for the behaviour that results.

# CLASSICAL CONDITIONING

Pavlov (1927) had little doubt that classical conditioning involved the learning of an association between the CS and the UCS that it predicted. This was supposedly mediated by an association between the centres of neural activity produced by the presentation of CS and UCS (1927, pp. 36–38). the conditioned response resulted from the CS thus becoming a *substitute* for the UCS (see p. 219ff.). This seems a very simple, clear account of classical conditioning, and one that is open to empirical test. But does it truly reflect the associations that are learnt and can it account for the form, magnitude and directionality of the conditioned response in the very wide variety of classical conditioning procedures that have been catalogued?

## What associations are learnt?

The main alternative to Pavlov's belief that classical conditioning involved the learning of CS–UCS associations is that the animal learns to associate the CS with the UCR that occurs in close temporal proximity to it; that is, it learns a stimulus–response relationship (Guthrie, 1935; Hull, 1943). Quite simply, the contiguous sequence CS : UCS →UCR is enough to ensure that an association is formed between CS and UCR – that the subsequent presentation of the CS will eventually elicit a facsimile UCR.

The evidence relating to a simple S–R account of classical conditioning is not encouraging for this theory. For instance, if the occurrence of the UCR is blocked in some way during conditioning, this does not prevent the occurrence of CRs when the block is lifted. Solomon and Turner (1962) cite a number of experiments in which dogs under the influence of the drug curare (which incapacitates the skeletal musculature) have been given pairings of a tone CS and an electric shock (UCS) to the leg. In the absence of curare this would have produced the response of leg-flexion. When these animals were subsequently presented with the tone CS after the effect of the curare had worn off, a CR of leg-flexion was immediately obtained. Thus a CR could be obtained even when there has been no previous opportunity to form an association between CS and UCR.

However, despite this initial blow, an S–R account of some description can still be salvaged. First, in experiments similar to those reviewed by Solomon and Turner (1962) it is possible to suggest that although the overt behaviour may have been blocked by response inhibiting drugs, for example, leg flexion by curare and salivation by atropine (Crisler, 1930; Finch, 1938) – the UCS may still have elicited activity in the central nervous system which was related to the overt response. It may have been this 'covert' response that became associated with the CS thus producing a CR when the blocking agent was removed. Secondly, a slightly different approach to this problem is to suggest that it is not the association between CS and specific responses that occurs during classical conditioning, but that it is the association between the CS and a central motivational state which in turn mediates the appearance of responses specific to the UCS (Konorski, 1967; Mowrer, 1960; Rescorla and Solomon, 1967). The argument goes like this: the UCS elicits a particular motivational state which in turn induces responses which are specific to the UCS (the sequence is UCS →motivational state → overt UCRs). Instead of forming an association between the CS and the UCRs, an association is formed between the CS and the motivational state relevant to the UCS (the sequence thus becomes CS → motivational state → overt UCRs). So, what evidence is there for the conditioning of central motivational states to classical CSs? The first piece of supporting evidence comes from the studies of conditioned suppression (see p. 142ff.). Conditioned suppression, you will remember, involves superimposing a pairing of a signal (CS) plus shock onto a baseline of responding maintained by operant reinforcement. This results in a suppression of operant responding during the CS. Now, the CS-shock pairing is basically a classical conditioning procedure so its effect on operant responding should tell us something about the classical conditioning process. The first intuitive explanation of the conditioned suppression effect is that competing responses normally elicited by electric shock (for example, prancing, freezing) are now elicited by the CS and thus interfere with responding. But if this were true superimposing a CS-shock contingency on a baseline of *avoidance* responding should suppress responding also; but it does not – it is actually observed to *facilitate* responding (see p. 144; Kamano, 1970; Lolordo, 1967; Sidman *et al.*, 1957). Thus, the effects of a CS predicting unavoidable shock seem to depend not on any overt responses that might be conditioned to the CS but on the motivational state maintaining instrumental responding. If this is appetitive, responding is suppressed; if it is aversive, responding is facilitated. A more plausible interpretation of these results is to suggest that what is associated with the CS is a motivational state similar to that elicited by the UCS, and that this conditioned motivational state will either conflict with or supplement the motivational state relevant to that maintaining the operant baseline.

It is quite clear that the association of the CS with a central motivational state is a more satisfactory explanation of some classical conditioning phenomena than is the association of the CS with specific UCRs. However, as Mackintosh (1974) points out, there becomes only a slim line of difference between an S–R interpretation and an S–S interpretation of classical conditioning when we allude to hypothetical central motivational states. If, for example, there is a good correlation between the central representation of the UCS and the central motivational state elicited by the UCS, then to say that an association is formed between a CS and the central motivational state is tantamount to saying there is an association formed between the CS and the UCS.

Perhaps the problem can be attacked from a different angle in an attempt to shed more light on the kinds of associations formed during classical conditioning. For instance, is there any evidence to suggest that an animal can, and indeed does, form associations between environmental stimuli? There are two sources of evidence here: (1) experiments which change the status or characteristics of the UCS, and (2) experiments concerning sensory preconditioning.

One important way of testing to see if the organism has formed an association between CS and UCR, or CS and UCS, is the following: first the CS and UCS are paired until a CR is formed, then the status of the UCS is altered in some way – usually by habituating the subject to the UCS by presenting it alone – and then the subject's response to the CS is tested again. If an association has been formed between CS and UCS, habituation to the UCS should eliminate any CRs to the CS on subsequent testing. If an association has been formed between CS and UCR, subsequently altering the status of the UCS should leave this association unaffected. Of the few experiments which have used this procedure some have produced support for CS–UCR associations (Harlow, 1937) and some for CS–UCS associations (Rescorla, 1973a).

A slightly different procedure can also attempt to tease out the answer to this problem, and this involves higher-order classical conditioning. Using a normal classical conditioning procedure, a stimulus $CS_1$ is paired with an appetitive UCS. After reliable appetitive CRs have been established, $CS_1$ is now paired with another stimulus $CS_2$ (the original UCS is discarded). In stage 3, $CS_1$ is now paired with a different UCS, this time an aversive one. If the animal has formed an association between $CS_1$ and $CS_2$ then altering the significance of $CS_1$ should also affect the CRs elicited by $CS_2$. For example, using dogs, Konorski (1948) paired $CS_1$ with food such that $CS_1$ produced salivation; he then paired $CS_1$ with $CS_2$ such that $CS_2$ also came to produce salivation; in stage 3, $CS_1$ was now paired with shock to the paw so that the CR to $CS_1$ was changed from salivation to leg-flexion; when $CS_2$ was subsequently tested alone after all this it continued to elicit salivation, suggesting that the animal had formed an association between $CS_2$ and salivation rather than between $CS_2$ and $CS_1$.

If there is any convincing evidence for the assertion that animals can form associations between two stimulus events it comes from studies of sensory preconditioning. The three stages of this procedure involve (1) pairing two neutral stimuli, $CS_1$ and $CS_2$; (2) following this by pairing $CS_1$ with a UCS until $CS_1$ produces a reliable CR; and (3) subsequently presenting $CS_2$ and assessing the nature of any response to it. Normally, the result is that the organism emits a measurable CR to $CS_2$ indicating that it has associated $CS_2$ with $CS_1$ (Brogden, 1939b; Prewitt, 1967; Tait, Marquis, Williams, Weinstein and Suboski, 1969).

What the evidence from these procedures seems to imply is that animals subjected to first-order and higher-order classical conditioning procedures can form both stimulus–stimulus associations and stimulus–response associations. However, as Mackintosh (1974) points out, there is no reason why an animal should not be able to form more than one type of association; whether he forms an association between CS and UCS or between the CS and his reaction to the UCS may well depend on more detailed points of procedure. Indeed, it has recently been suggested (and the evidence we have just discussed seems to support this interpretation) that animals may form different kinds of associations during first- and second-order conditioning (Holland and Rescorla, 1975; Rescorla, 1973b, 1974). Stimulus–stimulus associations (between CS and UCS) seem to be more involved in first-order conditioning, whilst stimulus–response associations appear to prevail when a first-order $CS_1$ is paired with a second-order $CS_2$.

## What determines the nature of the conditioned response?

It is one thing to say that an organism has formed certain associations during conditioning, but how, for example, is a CS–UCS association translated into the behaviour we observe. Pavlov's answer was quite simple: the CS becomes a 'substitute' for the UCS and thus elicits those responses that are linked to the UCS. But is the simple answer the correct one? This question basically boils down to asking the question 'What processes determine the nature of the conditioned response during classical conditioning?' A different answer will probably be given depending on what kinds of associations one believes are formed during classical conditioning.

### Stimulus substitution

Stimulus substitution theory claims that classical conditioning is a procedure which determines that the previously neutral stimulus, the CS, will come to be treated by the animal in the same way that it treats the UCS. The CS becomes a substitute for the UCS, and this results because the animal has formed an association between CS and UCS. The validity of this theory seems to depend on whether or not the CRs elicited by the CS are good replicas of the UCR, and often they are not, (see pp. 29–30). For instance, in the rat the CR to a CS predicting shock is usually freezing whilst the UCR is prancing and squealing; also, one of the UCRs to shock is an *increase* in heart-rate, whilst CRs involve sometimes an increase and sometimes a decrease in heart-rate (Black, 1971); similarly, CRs are usually of lesser magnitude than UCRs – in the case of eyelid conditioning they have a slightly longer latency (Hilgard and Campbell, 1936), and in the case of salivation, a slightly different chemical constitution (Gormezano, 1966). A further argument against stimulus substitution theory is one first proposed by Zener (1937). He claims that since the experimenter usually restricts the range of CRs he will record in a classical conditioning experiment he is virtually assured of producing a CR which resembles the UCR. Zener further argued that the CR was not a simple facsimile of the UCR, but if allowed to express itself fully could be seen to be anticipatory or preparatory rather than mere replication – in effect, it prepared the animal to receive the UCS. Zener's argument is partly true – many experimenters do restrain their subjects and so eliminate orienting and approach reactions, and they also only record behaviour which is identical to the UCR (for example, they record only salivation by a dog restrained in a harness). Such experiments are bound to support stimulus-substitution theory, but if one takes stimulus-substitution theory to its logical end, the CR to an appetitive UCS must be approached to the CS and, in the end, attempted consuming of the CS! However, strange though it may seem, this is often what is observed in unrestrained animals. A number of reports are available which have stated that unrestrained dogs in classical salivary conditioning experiments will often tend to approach and lick the CS if it is accessible. For example,

My late friend Howard Liddell told me about an unpublished experiment he did while working as a guest in Pavlov's laboratory. It consisted simply in freeing from its harness a dog that had been conditioned to salivate at the acceleration in the beat of a metronome. The dog at once ran to the machine, wagged its tail at it, tried to jump up to it, barked and so on.

(Lorenz, 1969, p. 47)

Further evidence comes from studies of autoshaping (see p. 180ff.). When an unrestrained organism in a conditioning chamber is subjected to pairings of a neutral stimulus and food, many different kinds of species will approach the CS and direct species-specific food-getting behaviours towards it. For instance, pigeons will peck at a lighted key which predicts grain (Brown and Jenkins, 1968), rats will bite and lick a retractable lever which is inserted into the chamber immediately prior to delivery of a food pellet (Peterson, 1975), many species of fish will direct species-specific food responses towards stimuli predicting food (Squiers, 1969), and so on. But perhaps the most convincing evidence for a stimulus-substitution explanation of classical conditioning comes from the study carried out by Jenkins and Moore (1973). When one key colour predicted food and a second key colour predicted water, pigeons attempted to 'eat' the food-paired key and 'drink' the water-paired key (see figure 5.7, p. 184). Although these animals do not go as far as to bite off and swallow appetitive CSs in such situations, there have been reports of such instances (for instance, turkeys will swallow coins paired with food, and porpoises have been known to swallow rubber balls under similar circumstances – Breland and Breland, 1966, p. 68). Such examples certainly favour an interpretation in terms of stimulus substitution rather than the 'preparatory' role suggested for CSs by Zener.

## Adventitious reinforcement

One process which has been suggested as a means for establishing conditioned responses is that of adventitious reinforcement. This account claims that CRs are developed, not directly because of the CS–UCS relationship, but because responses which are emitted during the CS are immediately followed by the UCS, and are thus adventitiously reinforced. Since the response elicited by the UCS (that is, the UCR) is likely to be the most dominant response in the experimental situation it is not inconceivable that this may occur during the CS and thus be strengthened by its temporal proximity to the UCS. However, the evidence against such an account of classical conditioning is quite overwhelming. If CRs were merely adventitiously reinforced operants, one could expect a reasonable level of transfer between reinforcers, that is, one would expect a leg flexion that is elicited during aversive classical conditioning readily to become an operant for food reinforcement. But this is extremely difficult to establish (Brogden, 1939a; Konorski and Szwejkowska, 1956). There are other arguments against an adventitious reinforcement account of classical conditioning, not the least being that although it can account for classical appetitive conditioning, how it can be applied to classical aversive conditioning is not fully clear. Finally, one has to return to the status of superstitious reinforcement itself; if, as the experiments of Staddon and Simmelhag (1971) suggest (see p. 163ff.), superstitious reinforcement is only a transitory phenomenon, it seems difficult to account for the robust character of classical CRs in terms of such a process.

## Pseudo-contingent reinforcement

Despite the impracticalities of an explanation in terms of adventitious reinforcement, a stronger version of this account does seem in part defensible. The CR may be operantly reinforced because it has a direct effect on the appetitiveness or aversiveness of the UCS – that is, it has a contingent effect on the palatability or painfulness of the subsequent UCS. For instance, salivation prior to receiving meat powder may make that food more tasty and digestible, flexing the leg prior to shock to the paw may reduce the intensity of the shock, as may closure of the nictitating membrane in rabbits prior to paraorbital shock (Prokasy, 1965; Perkins, 1968), and so on. Now, although it may be stretching the theory a little too far to suggest that this process *alone* can account for the CR in *all* classical conditioning experiments, some support can be mustered for the operation of pseudo-contingent reinforcement in many classical conditioning studies. The main evidence in its favour comes from studies on omission training. The argument goes thus: if a CR is established through its reinforcing consequences, then changes in the value of these consequences should affect the CR in a predictable fashion. For instance, if leg-flexion is established because it helps to reduce the painfulness of the electric shock, then it should be equally easy to establish witholding of leg-flexion to avoid shock, that is, an omission contingency states that the animal is only presented with the UCS if he *withholds* making the CR in appetitive conditioning, and avoids the UCS only if he withholds the CR in aversive conditioning. Using this procedure a number of CRs established by a classical conditioning procedure have been shown to be sensitive to their consequences; these include conditioned leg-flexion in dogs (Schlosberg, 1936; Wahlsten and Cole, 1972), conditioned running in guinea-pigs (Brogden, Lipman and Culler, 1938), and conditioned key-pecking in pigeons (Barrera, 1974; Schwartz and Williams, 1972). However, some other kinds of CRs are just *not* sensitive to an omission contingency, and these include appetitive CRs such as conditioned salivation in the dog (Sheffield, 1965), conditioned licking of a water-tube in rats (Patten and Rudy, 1967), and conditioned jaw-movement in rabbits (Gormezano and Hiller, 1972). In all of these last three experiments, if the subject emitted a CR in the presence of the CS, the appetitive UCS was withheld – in all cases the subjects still found it difficult to refrain from making the CR and fared no better than yoked control animals (see p. 468), a fact which makes it hard to believe that the CRs were established in the first place primarily because of their consequences. A further study which bears on this discussion is an elegantly designed experiment performed by Soltysik and Jaworska (1962). With dogs as their experimental subjects they presented shock to the forepaw (UCS) which was preceded by a buzzer CS. Eventually this established conditioned paw-flexion as the CR to the buzzer. They argued that this response could be established in one of two ways: (1) by the animal learning that the CS predicted the UCS and the CS thus eliciting a response identical to that elicited by the UCS, or (2) by the CR of paw-flexion being operantly reinforced because it helped to reduce the intensity of the ensuing electric shock. When conditioned responding was established they introduced a number of test trials on which the UCS was deliberately omitted. Now, if the CR were established because it reduced the intensity of shock, actually omitting the shock altogether should be even greater reinforcement. Therefore on the next trial an even stronger CR should be expected. If, however, the CR were established because of association between CS and UCS, omission of the UCS on trial $n$ should result in a weaker CR on trial $n + 1$, because the CS–UCS relationship would itself have been weakened. The results of their experiment supported this latter outcome: the CR was actually weaker on trials that followed omission of the UCS, and so did not support a pseudo-contingent reinforcement hypothesis.

What all this evidence seems to imply is that in situations where the UCS is aversive, many CRs do seem to be sensitive to their consequences and so pseudo-contingent reinforcement may play some role in determining the strength of the CR. There is little evidence, however, that CRs maintained by appetitive UCSs are sensitive to their consequences (but autoshaped respond-ing seems to be an exception to this). Finally, the experiment of Soltysik and

Jaworska suggests that even though a CR such a leg-flexion may under some circumstances be sensitive to its consequences it appears to be the CS–UCS relationship that is most important in determining its intensity on a trial-by-trial basis.

# What determines the rate and subsequent strength of conditioning?

Is there any way in which we can make predictions about the strength of conditioned responding to a particular CS, either in terms of the rate of acquisition of conditioning or in terms of the asymptotic level of conditioning that is eventually reached? In chapter 2 we noted that the acquisition curve for classically conditioned responses followed an S-shaped function with the speed of this acquisition appearing to depend on such factors as varied as the nature of the response, the intensity of the UCS, the clarity of the correlation between CS and UCS and so on. Until recently, however, there has been no compact theory of classical conditioning which has made predictions about the rate and eventual asymptotic strength of conditioning to particular CSs. The 'Rescorla–Wagner' model of classical conditioning has attempted to tackle some of these problems.

## The Rescorla–Wagner model

Successive pairings of CS and UCS eventually produce a negatively accelerated increase in the strength of the CR (see figure 2.3, p. 25). This curve has traditionally been taken to imply that repeated CS–UCS pairings produce successively smaller increments in associative strength until no more associative strength can accrue and asymptotic performance is reached. This effect can be characterised as a 'saturation-like' process, and traditionally it has been assumed that these increments get smaller and smaller because the associative strength of the CS is becoming less and less capable of being incremented (Hull, 1943). Thus, it had traditionally been considered that the 'saturation' process was due to some property possessed by the CS rather than the UCS. Wagner and Rescorla (1972) and Rescorla and Wagner (1972) point out that it is perhaps more valid to attribute the 'saturation' process to the continually changing status of the UCS rather than the CS.

As repeated CS–UCS pairings produce smaller and smaller increments in associative strength, it is easy to see this in terms of a saturation-like process. As the individual CS acquires more and more associative strength it becomes less and less possible for that CS to acquire further strength by pairing with a designated UCS. According to this view, it is reasonable to expect that when multiple CSs, e.g. A and B, are present on a conditioning trial that each should be independently incremented according to the distance from its own saturation level. This "saturation" viewpoint has apparently dominated previous theorizing, and would probably not have been challenged by us, had Kamin (1968, 1969) not called our attention to another intriguing viewpoint. Quite simply, *perhaps repeated CS–UCS pairings produce smaller and smaller increments, not because the associative strength of the CS is becoming less and less capable of being incremented, but because the UCS is becoming less and less effective as it is announced by a cue with increasingly greater associative strength.*

(Wagner and Rescorla, 1972, p. 303, my italic)

Thus associative strength only accrues to a CS if the UCS is not already signalled by a reliable CS. Furthermore, the amount of associative strength a CS will acquire on a given trial will be inversely proportional to how 'surprising' the occurrence of the UCS is to the subject. For instance, if $CS_1$ has already been paired with the UCS for, say, 50 trials, on trial 51 the increment in associative strength will be smaller than that on early trials. Similarly, if on trial 51 a compound of $CS_1$ and a new stimulus, $CS_2$, is presented prior to the UCS, the increment in associative strength to $CS_2$ will also be small since the UCS is well-predicted by $CS_1$ and so the power of the UCS to increment associative strength will have been greatly diminished. This model can be formulated more precisely in the following equation which predicts the increment in associative strength of a CS, A, on any one conditioning trial ($\Delta V_A$).

$$\Delta V_A = \alpha_A \beta (\lambda - \overline{V})$$

where

$\overline{V}$ = the aggregate associative strength of all the cues present on that trial (including background stimuli).

$\alpha_A$ = salience or learning rate parameter of CS, A (a constant related to the nature of the *CS*).

$\beta$ = scaling factor or learning rate parameter of unconditioned stimulus (UCS) (a constant related to the nature of the *UCS*).

$\lambda$ = asymptotic level of associative strength that the UCS will support.

So, as the value of $\overline{V}$ approaches $\lambda$ the amount of conditioning to any cue paired with the CS becomes less and less. This model is particularly successful at predicting the associative strength that will accrue to the components of a compound CS during classical conditioning, and indeed it predicts precisely the effects known as 'overshadowing' and 'blocking' (see chapter 2, p. 33). For instance, if a subject is given repeated pairings of a CS, A, with a UCS until the CS is eliciting a reliable CR, and following this the UCS is now paired with a compound CS, AX, for a further number of trials, subsequent testing of the potency of the elements A and X will reveal that while A elicits a CR, there is little if any conditioning to component X (see Kamin, 1969, for detailed procedural points). This is clearly what one would expect if there is a limited amount of associative strength available, and the UCS has already 'granted' this to component A during pretraining. Once the UCS is reliably predicted by some event there is little conditioning that takes place to any other event that may become paired with the UCS.

## Summary

It is clear that the principles of classical conditioning are not as simple as one might initially be led to believe. First, it seems that the pairing of environmental stimuli can have a number of effects on the kinds of things the animal will learn. If one of the two environmental events is a biologically important event (a UCS) which elicits predictable and consistent reactions, then the animal is more likely to form S–S associations than if neither of the two stimuli have intrinsic biological value (see pp. 218–219). S–R associations seem more likely to be formed during higher-order classical conditioning. Secondly, the role of operant reinforcement in the establishment of CRs has also been discussed. Its implication during classical aversive conditioning is more likely than during classical appetitive conditioning; but even in the former it does not seem that 'pseudo-contingent' reinforcement can be the whole answer to classical conditioning. Furthermore, the fact that in many cases animals have been reported practically to 'consume' a CS conditioned to an appetitive UCS lends support to Pavlov's stimulus substitution theory of classical conditioning. Finally, the Rescorla–Wagner model provides some predictions about the acquisition of conditioning by classical CSs – the degree of associative strength that will accrue to a given CS is inversely related to how reliably the UCS is already predicted by that or any other CS.

# OPERANT CONDITIONING

A theoretical analysis of operant conditioning contains many of the same problems as an analysis of classical conditioning; just because the experimenter arranges a contingency between a response and its consequences does not mean that this is what the organism in fact learns. Indeed, if it is what the animal learns we then have to try and account for those increasingly numerous occasions where it is reported that operant conditioning does not work (see chapter 5), and many theorists have insisted that these constraints are evidence enough for denying even the existence of an underlying operant conditioning process (Bindra, 1972, 1974; Moore, 1973). However, animals *do* frequently learn to adapt their behaviour successfully in operant conditioning situations, so how do they do this? In the remainder of this chapter we shall look at the evidence relating to the kinds of things animals can and cannot learn under contingencies of operant reinforcement, and the light that this might throw on the existence of an operant conditioning mechanism.

## The learning of stimulus–response associations

The traditional account of operant conditioning that goes back to the days of Thorndike and the 'Law of Effect', is that animals learn S–R associations when trained with operant conditioning procedures. Thorndike (1911), Hull (1943) and Guthrie (1935) all believed that operant reinforcement had its effect by strengthening associations between responses and antecedent stimuli. This solution seemed the simplest and most parsimonious to the problem of operant conditioning. We have already mentioned Thorndike's Law of Effect (chapter 2, p. 20), which stated that the role of reinforcement was to 'stamp in' the correct responses in order that they would be 'more firmly connected with the situation, so that, when it recurs, they will be more likely to recur' (Thorndike, 1911, p. 244). However, of all the S–R theories of operant conditioning, that of Guthrie is probably the most mechanistic. According to Guthrie's theory, whatever responses are occurring at any moment will become conditioned to the stimuli which are present at that moment, and after a number of reinforced trials the correct response will have become conditioned to virtually all the stimuli in the learning situation, so that the response is practically inevitable thereafter. In essence, Guthrie's theory boils down to claiming that the correct response occurs because the stimuli in the learning situation come to elicit the pattern of motor responses appropriate to obtaining reward.

A strong version of S–R theory can fairly easily be disproved. Such a version would claim that the S–R associations which are formed during operant conditioning are to all intents and purposes reflexive. For instance, if a rat has learned to turn left in a T-maze to receive food, it could be argued that stimuli along the path of the maze and at the choice point come to elicit the appropriate motor behaviour along the maze so that the goal-box is eventually reached. The fact that rats trained in this way still exhibit good goal-box oriented behaviour when normal running is prevented by cerebellar lesions (Lashley and Ball, 1929), and also exhibit excellent transfer from running in a T-maze to swimming in it when it is filled with water (MacFarlane, 1930), suggests that if S–R associations are formed during learning they are not of the rigid reflexive variety.

A further problem with the S–R account is that it implies that the animal learns little, if anything, about the nature of the reinforcer, nor does it learn to 'expect' or 'anticipate' the reinforcer – the reinforcer is merely an agent for establishing S–R associations. It is difficult to believe that animals do not learn something about the reinforcer itself. First, an experiment by Tinklepaugh (1928; see chapter 10, p. 280) has demonstrated that chimpanzees do appear to possess some knowledge of the nature of the reward used in an operant learning study. When a preferred food, such as banana, was replaced with a different food, lettuce, the subject exhibited both a disruption of discriminative responding and subsequent 'searching' behaviour. Secondly, if reinforcers simply stamped in S–R associations, then animals should exhibit good transfer between reinforcers if the response remains the same. In a number of instances this is patently not the case, and the nature of the reinforcer seems to play an important role in determining either the nature of the response or, in some cases, whether the response will be maintained at all. For instance, if a pigeon is reinforced with food for key-pecking and then transferred to a shock-avoidance schedule on which key-pecking avoids shock – the response is not maintained (Schwartz and Coulter, 1973; see pp. 193–194). Furthermore, the actual nature of the response can be altered by the nature of the reinforcer. Smith (1967, cited by Moore, 1973) has found that when pigeons are reinforced with food for key-pecking, the key-peck response resembles the consummatory action itself, with the beak opening slightly on impact as if it were picking up grain. Similarly, when reinforced with water, the key-pecking now resembles drinking rather than eating. So although the response requirement for both reinforcers is in principle identical, the reinforcer still determines more discrete aspects of the operant.

Finally, studies of irrelevant incentive learning also pose difficulties for a strict S–R interpretation of operant conditioning. Experiments by Krieckhaus and Wolf (1968) and Khavari and Eisman (1971) suggest that rats can learn that a response produces a particular reinforcer even though they have no need for that reinforcer. Thirsty rats were first trained to press a lever for sodium solution even though they were not sodium deficient. When they were subsequently made sodium deficient and satiated for water, these animals persevered with lever-pressing into extinction. Control animals which had received only water in the first phase did not persist in lever-pressing.

So, the evidence for a simplistic S–R account of operant learning is not encouraging. Animals do appear to learn something about the nature of the reinforcer, and the nature of the reinforcer can in turn influence details of the response. However, from classical conditioning studies we know that animals can form S–R associations so it seems unduly harsh to deny out of hand that they do not do so during operant conditioning. The question arises as to what associations are most *readily* formed and have most influence on behaviour during operant learning; we have seen that the animal's behaviour is quite sensitive to features of the reinforcer, so this must naturally lead us on to consider the role of stimulus–reinforcer and response–reinforcer associations during operant conditioning.

# The role of stimulus–reinforcer relationships

Under many of the conditions where operant reinforcement does not work, the animal's counterproductive behaviour appears to be under the control of other aspects of the conditioning situation which are closely related to the reinforcer. Prominent examples of this are instinctive drift (see p. 157) and negative automaintenance (see p. 181). In these cases, the animals direct their behaviour towards features of the conditioning situation which are highly associated with food, and as a consequence this interferes with the behaviour needed to obtain the reinforcer. It would not be too much of an assumption to claim that in these situations the behaviour of the animal has come under the control of stimulus–reinforcer relationships rather than the response–reinforcer relationships specified by the operant contingency. The fact that under some conditions a stimulus–reinforcer contingency can over-ride experimenter-imposed response–reinforcer contingencies had led a number of theorists to claim that operant learning *normally* occurs through the formation of stimulus–reinforcer associations. Indeed, the more radical of these theorists deny the need for postulating stimulus–response and response–reinforcer learning to account for operant performance – an adequate account of conditioning, they claim, can be based on the assumption that animals can only form associations between reinforcers and the stimuli that predict these reinforcers. One of the earliest accounts of this kind is incentive motivation theory, first suggested by Hull (1952) and Spence (1951, 1956), but since championed primarily by Bindra (1972, 1974).

## Incentive motivation

In its barest form this account of operant conditioning says that stimuli which are closely associated with reward also acquire incentive motivation – a property which energises behaviour. Bolles gives the following examples:

> If an animal is given food for running down an alley and into a goal-box, the presentation of food is said to reinforce running and also to attach incentive motivation to whatever stimuli are near to or similar to the goal-box stimuli; so that the animal will run faster whenever these stimuli are subsequently presented. Moreover, if a lot of incentive motivation is attached to one cue by giving reinforcement in its presence, and if none is attached to a second cue by withholding reinforcement in its presence, we may have an animal choosing or approaching the first cue, or responding more vigorously to it, even though there is no difference in the associative strength of the responses evoked by the two stimuli.
>
> (Bolles, 1972, p. 399)

There are three variations of incentive motivation theory which require discussion here: the first claims that incentive motivational stimuli merely energise behaviour in a non-specific way; the second claims that such stimuli increase motivation, but only appetitively or aversively depending on the kind of reinforcer they have been paired with; and the third claims that incentive motivational stimuli not only possess the property of motivating behaviour, but they also elicit orienting reactions; more precisely, behaviour is directed towards such stimuli.

*Incentive stimuli as a source of nonspecific 'drive'*

Early accounts of incentive motivation assumed that incentive stimuli provided a source of non-specific drive that motivated on-going behaviour (Hull, 1952; Spence, 1956). Support for this assumption came from the fact that stimuli that had been paired with reinforcers did appear to elicit increases in general activity levels (Slivka and Bitterman, 1966; Zamble, 1969). Nevertheless, incentive motivation theory needs more than this to support it; it needs to be shown that an incentive stimulus will increase the vigour or frequency of behaviours that are being performed when it is presented. (An incentive stimulus is one which is predictive of reinforcement; therefore, to all intents and purposes, it is synonymous with a classical CS.) There is quite some evidence which suggests that a CS will not do this. For instance, a stimulus which is paired with an appetitive reinforcer (an appetitive CS) should, according to this theory, possess motivating properties which will energise any on-going behaviour. However, experiments by Bull (1970) and Grossen, Kostansek and Bolles (1969) have shown that presenting an appetitive CS during avoidance responding does not invigorate avoidance responding but in fact suppresses it. This now leads us to consider whether incentive stimuli might not be specific to the motivational state under which they were conditioned. For instance, an appetitive CS (nominally an appetitive incentive stimulus in this theory) might only motivate on-going behaviour if that behaviour is maintained by an appetitive outcome.

*Incentive stimuli and specific motivational states*

An incentive stimulus may only provide motivation for an on-going behaviour if that behaviour itself is maintained by an identical motivational state. We have come across this theory in a number of guises before: it forms one account of the effects of conditioned suppression (see p. 142ff.), and it has also been cited by Konorski (1967) and Rescorla and Solomon (1967) as an explanation of many classical and operant conditioning phenomena. In chapter 4 it was noted that aversive CSs would generally suppress appetitively reinforced operant behaviour but enhance operant avoidance responding. If there can be shown to be a conversely symmetrical effect with appetitive CSs some support can be found for incentive stimuli as motivators of positively reinforced operant behaviour. Unfortunately, there is precious little evidence which suggests that an appetitive CS will enhance the rate of an on-going appetitive operant response. Meltzer and Brahlek (1970) found that presenting an appetitive CS for 2-min periods during lever-pressing for food in rats did produce a slight enhancement of lever-pressing during the CS, but the majority of other studies which have examined this procedure have found that an appetitive CS will actually *suppress* on-going appetitive operant performance (Azrin and Hake, 1969; Miczek and Grossman, 1971; J. B. Smith, 1974). This effect is known as positive conditioned suppression. The solution to this theoretical anomaly is perhaps found in a pair of experiments by Lolordo (1971) and Lolordo, McMillan and Riley (1974). Lolordo (1971), using pigeons as subjects, found that if key-pecking was maintained on a VI schedule and the appetitive CS was a change in colour on this pecking key, rate of key-pecking increased during the CS. If, however, the operant response was treadle-hopping but the CS was still projected onto a pecking key (Lolordo *et al.*, 1974), CS presentation resulted in a suppression of treadle-hopping *and* the development of key-pecking during the CS. Thus, the *location* of the CS is an extremely important factor: if the location of the CS is identical to the location of the operant manipulandum, response facilitation will occur; if it is some distance from the manipulandum, suppression of responding will occur. A further experiment by Karpicke, Christoph, Peterson and Hearst (1977) suggests that the suppression induced in the latter case may often be due to the animal orienting toward and contacting the appetitive CS. They trained rats to press a lever on a VI schedule of food reinforcement; then, at intervals during this training a visual CS preceded a free-food delivery. For one group of rats the CS was located some distance from the operant manipulandum, for a second group it was located relatively close to the lever. They found that the 'far' CS produced greater suppression of responding than the 'near' CS and this was mainly because most of the subjects in the 'far' group abandoned the operant lever and directed their attention to the distant CS when it was illuminated. In fact, they appeared to be autoshaping to an appetitive CS.

At this point, a restatement of the relevance of this account of incentive theory to operant responding is probably valuable. This account stresses that those stimuli which reliably precede or predict an appetitive reinforcer elicit a motivational state which facilitates – or in its strictest form, initiates – operant responding. The main problem here is that incentive stimuli may well leave the animal highly motivated, but they do not, in this account, suggest how the *form* of the operant response is established. Quite simply, why does the rat in the Skinner-box press the lever rather than turn rapidly round in circles, and why does the rat in the alleyway run to the goal-box rather than remain vigorously sniffing at the corners of the start-box? A possible answer is derived from the experiments cited in the preceding paragraph, and has been formulated into the third variation of incentive theory by Bindra (1972, 1974).

*Incentive stimuli and directed behaviour*

Bindra (1972, 1974) has argued that incentive stimuli have two functions: (1) they activate a central motivational state and (2) they elicit orienting responses such that the animal's behaviour is directed towards them. Bindra summarises his approach accordingly:

> There are two essential ideas in the present view of the principles underlying learned modifications of behaviour. The first is that the principle of contingency learning between stimuli (which is conceptually derived from the classical conditioning paradigm) is sufficient for explaining learned behavioural modifications; the principle of response reinforcement is unnecessary. The second is that performance – the production and form of an instrumental response – is determined jointly by the type of the prevailing organismic state and the spatio-temporal layout of the con-ditioned incentive stimuli in the situation.
>
> (Bindra, 1974, p. 207)

Consider a rat pressing a lever in a Skinner-box for food reinforcement. According to the present account he presses the lever, not because he has learnt that the *response* of lever pressing is followed by food, but because the lever has become an incentive motivational stimulus. That is, during the period in which he learns to press the lever for food (presumably his first few presses are accidental), food is always preceded by a close-up view of the lever (because he only receives food when he has pressed the lever). Thus, the lever eventually becomes reliably paired with food delivery and so acquires incentive value (or, in alternative terminology, it becomes a classical appetitive CS). Thus the lever not only induces a motivational state appropriate to the reinforcer, but, because it is an incentive stimulus, the rat's behaviour is directed towards it. The response–reinforcer contingency that is set up by the experimenter ensures that the lever remains the only part of the environment which is reliably paired with food.

This theory has in its favour the fact that it accounts not only for the basic phenomenon of operant conditioning, but also predicts many of the constraints on conditioning we talked about in chapter 5. For instance, instinctive drift can be explained by the assumption that some aspect of the conditioning situation other than that where the operant response is to be made, has acquired incentive motivational value. In the examples cited on page 157, the coins that the pig and racoon had to drop into the piggy-bank were the stimuli most reliably paired with food; therefore, these animals directed their attention – in this case their species-specific food-getting behaviours – to these coins rather than the piggy-bank. Autoshaping is a further phenomenon which fits comfortably into this explanation. The pigeon pecks the CS because it is the only aspect of the environment reliably paired with food – it therefore acquires incentive motivational properties and behaviour is directed towards it. The fact that an omission contingency fails to suppress this behaviour is further predicted by this account. Since it is claimed that an animal cannot directly learn about the consequences of its behaviour, only about those stimuli which predict reward, then even when an omission contingency is imposed, the CS is still the only reliable predictor of food and so behaviour is still directed towards it.

One further advantage of this theory is that it takes good account of the interactions between conditioned incentive stimuli. In the previous section it was noted that, in general, superimposing an aversive CS on behaviour maintained by an appetitive reinforcer suppressed responding, and super-imposing the aversive CS on behaviour maintained by an aversive reinforcer (that is, avoidance responding) facilitated responding. There appeared to be a predictable interaction between the motivational state elicited by the aversive CS and the motivational state which was maintaining responding. However, the converse is not generally true when an appetitive CS is superimposed on operant responding (see p. 230). The reason for this, according to Bindra (1974), is because the effect of a classical CS on behaviour depends not only on the nature of motivational states elicited by the CS and maintaining on-going operant behaviour, but also on the spatio-temporal relationship of the CS to the place of operant responding. Organisms orient towards appetitive CSs and also orient *away from* aversive CSs. The experiments of Lolordo (1971) and Lolordo et al., (1974; see p. 230) suggest this to be the case with appetitive CSs and Bindra describes how the spatial positioning of an aversive CS can influence avoidance responding:

In a typical avoidance experiment in a unidirectional alley, the avoidance-response performance may be altered by the introduction of a conditioned stimulus, say a tone that has previously been correlated with an electric shock or some other aversive unconditioned stimulus. The instrumental response of running from the dangerous start box to the safe box at the end of the alley would be facilitated by the introduction of the tone. This could occur because the tone would enhance the motivational state and thus increase the conditioned incentive (aversive) properties of the start box (in the absence of shock). But suppose that the tone, instead of being presented as an unlocalizeable encompassing sound, as is usually the case, is presented through a loudspeaker placed in the safe box at the end of the runway. According to the present view, the tone would suppress the instrumental response (of running in the direction of the aversive tone). Certain experimental findings (e.g., Bolles and Grossen, 1970; Katzev, 1967) support this type of prediction.

(Bindra, 1974, p. 209)

Moore (1973) is another writer who has outlined the advantages of considering operant behaviour primarily as approach to appetitive CSs and withdrawal from aversive CSs. He suggests that in the pigeon, as one example, operant responding can be more profitably conceived of as determined by a Pavlovian conditioning process, in which the bird learns that certain aspects of the environment are more reliably associated with reinforcement than others. This aspect of the environment becomes a classically conditioned CS eliciting consummatory behaviour which is directed towards that CS. Moore suggests that when a pigeon is shaped to peck the response key using successive approximations (pp. 40–41), it is not a response that is being strengthened but a classically conditioned relationship between the response key and food: the response key predicts food, therefore the bird directs consummatory behaviour (pecking) towards it.

> The shaping process is said to depend upon the direct operant strengthening of the reinforced movements, the "successive approximations" to pecking. However, we have seen that the same progression occurs without selective reinforcement; autoshaped birds, too, tend first to begin to face the key, then to begin to approach it, then to peck towards it and so forth. Apparently, successive approximations to key pecking can arise through the mere strengthening of a Pavlovian association between the key and grain.
>
> (Moore, 1973, p. 176)

This explanation has many points in its favour. It claims that because operant responding is developed by a classical conditioning process: (1) the resulting topography of the operant response should usually resemble the behaviour elicited by the reinforcer; this has been found to be the case under appetitive operant reinforcement where the form of the pigeon's key-peck is determined by the type of reward (Smith, 1967; see also Breland and Breland, 1966, p. 104); and when avoidance key-pecking can be established in pigeons this is very often accompanied by full-blown pseudo-aggressive behaviours normally elicited by shock (Rachlin, 1969); (2) if the behaviour elicited by the reinforcer is incompatible with the required operant, then conditioning should fail to occur. Numerous examples of this can be found in both appetitive and aversive instrumental conditioning (see p. 189ff.); and (3) if approach to the CS which most reliably predicts food is contrary to the operant requirement, the operant contingency will fail to have effect. This is, of course, what happens when an omission contingency is imposed on autoshaped responding (see p. 181), and also explains why positive conditioned suppression occurs when an appetitive CS is presented some distance from the operant manipulandum.

Although this explanation of operant conditioning has the benefit of parsimony – it can explain quite a number of diverse phenomena with just a few simple principles – there are some facts that a classical conditioning account of operant performance has difficulty in handling. First, this account requires that operant responding can be developed only if there is a classical CS towards or away from which behaviour can be directed: in essence, it maintains that the only responses that can be learnt under operant contingencies are those which require manipulation of part of the environment, or are directed towards some part of the environment which has been established as a CS predicting reinforcement. There are a number of important exceptions to this claim; the most notable being the work of Miller and his associates on the operant conditioning of autonomic and visceral responses (Miller, 1969; see chapter 2, pp. 55–57). These experiments demonstrated that operant contingencies could alter characteristics of heart-rate, blood-pressure, urine-formation, etc., in animals paralysed with curare. It is difficult, if not impossible, to envisage these responses as directed at a classical CS, and the bidirectional controls adopted by Miller (see p. 56) seem to rule out the possibility of mediation by implicit classical conditioning contingencies. Other successfully reinforced operants which fall into the category of being non-directed are wheel-running in rats (Bernheim and Williams, 1967; Bolles, Stokes and Younger, 1966), leg-flexion in the dog (figure 7.1) (Konorski, 1948; Wahlsten and Cole, 1972), and 'head bobbing' in the pigeon (Jenkins, 1977).

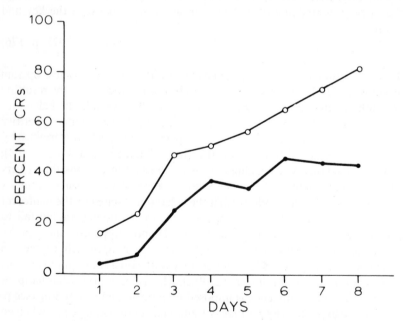

Figure 7.1    Percentage leg-flexion responses elicited in dogs during classical conditioning (filled circles) and omission training (open circles). In the latter condition the UCS (shock) was omitted if a leg-flexion CR occurred on that trial. (After Wahlsten and Cole, 1972).

The second problem with this account arises from one of the implicit assumptions of a Pavlovian account of operant conditioning: if appetitive operant responding is elicited by stimuli established as predictors of food, then operant responses should be accompanied by those responses which indicate the anticipation of food. For example, if lever-pressing in a dog is elicited because the lever has been established as a classical appetitive CS, then lever-pressing should also be accompanied by responses more traditionally associated with appetitive CSs – for example, salivation. An experiment by Williams (1965) suggests that in some instances the two are not well correlated. He trained dogs to press a panel for food reinforcement on either an FI or an FR schedule. During each session he recorded both the rate of panel-pressing and the rate of salivation at different times during each trial (see figure 7.2). Although there was quite a good correspondence between probability of salivation and probability of panel-pressing on FI, on the FR schedule panel-pressing was initiated much earlier in the trial than salivation – suggesting that the two responses are under the control of different factors. Since the two schedules were matched for reinforcement frequency, it seems difficult to avoid the conclusion that while salivation may have been under the control of stimuli predicting the delivery of food (that is, appetitive CSs), panel-pressing was being more directly influenced by the response requirements of the operant schedule.

In summary, an account of operant conditioning which stresses the importance of the learning of stimulus–reinforcer relationships has many advantages. It not only provides an explanation of how conventional operant responding might be developed, it also adequately predicts those circumstances in which operant conditioning fails. The strong form of this theory stresses that an animal only learns stimulus–reinforcer associations and either does not need to learn directly about response–reinforcer relations (Bindra, 1974), or is simply just incapable of doing so. However, other weaker forms of this type of account claim that although stimulus–reinforcer learning may represent an animal's primary adaptive mechanism, it is not unable to form other types of associations when this is required.

**Expectancy and Learning**

Bolles (1972) has tempered the Pavlovian account of operant conditioning a little by suggesting that the formation of stimulus–reinforcer associations alone is not adequate to explain all operant conditioning. He claims that one also has to permit the animal the ability to form response–reinforcer (outcome) associations. Thus, in effect, he is claiming that animals learn about 'expectancies', that is, they learn what event to expect in the presence of certain stimuli (S–S associations) and what outcomes to expect after certain responses (R–S associations). Although this theory possesses the same ability to cope with constraints on operant conditioning as pure stimulus–reinforcer

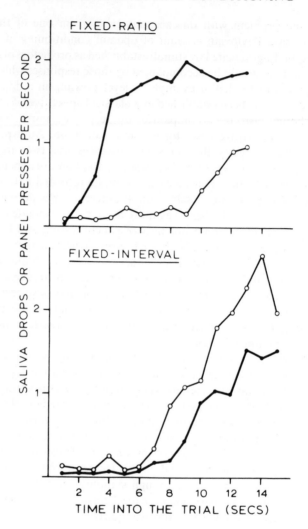

Figure 7.2 Simultaneous measures of salivation (open circles) and panel-pressing (filled circles) in a dog trained on FR and FI schedules of food reinforcement. (After Williams, 1965).

accounts, it has the added bonus of not leaving the form of the operant response to interactions between the site of the CS and the nature of behaviour elicited by the reinforcer. The operant contingency has a *direct* effect on establishing the form and frequency of the operant response. Bolles summarises it in the following way,

Any theoretical account of behaviour must embody some rule for translating what is learned into observable behaviour. The following is sufficient: Two expectancies of the form S–S* (stimulus–reinforcer) and R–S* (response–reinforcer) are 'synthesized', or combined in a 'psychological syllogism' so that in the presence of the cue S the animal is likely to make the response R.

(Bolles, 1972, p. 404)

As comprehensive as this account may seem it still begs an important question. There is adequate evidence to suggest that animals can form stimulus–reinforcer associations, but where is the evidence that they can also form response–reinforcer associations? In effect can animals learn about the consequences of their behaviour?

## The learning of response–consequence relationships

The main problem involved in establishing whether animals can form response–reinforcer associations is to rule out unequivocally whether they are learning stimulus–response or stimulus–reinforcer relationships. As we have seen in the preceding pages, behaviour that superficially looks as though it is being controlled by response–reinforcer contingencies can often be explained in terms of stimulus–response or stimulus–reinforcer learning. Paradoxically, some of the best evidence in favour of a process of operant reinforcement comes from studies using responses which had traditionally been considered out of the range of operant control. There are the studies of the operant reinforcement of visceral responses (Miller, 1969; see pp. 55–57). The bidirectional and yoked controls used by Miller and his associates during these experiments make it difficult to conceive of visceral control being mediated by implicit classical conditioning contingencies or, because of the curarised state of the animal, by behaviour directed at parts of the environment predictive of reinforcement. Thus, support for an operant reinforcement process which is independent of a classical conditioning process can perhaps be mustered. However, these experiments really lend themselves as much to an interpretation in terms of stimulus–response learning as response–reinforcer learning. In true Thorndikian fashion, the reinforcer may simply have established associations between environmental stimuli and changes in autonomic responses; the animal need not necessarily have directly associated response with consequences.

Even so, there is evidence enough that animals are at least sensitive to the consequences of their behaviour. One important source of this comes from studies which have imposed an omission contingency on behaviour generated by an autoshaping procedure. If autoshaped key-pecking results in omission of food on that trial, the frequency of key-pecking – although not reduced to zero – is eventually suppressed (see p. 181). On a very superficial level this means that the subject's behaviour is sensitive to the experimenter imposed response-consequence contingency. What it does not tell us is how this contingency has its effect. For instance, it is possible to conceive of this suppression in terms of a Pavlovian process without having to grant the subject the ability to learn about the consequences of its behaviour. This account suggests that the omission contingency suppresses responding because it also weakens the contingency between key-light and food. That is, if the pigeon pecks the key on any given trial, that trial is terminated *without food* so the autoshaping procedure becomes a partial Pavlovian schedule on which the key-light is sometimes followed by food, and on some occasions not (see figure 7.3). Therefore, the subject's behaviour is indirectly weakening the strength of the contingency between key-light and food. This being the case, if it is assumed that autoshaped responding is determined primarily by the key-light–food contingency, one would also expect the strength of key-pecking to be weakened if the CS–UCS contingency is weakened. However, partial

Figure 7.3 Schematic representation of normal autoshaping and autoshaping omission procedures. In the former, food terminates each trial regardless of whether the pigeon pecks the key. In the latter, food only follows those trials on which no key-peck occurred.

Pavlovian conditioning cannot account for all of the suppression of responding during omission training. Schwartz and Williams (1972) carried out an experiment using a within-subject yoked control procedure designed to assess the relative roles of the omission contingency and partial Pavlovian reinforcement in suppressing autoshaped responding. Pigeons were presented periodically with one of two different key colours; red key trials were associated with an omission contingency – if the subject pecked the key food was witheld at the end of the trial; on white key-trials food delivery was unrelated to responding. However, reinforcement frequency on white-key trials was yoked to the obtained reinforcement frequency in response-related red-key trials so that both key colours had the same number of food pairings but only responding to the red-key had negative consequential effects. Schwartz and Williams found that all their subjects pecked substantially more on the non-contingent white-key than on the negative red-key; these are results which suggest that although the weakening of the CS–UCS contingency does suppress responding to some degree, the negative response–consequence relationship suppressed responding even more. In short, the pigeons behaviour was being directly affected by its consequences. Even so, this does not necessarily mean that the animal is learning directly about the consequences of its behaviour; for instance, the omission contingency could be instrumental in making some other aspect of the environment more predictive of food than the response-key, so the subject now directs his behaviour elsewhere than the pecking-key. Barrera (1974) has observed that key-pecking occurred at a lower rate under a negative contingency than under standard autoshaping conditions; however, pecking was occurring at as high a rate as ever but it was directed to the side of the response key. In this example it could be argued that the omission contingency merely established the area around the lighted pecking-key as a better predictor of food than the lighted pecking-key itself. Nevertheless, it does seem difficult to fit the results of all omission studies into this account. For instance, omission studies which have used non-directed classically conditioned responses such as leg-flexion in dogs (Wahlsten and Cole, 1972, see Figure 7.1) and nictitating membrane responses in rabbits (Gormezano, 1965) have found that the omission schedule produces slightly greater suppression of responding than that found in yoked-control subjects. If the animal is restrained and inactive it is difficult to explain how covert classical contingencies could have produced greater suppression of responding in the omission group but not the yoked group, who received an equal number of CS–UCS pairings; the omission subjects appear to have been directly affected by the operant contingency. Similarly, omission contingencies can sometimes alter subtle aspects of behaviour without affecting the directedness of that behaviour. For example, Atnip (1977) found that when a retractable lever (CS) was inserted into the conditioning chamber as a signal for food (UCS), rats rapidly learned to contact the lever by licking, pawing and biting it. However, when an omission contingency was imposed on lever-presses the rats still approached and contacted the lever but now only sniffed or nosed at it with insufficient force to cause a lever depression. The omission contingency had not changed the directedness of the behaviour, but had altered the nature of the CS contact. Again it is difficult to conceive of this effect as being due to anything other than a direct effect of the operant contingency. Response suppression in this case is definitely not due to the subject approaching a new CS which has been indirectly established through the effect of the omission contingency.

What these studies tend to suggest is that there is evidence for an operant conditioning process; that is, it is not possible to account for all of those occasions where operant contingencies have their effect by suggesting that an implicit classical conditioning process is responsible (Bindra, 1974; Moore, 1973). But what this evidence does not tell us is what associations the animal forms during this operant conditioning process: if he does not form stimulus–reinforcer associations then he presumably must form either stimulus–response or response–reinforcer associations. In this section we have been concerned with discovering whether or not animals can form response–consequence associations. Unfortunately, there is little evidence that animals normally adapt to operant contingencies simply by forming response – consequence associations, although the evidence is not contrary to such a possibility – where data require the postulation of an underlying operant reinforcement process, the behaviour could as likely be mediated by stimulus–response learning as response–reinforcer learning. Finally, one added point of difficulty with an interpretation of operant conditioning purely in terms of response – consequence learning is that suggesting that the animal learns only response–reinforcer associations does not explain how this association becomes translated into behaviour. Accounts of conditioning which stress the formation of stimulus–response and stimulus–reinforcer associations have little problem in this respect. In the case of S–R learning, antecedent stimuli 'elicit' the associated response, and in the case of stimulus–reinforcer learning the resultant behaviour can be accounted for by making any one of a number of fairly simple assumptions about the learning process (see p. 231 and p. 233). But how is a response–consequence association translated into responding, because, as Mackintosh points out,

> to say, for example, that a subject has learned that a response is followed by a particular reinforcer is not to say that the probability of that response will increase
>
> (Mackintosh, 1974, p. 222)

It seems difficult to account for the emission of responses unless one postulates the formation of other associations in addition to response–reinforcer associations. Bolles (1972) is one who has taken this line of reasoning by suggesting that although animals can form response–reinforcer associations, the behaviour observed in an operant conditioning situation is more likely to result from a synthesis of response–reinforcer and stimulus–reinforcer associations (see pp. 236–237).

# CONCLUSIONS

As a general principle it seems unreasonable to asset that animals learn only one type of association. If they can learn stimulus–stimulus associations then why not stimulus–response and response–stimulus associations, since responses can quite easily be considered as special kinds of stimuli, especially if the sensory feedback from the response is consistent and intense enough. The real theoretical problems, however, appear in two forms: (1) do some conditioning procedures establish certain kinds of associations more readily than others? and (2) how do these associations generate the behaviour we observe in the conditioning situation? We have reviewed these questions in detail in the preceding pages and the answers are not without their ambiguities. However, in recent years, theories of both operant and classical conditioning have increasingly opted for explanations which stress the importance of stimulus–reinforcer associations. These theories can account fairly well for the form of the response in operant conditioning (Moore, 1973), and can also predict many of those occasions when operant reinforcement fails to work (constraints on conditioning, Moore, 1973; Bindra, 1972, 1974). But there is perhaps yet another dimension to the problem. So far in this chapter we have worked under the assumption that all animals learn similar things under similar contingencies of reinforcement; that is, there are underlying mechanisms of conditioning which are common to a very wide variety of species. This need not be, and in all probability is not, the case. Through evolution different species may have adapted to solve similar problems in different ways. For example, one species may learn in an operant conditioning procedure by processing stimulus–reinforcer relationships; another species may have developed mechanisms appropriate for directly processing response-consequence relationships. For instance, the operant behaviour of the pigeon can in theory be explained quite thoroughly merely by appeal to processes of classical conditioning (Moore, 1973; see p. 233), but the greater flexibility of mammalian behaviour may require more sophisticated contingency processing abilities. As Bolles (1972) points out:

> Flexibility is characteristic of much mammalian behaviour (although defensive behaviour in rats seems to be seriously constrained to innate behaviour patterns: e.g. Bolles, 1970). I believe some mechanism is needed to account for this contrast, as well as the contrast in flexibility between birds' and mammals' behaviour. Bindra's analysis appears appropriate for the learning of birds, which consists predominantly of responding in old

ways to new stimuli. But to account for the plasticity of mammalian behaviour, it may be useful to hypothesize the possibility of both stimulus (S–S) and response (R–S) learning.

(Bolles, 1972, p. 406)

Bearing this in mind, it might not just be the conditioning procedure (that is operant or classical) which determines the kinds of associations that will be formed, but also the nature of the organism undergoing conditioning; and certainly, the recent evidence that has accumulated on constraints on conditioning has highlighted species-differences in adaptivity to learning contingencies. Certain species appear to be more reflexive in their coping with conditioning contingencies, that is, they prefer to learn by forming stimulus–reinforcer assocations – even in operant conditioning situations; while others may process response related contingencies more directly, and instead of 'behaving to new stimuli in old ways', they possess the mechanisms capable of generating new behaviour forms in direct response to conditioning contingencies.

# Section Two

# Non-associative Aspects of Learning in Animals

# 8 Critical Periods: Perceptual Learning and Imprinting

We very often take it for granted that we are able to perceive so much of our environment. In the visual modality alone we can discriminate such facets as depth, figure versus background, movement, colour, etc., as well as identifying subtle changes in the texture, pattern and orientation of objects. We have mentioned in previous chapters (5 and 6) that the qualitative and quantitative characteristics of perception differ from species to species, and that these differences in perceptual abilities often reflect the differing requirements of different species for coping with their environmental niche. For example, nocturnal animals require more monochromatic than colour receptors in their retinae; upland animals, such as mountain goats, who frequently graze on precarious rock ledges, require an exceptional sensitivity to depth cues, and so on. Perhaps because the early behaviour of a neonate organism tends to reflect relatively sophisticated knowledge of its surroundings, and perhaps because different species appear to be born with perceptual abilities suited to their eventual survival, we tend to assume that a large proportion of the perceptual capabilities of animals is inbuilt or innate. Some of it is: the visual system of the frog provides an example of the extent of built-in analysers which selectively 'tune-in' the animal's attention to relevant aspects of its environment (chapter 6, p. 203). Nevertheless, the distinction between learnt and innate is a tenuous one: in reality, no behaviour can be thought of as free from environmental influences, and from the moment of conception the developing organism is undergoing stimulation of various kinds, whether *in utero, in ova* or during the immediate post-natal period. It is quite clear now that animals require certain kinds of perceptual and sensorimotor experience during periods of their early development in order to acquire the healthy and versatile perceptual processes they possess during adulthood. This early experience determines not only *what* an animal can perceive (the quantitative aspects of perception), but also *how* he perceives objects in his environment; in short how he will react to them (the qualitative aspects).

Perceptual learning differs in two important ways from the learning we have discussed in section one. First, it is difficult to conceive of it as associative learning – it often occurs in the absence of unconditioned stimuli or operant reinforcers. Merely experiencing certain environmental conditions seems adequate for normal development. Secondly, perceptual learning often involves what are known as sensitive or critical periods. That is, the experience required for learning must occur during certain specified periods of the animal's development: encountering this experience either before or after the critical period has little or no effect. The notion of critical periods has been most closely examined in the context of imprinting (the learning of attachments to specific objects or places), and although it is a matter for conjecture whether this can truly be considered as perceptual learning (see p. 262ff.) it has been included in this chapter to further illustrate the importance of critical periods in a number of aspects of learning.

## THE CRITICAL PERIOD

The concept of a critical period is best illustrated by an example from embryology. In the early stages of embryological development, tissue is usually non-specific, its subsequent function is undetermined and it can, given the appropriate conditions, develop into nerve tissue, skin tissue, or muscle tissue. However, there is a critical point at which the nature of the tissue does become determined and subsequent influences cannot change this. This example emphasises the importance of maturational factors and physical change in determining critical periods, and certainly the notion of maturation is one that is closely linked with its definition. Nevertheless, a descriptive definition of critical periods is probably more in order in this chapter. Even though it is reasonable to suggest that many of the learning effects we are going to discuss involve critical periods determined by physiological or maturational factors, the evidence suggests that in many instances experiential and maturational factors combine to determine the duration of a critical period. However, a number of types of learning are characterised by two common features: these are (1), that the response in question is learnt only during a very brief and specified period of the animal's lifetime and (2), the learning appears to be relatively permanent and impervious to subsequent influences. It is these two features that we will maintain as important in defining a critical period for learning.

## EARLY EXPERIENCE AND PERCEPTUAL DEVELOPMENT

Before we look at what kinds of early learning are necessary for perceptual development, we should first look at the problems involved in investigating this topic.

## Methodological considerations

The most common techniques for assessing the effects of early perceptual experience on later perceptual ability involve either depriving the organism of normal sensory stimulation during early development (deprivation studies), or during this period, to subject the animal to specialised sensory conditions, that is, to allow it to experience only limited kinds of stimulation. Given this kind of preparation, there are three important problems to be borne in mind when interpreting the eventual data obtained from such experiments. First, if we attempt to assess the breadth of an organism's visual acuity after having selectively deprived it of early visual experience, it must be ensured that any deficits observed are not simply due to stunted physiological maturation of the peripheral visual system. For example, depriving an organism of all light stimulation for a period of time immediately after birth will probably have severe detrimental effects on subsequent visual ability. However, this need not be directly due to the lack of visual experience on the development of 'central' discriminative abilities, but to peripheral physiological factors such as retinal degeneration due to light starvation. It certainly seems the case that in most visually sophisticated organisms light stimulation during the early period of life is essential for healthy development of the retinae. For instance, Mowrer (1936) subsequently found gross deficiencies in the optokinetic responses of pigeons after their eyelids had been sewn together for 6 weeks after hatching. However, when the eye is allowed diffuse but unpatterned light, by using translucent hoods, little or no deficiencies in the optokinetic responses are found (Siegel, 1953).

Secondly, merely because we have deprived the animal of sensory experience may not simply leave a perceptual deficit, it may totally alter the course of perceptual development. Fantz (1967) has been prominent in pointing out this possibility; in fact, he goes as far as to suggest that the performance of an animal that has been visually deprived since birth may provide little insight into the abilities of the newborn. Simply because we have deprived an organism of stimulation from birth does not mean that when we do come to test its abilities these will have been held in 'suspended animation' and thus faithfully reflect the abilities that the animal is born with. For example, Fantz (1965, 1967) found that rhesus monkeys reared in darkness for the first 3–4 weeks of life, showed on subsequent testing a preference to fixate on patterned rather than plain objects. This was a trend that continued if unrestricted visual experience was given before 2 months of age. However, if this was not given and the animals were further deprived, an opposite tendency appeared at about 1 month of age and continued irreversibly: the long deprived monkeys showed little differential response to configurational variables (for example, patterned versus plain, centred versus uncentred, solid versus plane) but did show differential responses towards non-configurational variables (that is, they preferred to fixate on one of two seemingly similar geometrical objects, see figure 8.1). So a period of dark rearing appeared not only to retard the development of visual perception but also – if prolonged enough – to actively change its course.

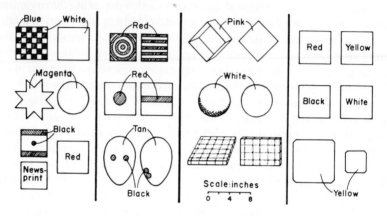

Figure 8.1   Stimuli used in differential-fixation tests of infant rhesus monkeys: (1) patterned vs plain, (2) centred vs uncentred, (3) solid vs plane, (4) non-configurational variables. (From Fantz (1965), copyright 1965 by Academic Press Inc. Reprinted by permission).

The third problem is one of selecting a response by which to test an organism's perceptual abilities. Because an animal does not respond differentially in a perceptual task it does not necessarily mean it does not have the perceptual apparatus necessary for making the discrimination, it may simply be that it does not have the neuromuscular development necessary to make the required response. For example, chicks reared in darkness for a number of days after hatching, show a low level of pecking accuracy when first presented with food in the light. This is in all probability due at least as much to postural instability as to perceptual deficits (Hinde, 1970, p. 472).

These three problems are ones which are particularly pertinent in the experimental study of the nature of perceptual experience. More recently, the development of sophisticated recording techniques (such as single-cell recording, see p. 250ff.) has stimulated a revival in the study of perceptual learning, and the kinds of perceptual experience necessary for healthy cortical development are becoming clearer. But, even so, studies vary considerably in procedural detail, and controversy still surrounds the interpretation of many phenomena (Barlow, 1975)

## Form perception

Although there is some evidence that birds will often develop form perception independently of visual experience (Pastore, 1962), many mammals require early visual stimulation for the development of this skill. For example, early visual experience is necessary for the development of size constancy in rats (Heller, 1968); dark-reared cats take longer than normally reared cats to discriminate patterns (Riesen and Aarons, 1959); and rhesus monkeys deprived of pattern vision for 20–60 days after birth learn a subsequent visual discrimination at the same rate as newborn monkeys (Wilson and Riesen, 1966). Similarly, the studies of Fantz (1965, 1967) noted earlier, suggest that early experience with patterned light is necessary for adequate pattern discrimination in rhesus monkeys. More interestingly though, many of the effects of early experience come about not as a result of conventional reward or punishment. It is as if there is a 'latent learning' effect; early experience without differential reinforcement normally increases the ease with which the learning of subsequent discriminations occur (Gibson, Walk and Tighe, 1959). However, some studies have noted that differential reinforcement can retard form discrimination under certain conditions (for example, Bateson and Chantrey, 1972).

## Depth perception

Perhaps the most famous piece of apparatus for assessing depth perception in animals is the 'visual cliff' apparatus developed by Walk and Gibson (1961). This normally consists of a piece of glass raised above the floor with a chequered board across the centre of it. On one side of this board a sheet of similar chequered material is placed beneath the glass (the 'shallow' end) while on the other side of this board a sheet of chequered material is positioned on the floor (the 'deep' end, see figure 8.2). The animal is placed on the central board and his movements to either the 'deep' or 'shallow' side recorded. When tested at an age at which they first begin to move about freely, most animals avoid the 'deep' end and leave the board on the 'shallow' end: chicks at 24 hours after hatching, goat kids and lambs at 1 day of age, kittens at 4 weeks, rhesus monkeys at 3 days, and human babies at 6–14 months. However, even though a wide variety of organisms begin to perceive depth as soon as they start to interact with their environment, the cues by which they perceive depth can differ greatly between species. Some organisms require binocular vision to perceive depth and the ability to utilise binocular disparity appears to be directly related to the amount of early binocular visual experience (for example, in cats, Pettigrew, 1974). However, other organisms can utilise such sensory cues as pattern density, relative retinal size, motion parallax, etc. Both chicks and day-old rats appear to fall into the latter group (Walk and Gibson, 1961), and there is some evidence that kittens can react appropriately to depth cues on the visual cliff when using only one eye (Held and Hein, 1967). The results of this latter study imply that although binocular vision appears to be necessary for fairly fine depth perception in cats, subjects that are forced to

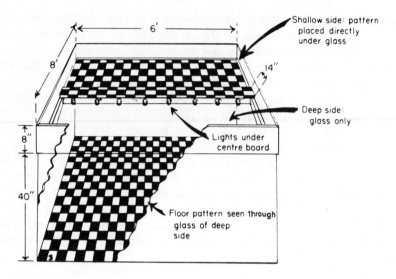

Figure 8.2   A model of a visual cliff for testing animals such as kittens, rabbits and goats. Lights mounted under the centre board permit adjustment of illumination on the deep side. Patterns on the deep and shallow sides, the width and height of the centre board, and the distance of the deep side below the glass can all be varied. (From Walk, 1965, copyright 1965, by Academic Press Inc. Reprinted by permission).

make monocular judgements of depth can develop *ad hoc* strategies for perceiving depth. Held and Hein suggest that their kittens were able to judge 'deep and shallow' by comparing the displacement of the retinal image during self-induced movement.

## Visual experience and cortical development

A greater understanding of the role of visual experience in perceptual development has been brought about through the use of single-cell recording techniques pioneered by Hubel and Wiesel (1963). It is clearly the case that a majority of the neurones in the visual cortex acquire specific functions; that is, they 'fire' only when certain characteristics are present in the visual field. For example, some may be oblique detectors (that is, fire only when lines of certain orientation enter the visual field). Some may be light/dark detectors, some may be edge detectors, and so on. By implanting electrodes within specific neurones of the visual cortex, and by recording their reaction to various visual stimuli, the location and percentage of neurones which fire to certain types of stimulation can be roughly assessed.

The role of experience in moulding the function of cortical neurones appears to be extremely important. For instance, the cortical neurones of kittens, prior to patterned visual stimulation, are largely unspecified (Barlow and Pettigrew, 1971; Imbert and Buisseret, 1975), although those neurones which are selective according to (1) the position of the stimulus, (2) whether lighter or darker than background, or (3) direction and velocity of movement, do occur in very young kittens without visual experience (Barlow, 1975, p. 200). Nevertheless, a hint of the degree of specificity obtained by some neurones can be illustrated by the fact that some neurones have been found in the monkey's inferotemporal cortex that respond optimally to the shape of a monkey's hand (Gross, Rocha-Miranda and Bender, 1972)!

Perhaps the most informative studies of this kind are those that have reared kittens in controlled visual environments and subsequently looked at neuronal specificity to types of visual stimulation.

## Binocular and monocular deprivation

Wiesel and Hubel (1965) closed both eyelids of newborn kittens. After periods of $2\frac{1}{2}$–$4\frac{1}{2}$ months the sutures were removed and the kittens exposed to varied visual stimulation. They found that only about one-third of the neurones they tested exhibited orientation selectivity (that is, 'fired' selectively to just a small number of line orientations). When just one eye is closed at birth and normal visual experience allowed with the other, monocular deprivation during weeks 4 and 5 of life appears to leave the cortex totally disconnected from the closed eye (Hubel and Wiesel, 1970). This rather dramatic effect appears to be the result of lack of stimulation during a critical period of about 3–12 weeks of age. Monocular deprivation from 9 to 19 days and from $3\frac{1}{2}$ to $6\frac{1}{2}$ months has no detectable effect on binocularity.

## Neuronal specificity and learning conditions

Further evidence for the role of experience, and indeed, of critical periods, in developing neuronal specificity has accrued from studies in which kittens have been reared in specialised visual environments. For example, Blakemore and Cooper (1970) reared kittens for the first 5–6 months of life in a totally dark environment, but allowed some kittens to experience vertical stripes and some to experience horizontal stripes, for 4–5 hours a day. This was achieved by placing a collar round the kitten's neck to prevent it from seeing its own body and placing it in a large vertical tube whose walls were painted with either horizontal or vertical stripes (see figure 8.3). At $7\frac{1}{2}$ months of age Blakemore and Cooper recorded from cells in the kitten's cortex. They found the neurones to be perfectly normal in every way except one: they responded only to an orientation of line similar to that experienced by the kitten when young. The kinds of orientation that the neurones would respond to are illustrated in figure 8.4. A randomly selected group of neurones from a normally reared cat would respond to a great variety of orientations: this was not so with the cats reared in horizontally or vertically striped tubes. Blakemore summarises their plight accordingly:

Figure 8.3   Apparatus to rear kittens in an environment of vertical stripes. (From Blakemore and Cooper, 1970. Reprinted by permission).

In short, they could not respond to edges of an orientation perpendicular to that which they had experienced. Probably the best test was to watch the way that the kittens reacted to a thin rod held and shaken by the experimenter. If it was held vertically the kitten raised in verticals would orient to it, run to it and play with it, while its fellow kitten was completely disinterested. If the rod was now turned horizontally the first kitten behaved as if it had disappeared and the horizontally experienced kitten would now come and play.

(Blakemore, 1973, pp. 65–66)

Two further points are of interest from this study. First it does not appear that 'busy' vertical neurones are activated by vertical experience and 'inactive' horizontal neurones simply degenerate. Blakemore and Cooper found 'no regions of silent cortex, and no decrease in the density of neurones'. The implication is that the visual cortex is to a large degree a *tabula rasa* at birth, and the subsequent function of the majority of cells is under environmental control. There is some evidence, however, that orientation selectivity might be, to some degree, under genetic control. Leventhal and Hirsch (1975) found that kittens reared in horizontal or vertical environments had cells which fired

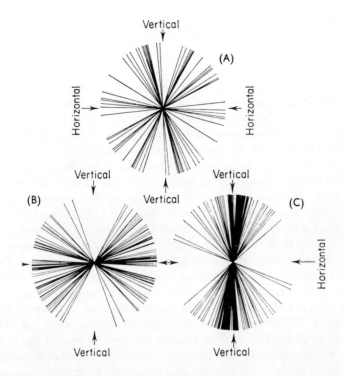

Figure 8.4   A. A polar diagram of the range of preferred orientations for 34 neurones from a normal adult cat. Each line represents the optimal orientation for one cell. B. A kitten reared in horizontal stripes from 2 weeks to $5\frac{1}{2}$ months, then allowed a little normal vision spread over 2 months. C. A kitten like that in B, but reared in vertical stripes. (From Blakemore, 1973. Reprinted by permission).

only to horizontal or vertical orientations. However, when kittens were reared with either left or right obliques, they predictably had cells that fired to left or right obliques but they also had many cells which responded to horizontal and vertical. They concluded that although experience is required for the establishment of oblique detectors, genetic factors may play an important role in developing horizontal and vertical detectors. Secondly, there does appear to be a critical period for this selectivity in kittens. Outside of a period between 4–12 weeks after birth no amount of stimulation will bias the distribution of preferred orientations.

Since this, and similar initial studies (Hirsch and Spinelli, 1971), further research has revealed that the proportion of neurones of different types can be manipulated by rearing in environments consisting only of stroboscopic light (Olson and Pettigrew, 1974), moving stripes (Pettigrew and Garey, 1974; Tretter, Cyander and Singer, 1975), or dots without any stripes or lines (Pettigrew and Freeman, 1973; Van Sluyters and Blakemore, 1973). In fact, some of the treatments not only modify the proportions of different neurones, but in some circumstances produce neurones with properties totally unlike those found in a normal cortex (Hirsch and Spinelli, 1971).

Finally it is important that we should look at some of the single-cell studies carried out in primates. This is quite necessary because the visual system of newborn monkeys and cats are quite different: the macaque monkey, for example, can observe and follow objects immediately at birth, but the cat is visually very slow to develop – it does not open its eyes for the first 10 days, and the visual cortex is anatomically immature at birth (Craggs, 1972). This difference has led to speculation that orientation specificity may indeed be moulded by experience in the cat, since its visual cortex is still developing even after birth. However, orientation specificity in the monkey may have a significantly larger innate component since the visual system is relatively sophisticated at birth. This hypothesis has been lent some support in a study by Wiesel and Hubel (1974). The eyelids of a macaque monkey were sutured together immediately after it was born by Caesarean section; visual experience could be considered virtually nil. However, Wiesel and Hubel found that when the sutures were removed 4 weeks later, there were neurones in the visual cortex which responded in an ordered way to lines of different orientation. To all intents and purposes, the results were not obviously different to those obtained from an adult monkey. So bearing this in mind, it seems to be the case that neuronal specificity may be determined by experiential factors – but perhaps only in animals with visual systems which are still relatively immature at birth.

## Summary

Early attempts to determine the role played by experience, in developing the perceptual systems of animals were unsatisfactory, primarily because methodological problems often made the interpretation of results equivocal. However, although techniques of single-cell recording are not without their methodological problems (Barlow, 1975, pp. 200–201), they do allow us to get more directly to the seat of perceptual development. The evidence that is available from kitten studies does stress the plasticity of the cortex to perceptual experience and also points to the importance of critical periods during which appropriate experience is necessary for subsequent normal development.

## IMPRINTING

We have touched briefly on the nature of critical periods and the role they play in the development of the visual system. However, critical periods appear to be characteristic of a number of types of learning, and perhaps the most widely studied example is that of *imprinting*. Imprinting has received a great deal of attention from animal behaviourists for well over half a century but there is still even argument as to what phenomena should actually be contained under the auspices of the term imprinting. The Austrian ethologist Konrad Lorenz was the first to subject this type of learning to extensive field and laboratory study. What he observed was that the young of certain precocial birds, (birds hatched with a complete covering of down and able to leave the nest at once and seek food) such as duckling and chicks, on emerging from the egg, will learn to approach and follow the nearest moving object. This attachment is strong, subsequently difficult to break, but most importantly this bond-formation occurs only during a fairly specific post-hatching period. The traditional view of this phenomenon, as expounded in the writings of Lorenz, (Lorenz, 1935), was that (1) imprinting was essentially a form of social learning (it familiarised the hatchling with the nature of its conspecifics) – without this 'social imprinting' the subsequent emergence of socially appropriate behaviours was adversely affected. For example, the attachment or imprinting of a young mallard duckling to its mother would ensure that at sexual maturity it would copulate successfully with other mallard ducks, and not attempt to direct sexual behaviour at any other animal or object; (2) Lorenz considered that the critical period during which imprinting occurred was genetically programmed. That is, the early experience of the young bird had no effect on the timing of the critical period, it was determined solely by maturational factors. In fact, 'it was as though a window opened on the external world and then closed again. While the window was open the young animal was affected by certain types of experience, at other times it was not'. (Bateson, 1973, p. 103). As we shall see shortly, both of these assertions of the traditional imprinting theorists have been subsequently questioned. A more recent and liberal view of imprinting is that it is a type of learning which occurs not just in a social context. Many other learning phenomena share the characteristics of Lorenz's 'social imprinting'. We shall discuss such examples as the formation of food preferences, song learning in birds and the formation of preferences for nest building sites (environmental imprinting). So imprinting can be considered as a special kind of learning which simply develops preferences for certain objects or places. As Hess summarises this:

. . . imprinting is a type of process in which there is an extremely rapid attachment, during a specific critical period of an innate behaviour pattern to specific objects which thereafter become important elicitors of that behaviour pattern.

(Hess, 1973, p. 65)

Whether it should be considered as 'social learning' or 'preference formation', imprinting is important to us as learning theorists because it has a number of characteristics which differentiate it from associative learning or conditioning. First, it is apparently characterised by a critical period – the learning only occurs during a brief and fairly specific period of the organism's lifetime (but see pp. 263–264). Secondly, conventional rewards and punishments do not appear to play any major role in the development of imprinted behaviour (see p. 265). Thirdly, the resistance to extinction of this type of learning is unlike that in conventional associative learning – once an attachment is established it is much more difficult to break (but see Salzen and Meyer, 1967, 1968, for procedures which have successfully reversed the imprinting process).

## Variables affecting imprinting

### The nature of the imprinted object

In most laboratory studies of imprinting, young precocial birds are incubator hatched and then, at various post-hatching periods, exposed to an object on which they can imprint. The range and variety of objects to which these hatchlings will imprint is quite enormous. During the critical period they will approach and follow such imprinting objects as people, boxes, cylinders, decoy ducks, model hens, rotating discs and flickering lights. However, given the sensitivity of imprinting to such a wide range of objects there are certain characteristics of stimuli which make them more likely to become imprinted objects: (1) Ducklings, for example, will imprint more readily to moving rather than stationary objects (Klopfer, 1971) and objects which emit animated or 'lifelike' movements are preferred to objects which emit smooth 'gliding' movements (Fabricius, 1951); (2) a vocal or 'noisy' object is approached and followed more readily than a silent object (Collias and Collias, 1956), and short, rhythmic sounds appear to be more attractive than long high-pitched notes (Weidmann, 1956). However, at later post-hatch ages visual rather than auditory cues become more effective (Gottlieb and Klopfer, 1962); (3) chicks are most responsive to lights at the red end of the spectrum and particularly unresponsive to green light (Bateson, 1966; Kovach, 1971); (4) size of the object is extremely important; if it is too small (smaller than a match box) it tends to be treated as food (Fabricius and Boyd, 1954), and ducklings prefer objects around 10 cm in diameter (Schulman, Hale and Graves, 1970).

### Age parameters

The age at which an animal will socially imprint is usually species-specific, but as a general rule it becomes increasingly difficult to imprint or obtain following reactions in young precocial birds as they get older. However, imprinting has been obtained with ducklings who were not exposed to an imprinting object until 10 days after hatching (Boyd and Fabricius, 1965; Smith and Nott, 1970). Jaynes (1956) exposed New Hampshire chick hatchlings to cardboard cubes at different times after hatching. He found that five-sixths of those tested at 1–6 hours met the criterion for imprinting to the cardboard cube, five-sevenths of those tested at 6–12 hours, two-fifths of those at 24–30 hours, three-fifths of those at 30–36 hours, and only one-fifth of those at 48–54 hours. Thus, there was an inverse relationship between post-hatch age and tendency to imprint.

However, more recent studies have suggested that this inverse relationship between post-hatch age and imprinting may result from the kinds of procedures used to implement imprinting. For instance Brown (1975) found that ducklings ranging in age from 20 to 125 hours showed an equal degree of following behaviour to an imprinted object when all age groups were trained to a criterion of following behaviour regardless of the time this took. Age did not correlate with the time to criterion and nor did it correlate with the tendency to choose the imprinted object over a novel object. Since most studies give their subjects a fixed period of exposure to the imprinting object this implies that the inverse relationship between age and imprinting may be due to attentional or arousal differences which are age-related, rather than to age differences in discrimination learning or imprinting *per se* (Fischer, 1966, 1967).

## Effects of early visual experience

Early experience, especially visual experience, does seem to influence the readiness with which hatchlings will approach and imprint to certain objects. For example, imprinting is enhanced if, prior to exposure to the imprinting object, ducklings are allowed to see only diffuse as opposed to patterned light. Moltz (1961) and Moltz and Stettner (1961) found that Pekin ducklings who wore a latex hood from hatching (allowing only diffuse visual stimulation) imprinted to a cardboard box more readily than control subjects whose hoods allowed them to experience patterned light. A study by Bateson (1964) suggested that the nature of the patterned light stimulation was also important. He reared chicks in isolation with one of three differing patterns: dark grey alone; dark grey and red horizontal stripes; white with red stripes and red circles between the stripes. After this early experience they were presented with a model painted with one of the three designs. A positive relationship was found between rearing conditions and approach to the model pattern: those reared in grey conditions were the quickest to respond while those reared in red and white walls were the slowest. Bateson interpreted these and the results of similar experiments as suggesting that the chicks reared in the more complex environment had learned the characteristics of their environment more thoroughly and therefore showed a stronger dis-crimination between the familiar and unfamiliar. In a study which investigated slightly different aspects of early visual experience on imprinting, Moltz (1963) found that if restrained ducklings were allowed to watch a cardboard box (1) approach them, (2) move away, (3) alternately move away and approach, or (4) remain stationary, and were subsequently allowed to follow the box, the group which had seen the box move *away* followed it the most. So, although ducklings will imprint more readily to moving objects than stationary ones, they are even quite sensitive to the direction of this movement.

## The role of the following response in establishing imprinting

A number of studies have purported to demonstrate that the expenditure of effort during the imprinting experience is essential to produce strong imprinting (Gottlieb, 1965; Klopfer, 1971). A study by Macdonald and Solandt (1966) illustrates this effect. They exposed 12- to 20-hour-old White Rock Cornish chicks to a flickering light and a metronome sound. Some of the chicks were exposed while under the influence of Flaxedil, a drug which blocks overt motor responses, while others were exposed without any drug influence. Their results suggested that although the drugged group did subsequently show some signs of imprinting to the light and metronome, the group that had been allowed to approach the imprinting object during training showed a significantly higher level of imprinting behaviour in the subsequent test. This relationship between the responses of the hatchlings and the subsequent degree of imprinting can be loosely summarised in a 'Law of Effort' (Hess, 1973, p. 392): the greater the amount of effort expended by the hatchling during initial exposure to the imprinting object, the more probable it is that following behaviour will be elicited by further presentations of that object. Although there are numerous methodological problems that need to be overcome when studying this relationship, it seems reasonable to assume that the Law of Effort is not an experimental artifact. One possible alternative interpretation of such results has been put forward by Bateson (1966). He suggests that it is not the amount of locomotor activity *per se* that influences imprinting, but the amount of visual experience with the imprinting object that this activity entails. At any rate, it certainly seems to be the case that hatchlings do strive to obtain as much visual experience of the imprinted object as possible and they will even perform an operant response in order to present themselves with different angled views of the imprinted object (Bateson and Reese, 1969). So, whether the law of effort reflects the direct relationship between amount of effort and degree of attachment, or whether it reflects the fact that greater activity allows the animal greater and more diverse visual experience with the imprinted object remains to be elucidated.

## Imprinting and the formation of behavioural preferences

In the introduction to this section on imprinting we suggested that imprinting can perhaps be considered as a kind of learning which extends beyond simple social learning. There is evidence that a learning process which closely resembles imprinting is involved in the formation of a broad range of preferences during the animal's early life.

### Food preferences

Nearly all animals develop preferences in their feeding habits. In some organisms the process for developing this preference can be quite complex – involving the interaction of a number of different basic mechanisms – but in some the process is much more simple, and long-lasting feeding preferences develop on the basis of mere experience with food objects during a critical period of early development. The chick appears to be one animal whose food preferences are defined by experience with food objects during a critical period in the first 3–4 days after hatching. Hess (1962, 1964) has conducted experiments which throw some light on the duration of the critical period for food preference formation in the chick. He found that, on hatching, chicks would innately prefer to peck at a white circle on a blue background rather than at a white triangle on a green background. He then gave different groups of chicks food reward for pecking at the less-preferred (green-triangle) stimulus at different post-hatch ages. Figure 8.5 shows the results he obtained. The food reward for pecking the non-preferred stimulus appeared to have an

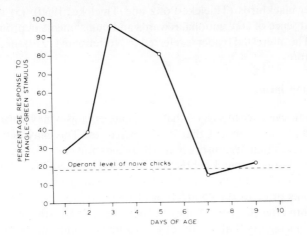

Figure 8.5  Percentage responses to the triangle-green stimulus during test sessions carried out during extinction. These data show that the reinforcement session only had a lasting effect on those chicks that were reinforced at the ages of 3–5 days. (After Hess, 1973).

effect only when it was given to chicks who were 3–5 days old. Hess concluded that there was a definite critical period during which food reward was most effective in modifying innate preferences of chicks for pecking at certain objects. He further suggests that a putative critical period for learning food objects is reasonable since (1) at the age of 3 days a chick can no longer depend on the yolk sac for nutritional resources but must look elsewhere, and (2) at the age of 3 days the chick has also achieved its highest level of pecking accuracy. A further fact which suggests that pecking preference is determined by a process resembling imprinting is that modification of the pecking response during the critical period is apparently permanent.

### Song-learning in birds

Although in many birds the species-typical song is innate, a large number still develop the song through experience. That is, they either learn the rudiments of the song by imitating experienced conspecifies or they learn to embellish a basic tune–a process which develops 'song dialects' within a given species of bird. Thorpe (1961) has suggested that in some species of bird this song learning may proceed by a process resembling imprinting. For example, song-learning in the chaffinch has characteristics which resemble imprinting: (1) there appears to be a critical period for the song-learning which has a maturational basis, (2) there is an inability to learn new material after this critical period, (3) there is a preference for certain classes of events, (4) short exposure to the imprinted event has long-lasting effects – for example, only a one-minute exposure during the sensitive period is sufficient to cause song-learning in blackbirds (Thielcke-Poltz and Thielcke, 1960), (5) there is an apparent absence of conventional rewards and punishments in promoting this learning. The interested reader is referred to Nottebohm (1970) for a review of this literature.

### Locality imprinting

There is some evidence to suggest that a neonate organism may learn to 'prefer' the immediate environment that it first perceives. This may determine the locality in which that organism eventually breeds and rears its young (Thorpe, 1944, 1945). For example, Loehrl (1959) found that flycatchers returned after the winter to the locality in which they had been reared if they were released at least 2 weeks prior to the completion of their first moult. If they were not released until after this period, they failed to return. A further experiment by Hess (1972) suggests that a kind of environmental imprinting can occur in mallard ducklings on the 2nd day of life. A group of ducklings were hatched in a simulated open ground nest and a second group were hatched in a simulated nest box. Some were kept in this environment for 1 day, some for 2 days. Hess found that when these ducks eventually came to breed and nest in the wild, those that had remained in the simulated nest box for two days all subsequently used elevated nest boxes while a majority of those that were reared in a simulated open ground nest, eventually used natural ground nests. However, we must conclude this section by emphasising that although the evidence is consistent with an imprinting interpretation of locality learning, it is still by no means conclusive in this respect.

## Imprinting in mammals

So far we have discussed imprinting mainly in relation to birds, and primarily precocial birds. What evidence is there for imprinting in classes other than Aves? Psychologists, for instance,. have often postulated the existence of critical or sensitive periods for healthy psychological and cognitive development in man. Piaget's developmental stages, Freud's psycho-sexual stages and the effects of early maternal deprivation on the development of social responses (Bowlby, 1958) are all examples of the way in which the concept of sensitive or critical periods have been adopted to explain human behaviour. However, simply to suggest that there are critical or sensitive periods for human development does not, of course, imply that such learning proceeds via imprinting. Imprinting is characterised by a relatively *short* critical period, its effects are fairly permanent and it does not seem to involve learning by conventional reward or punishment; on these criteria the learning stages of Piaget, Freud and Bowlby should not be construed as imprinting. However, a process which closely resembles imprinting can be identified in the formation of the mother-young bond in sheep and goats. Collias (1953) noted that if a lamb is removed from the ewe at birth and kept away for 2 or more hours the mother will subsequently refuse to accept it. However, young which are removed 1 hour after birth and then returned to the mother 2–3 hours later will be accepted by the ewe. Thus there appears to be a critical period of 1–2 hours after birth in which the mother–young bond is formed; this appears to be the case with both sheep and goats (Klopfer, Adams and Klopfer, 1964); and the most likely medium for this attachment is olfactory cues (Smith, 1965). If this mother–young attachment can be considered as a form of imprinting it is interesting in that it is the mother who comes to recognise her young, rather than the young coming to recognise the mother, as in the case with precocial birds. Similarly, because goats and sheep usually give birth more than once in their lifetime, the imprinting process can occur on a number of occasions. This is again different from the once-and-for-all characteristic of imprinting in birds.

## Theories of imprinting

Two important questions must be raised before we talk about theories of imprinting. First, what phenomena should be considered as being examples of imprinting? As a general policy it is perhaps best to adopt a fairly broad definition of imprinting, one similar to that espoused by Hess (1973). This definition emphasises that an organism forms preferences for objects or places on the basis of early experience with these events; this experience must occur during a relatively short critical period in the organism's lifetime; the resulting preferences should be relatively permanent and seemingly established by a process other than conventional associative learning. Many phenomena seem to fit this definition, including social learning, food preference formation in chicks, song-learning in birds and some aspects of locality imprinting. However, simply because we are able to subsume all of these phenomena under one umbrella term does not mean that they share a common underlying mechanism. They certainly share common features but may differ in their *modus operandi*. For example, the factors which determine the duration of a critical period for social learning may be entirely different from the factors that initiate and terminate a critical period for the formation of food preferences. However, if imprinting is a generalised, non-specific learning process, we should be able to discover principles of operation which are common to different responses, environmental objects and – hopefully – different species. Secondly, what criteria do we use for maintaining that imprinting has occurred? In effect, what response is learnt? This is a difficult question to answer since, if we assume that the animal merely learns a 'cognitive set' or 'preference' for an object or place, we still have to inter- pret how this preference is translated into observable behaviour. It is all very well Salzen (1970) suggesting that 'imprinting is a process of goal or object acquisition and not response acquisition' but how should we measure the degree of goal acquisition? What kinds of behaviours reflect the influence of this 'cognitive set'? These problems are pertinent ones in a theoretical analysis and should be borne in mind when we discuss the prominent theories of imprinting.

## Imprinting as associative learning

The principle of scientific parsimony urges us to seek explanations of new or special phenomena in terms of accepted principles. In deference to this principle some theorists have attempted to subsume imprinting under the mantle of associative learning. For example, Moltz (1960, 1963) has claimed that social imprinting could be thought of as a simple sequence of classical and then operant conditioning. He suggests that as soon as the chick is able to move about, any relatively large object in its environment is attention- provoking and the bird orientates towards it. Since the hatchling has little fear of its environment at this stage (and thus a low anxiety drive), autonomic components of this low anxiety drive become classically conditioned to the object that the bird orientates towards. Eventually, when the chick does acquire a fear of unfamiliar objects in its environment, the imprinted object becomes a source of 'relief' – approaching and following the imprinted object is operantly reinforced by fear reduction. Although there is some evidence that the imprinted object does seem to have anxiety or arousal reducing effects (Bateson, 1969; Hoffman, 1968) there is evidence which is not easily integrated into this account. First, chicks do tend to have apparently innate preferences for the kind of objects they will follow or imprint to (see p. 256); although this might reflect the different attention-gaining properties of certain objects, it is interesting that some organisms imprint more readily to objects which possess the characteristics of an adult conspecific (Gottlieb, 1965, 1966; Hess and Hess, 1969) – a fact which might suggest that there is an inbuilt tendency to orient not towards attention-provoking stimuli *per se* but more towards stimuli which resemble the natural mother. Secondly, this theory has difficulty explaining why chicks tend to imprint to objects moving away from them and not so readily to objects moving towards them (Moltz, 1963). Intuitively one would assume that objects moving towards the chick would be more attention-provoking than those moving away.

## Imprinting as perceptual learning

A second account of imprinting emphasises the role of perceptual learning and in particular, visual experience in determining both the establishment of preferences and the length of the critical period for this preference formation (Bateson, 1966; Sluckin, 1964, 1973). This approach suggests that, on hatching, chicks have a simple preference for 'conspicuous' objects; the imprinting process merely refines and sharpens these preferences to the point where the chick can now make distinctions between familiar and unfamiliar aspects of its environment. It is at this point in its perceptual organisation that the critical period for imprinting comes to an end – the hatchling becomes able to learn the characteristics of the environment in which it is reared and hence avoid dissimilar objects. Certainly, this theory does have a number of factors in its favour. First, as one might expect in a perceptual account of imprinting, the amount of visual experience that a hatchling has with the imprinting object will determine the strength of the imprinted response (although differentiating this from actual effort expended is difficult, see p. 258). Secondly, the amount and type of early experience can affect the length and timing of the critical period for imprinting. For example, MacDonald (1968) reduced early visual experience in chicks with injections of sodium pentobarbital – which effectively anaesthetises the birds. When tested 5 days after hatching these birds showed significantly more approach responses to a strange moving object than did control birds. Indeed some studies might suggest that the concept of a 'critical period' is indeed misleading in the context of imprinting. We have already mentioned that the relationship between age and the degree of imprinting seems to depend on the kind of training given (Brown, 1975) and this has led proponents of the perceptual learning hypothesis to suggest that imprinting might more profitably be considered as occurring during a *sensitive* period, since many studies suggest it is *not* critical. It is not critical in the sense that conditions during rearing can have a marked effect on the age at which this 'critical' period terminates (Moltz and Stettner, 1961; Sluckin and Salzen, 1961) and this can also be affected by the nature of the rearing, training and testing conditions (Brown, 1975; Brown and Hamilton, 1977). So although one might be able to argue that maturational factors are important in determining the time of onset of the critical period, then it clearly seems that experiential factors are important in determining its duration.

## Imprinting as a special learning mechanism

Lorenz (1935) was perhaps the first to characterise imprinting as a specially adapted form of learning. He claimed that imprinting was irreversible and confined to a brief critical period of the organism's life – the duration of this period was, he further suggested, genetically programmed and thus determined by endogenous changes rather than experience. Some of these claims have since been shown to be excessive – for example, imprinting can be reversed under some conditions (Salzen and Meyer, 1967, 1968) and experiential factors can influence the duration of the critical period (MacDonald, 1968; Sluckin and Salzen, 1961). Nevertheless, there is some evidence still to suggest that imprinting might have some genetically programmed components; a case can be made for a weaker version of Lorenz's original formulation. This case is based on two main facts: (1) even though hatchlings will imprint to a wide variety of animate and inanimate objects they still tend to exhibit seemingly innate preferences for some objects over others. Although factors such as size, colour, sound etc. are important in this respect (see p. 256), there is also a tendency to imprint more readily to objects resembling the natural mother. Hess and Hess (1969), for example, found that mallard ducklings when imprinted on the first day of life will show a 100 per cent preference for the imprinted object in a choice situation. However, the speed with which they join the imprinted object depends on the nature of the object: they will immediately rush over and join a mallard decoy but are considerably slower to approach and join a human being. Hess suggests that:

> we must consider that young ducks innately possess a schema of the natural imprinting object, so that the more a social object fits this schema, the stronger the imprinting that occurs to the object. This innate disposition with regard to the type of object learned indicates that social imprinting is not just simply an extremely powerful effect of the environment upon the behaviour of an animal. Rather, there has been an evolutionary pressure for the young bird to learn the right thing – the natural parent – at the right

time – the first day of life – the time of the sensitive period that has been genetically provided for.

(Hess, 1973, p. 380)

(2) the learning that occurs during imprinting appears to occur via a process which is so markedly different from conventional associative learning. The points of disparity are the following: (a) imprinting occurs only during a relatively brief specified period of the organism's lifetime, (b) the imprinted object is not initially a neutral stimulus as it would be if it were a conventional Pavlovian CS; the imprinting object by itself elicits the behaviour in question, (c) imprinted behaviour does not show a readiness to extinguish when rewards are subsequently withheld. For example, altering the pecking preference of a 3–4-day old chick with food reinforcement appears to have a permanent effect, even when subsequent food reinforcement is discontinued (Hess, 1962, 1964), (d) the imprinted response appears to be insensitive to punishment during the critical period. Actually subjecting the young chick to electric shocks as it approaches the imprinting object during the critical period fails to prevent the approach behaviour (Kovach and Hess, 1963), (e) during imprinting it is *primacy* of experience that has the greatest effect on future behaviour, in conventional operant or classical conditioning studies it is usually recency of experience which affects subsequent behaviour the most.

Given this catalogue of differences it is certainly difficult to think of imprinting as a form of associative learning. But equally so, simply because of this we should not directly opt for the polar view that the imprinting process is primarily genetically determined. Certainly, components of the imprinting process do appear to be innately determined, for example, the approach behaviour in social imprinting appears to be an unconditioned response to environmental stimulation. However, the process appears to be more flexible than was originally believed and experiential factors do play an important role in modifying many facets of imprinting.

**Summary**

It is surely safe to say that imprinting is worthy of study as a specialised form of learning, whether it be a 'genetically programmed learning process' or merely a result of 'perceptual development'. Nevertheless, it is still difficult to know confidently what phenomena should be classed as imprinting. We have defined imprinting here as the development of long-lasting preferences during a brief 'critical period' of an organism's lifetime; however, this does not presuppose that phenomena which share these characteristics also share a common underlying mechanism, and, surely, some aspects of the theories we have discussed can really only be applied to social imprinting. However, the fact that such phenomena as social imprinting, food preference, song-learning, etc. cannot be easily interpreted in terms of associative learning, and that they are all characterised by critical periods suggest that their similarities might reflect at least some common principles of operation. If this is the case, these principles should have predictive significance independently of organism, nature of response, or type of imprinted object.

# 9 Imitation and Observational Learning

If we compare the potential adaptive utility of imitation learning and the amount of research it has engendered, then imitation learning has been sorely neglected by animal behaviourists. There is a great deal of anecdotal evidence that animals successfully learn to adapt to their environments by observing experienced conspecifics – we shall discuss some examples shortly – but what should interest us here is the mechanics of imitation: Are there different types of observational learning? Is imitation simply 'copying'? Is imitation facilitated by non-specific social factors? But before all this, it is not difficult for one to envisage the adaptive benefits of observational learning. First, it can facilitate social cohesiveness by maintaining a group 'identity' – in most circumstances a group of conspecifics is more likely to survive than a single animal foraging alone for food whilst also having to remain alert to predators. Secondly, as an adaptive process, imitation is a vehicle for communicating useful information between conspecifics. For example, birds, by observing the food preferences of experienced conspecifics, can quickly appreciate which foods are palatable and which should be avoided – without having to 'gamble', in a trial-and-error fashion, with possibly toxic foods (Turner, 1964). On the other side of the coin, social facilitation – as a special kind of observational learning (see below) – can enable aversive or dangerous situations to be avoided without prior experience with the aversive agent. For example, if one of a flock of birds takes to the air, the rest will immediately follow, thus avoiding the possible dangers which aroused the original bird to flight (Klopfer, 1961). So, in their natural environments it is not unreasonable to expect that great importance should be attached to learning by observing experienced conspecifics. However, this importance has not been reflected in the proportion of research time alloted to imitation learning and associative learning respectively. We seem to know a great deal about how an animal in isolation learns to respond to obtain food or avoid electric shock, but how often is the animal in the wild confronted with a novel learning situation which it must cope with purely on its own? Rats and pigeons, for example, are fairly social animals and we might expect that much of what they learn about their environment is learnt with a helping hand from the observation of others. Bearing this in mind, there are two theoretical views we can take of observational learning; (1) it can be seen as an adaptive process owing nothing to associative learning (for example, which possesses its own mechanisms and principles), or (2) as a process which facilitates the learning of appropriate associations; that is, observation of an experienced conspecific draws the animal's attention to the important inter-relationships in the environment (for example, between stimuli and reinforcers, or behaviours and consequences, etc.) and thus enables them to be learnt more quickly. However, how we view such effects theoretically will depend on what phenomena we class under observational learning. A few definitions are necessary here. *Observational learning* will be used as a blanket term to denote all those situations in which successful behavioural adaptation is facilitated by observing another experienced animal (conspecific or not). This is quite difficult to define operationally, but the majority of the experimental studies we shall discuss assume that observational learning has occurred if an animal that has viewed an experienced 'demonstrator' learns the response in question more rapidly than one that has either viewed a 'naïve' demonstrator (one that is not skilled at executing the appropriate response) or has had no previous observational experience. As Zajonc puts it, a study of observational learning constitutes 'the study of cues which are generated by the behaviour of one organism and which benefit the learning and performance of another' (Zajonc, 1969, p. 62).

So, observational learning does not imply that an organism *copies* the behaviour of another organism, but simply that it benefits in some way from the observing experience. When the observing experience does eventually culminate in the demonstrator's behaviour being replicated by the observer, this is more specifically designated as *imitation learning*. Another category that should be mentioned in this context is *social facilitation*. Although it is perhaps stretching a point to suggest that this phenomenon can truly be labelled learning (see p. 272), it has traditionally been closely linked with observational learning. Social facilitation can be thought of as 'contagious' behaviour; it is characterised by the action of one animal appearing to act as a *releaser* for identical behaviour in another (Thorpe, 1956). However, whereas learning effects should be considered as relatively permanent, social facilitation effects are fairly transitory (Klopfer, 1961). Nevertheless, it is relevant in this context because explanations of observational learning often implicate the mechanism of social facilitation.

Anecdotal accounts of observational learning in the wild are numerous. For example, Fisher and Hinde (1949) report that the first account of birds opening milk bottles left on doorsteps occurred in 1921 in Southern England. Since that time the habit has become widespread all over the United Kingdom – its current popularity among birds probably owes more to imitation than to birds stumbling independently on this rich source of nourishment. A second set of examples of the way in which observational learning can facilitate changes in the behaviour of whole populations of animals comes from Japan (Imanishi, 1957; Miyadi, 1964). For instance, while observing the feeding habits of Japanese monkeys it was found that one young female developed a novel method of clearing the sand off sweet potatoes by washing them in the sea, this new behaviour was also useful for separating grains of wheat from sand by floating the wheat in a stream. In both instances, the new behaviours eventually spread to all members of the group, first being taken up by close relatives and associates of the originator and then spreading to other members.

Apart from these relatively casual accounts, observational learning has received some experimental examination. The paradigmatic example is to allow one animal (the observer) to watch an experienced conspecific indulging in the behaviour to be learnt (the demonstrator). In most cases this involves making a relatively unsophisticated response (for example, bar-pressing or alley running in the rat; key-pecking in the pigeon) to procure food, or to avoid electric shock. The observer is subsequently placed in the conditioning chamber and the rate at which he acquires the response is compared with animals who have had no observing experience, or who have observed only naïve conspecifics. Using this or relatively similar assessment techniques, observational learning has been demonstrated in a variety of species including rats (Angermeier, Schaul and James, 1959; Bankart, Bankart and Burkett, 1974; Gilbert and Beaton, 1967), some species of monkey (Darby and Riopelle, 1959; Presley and Riopelle, 1959; Warden, Fjeld and Koch, 1940), cats (John, Chesler, Bartlett and Victor, 1968; Tachibana, Yamaguchi and Huruki, 1974), birds (Klopfer, 1957; Turner, 1964; Zentall and Hogan, 1976), and, of course, humans (Bandura, 1969, for an interesting account of imitation learning in man). It occurs not only under conditions of appetitive motivation but can also be demonstrated with avoidance learning tasks (Del Russo, 1975; Kohn, 1976; Kohn and Dennis, 1972).

# THEORETICAL ACCOUNTS OF OBSERVATIONAL LEARNING

By and large, theoretical accounts of observational learning can be packaged into two main groups: (1) those that consider 'social' or motivational variables to be important, and (2) those that consider the demonstrator animal to be a vehicle for focussing the observer's attention on the important relationships in the environment. The former category assumes that observational learning comes about in an 'indirect' fashion – the particular behaviour is acquired by the observer not because the demonstrator provides information about the learning task, but because the presence of a conspecific induces non-specific behavioural changes which are conducive to the acquisition of the response. The latter of the two categories emphasises associational learning and that the behaviour of the demonstrator merely accelerates the learning of associations necessary for efficient performance.

# Social or motivational accounts of observational learning

### Imitation as an intrinsic reinforcer

It has often been suggested that imitation is intrinsically reinforcing; that is, copying the behaviour of a conspecific does not need any extra reinforcers to shape or maintain it. Now, what this statement actually means in practice is a difficult matter but, this aside for now, there is another way of formulating this account which stresses that merely *viewing* a conspecific has reinforcing properties; put another way, animals exhibit a tendency to observe conspecifics, and this has subsequent adaptive value for the organism. For instance, pigtail monkeys will respond not only to viewing one another but also to observing themselves in a mirror (Gallup, 1966); zebra finches will also respond to view a conspecific, and more specifically their own mate (Butterfield, 1970); similarly, rats will press a lever to obtain a period of either visual or physical access to another rat (Angermeier, 1960, 1962; Thach, 1965). Now, having established that viewing a conspecific does act as a reinforcer in a variety of species, how does this promote observational learning? One possibility is to conceive of observational learning as a two-process phenomenon: the first process involves an 'attention-focusing' mechanism, that is, one which merely facilitates the observation of conspecifics; the second process assimilates what is observed into the behavioural repertoire of the observer. In this sense the factors which facilitate viewing of a conspecific do not constitute a full account of observational learning because they only address themselves to the first process; they do not tell us anything directly about 'how' the animal imitates, or what he 'learns' during observational learning. Nevertheless, the importance to observational learning of viewing or following a conspecific has recently been illustrated in experiments by Neuringer and Neuringer (1974). They first of all taught a number of pigeons to feed from their hand; subsequently these pigeons observed the hand 'peck' a lighted key in a conditioning chamber. This group of birds learnt to peck the key significantly faster than another group of pigeons who also received the demonstration but were never fed from the experimenter's hand. Although the 'hand' in this case was not a conspecific, the Neuringers suggest that animals will follow a *food source*, and learning to acquire food is facilitated by this tendency. Although this account does not specify *what* the pigeons learnt from observing the food source, it does suggest a way in which animals might be primed to benefit from observing conspecifics. In the wild, the mother is obviously the first food source that altricial animals will encounter; eventually following this mobile source of food will not only enable the animal to discover important food sites, but also to avoid potentially toxic foods.

## Reduction of fear by a conspecific

A second account of observational learning suggests that animals may learn a response more quickly when observing a conspecific because the conspecific reduces the effects of 'fear' or anxiety in the observer. Most naïve subjects, when placed in a novel environment such as a Skinner-box, show signs of nervousness; it is suggested that an adjacent conspecific might reduce that fear which would ordinarily interfere with the learning process. To support this claim there is evidence that the presence of another rat will both reduce fear (Davitz and Mason, 1955) and facilitate exploration (Hughes, 1969) in an observing conspecific. Nevertheless, although fear reduction might be implicated in some observational learning studies there are many where it cannot. First, many studies purport to show observational learning when the observer is put into the learning situation *alone* after the observing experience (Del Russo, 1975). Secondly, observers watching a naïve demonstrator do not learn the task as quickly as those observing an experienced demonstrator (Bankart *et al.*, 1974; Zentall and Hogan, 1976); presumably the fear-reducing properties of the conspecific should be identical whether they are naïve or experienced demonstrators. Thirdly, some observational learning has been observed in rats when they are allowed to watch an experienced gerbil execute the response (Benel, 1975). Without making further assumptions, it is difficult to conceive of how a rat's anxiety might be reduced simply by being placed adjacent to an animal that is not a conspecific. These three reasons, when considered together, make a strong argument against an interpretation of observational learning merely in terms of fear reduction.

## Arousal properties of the demonstrator

The presence of a conspecific appears to stimulate an animal to increased levels of activity. That is, it acts as a source of arousal. Wheeler and Davis (1967) have demonstrated that the close presence of another rat would disrupt stable DRL performance by a second hungry rat. This disruption was not a result of the DRL rat's curiosity over his adjacent partner causing him to cease responding – it was due to the DRL rat pressing the lever too *often* to meet the DRL criterion. If we make the assumption that this increased rate of lever-pressing was brought about by an increase in activity or arousal level then, in situations where the observer has to learn an active response (for example, pressing a lever, running in a shuttle-box, etc.), increased activity levels should hasten the first occurrence of the response and thus facilitate learning. This is not an unreasonable suggestion in many observational learning studies, except that the first two objections that were raised to the previous account also apply here. Given these constraints, induced activity also cannot provide a full account of observational learning.

**Social facilitation**

It has been suggested that much of what is called observational learning closely resembles the more general phenomenon of social facilitation (Klopfer, 1961). Social facilitation, as we mentioned earlier, is 'contagious' behaviour: in a group of conspecifics, a behaviour executed by one of the members will often be copied by other members until the whole group is involved. Common examples of this phenomenon include a flock of feeding birds taking to flight when one of the members takes to the air (Davis, 1974), or the induction of pecking in a group of chicks by the pecking of one individual – even when there is no food around (Turner, 1964). An account of observational learning in terms of social facilitation implies two things: first, that observational learning can only involve 'copying' or imitating, and secondly, that the mechanism involves an 'instinctive tendency' (Thorpe, 1956). However, although social facilitation can be thought of as imitation and must be considered as a factor in observational learning experiments (Zentall and Hogan, 1976), it is doubtful if the mechanisms underlying social facilitation can provide a thorough account of observational learning. There are two important reasons for this. First, social facilitation is only a *transitory* social phenomenon, it does not produce the permanent behavioural changes that many examples of observational learning do. Secondly, one or two studies have reported instances of observational learning which do not involve imitating the demonstrator. Darby and Riopelle (1959) and Riopelle (1960) report experiments involving two rhesus monkeys. One monkey observed the other choosing one of a number of stimulus objects and subsequently getting reinforced or non-reinforced for the choice. After each trial the observer had to make a similar discrimination. Although observer monkeys learnt to make the correct response in the first trial in almost 75 per cent of the cases, what was more interesting was that the observer's performance was more accurate when the demonstrator's first trial choice was incorrect; thus, the observer was not merely imitating the demonstrator, he was actually learning something from the demonstrator's mistakes which facilitated his own performance. Clearly, this cannot really be labelled social facilitation since aspects of the observational learning do not involve the mere imitation that social facilitation implies.

Although instances of what appear to be observational learning can be accounted for in terms of the four factors we have just discussed, there appear to be examples of observational learning which require that we attribute to it the status of an individual learning process. Although this process may under certain conditions be facilitated by the non-specific factors we have discussed, it seems that a full account must elucidate other principles which will point more specifically to 'what the animal has learnt' from his observing experience. It is the second class of explanations which address themselves to this problem.

# Associative accounts of observational learning

Associative accounts of observational learning stress that what an animal gains from the observing experience is an insight into the relationships either between behaviour and environmental events, or between different environmental stimuli.

## Learning about the significance of the response

It may be that the observer learns that the demonstrator's response has certain consequences. The experiments of Darby and Riopelle (1959) and Riopelle (1960) suggest that such information might be gleaned from the observation of a demonstrator – their observer monkeys adjusted their behaviour according to the relationship between the demonstrator's response and its consequence (reinforcement or non-reinforcement). Similarly, other studies have found results which suggest that the observer is sensitive to such relationships. For example, Lore, Blanc and Suedfeld (1971) found that punished approach responses to a lighted candle decreased in rats if they had earlier observed a conspecific getting burnt; approach responses did not decrease, however, if they had previously observed a conspecific unsuccessfully trying to approach the candle. Apart from these two examples there are a number of experiments whose results suggest that the observer learns something about the inter-relationships in the learning task. Unfortunately it is difficult to isolate whether the observer is attending to S–S, S–R or R–S relationships, or some combination of these (Del Russo, 1975; Kohn, 1976; Kohn and Dennis, 1972). For example, although the observer monkeys of Darby and Riopelle behaved as though they had grasped the relationship between the response and its consequence, what they actually learnt from the observing experience may have been something quite different. For instance, they could have learnt that particular stimulus objects were related to particular consequences (S–S relationships), or that certain stimulus objects required certain responses to be performed (that is, choosing that object or not choosing that object; S–R relationships). Experimentally teasing out what associations the observer learns is a problem that is not specific to observational learning – as we saw in chapter 7 it is one that is at the core of associative learning. This being the case, a better understanding of observational learning would benefit from a finer knowledge of what an animal learns in a simple associative conditioning experiment. However, these problems apart, some theorists have indicated the possible importance of certain environmental relationships to observational learning.

## Learning the significance of places and events

Both Thorpe (1956) and Klopfer (1959) stress the importance to observational learning of what is called *local enhancement*. This is 'apparent imitation resulting from directing an animal's attention to a particular part of the environment' (Thorpe, 1956, p. 133f); for example, observing a conspecific feeding in a particular part of the experimental chamber has been shown to facilitate approach to the food site (Tachibana *et al.*, 1974; Zentall and Hogan, 1976). More generally, Spence (1937) has suggested that the main contribution of the demonstrator to the learning of the observer was the 'enhancement' of stimuli that were critical to the observer's learning. If a demonstrator rat spends much of its time contacting a response lever, the importance of this part of the environment will in some way be conveyed to the observer. Enhancement accounts such as these can be considered as stressing the importance of the demonstrator's 'attention-focusing' properties: that is, the demonstrator merely helps the observer to isolate the relevant aspects of the environment for exploration.

A more detailed associative account of observational learning has been provided by Bandura (1969). He claims that during the observing experience, a sequence of observed sensory events becomes integrated through a process of sensory–sensory conditioning. When an observer is repeatedly exposed to the pairings of sensory events (for example, watching a tone elicit bar-pressing in a demonstrator rat, or watching a behaviour followed by food, etc.), the subsequent presentation of one of the stimuli to the observer acquires the ability to elicit in the observer centrally aroused perceptions of the associated sensory events. For example, repeated exposure to a demonstrated stimulus–response sequence of tone followed by running (Del Russo, 1975) results in the formation of a sensory connection between tone and running in the observer. Subsequent presentation of the tone to the observer during his acquisition phase would elicit centrally aroused perceptions of the associated running response and therefore facilitate the acquisition of the association. Tortuous though this explanation may be in comparison with other accounts of observational learning it does at least attempt to get to the core of the problem, and, although not necessarily indicating 'what' is learnt, it does try and deal in more detail with 'how' observational learning occurs. Although many experimental studies yield results which are consistent with this sensory–sensory conditioning account, how easy it will be to verify the existence of sensory–sensory associations and subsequently to assess their contribution to observational learning is another matter.

## SUMMARY

There is little doubt that observational learning is an important part of an animal's adaptive armoury. In the wild it can aid group cohesiveness and it provides a psychological 'short-cut' to finding new food sites and learning new methods of obtaining food or avoiding predators. We have defined observational learning in very general terms and because of this it is quite likely to be the case that we have talked about phenomena which have very different underlying mechanisms. Social facilitation, for example, is a form of imitation whose mechanism is in all probability quite different from that which operates in an instrumental observational task (such as learning to bar-press through observation). Similarly, more than one process may be operating in an observational learning task. These appear to include processes which facilitate observing of conspecifics, non-specific processes which produce general arousal and motivational changes, and more specific processes which facilitate the formation of associations during observational learning. The evidence that is available suggests it is wrong to try and boil down all instances of observational learning to one of these processes, and more reasonable to conceive of observational learning as the result of the integrated action of all three.

# 10 Animal Memory

Two psychological processes which seem to be inextricably bound together are those of learning and memory; without memory there could be no learning and without learning, organisms would have nothing to remember. Yet, in the case of animals, psychologists have concentrated almost entirely on the process of learning to the detriment of memorial processes. We know quite a lot about how associations are formed but very little about how they are stored and retrieved. One reason for this bias immediately springs to mind; if one were to adopt directly the methodology of those who study human memory, then it is evident we would find out very little about animal memory. Until recently the study of memory processes was heavily biased towards semantic and verbal memory in man; the questions that were asked were of the type 'How is information coded in memory?'. 'How does structure of recall reflect the structure of storage?'. 'Is recall affected by semantic or acoustic similarity?'. The experimental procedures involved free recall of word lists or paired-associate learning. It is fairly obvious that in the case of animals, both the questions and methodology are largely inappropriate. Yet if these are inappropriate questions, what are the right questions to ask? Do animals have memories? We are told that elephants never forget, so by what process do they remember?

To search out the appropriate questions, let us go back to the distinction between learning and memory. By its traditional definition learning is a 'relatively permanent change in behaviour (resulting from practice)' (Kimble, 1961, my brackets). To measure learning we usually expose the animal to specified environmental contingencies and then, *at a later time or date*, assess what changes in behaviour have occurred. Thus, the only evidence we ever have of learning is a subsequent change in behaviour which is related to some earlier environmental events. The mere fact that an animal's current behaviour is controlled by an event or contingency which is no longer present implies an altered internal state of some sort (given that one does not believe in action at a distance!). This might be central or peripheral, representational or merely sensory. To give you examples of the extremes of these possibilities, consider the following two examples. We can present a pigeon with a brief illumination of a key-light followed by a delay of 1 or 2 s; after this delay two further keys are illuminated, one with light of the same wavelength as the original sample and one different. To obtain food the bird must peck the key whose colour is identical to the original sample. With relatively short delay intervals this is something that pigeons can do quite successfully (Roberts and Grant, 1976). Yet, rather than some central storage processes being needed to account for these results, the 'memory' in this case could more readily be likened to a 'sensory trace' which, at its simplest, might be a retinal after-image that decays with time. The pigeon simply matches the stimulus trace to the appropriate key colour. On the other side of the coin consider the phenomenon of 'learning sets' (Harlow, 1949). First, an animal is taught to discriminate between two objects, A and B, until it has reached a criterion of discrimination. After this it is given another discrimination, but with different objects, C and D; after this it is taught a further discrimination with E and F, and so on. What is often found is that the animal becomes 'learning-set sophisticated'. That is, it takes fewer and fewer trials to reach the criterion with successive problems. In this case, the animal is certainly learning from prior experience – what has happened in that past is influencing his present behaviour. But in what form can the animal retain this information? It can hardly be isomorphic with a simple stimulus trace, but might be thought of as reflecting some attribute of an organised memory (see p. 279ff. for elaboration of this point). The animal appears to have abstracted the relevant information from the early discrimination, stored it, and eventually combined it with the logistics of the present problem in order to arrive at a successful response strategy. These examples, however, reflect the extremes of what we understand by memory: the first is memory in perhaps its most trivial sense, while the latter is memory in its most sophisticated form. But given that we can identify memorial processes in a wide variety of animal learning situations, why have these processes received so little attention? One main reason was the linguistic straight-jacket applied to the explanation of behaviour by the early conditioning theorists. The rhetoric of the behaviourist manifesto demanded that only observable behaviour should be the subject matter of psychology and that the fabric of explanation was to consist of environment–behaviour interactions. Memory unfortunately could only be inferred from behaviour and not directly observed: if behaviour could be explained solely in terms of S–R connections, or, more fully, in terms of environmental contingencies, there seemed no need to invoke concepts of memory. Behaviour, it was assumed, was controlled by the environment, and so to fulfil the behaviourist's aims of prediction and control it was unnecessary to invoke either the concept or process of memory. Spear emphasises this point particularly well.

As a concept 'memory' has not always been viewed as particularly useful for understanding animal learning. If the existence of such a representation of the events of learning cannot be defined independently of contemporary responding by the animal, then a description of the antecedents of that behaviour does not really benefit from invoking the term 'memory'. Thus, rather than attributing the decrements in learned behaviour found after a retention interval to 'forgetting' or 'loss of memory', one might simply attribute this to 'loss of experimental control' . . .(Sidman, 1960, p. 310)

(Spear, 1973, p. 164)

As an instance of the application of Lloyd-Morgan's canon this is probably a praiseworthy attitude, but before the reader should write-off the concept of animal memory altogether let us consider the following. To opt for an explanation of behaviour in terms of environment–behaviour associations (for example, S–R links) does not mean that animals do *not* store information about their environment. Many early theorists were loathe to admit that animals could store information, and even when they did admit the possibility, this 'store' was usually the crudest of neural traces which replicated the bland physical attributes of environmental events (Hull, 1943). As students of animal learning we have inherited a legacy which implores us to attribute behavioural changes and behavioural decrements to environmental control rather than to memorial processes. But consider what it might mean to say this. Does it mean that all instances of 'relatively permanent behavioural change' can be attributed to principles of learning? Or that all instances of what we might call 'remembering' or 'forgetting' can be adequately explained by describing the environmental conditions under which they occur? In the light of these questions now consider the following examples. Having taught a rat a left–right discrimination in a T-maze we now leave him for some months before returning him to apparatus; his ability to execute the discrimination correctly has greatly diminished. Has he forgotten the discrimination? We carry out further studies which show that the decrement in performance after a 'lay-off' is proportional to the length of the 'lay-off' period. Couched in strict environmentalist terms the correct response becomes 'unlearned' in proportion to the time since the last practice trial. But this tells us very little about the reasons for the discrimination becoming 'unlearned', in fact it merely redescribes the empirical evidence. As animal psychologists there is much more than this we would like to know about the process. Some information must be stored in some form by the animal (since even with a long lay-off between training and testing, the 'lay-off' rats would probably perform better at the discrimination than totally 'naïve' animals), so does this drop in performance simply reflect the disintegration of previously formed associations (unlearning), the decay of memory traces over time, interference of old learning by new learning during the retention interval, or the inability of the animal to retrieve information from storage after long periods of disuse, etc.? If, as workers have recently tried to do, we attempt to pin down the forgetting phenomenon to one of these processes, how likely is it that the principles we discover will be related directly to principles of learning? There are two points to be made here. First, there are a number of behavioural phenomena whose recent discovery has stretched the credibility of established learning principles, but whose occurrence might be more profitably interpreted in terms of memorial factors. One example we have already discussed is conditioned taste aversion (chapter 6, p. 206ff.) where long-delay learning appears to violate fundamental associative laws. Other instances include examples we shall discuss shortly and thus only mention by name here: the 'Kamin effect', warm-up, state-dependent learning, and a number of phenomena related to delayed matching to sample (DMTS). As we shall see, numerous examples of this kind fit more comfortably into a framework of memory rather than being integrated awkwardly and incompletely under the auspices of learning principles. Secondly, there is some evidence to suggest that memories can be retrieved and even modified in the absence of the reinforcement contingencies originally used to establish the memory (Spear, 1973). After lengthy 'lay-off' periods quite dramatic behavioural changes can be documented by presenting certain 'retrieval cues' to animals (see pp. 292–293). It seems extremely fortuitous to attribute these changes in behaviour to 'fluctuations in stimulus control' or to 'relearning' since the contextual cues presented in order to facilitate learned performance after long retention intervals have immediate effects and are non-critical and often trivial aspects of the learning situation – certainly not aspects that one would have said had acquired any degree of control over the behaviour in the first place!

As more and more evidence becomes available from studies of retention in animals it is becoming apparent that the concept of memory is more valuable than learning theorists originally conceived. Although the methods that we use to study them seem to intertwine inextricably both memory and learning it appears that both processes command their own principles of operation. The problem for contemporary students of animal learning is not only to define more precisely the principles of remembering and forgetting, but also to tackle the more fundamental and 'grass-roots' problem of whether perplexing behavioural phenomena reflect attributes of the learning or memorial processes. In this chapter we shall discuss some of the phenomena which have been attributed to memory processes and the methodological problems involved therein. Similarly, we shall address ourselves to the more funda-mental question of 'why do animals forget?'.

## RECOGNITION VERSUS RECALL: S–R MEMORY VERSUS REPRESENTATIONAL MEMORY

It was the classical thinker Seneca (54 B.C.–A.D. 39) who is attributed as saying

> Animals are unable to recall their past experience, their memory capacity being limited to recognition. Thus the horse may recognise a road over which it has travelled before, but remembers nothing of it when in the stable afterwards.
>
> (from Warden, 1927, p. 77 in Winograd, 1971)

Has the study of animal learning thrown any new light on this question? Certainly animals exhibit recognition, virtually all experiments in the S–R framework suggest this much – appropriate stimulus conditions are presented after a retention period and the animal either does or does not behave appropriately. The implication of such S–R studies is simply that animals retain some memory of the learned response, and presentation of the appropriate stimulus subsequently elicits this response. But this is hardly a sophisticated form of memory, the utilisation of a sensory memory (for example, retinal after-image) is an example of recognition in its crudest form; and since even the lowly planaria can be classically conditioned (Thompson and McConnell, 1955), it too must exhibit recognition of a kind. The leap in sophistication of memorial abilities comes when it can be shown that an organism possesses what is known as *representational memory*. It is easier to define representational memory by outlining what it is not: in essence it reflects the ability to respond appropriately in a situation when controlling stimuli are *not* present or when there has been no opportunity to form an S–R bond. The type of phenomena which fall under the auspices of representational memory are quite varied and the reader will best acquire the flavour of the term as we discuss some examples.

An experiment performed by Tinkelpaugh (1928) illustrates the first example of representational memory in chimpanzees. The animal first watched the experimenter bait one of two cups with a piece of banana. After a delay period, during which the chimp was not allowed to view the two cups, it was given the opportunity to look under one of the two cups. The animal's responding was well above chance even at delay intervals of 15–20 h. Now this in itself could simply reflect the running-off of an S–R habit – the chimp merely recognises the appropriate cup and reels off the appropriate motor responses. However, a further manipulation by Tinklepaugh suggested that his chimps were exhibiting more sophisticated memorial abilities than this. During one of the delay intervals he substituted a non-preferred food (lettuce) for the piece of banana. When subsequently allowed to respond, the chimp 'jumps down from the chair, rushes to the proper container, and picks it up. She extends her hand to seize the food. But her hand drops to the floor without touching it. . . . She looks around the cup and behind the board. She stands up and looks under and around her. She picks the cup up and examines it thoroughly inside and out. She has on occasions turned towards observers present in the room and shrieked at them in apparent anger' (1928, pp. 224–225) and so on. What this suggests is not the simple running-off of a motor response because the lettuce would have been consumed by the hungry chimp if this were the case. What it does suggest is that the chimp had some *expectancy* of what was under the cup; it appears she had specifically expected a piece of banana rather than just any morsel of food. Her subsequent agitated searching implies some representation of the reward in memory.

A further example of representational memory in animals is provided in studies by D'Amato and Worsham (1974; see also D'Amato, 1973). Their first set of experiments with capuchin monkeys (*Cebus apella*) utilised an experimental procedure known as delayed matching to sample (DMTS). In this procedure the animal is presented with a sample stimulus for a brief exposure period, then after a delay interval the animal is simultaneously given two 'choice' or comparison stimuli, one of which is identical to the sample and one different. The animal's task is to respond to the 'choice' stimulus which resembles the original sample stimulus. This is an extremely useful paradigm for assessing the characteristics of short-term memory in animals, and earlier studies by D'Amato and colleagues (D'Amato, 1973) had shown that capuchin monkeys could respond quite accurately on DMTS tasks with delay intervals of up to 2 to 3 min. However, the reader will probably have realised that the DMTS task is basically a recognition task: the memory processes invoked to explain efficient performance need be no more complex than sensory traces – the animal retains a sensory trace of the sample stimulus and matches this to the appropriate comparison stimulus. To discover whether monkeys could perform successfully on a delayed matching task which required a higher level retention mechanism than simple sensory traces, D'Amato and Worsham developed the delayed conditional matching (DCM) task. This created a situation which is analogous to recall in human experiments. The subject is first shown a sample (either a red disk or a vertical line); after the delay interval two comparison stimuli are simultaneously presented (an inverted triangle and a small circle). When the red disk had appeared as the sample the correct comparison was the inverted triangle; when the vertical line was the sample, the small circle was correct. As D'Amato points out 'the stimuli at the time of choice provide no information whatever regarding the identity of the standard stimulus. Or in somewhat different terms, the comparison stimuli serve no differential retrieval function' (D'Amato, 1973). Efficient performance on this task cannot be accounted for in terms of the matching of a stimulus trace to an identical comparison stimulus and the fact that D'Amato and Worsham's capuchin monkeys performed as accurately on this task as on the conventional DMTS task suggests that these animals do indeed have representational memory of a kind. The DCM task clearly requires the animal not only to respond on the basis of a stimulus that is not present, but also to use this information in conjunction with information presented later in order to respond correctly.

A further series of experiments performed by Flagg (1975) illustrates how rhesus monkeys can hold stimulus information in some kind of representational memory. Figure 10.1 describes the basic procedure and rationale for the experiment and presents the important results. The 16-cell matrix was presented to the monkey and a row would be immediately illuminated with one of 4 possible colours of light (figure 10.1(i)). One second later, one of the cells in the coloured row would be illuminated with white-light for half a second (figure 10.1(ii)). The row of coloured lights remained on for one second

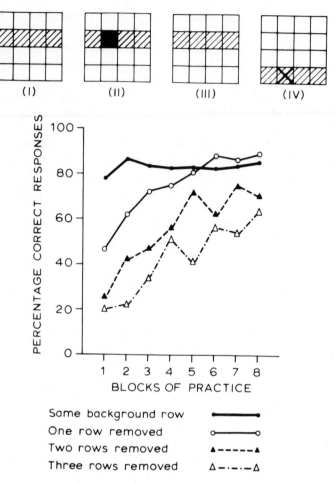

Figure 10.1 Stimulus display conditions and results of Flagg's (1975) transformation of memory experiment. One line of the 16-cell matrix is illuminated with coloured light (i): One of the cells on this line is briefly illuminated with white light (ii) and then reilluminated with the original background colour (iii); Finally another line is illuminated in the same colour as the original target line (iv) and the correct response is to the cell marked X. The results show the percentage correct responses as a function of practice when transformations of 0, 1, 2 and 3 rows are required. (After Ruggiero and Flagg, 1976).

after the white stimulus cell had been illuminated, (figure 10.1(iii)). After a short delay either the same row or a different row was illuminated by the same colour as the original background (figure 10.1(iv)). The monkey's task was to press the cell on the newly illuminated row which corresponded to the cell on the originally illuminated sample row. Figure 10.1 also shows the results of this task, analysed in terms of displacement of new background from original background (1 row, 2 rows or 3 rows). When there was no displacement (that is, the original and new rows were identical) the subjects performed with great accuracy. The other three conditions require him to perform an operation on the stimulus information in order to arrive at the correct response. The results show that the subjects were able to accomplish this, given practice, and operate efficiently with a delay interval of up to 12 s.

What these examples and a number of others purport to show (Ruggiero and Flagg, 1976, pp. 3–19) is that certain animals seem to retain representations of stimulus information in memory, and, more specifically, to abstract and retain information from these stimuli which they can apply to later problems. That is, they appear not only to be able to respond appropriately in the absence of the stimulus itself, but they can also retain and utilise information which is not directly related to the gross physical properties of previous events (for example, 'learning sets' – see p. 309, delayed conditional matching; the displacement matching experiments of Flagg). Now, if animals can abstract and retain useful information in this way, can they organise their memories in such a way as to accumulate a generalised body of knowledge? The examples we have discussed suggest that animals can make a start at this, but the extent of their abilities to abstract and apply memorial representations is yet to be assessed. Although we might never be able to know whether Seneca's horse is capable of reminiscing the day's events in his stable, we are beginning to realise that some animals can do more than just recognise and react to familiar stimuli.

# SOME CHARACTERISTICS OF ANIMAL MEMORY PROCESSES

So far we have talked in fairly general terms about the possibilities of different types of memory in animals. But to get down to the core of the problem, what hard facts are known about memory processes in animals? Already a number of phenomena have received intensive study and have relevance to the theories of forgetting that we shall discuss shortly. In the meantime, this section will deal with the facts of some of these phenomena and the methodological problems associated with them.

## Methodological considerations

The most important experimental paradigms in animal memory can be divided into two groups; (1) those which attempt to test long-term retention of information and learned responses; and (2) those which test the capacity of short-term retention on a trial-by-trial basis. The former usually involves training an animal to a criterion of efficient responding and then, after a lengthy retention interval (weeks, months, even years), again retraining the animal (Gleitman, 1971). Retention of learned responding is indicated in a 'savings score' – if the amount of practice that is needed to reach criterion during retraining is significantly less than that originally required to reach criterion, then the animal has retained some information about the task over the delay interval. Using this type of paradigm, Skinner (1950) has reported the retention of a discrimination by a pigeon with a lay-off of several years.

The studies of short-term memory in animals are more varied in their format, but one particularly popular procedure is the delayed matching to sample (DMTS) procedure we discussed earlier (also p. 285ff.). This is ideally suited to studies of decay and interference in short-term memory (D'Amato, 1973; Roberts and Grant, 1976); topics we shall discuss shortly. (The 'long-term'–'short-term' distinction is considered to be more operational than theoretical. The argument over one or two types of memory is not yet an appropriate one in animal memory.)

Perhaps the biggest problems arise, however, when considering how one should *measure* retention. In a single retention task some measures will show evidence of retention while others will show none. For example, in studies of long-term retention of choice-responding (for example, left–right or GO–NOGO discriminations) forgetting is often hard to find if the dependent variable is the number of correct choices (Chiszar and Spear, 1968; Gleitman and Jung, 1963; Maier and Gleitman, 1967). However, when latency measures are taken, such as running speed in the T-maze, significant decrements in performance are obtained over time (Hill, Cotton, Spear and Duncan, 1969). Thus, forgetting may be reflected more accurately in some measures than others. Winograd illustrates how forgetting might selectively influence particular variables

> Let us say that I have returned to my boyhood neighbourhood after an absence of many years and am assigned the task of retracing my daily walk from home to elementary school. The omnipotent psychologist testing me records my speed of locomotion as well as any errors I commit at the several streetcorner choice points (fortunately, he has such data for the last time I performed this routine years ago). The data would probably show that, while I made no errors at the choice points, I walked significantly slower while exhibiting much orienting behaviour. Does this change in behaviour represent a loss of stored information? Again, certain changes in behaviour correlated with the passage of time away from the situation seem to be reasonable and even adaptive.
>
> (Winograd, 1971, p. 268)

Memory effects are fraught with this problem of measurement, probably more so than most behavioural phenomena. On a related point, the method of testing retention is also critical. For example, after a delay period retention can be assessed by relearning (re-exposure to the original reinforcement contingencies), extinction (performance may be indexed only on the first trial prior to presenting any reinforcement contingencies, or reinforcement contingencies can be omitted altogether), or transfer (assessing how the animal can transfer the knowledge he has acquired to new reinforcement contingencies). One testing method may be more conducive to recall than others. For example, Spear (1973) has pointed out that aspects of the reinforcement condition (for example, the consummatory and ingestive properties of the reinforcer) are very often potent retrieval cues. Thus, on this basis one might expect to index better retention when the reinforcement contingencies are presented during testing (for example, relearning) than when they are not (for example, during extinction).

## Some memory phenomena

### Delayed matching to sample (DMTS)

We have already touched upon the usefulness of the DMTS paradigm for studying short-term memory in animals and the use of this paradigm has spotlighted a number of interesting memory phenomena. The procedure is illustrated schematically in figure 10.2 and it is a method which has been used to study short-term memory in pigeons (Grant and Roberts, 1973; Shimp and Moffit, 1974; Zentall, 1973), rats (Roberts, 1972, 1974), and monkeys (D'Amato, 1973; Jarvik, Goldfarb and Carley, 1969; Moise, 1970).

First, D'Amato and O'Neill (1971) found that DMTS recall in capuchin monkeys was highly efficient with delay intervals ranging from 16 to 120 s. Moreover they found that recall was facilitated when the delay interval was spent in darkness rather than in moderate illumination. Further studies suggested that this facilitation was directly due to *reduction* rather than change in delay interval illumination. This being so, it might be suggested that darkness during the retention interval leads to less motor activity and hence to better preservation of the memory trace. Studies by Etkin (1972) and Jarrad and Moise (1970) discount this possibility. For example, Etkin found his animals were *more* active during dark than during illuminated retention intervals. So, by what mechanisms does darkness facilitate retention? It may be that a period of darkness provides much more opportunity for the consolidation of a memory trace than does a period of illumination, during which the organism is busy processing visual information (a simplified form of this is that darkness may allow the survival of the after-image of the sample whereas an illuminated delay interval would not). This account suggests that illumination introduced at the *beginning* of the delay interval (when the memory trace is at the critical stage of consolidation), should have more damaging effects than illumination presented at the end of the interval. Etkin

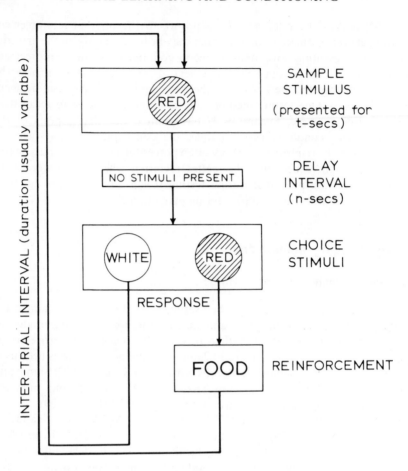

Figure 10.2    Schematic representation of the delayed matching to sample (DMTS) paradigm. See text for further explanation.

(1972) carried out an experiment which assessed the effects of introducing light probes at various periods into a dark delay interval. Figure 10.3 shows that it is the duration of the light probe rather than its location in the delay interval that influences recall decrement. So this evidence weighs against the interpretation that illuminated delay intervals damage recall by disrupting any consolidation process or by 'smudging' or 'deleting' any rapidly decaying sensory after-images. At this point we must leave discussion of the actual mechanism underlying the facilitation of recall by darkened delay intervals – simply because there is little further empirical evidence on this subject to point us in any one direction. One further point which is of interest here, however, is the effect of sample duration on DMTS recall. If one assumes that longer

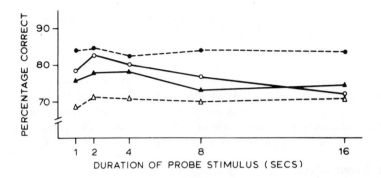

Figure 10.3   Delayed matching to sample performance as a function of the duration of a houselight-on-period and its location in the 18-s delay interval. This was either at the beginning (unfilled circles) or end (filled triangles) of the interval. Filled circles and unfilled triangles indicate conditions where the houselight was off and on, respectively, during the delay interval. (After Etkin, 1972).

sample durations will allow the reception of more features of the sample stimulus than can be attended to at brief exposures, then there should be a direct relationship between sample duration and accuracy of recall. D'Amato and Worsham (1972) carried out such a study on capuchin monkeys. However, to their surprise, they found that sample duration had no effect on recall accuracy (using sample durations of 0.075, 0.10 and 0.15 s); all of their subjects showed a remarkable ability to retain information over long periods even when it was presented for the briefest of intervals. Since this finding was contrary to the effect of sample duration found with human recall, D'Amato (1973) suggests two possible reasons for the negative findings with monkeys. First, he points out that his subjects were chronically overtrained in DMTS tasks; thus, their familiarity with the task and the stimuli involved may have produced a 'ceiling' effect long before they were tested with sample duration. Secondly, it may be that sample duration only has relevance when the subject has the ability to code information verbally – if longer sample durations allow the subject to process a large number of the sample's attributes he must have some method of coding and storing this information. Coding this material verbally is certainly helpful and may explain why sample duration is an effective variable with human subjects, and not with non-human primates. However, although verbal coding of attributes may be a useful strategy for human subjects, it does not imply that sample duration might not influence recall in other ways. For instance, the eminently non-linguistic pigeon is a subject whose DMTS recall is facilitated by longer sample durations (Roberts, 1972; Roberts and Grant, 1974). The problem we have encountered here is one that constantly crops up in memory studies: the effect of a simple variable such as sample duration does not necessarily tap an invariant attribute of memory. For instance, different animals may adopt different strategies of recall depending on their ability to code information. Verbal organisms, such as man, may verbally encode attributes of the stimulus whereas more 'lowly' organisms, such as the pigeon, may have to rely on a rapidly decaying sensory trace. Those non-human primates, who appear to exhibit some kind of representational memory, may have entirely different strategies; these may involve neither simple trace decay nor symbolic encoding of attributes, or alternatively may involve even a combination of both. Given differential storage strategies a variable such as sample duration is unlikely to have a constant effect across species, or even across different recall tasks.

### The 'Kamin effect'

The 'Kamin effect' is a phenomenon usually found with the retention of aversively motivated behaviour (Kamin, 1957b). It is a V-shaped function with peak retention at 0 and 24 h after acquisition and a low retention between 1 and 6 hours after acquisition, (see figure 10.4). Thus, retention deficit is greatest if it is measured between 1 and 6 h after acquisition, if tested later retention appears to exhibit a 'spontaneous recovery'. Traditionally this

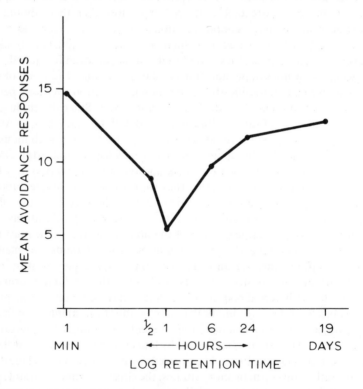

Figure 10.4  The 'Kamin' effect. Avoidance performance as a function of the delay between training and subsequent testing. (After Kamin, 1957b).

effect has been interpreted as reflecting motivational rather than memorial processes (see Brush, 1971, for a review), however recent studies by Spear and his colleagues suggest that the 'Kamin effect' may indeed reflect features of memory reactivation (Spear, 1973, pp. 174–176). First, Klein and Spear (1970) employed a negative transfer procedure: they trained rats on an active avoidance task which was followed at various intervals by training on a passive avoidance task (or vice versa for other subjects). They found that retraining on a passive avoidance task either 0 or 24 h after training on an active avoidance task resulted in massive negative transfer – the subjects had great difficulty learning the new passive avoidance task. However, at intermediate retraining intervals (corresponding with the minima on the Kamin V-shaped function) there appeared to be only minor deleterious effects of the previous learning and the new passive avoidance task was learnt relatively quickly. These results imply two things: (1) non-specific motivational levels could not be low at intermediate intervals because the animal readily learnt a new avoidance response. It is therefore unlikely that the low in the Kamin function results from a motivational deficiency. (2) The negative transfer observed at relatively short and long relearning intervals suggests that recall of the active avoidance response was interfering with acquisition of the passive avoidance task at these intervals. What has to be accounted for now is why the animal has trouble remembering the first avoidance response at retention intervals of 1–6 h after acquisition. Klein (1972) has pursued this question and obtained data which suggest that internal physiological factors may act as 'retrieval cues' for the learned response and that at intermediate intervals of 1–6 h after learning these internal cues become inaccessible. More specifically it is known that for a few hours after a stressful experience the immediate release of adrenocorticotropic hormone (ACTH) is actively inhibited. Experiments conducted by Klein, although not positively identifying ACTH as a retrieval cue, suggest that it may serve as such.

**Warm-up**

'Warm-up' is a behavioural phenomenon commonly encountered by animal experimenters. Essentially it is defined as 'poor performance on the initial trials of a daily training session relative to both the terminal trials of the immediately preceding session and the subsequent trials of the current session' (Spear, 1973, p. 169). Again this is a phenomenon which until recently had been interpreted in motivational terms and there is some evidence that warm-up may be attributed to relatively low motivation early in a daily session (Hoffman et al., 1961), but neither sensory adaptation (Azrin et al., 1963) nor muscular practice (Powell and Peck, 1969) seem to be important determinants of warm-up decrement. More recently, however, it has been suggested that warm-up decrement can more readily be associated with memorial processes than with motivation or the practising of the motor activity necessary for execution of the response. Spear (1973) suggests that warm-up decrement occurs because a 'sufficient number of retrieval cues have been absent or not noticed during the early stages of a training session' (p. 170). There is some evidence to support this interpretation. First, Lyon and Ozolins (1970) found that the presentation of 50 inescapable shocks prior to each avoidance session was ineffective in alleviating warm-up decrement: the assumption is that if warm-up is purely a motivational decrement then the shocks prior to each session should suitably remotivate the animal before he re-enters the experimental chamber. A study by Spear, Gordon and Martin (1973) further supports the memorial interpretation of Spear (1973). Twenty-four hours after learning an active avoidance response each rat was given training on a conflicting passive avoidance response. At the beginning of passive avoidance training a number of 'reactivation treatments' were presented to the animals (in the form of pre-test exposure to either the UCS alone or to pairing of the CS and UCS); if these enhanced retrieval of prior learning (the active avoidance response) then negative transfer should increase; but if only the level of motivation accompanying aversive conditioning were enhanced then passive avoidance learning should be facilitated. The results were clearly in favour of the former proposition.

**State-dependent learning**

Decrements in learned performance are often observed after a shift in the subject's physiological state. If a response is learnt under the influence of a particular drug, its subsequent performance when the animal is not under the influence of the drug will often be inefficient; it is as though the drug acts as a discriminative stimulus for the learned response ( Schuster and Balster, 1977; Omello and Stolerman, 1977). Similarly, administration of a drug which changes the animal's physiological state immediately after acquisition of a learned response will often cause a decrement in performance, although in simple physiological terms it should not prevent the animal from executing the response. Both of these instances provide examples of what is called state-dependent learning (Overton, 1964, 1966; Pusakulich and Nielson, 1976) and the latter example, when decrement in performance is almost total, defines what is known as an 'amnestic treatment'. On such occasions the drug is often labelled an 'amnestic agent'. One example of such a drug is the protein inhibitor cyclohexamine whose biochemical effects were originally thought to inhibit learning and thus prohibit the laying down of memory traces (Agranoff 1967; Agranoff, Davies and Brink, 1966). However, recent evidence has suggested that cyclohexamine and other similar acting drugs do not have truly amnestic effects. For instance, spontaneous recovery of the 'forgotten' learned behaviour has on some occasions been observed at varying time intervals after cyclohexamine administration (Quartermain and McEwen, 1970; Serota, Roberts and Flexner, 1972). To the extent that important amnestic agents such as cyclohexamine, potassium chloride and electroconvulsive shock (ECS), have been observed to have transitory amnestic effects, it suggests that they may impair memory retrieval rather than have direct deleterious effects on the memory trace itself. Although the evidence here is still fragmentary, Spear suggests one plausible account of the effects of these agents.

> The consequences of certain amnestic agents, such as potassium chloride and antibiotics, may persist in modifying the animal's internal state long after their administration. To the extent that the resulting internal state is noticeably distinct from the state represented as one or more memory attributes during original learning, retrieval of the target memory will be retarded. Alternatively, the retrieval process itself may be temporarily impaired at a hormonal, neurotransmitter, or molecular level by these agents. But as these ancillary consequences of the 'amnestic' agents dissipate with time the probability of retrieving the target memory will increase and 'spontaneous recovery' will be said to occur.
>
> (Spear, 1973, pp. 177–178)

So, although there is not enough evidence yet to enable us to make firm conclusions about the actions of amnestic agents, they may well reflect instances of state-dependent learning and failure of memory retrieval, rather than permanent amnesia.

# FORGETTING

Having talked about some of the characteristics of memory in animals we should now turn our attention to why animals forget. Forgetting in animals has often been thought of as inevitable and hence has received little attention until recently. But in itself forgetting provides an interesting enigma: if an animal can remember something for 2 min, why not 20 min, or 20 h, or even a lifetime (D'Amato, 1973)? Thus, the simple but important question for us to ask is 'what accounts for this sometimes sudden, sometimes gradual, information loss?' The terminology and theories of human forgetting have been readily applied to this problem with – as we shall see – varying degrees of success. But what is interesting is that some of the characteristics of animal memory differ quite markedly from those of human memory. This is something we shall come across quite frequently in this section. Before progressing to discuss accounts of animal forgetting in detail it must be pointed out that theorists have in general been unwilling to make a theoretical distinction between short-term and long-term memory in animals; when these terms are used they usually refer to procedural differences rather than hypothetical storage systems. This being the case, accounts of forgetting in animals have tended either to be very general (that is, attempting to account for as many memory phenomena as possible), or particularly specific (that is, attempting to account for forgetting in one experimental paradigm such as DMTS). Thus, different theories need not necessarily be incompatible, and perhaps rather than thinking of each of the following as integrated and comprehensive theories of forgetting it might be better to conceive of them simply as factors which might contribute to forgetting.

## Forgetting as retrieval failure

One way to view forgetting is to conceive of it not as the loss of memory traces from storage but as a failure to retrieve information from storage. Retrieval failure may occur because the necessary retrieval cues are absent. This is a view of animal forgetting which has been championed by Spear (see Spear, 1973; Spear and Parsons, 1976 for reviews) and he has attempted to interpret a number of animal learning phenomena in terms of it. We have already discussed warm-up, the 'Kamin-effect' and state-dependent learning, all of which can be considered in terms of retrieval failure. In the case of warm-up decrement it is suggested that it occurs when a sufficient number of retrieval cues have been absent or not noticed during the early stages of a training session, (Spear, 1973, p. 170); and both the 'Kamin effect' and state-dependent learning are postulated as reflecting the retrieval cue function of internal physiological states – in these cases, performance decrement is said to result from the absence of the physiological states which accompanied training. Spear further emphasises the importance of the learning 'context' for memory retrieval. Those cues which can function to reactivate memories of learned responding need not be crucial aspects of the original learning situation. They can be as unimportant as the level of ambient noise in the conditioning chamber, the colour of the houselight or, as we have just noted, the momentary physiological state of the animal. The assumption which stems from this account of forgetting is that retention decrement can be alleviated by presenting the appropriate retrieval cues – this is what Spear calls a *reactivation* treatment (Spear, 1973, p. 168). Of course, it is not obvious what aspects of the experimental set-up are going to become effective retrieval cues, but re-exposure to either the CS or UCS does appear to alleviate retention decrement. For example, brief exposure to electric shocks prior to aversive reconditioning has been shown to reduce the magnitude of warm-up decrement (Hoffman *et al.*, 1961; Hake and Azrin, 1965) and when appetitive conditioning is tested, performance decrements which have been caused by extinction or counterconditioning disappear quite rapidly if the reinforcer associated with original learning is presented (Spear, 1967, pp. 226–232). Although CSs and UCSs might intuitively be assumed to become the most potent retrieval cues, more subtle aspects of the learning situation such as reinforcement· schedules (Gonzalez, Fernhoff and David, 1973) have been shown to alleviate retention decrement. During this discussion of contextual stimuli as retrieval cues the reader may have noted one paradox: presenting contextual cues, discriminative stimuli or conditioned stimuli, without their associated reinforcers or UCSs defines extinction, yet this account of forgetting suggests that just such presentation should actually *enhance* responding. However, there is recent evidence to suggest that limited exposure to a CS alone can actually increase response probability (Miller and Levis, 1971) whereas prolonged exposure does in fact result in subsequent extinction of the response. Further elucidation of the temporal limits of these two effects of unpaired CS presentation should shed further light on the validity of the 'retrieval failure' account of forgetting in animals.

## Trace decay

Trace decay theory of forgetting simply states that the memory trace deteriorates with time (Broadbent, 1958; Brown, 1958). This is a fairly straightforward and simple proposition but it has often been criticised on the grounds that without any physiological evidence to back it up it is little more than a restatement of the facts of forgetting. However, there are two possible approaches which would give trace decay theory some kind of validity. The first is to make some assumptions about the biochemical processes underlying trace decay and test these empirically. The second is to make some assumptions about the factors which determine the strength and rate of decay of the memory trace, and test these empirically. An example of the first approach is provided by Gleitman:

> If the decay process is viewed as part and parcel of the normal biological life functioning of the organism (e.g. somehow related to metabolic mechanisms), then whatever speeds up these overall normal patterns (of which the hypothesized decay process is somehow a part) must necessarily increase forgetting; whatever slows them down, will slow down memory loss.
>
> (Gleitman, 1971, p. 37)

Since metabolic rate is a function of temperature, a number of studies have looked at the effect of retention interval temperature on recall – the assumption being that higher temperatures should decrease retention because the increased metabolic rate it entails should increase the rate of decay of the memory trace. Studies by French (1942) on fish and Alloway (1969) on grain beetles (*Tenebrio molitor*) have been consistent with this account (although it must be pointed out that alternative interpretations of their results are possible). A slightly different approach is to assume that rate of decay of the memory trace is directly related to the amount of overall neural activity. Studies by Rensch and Dücker (1966) and Dücker and Rensch (1968) found that goldfish retained a visual discrimination better if treated with chlorpromazine or kept in darkness. Both treatments were assumed to reduce the overall excitatory state of the relevant modality system and thus help to preserve the memory trace.

Adoption of the second approach has been made by Roberts and Grant (1974, 1976) in an attempt to explain aspects of DMTS performance in pigeons. In constructing a model of short-term memory they asserted that successive stimulus events form separate and independent memory traces which do not interfere or compete with each other. The strength of a trace is assumed to be a negatively accelerated function of sample duration time, and decay of the trace, a negatively accelerated function of the decay period. Without going into the complex details of their experiments (see Roberts and Grant, 1976, pp. 96–110) this account had a relatively high level of success in accounting for the effects on DMTS recall of such variables as presentation time, delay and spacing of sample stimuli.

Given these attempts to validate the trace decay theory of forgetting in animals it must still be remembered that alternative explanations are far from ruled out. For example, interference effects from retentional interval learning cannot readily be ruled out of any study if the animal is active, or even merely conscious, during the retention interval. So, although on the face of it trace decay theory can account for many of the facts of forgetting, its validation as a viable mechanism underlying forgetting must depend not only on more specific evidence for its biochemical under-pinnings, but also on experimental techniques which can adequately rule out interference effects or alternative explanations.

## Interference

Since animals are continually processing information, both before and after a learning task, it is not unreasonable to assume that this processing might interfere either with established memory traces or with the laying down of future memory traces. There are two major types of interference effects which have been postulated to contribute to forgetting: retroactive inhibition (RI) and proactive inhibition (PI). RI entails the forgetting of responses by the subsequent acquisition of competing responses: the greater the similarity between stimuli and activities learned during the retention interval and the original learned conditions, the greater will be the deleterious effects in the subsequent retention test. Gleitman (1971, pp. 21–22) cites an experiment designed to assess the effect of similarity between training environment and retention interval environment. After learning a runway response for food, one group of rats was housed during the retention interval in their original home cages, while a second group was housed in completely new cages (animals in the latter group were moved back to their original cages 3 days prior to the retention test). On the basis of these manipulations interference theory would predict that the second group should perform better on the retention test than group one, since the former group were housed in conditions which should minimise interference. Unfortunately, Gleitman found no difference in test performance between the two groups.

Although studies have had difficulty finding evidence for the deleterious effects of RI, more success has been achieved in identifying the contribution to forgetting of PI. The traditional view of PI is that forgetting of a learned response is caused by interference from *previously* learned habits. That is, the animal brings a set of learned responses with him to the experimental situation; these old responses are extinguished during the course of new learning but they reappear due to spontaneous recovery and interfere with the correct response during the retention test. The most appropriate way to test such an account is to consider what responses might suitably interfere with a learning task and teach this to the animal prior to training on the task, whose retention is to be tested later on. A study by Gleitman and Jung (1963) provides an example of experimentally determined PI. Rats were first trained to press the right-lever in a Skinner-box for food, then retrained to press the left one; they were then rested for 44 days and again tested on the second discrimination. Figure 10.5 illustrates that subjects trained with two prior discriminations showed a massive retention loss compared with subjects who had not received the 'interfering' right-lever training. However, while some studies have supported Gleitman and Jung in observing the deleterious effects of PI on long-term retention (Chiszar and Spear, 1968; Maier and Gleitman, 1967), others have found no evidence of PI (Crowder, 1967; Gleitman and Steinman, 1963; Kehoe, 1963). Details of design and procedure of experiments differ quite markedly between studies, and because of this it is quite reasonable to suppose that some studies will have more success in observing PI than others. Such factors as type of learning task (for example, visual discrimination versus GO–NOGO) and distribution of practice (Spear, 1971, pp. 68–78) are variables which have been singled out as determining the appearance of PI effects. But given that some studies have observed PI effects, how can these be explained? As we mentioned earlier, the traditional account of PI effects in animal studies is in terms of spontaneous recovery of interfering response tendencies after long retention intervals. An insightful experiment by Maier, Alloway and Gleitman (1967) addressed itself to this problem. In the study of Gleitman and Jung there are two reversals to be made on the shift from first to second discrimination – S + becomes S −, and S − becomes S +. The spontaneous recovery hypothesis would pinpoint the first reversal as being the crucial one. It is the responding to the original S + which recovers to interfere with the new discrimination. In their study, to look more

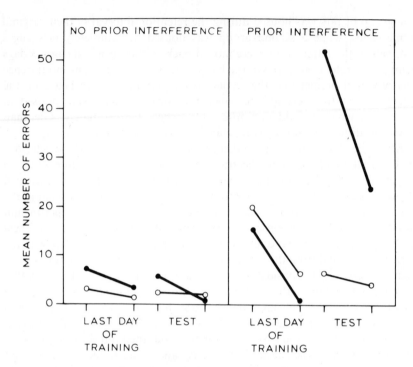

Figure 10.5 Error rate on a spatial discrimination task as a function of retention interval and proactive inhibition. Of the 'no prior interference' groups one group was tested for retention one day after criterion had been reached (unfilled circles) and the other tested 44 days after criterion (filled circles). Both of the 'prior interference' groups were trained to reverse the discrimination immediately after having achieved criterion. When the discrimination had been successfully reversed they were tested on the *original* discrimination either one day (unfilled circles) or 44 days (filled circles) later. (After Gleitman and Jung, 1963).

closely at the importance of the S + to S − shift, Maier *et al.* first trained rats on a visual discrimination S1 + v. S2 −. After this they were divided into two groups and given partial reversal training. One group was trained on S3 + v. S1 − (a plus–minus shift), while a second group were trained on S2 + v. S3 − (minus–plus shift). Half of group one and half of group two were then tested on the second discrimination the day after reaching a learning criterion, the rest were tested on this discrimination 32 days later (see table 10.1 for a summary of the procedure). If spontaneous recovery of old responses is implicated in PI effects then retention loss should be greater in the case of group one animals who had a plus–minus shift. The results, however, indicated that this was not the case – those animals given the long retention interval (32 days) showed an equal amount of retention loss whether their second discrimination had involved a plus–minus shift or a minus–plus shift.

**Table 10.1  Experiment to Test the Spontaneous Recovery Hypothesis of Proactive Inhibition Effects (Maier, Alloway and Gleitman, 1967)**

| Group | PM-1 | PM-32 | MP-1 | MP-32 |
|---|---|---|---|---|
| Discrimination 1 | Triangle (+) v. vertical stripes (−) | | | |
| Discrimination 2 | Horizontal stripes (+) v. triangle (−) | | Vertical stripes (+) v. horizontal stripes (−) | |
| Retention test on Discrimination 2 | 1 day later | 32 days later | 1 day later | 32 days later |

Hypothesis:  Group PM–32 is the treatment that should exhibit most proactive inhibition since it is the only group that experiences a change from a particular stimulus being plus in discrimination 1 to it being minus in discrimination 2. According to the spontaneous recovery account of PI the retention interval of 32 days should enable responding to that stimulus to recovery thus producing errors on the retention test.

Results:  There was no significant difference in the number of errors produced by groups PM–32 and MP–32 on the retention test.

Conclusions:  A plus-to-minus reversal does not generate any more proactive inhibition than a minus-to-plus reversal, thus suggesting that PI is not wholly explained by spontaneous recovery of previously learned responses.

This argues strongly against a spontaneous recovery interpretation of PI effects.

So what are the important factors underlying PI effects? A number of theorists suggest that factors associated with temporal discrimination may be important. For example, Gleitman (1971) proposes quite simply that 'S forgets which of the two discriminations he has encountered more recently, an effect which grows as the retention interval increases and the relative recency of the two situations become more alike' (p. 30). Similarly, Spear (1971) also proposes that animals tend to respond as they have done most recently, thus the animal's ability to discriminate which task was last in the temporal sequence of tasks he had encountered is an important determinant of remembering. This leads us on to discuss the temporal discrimination account of forgetting.

## Temporal discrimination

Since forgetting is very much a function of temporal variables, it seems reasonable to suggest that forgetting might be related to temporal discrimination in some way. For example, at relatively short retention intervals an animal can fairly readily identify which stimulus was presented last, or which response was learned last – his discrimination of short intervals should be fairly precise. At relatively long retention intervals, when the animal's discrimination of the interval becomes less accurate, he may have more trouble discriminating which of a number of stimuli were presented last (in a DMTS task) or which response was the last one to be learned (in a long-term retention study) – hence the appearance of PI and RI-type effects – which impair retention. This type of approach, pioneered by D'Amato (1973; see also D'Amato and Cox, 1976) is characterised thus:

> (It appears that) the kind of memory studied in DMTS and related animal retention tasks was more likely based on temporal discrimination processes than on the limited-capacity storage mechanisms postulated for human STM. The argument in brief is that when an animal is confronted with the sample stimulus and one or more comparison stimuli at the end of the retention interval, its task is to decide which of the various stimuli it has seen most recently. Essentially this amounts to forming a temporal discrimination between the 'time to last seen sample' for each of the choice stimuli.
>
> (D'Amato and Cox, 1976, p. 50)

The approach has a number of advantages, both conceptual and empirical: (1) It collapses two processes into one; that is, rather than thinking of memory as a capacity separate from discrimination, the two can be considered isomorphic. This being the case, the knowledge we have about discrimination processes – and temporal discrimination in particular – should be applicable to memory processes; (2) this interpretation can account for the fact that DMTS performance generally improves as the number of stimuli in the sample set increases (Herman, 1975; Worsham, 1975), that performance improves with longer inter-trial intervals (Herman, 1975), and also explains certain trial sequence effects in DMTS (Herman, 1975; Worsham, 1975); (3) species-differences in retention can be considered in terms of their temporal discriminative abilities rather than in species differences in storage capacity or retrieval processes; (4) when holding delay interval constant performance on DMTS tasks improves with practice; if this behaviour can be considered as a discrimination (albeit a temporal discrimination) this is exactly what one would expect (D'Amato, 1973).

The temporal discrimination account of performance on memory tasks is interesting in that it can, on a qualitative level, account for many facts from both short-term and long-term retention studies; it can provide a possible explanation for PI and RI effects, and is not inconsistent with some trace decay accounts of short-term memory (Roberts and Grant, 1976). Also, unlike some theories of forgetting, it does generate predictions which can be empirically tested quite easily. However, we should still emphasise that there is some evidence which does not fit readily into this framework (D'Amato and Cox, 1976), and, although failure of temporal discrimination might well contribute to poor performance on retention tasks it is quite wrong to think that it will provide a full account of forgetting. Although it might be a more fruitful way of looking at phenomena which have previously been thought of in terms of interference or trace decay, it might only be one of many contributions to forgetting.

# 11 Concept Formation Problem Solving and 'Intelligent' Behaviour in Animals

'Intelligent' is an adjective we use almost daily to describe the behaviour of other people and, in some cases, other animals. Yet it is notoriously difficult to pin down what we understand by that word. On one occasion we may mean that someone is particularly good at solving problems, on other occasions that someone has a good memory, or simply that someone is particularly good at arguing their point of view. Intelligence obviously means different things to different people, but in general most of us would probably agree that it involves the ability to abstract relevant information from a testing problem and, through processes of reasoning, to arrive at a relevant and efficient solution. Most of us at some time have attributed the adjective 'intelligent' to the behaviour of animals. It may have occurred when watching a pet cat open the cupboard where food is stored; or at the zoo, when watching the antics of playful chimpanzees. In such cases we tend to see the animal's behaviour as purposive or intentional, in that we assume the animal has 'reasoned out' what behaviour is appropriate to obtain its goal. But how far is this attributable to us merely projecting human traits onto behaviours which at a formalistic level appear intentional? Can animals reason, can they abstract relevant inform- ation from tasks that aid them in finding a solution? Darwinians of the late nineteenth century certainly thought so. As we pointed out in chapter 1, Darwin's theory of evolution opened up the search for human traits in animals. Since animals were only quantitatively different from human beings in their intelligence, then it was quite proper to assume that they possessed the ability – albeit at some lower level – to think, reason, abstract, communicate, etc. However, much of the evidence towards these ends was anecdotal, and based on casual observation, a fact which can lead to one 'seeing' human traits when in fact there may be none there. Two examples will serve to illustrate the need for caution in interpreting seemingly intelligent behaviour in animals. The first is known as the 'Clever Hans error' (Pfungst, 1911). Clever Hans was a horse who had apparently learnt to carry out complex arithmetical computations and communicate the answer by tapping his foot. A human observer would write the sum on a blackboard and wait eagerly as Clever Hans tapped out the answer. Many scientists were convinced that Clever Hans was demonstrating an ability to reason. However, closer scrutiny of the horse's behaviour revealed that he was in fact watching the behaviour of the human observer who had set the task. Since this observer also had to work out the answer to the problem he unconsciously conveyed information to Clever Hans. His computations might be accompanied by nods of the head which the horse would follow; when the observer stopped nodding, the horse stopped tapping. Both had arrived at the correct answer, the observer by arithmetic computation, Clever Hans by simply picking up and reacting to subtle cues exhibited by the observer. A behaviour that looked like reasoning in an animal was in fact produced by a much less sophisticated process. A second example shows that, although an animal may have solved a problem in a seemingly 'intelligent' way, his behaviour may be very rigid, ritualistic or inflexible to such an extent that it seems the animal is merely executing a sequence of motor responses in an 'unthinking' way. Razran (1971) relates the behaviour of a chimpanzee called Raphael in an experiment conducted by Vatsuro (1948):

In a laboratory Raphael was taught to extinguish a flame, which was barring a visible fruit in a fruit dispenser, by inserting a stick in a hole of a box to obtain a cup, filling the cup with water by turning on a faucet in a water jug, and pouring the water on the flame. At first, the jug was on a platform above the flame, but later it was placed at some distance on a separate stand. On hot summer days, Raphael was taken to Lake Ladoga, where he stayed on a float 5 meters away from another float. He quickly learned to cool himself by pouring over his body water which he obtained from the surrounding lake by means of the cup in the box. Now, the fruit dispenser was brought to the lake and placed on the float on which Raphael was staying, while the water jug was put on the other float. The flame barring the fruit was lit, and two bamboo poles were made available . . . Raphael finally solved the problem of securing the fruit in a very laborious and 'idealess' way. He obtained the cup, joined the poles, threw them to the other float, crossed over, filled the cup with water from the jug, recrossed and extinguished the flame. . . . The water-in-the-cup-obtained-from-the-lake-to-cool was not 'ideated' – combined or abstracted-with   the   water-in-the-cup-obtained-from-the-jug-to-extinguish-the-flame.

(Razran, 1971, pp. 274–275)

When you consider examples like this it illustrates the need for caution when attributing higher mental processes to animals. We cannot directly ask the animal what he is thinking or if he is reasoning, so we must find other ways of getting this information. We can, however, infer the kinds of mental processes necessary for solving particular learning tasks. Using this logic, experiments have suggested that some animals do possess the ability to form concepts, to reason at a rudimentary level, and to utilise the fundamental principles of language in order to communicate.

# CONCEPT FORMATION

To have formed a 'concept' an animal must have developed a response to a *class* of objects or events; or as Razran (1971, p. 270) puts it, it is 'the capacity to integrate aspects of different portions of the environment and not just of one particular portion'. Concepts, of course, can differ in the degree of abstraction necessary to form them. To respond to an object on the basis of 'squareness' of the object does not require as much abstraction of features as, say, responding on the basis of dissimilarity to surrounding objects ('oddity'). One of the more comprehensive early studies of concept formation in animals was carried out by Fields (1932). Using a Lashley jumping stand he attempted to teach rats the concept of 'triangularity'. They were rewarded for jumping towards pictures of triangles but were punished for jumping towards figures such as circles, rectangles, crosses, and dots. They eventually jumped towards triangles even when the size of the triangle was manipulated. However, when the triangle was tilted the discrimination broke down until further training with tilted triangles corrected this. Eventually, after still further training procedures, rats were able to respond correctly to triangles regardless of size, tilt, whether they were outline triangles or solid triangles, or whether the triangle was composed of dots. The rats could be said to have formed the concept of 'triangularity'. However, the step-by-step training procedure is fairly laborious and it seems from this study in particular that the concept has to be formed by learning to discriminate triangles from non-triangles on the basis of one feature at a time. Given adequate discrimination training a great variety of animals can be taught to respond on the basis of some abstracted feature of a stimulus array. This is certainly not a startling revelation since it should be quite obvious from everyday observation of animals that their behaviour is often determined not by specific features of environmental stimuli but by more general characteristics. For example, many wild animals, especially birds, will flee at the sight of a human being, regardless of the size, shape or colour of the person, while they will happily remain in the presence of other animals. This example could be explained by discrimination of a single specific feature, such as odour, but it is more likely that animals have formed the 'concept' of a human being and react to this rather than to individual features. A number of studies have already demonstrated that pigeons, for example, can be trained to respond to the presence or absence of specific individuals (Herrnstein, Loveland and Cable, 1976).

Perhaps a more taxing conceptual task is to respond on the basis not of the physical characteristics of the stimuli *per se*, but on the basis of the relationship between them. One such example is an 'oddity' task. Three stimuli are presented to the animal, of which two are similar and the third differs in some important way. The animal has to learn to respond by choosing the 'odd-man-out'. Although non-human primates such as chimpanzees are best at this kind of problem (Nissen and McCulloch, 1937), it is a task which can also be mastered by rats (Wodinsky, Varley and Bitterman, 1953), and even canaries (Pastore, 1954). A similar kind of concept is that of 'middleness'. Rohles and Devine (1966, 1967) have been able to train chimpanzees to select from a row of objects the one in the middle position (that is, equal numbers of objects were on its left and right); this has been successful with as many as 17 objects simultaneously present – an ability which is roughly equivalent to that of a 4- to 6-year-old child.

Non-human primates are also able to solve what can be called 'two-dimensional' oddity problems. For example, rhesus monkeys are able to choose correctly one of three objects when one background signals they should choose on the basis of odd *form* and another background signals they should choose on the basis of odd *colour* (see figure 11.1) (Harlow, 1943). Some studies have been indicated that chimpanzees can acquire relatively sophisticated number concepts (Ferster, 1964).

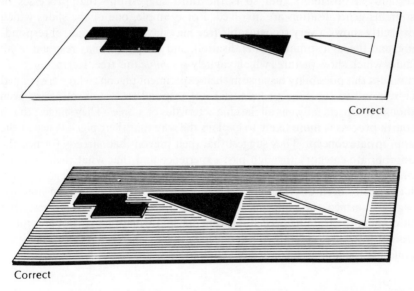

Figure 11.1 A two-dimensional oddity problem. Odd colour is correct on the white tray, odd form is correct on the dark tray. (From Maier and Maier, 1970, copyright © 1970 by Wadsworth Publishing Co. Inc. Reprinted by permission).

However, knowing what conceptual tasks animals can master does not readily indicate *how* they solve the problem, or more specifically how they develop the necessary concept. We can perhaps indicate some of the possibilities by discussing an experiment by Herrnstein *et al.*, (1976). Herrnstein *et al.* presented pigeons with slides portraying a wide variety of objects and scenes. Some pigeons received food for pecking a key only when a tree or part of a tree was in the picture, others when a body of water (for example, a puddle, river, stream, rain, etc.) was contained, and others only when a particular person was portrayed. They found that the majority of pigeons in each group learnt the appropriate concept. They suggest a number of ways in which the discrimination could be formed. First, they suggest that the pigeon may discriminate on the basis of common elements between the reinforced slides (for example, in the case of trees they may select out a feature such as green, leafy, vertical, woody, branching, etc. and respond when one of these is present). However, they point out quite rightly that 'to recognise a tree, the pigeons did not require that it be green, leafy, vertical, woody, branching and so on . . . Moreover, to be recognisable as a non-tree, a picture did not have to omit greenness, woodiness, branchiness, verticality, and so on'. Secondly, we could modify this first account by suggesting that the birds respond to *clusters* of features. For example, the green should be on the leaves, the vertical branching parts should be the woody parts, and so on. This is certainly a possibility, but so is the third suggestion – that processes of stimulus generalisation are involved. For example, one of the slides which explicitly shows a very recognisable tree may have come to control responding, and through stimulus generalisation, and the pigeon may respond to all slides which show pictures which vaguely *resemble* this tree. Herrnstein *et al.* discount this possibility because in their experiment pigeons who were trained to respond to the concept 'tree' did not respond to a stalk of celery even though it superficially has all the characteristics of a 'tree'. They suggest that a fourth process is more likely to explain the way that their pigeons learnt the appropriate concept. They suggest that their pigeons had already formed the appropriate category through past experience and that what their training procedure did was not to refine the category but to *activate* a category which had already been discriminated. This, of course, does not explain how the pigeons formed the concepts in the first place, but it does suggest that discrimination training procedures for concept formation may not give much insight into how animals *form* concepts, it may merely serve to reactivate concepts the animal has already acquired.

## REASONING AND 'INSIGHT' LEARNING

Proponents of traditional associative learning assumed that the most complex problems could be solved by animals on the basis of forming simple S–R associations. This process was known as 'trial-and-error' learning since an animal was supposed to solve each step of a problem by 'stumbling' onto the correct solution by accident; the more remote the possibility of the correct response being emitted by chance, the less likely it was that the problem would be solved. Once the correct response was emitted the appropriate S–R associations were 'stamped-in' by the rewarding consequences of the action (see chapter 2, pp. 19–21). However, this kind of analysis denied that animals possessed the capacity to 'reason out' the correct response to a problem. The possession of higher mental abilities in animals continued to be denied until the publication in 1925 of an influential book by the Gestalt psychologist Köhler. Köhler spent a number of years studying the problem solving behaviour of chimpanzees, and in his book, called 'The Mentality of Apes', he strongly suggested that chimpanzees do 'reason out' the answers to problems. Whilst observing his chimps solve fairly complex problems he noted one peculiarity of their behaviour which was common to nearly all of them: they appeared to hit on the correct solution to the problem very suddenly – one moment they would be completely baffled by the problem, the next moment they would very suddenly and spontaneously execute the correct sequence of responses. Before we look at his analysis of their problem solving behaviour let us look at some of the problems he set his chimps; (1) the chimp must arrange a number of boxes in a 'tower' and climb on the 'tower' in order to obtain a bunch of bananas suspended from the ceiling of the cage (figure 11.2); (2) to obtain a banana which is out of its reach, the chimp has to pick up a short stick in order to reach and pull towards it a second stick which is long enough to reach the banana; (3) to rake in a banana which is out of its reach the chimp must join two 'fitting-sticks' end-to-end and use this (figure 11.3); (4) to obtain a basket of fruit that is hanging from the ceiling of the cage, the chimp must untie the suspending rope which is fixed to a nearby tree. All of these problems are characterised by a number of common features. First, the goal is directly visible but not directly obtainable; secondly, the problem consists of a number of independently perceptible elements; and thirdly, each element of the problem serves a specific 'function' in the route to obtaining the goal. Most of Köhler's chimps were able to solve these problems, but far from appearing to do so in a trial-and-error fashion, Köhler suggested that they exhibited what he called *insight* learning. The important features of insight learning are: (1) the animal is able to perceive individually each element in the problem situation; (2) it becomes aware of the functional relationships existing between each element, and (3) it is able to 'shuffle around' cognitively the elements of the problem until the correct solution is arrived at. Thus, in the 'fitting-sticks' problem (no. 3), the chimp is able to perceive each element of the problem (the two sticks and the banana), and it becomes aware of the functions that the two sticks can serve (in this case to rake in the banana) and how they might be inter-related (for example, by fitting the two together). Once the chimp has become aware of these functions and inter-relationships, it can 'reason out' all the possible combinations until he comes across one that solves the problem. This is considered to be the moment of 'insight', when the chimp suddenly

Figure 11.2  A chimpanzee having successfully stacked a tower of boxes in order to obtain bananas hanging from the ceiling of the room. (From Köhler, 1973, copyright © 1973, by Routledge and Kegan Paul Ltd. Reprinted by permission).

grabs the two sticks and . . . 'quickly puts one into the other, and attains his objective with the double stick'. (Köhler, 1973, p. 128).

Insight learning, however, is only possible when all the relevant aspects of the problem are open to inspection by the animal. For example, the cats in Thorndike's puzzle boxes could never have formed an 'insightful' solution to the problem because the mechanism which operated the latch was out of the animal's view: thus, '. . . if essential portions of the experimental apparatus cannot be seen by the animals, how can they use their intelligence faculties in tackling the situation?' (Köhler 1973, p. 22). Köhler was quite convinced that when all the relevant elements of a problem were available for scrutiny, many animals could 'reason out' the solution.

Apart from chimpanzees a number of non-human primates seem capable of solving spacial-relations problems with what appears to be processes akin to what we call reasoning; these include gorillas (Riesen, Greenberg, Granston

Figure 11.3   The 'fitting-sticks' problem being attempted by one of Köhler's chimpanzees. (From Köhler, 1973, copyright © 1973, by Routledge and Kegan Paul Ltd. Reprinted by permission).

and Fantz, 1953), and gibbons (Yerkes and Yerkes, 1929). Yet how sure can we be that processes of reasoning are the vital contributors to solving a problem? Does 'insightful' behaviour in chimpanzees represent a new performance never before stereotyped in the course of previous experience? Whereas some appreciation of spatio-functional relationships does appear necessary to solve many of Köhler's problems some experiments have suggested that there may be a number of innate elements in the problem solving behaviour of his chimps. For example, Schiller (1952) gave a number of chimps of different ages sticks and boxes merely to play with. He made two interesting discoveries. First, he gave 48 chimps two 'fitting-sticks' to play with (that is, there was no problem to solve). Of these, 32 fitted them together within an hour, and of the 20 adults in the group 19 fitted them together within 5-min and repeated the performance several times. Secondly, Schiller gave his chimps a number of boxes to play with. He found that they

dragged the boxes along the floor, sat and stood on them, rolled them over, carried them carefully balanced to some preferred corner, and used them as pillows. Six of the animals actually stacked them and climbed on the tower jumping upward from the top repeatedly with arms lifted above the head and stretching towards the ceiling. For the human observer it was hard to believe that there was no food above them to be reached.

(Schiller, 1952)

This strongly suggests that a chimp's ability to solve a problem may well depend on how compatible his naturally preferred playforms are with the behaviour needed to solve the problem. It seems that during play chimps will spontaneously fit together sticks and stack boxes, so to solve a problem we might suggest that rather than reasoning out the solution 'all that is required in the simplest instance is that these activities (natural playforms) should be brought into juxtaposition to a lure and a solution will soon be hit upon' (Chance, 1960). So what even appears superficially to be quite sophisticated behaviour might contain a large trial-and-error component.

A further study by Birch (1945) found that previous experience with some of the objects in a problem solving situation aided solution of the problem. Birch gave six chimps a relatively simple problem to solve: they had to reach out and obtain a hoe with which they could rake in a morsel of food which was beyond arm's length. Only two of the subjects managed to solve the problem in an hour, and one of these did so after persistent reaching with his arm and soliciting the experimenter – this behaviour resulted in the chimp accidentally knocking the stick on to the food. After this first test the chimps were returned to their home cages for 3 days in which time they were provided with the opportunity to handle sticks in their play. All subjects were observed to chew, handle and reach out with the sticks, if only to hit another chimp. They were not used to sweep objects into reach, but were used to establish contact from a distance. On return to the problem situation after this 3-day interlude, every one of the 6 chimps was able to solve the problem without any sign of difficulty.

These studies raise some reservations about Köhler's notion of 'insight'. First, it seems there is an important innate or maturational component to problem solving behaviour. Some animals, such as chimpanzees, generate quite sophisticated forms of behaviour during their play; components of these behaviours which are relevant to a solution may well occur 'by chance' in a problem solving situation. Secondly, direct physical experience with elements of a problem appears to be an extremely important contributor to finding a solution. Chimps have to learn that a stick can act as an 'arm extension' and need to discover its physical properties by direct manipulation. Thus, to say that chimps can solve problems at a purely cognitive level of reasoning is perhaps an overstatement. Of course, a notion like 'insight' is a difficult one to verify and depends for its acceptance 'not upon demonstrable fact, but upon faith in the validity of intuitive judgement' (Thorpe, 1956). Again, we return to the problem of the human observer perceiving in animal behaviour higher mental processes when none can be objectively ratified.

Nevertheless, however difficult it might be to pin down processes of reasoning in animals, the notion has been retained in a 'weaker' form by Harlow (1949). He suggests that rather than being different explanations of one phenomenon, trial-and-error learning and insight learning are but two different phases of one long continuous process. They are not different capacities, but merely represent the orderly development of a learning and thinking process. Animals, it seems, have to 'learn to learn': in solving a problem the organism first selects from unlearned responses or previously learned habits. As his experience increases, responses that do not help in the solution drop out and useful responses become established (trial-and-error learning). After solving many problems of a certain kind, the organism develops organised patterns of responses that meet the demands of the problem, and these patterns, which are called *learning sets*, can be applied to the solution of still more complex problems. The paradigmatic example of the formation of learning sets is in reversal learning situations. An animal is taught to choose one of two stimuli until he reaches a predetermined criterion of successful discrimination. At this point the discrimination is reversed so that the previous S − now becomes the S +, and vice versa. When criterion has been achieved on this discrimination, the discrimination is reversed back again, and so on. In an experiment of this kind, an animal such as the rat typically shows a dramatic improvement in performance. It may make a number of errors in the early reversals, but as the number of reversals increases the animal learns to switch its behaviour appropriately sooner and sooner. Thus, it forms a *learning set*: the experience from previous reversals is retained and applied to later discrimination problems. Bitterman and his colleagues (1965a, b,) have extensively investigated the ability of different species to form reversal learning sets, and have found interesting differences in the phylogeny of learning. On spatial reversal problems, learning sets were formed by monkeys, rats, pigeons and turtles, but not by fish (African mouthbreeders − Bitterman *et al.*, 1958). The fish took about the same number of trials to learn each reversal; it was as if each reversal were a completely new discrimination. On visual reversal problems, however, both fish and turtles were unable to form reversal learning sets (see table 11.1 for a summary of Bitterman's results). Bitterman also investigated probability learning in animals (see chapter 3, pp. 98–99) and found differences in the way that different species coped with the problem. This list too had some kind of evolutionary continuity to it, with animals which possess relatively underdeveloped brains exhibiting 'matching' behaviour, and those with more sophisticated brain structures exhibiting 'maximising' behaviour (see p. 99). These studies are important for a number of reasons. First, they provide techniques for comparing the 'intelligence' of different species; and secondly, they tell us something about the evolution of intelligence. Certainly, we must be cautious when relating intelligence to either phylogenetic development or to the

**Table 11.1   The behaviour of five different animals in four different test situations. Their behaviour is characterised either as ratlike (the animal exhibits progressive improvement on reversal learning tasks and maximising on discrete-trial probability learning tasks), or fishlike (no improvement on successive reversals and random matching on discrete-trial probability learning tasks).**

|  | Spatial Problems | | Visual Problems | |
|  | Reversal | Probability | Reversal | Probability |
|---|---|---|---|---|
| Monkey | Rat | Rat | Rat | Rat |
| Rat | Rat | Rat | Rat | Rat |
| Pigeons | Rat | Rat | Rat | Fish |
| Turtle · | Rat | Rat | Fish | Fish |
| Fish | Fish | Fish | Fish | Fish |

(After Bitterman, 1966)

possession of sophisticated brain structures, since different species may have developed different brain structures to solve essentially similar problems – that is, they can learn similar things but in different ways. However, one conclusion that Bitterman does come to is that his '. . . studies of habit reversal and probability learning in the lower animals suggest that brain structures evolved by higher animals do not serve merely to replicate old functions and modes of intellectual adjustment but to mediate new ones" (Bitterman, 1965b, pp. 99–100). The ability to form learning sets appears to be one of these new functions.

# LANGUAGE AND COMMUNICATION

## Animal communication

A great variety of animals can communicate to one another using visual, auditory, tactile or olfactory signals. Usually these signals can only transmit the very minimum of information, but it is information that is particularly pertinent in the animals' world. For example, it can convey that the transmitter is of a particular species, sex or age, or that the transmitter is in a particular biological state, such as a readiness for fighting, fleeing or mating (Hinde, 1972; Sebeok and Ramsay, 1969). Under certain circumstances, however, quite sophisticated communication systems have been discovered in the most unlikely animals, and it is when we look at some of these enigmatic examples that we start to ask what level of 'intellectual' ability is necessary for an efficient and flexible communication system.

Perhaps one of the most dramatic examples of a sophisticated and flexible communication system in the animal world, is that of the 'dance speech' (*Tanzsprache*) of honeybees (Von Frisch, 1967, 1972, 1974). Von Frisch found that honeybees perform several communicative dances, the most important being the 'waggle dance'. This involves a figure-eight movement usually carried out inside a hive by bees crawling rapidly over the surface of the honeycomb. They are usually performed when a bee has returned from a profitable source of food. After an ingenious and extensive series of studies, Von Frisch found that features of this dance enabled the returning forager to communicate to other bees the distance and direction of the nectar source, as well as its quality. This 'language' has a great deal of flexibility, precisely transmits very sophisticated information, and allows an individual animal to be both a transmitter and a receiver. But is it a language as we understand that term? What is the ontogeny of this sophisticated system? And is it not a symbolic process which requires at least some degree of higher 'intellectual' apparatus? In answer to the last question it is certainly difficult to conceive of higher mental faculties in something like a honeybee. As Lewis Thomas points out in the case of the ant

> A solitary ant, afield, cannot be considered to have much of anything on his mind; indeed, with only a few neurones strung together by fibres, he cannot be imagined to have a mind at all, much less a thought. He is more like a ganglion on legs.
>
> (Thomas, 1974, quoted in Griffin, 1976, p. 44)

This apparent paradox – of a sophisticated communication system in a neurologically impoverished organism – raises a number of important points for consideration. First, an observed correlation between the behaviour of one organism (the 'transmitter') and the subsequent behaviour of a second conspecific (the 'receiver'), does not imply that the initial behaviour serves for communication (Gould, 1976). For example a 'transmitter' bee may bring back odours which are characteristic of the nectar site; these odours may serve both to induce a particular dance pattern in the 'transmitter', and in some way direct the 'receiver' to the food site. There is thus not a causal relationship between the behaviours of the 'transmitter' and 'receiver', they are both mediated by the odour. (Recent studies by Gould (1974, 1975, 1976) have suggested that the dances of bees seem to be directly communicative rather than spurious correlations. However, this hypothetical example still serves to illustrate the dangers of jumping from correlation to causation.) Secondly, it raises questions about the kinds of information that need to be transmitted in order for a communication system to be called truly linguistic. Many psychologists and zoologists have pointed out that animal communication is distinguished from human language by the fact that the former involves a rigid response to specific external or internal stimuli (Black, 1968). For example,

Langer (1972) interprets the waggle dance of the honeybee not as actual communication about external objects, but rather as a function of the internal state of the 'transmitter'. This being the case honeybees 'cannot tell lies' (Black, 1968) and though they use signs 'they do not know that these are signs. The whole thing belongs to the realm of conditioned reflexes' (Maritain, 1957). Bearing these points in mind it seems that organisms could quite well transmit what appears to us to be relatively sophisticated information, using relatively fundamental and unsophisticated mechanisms – the need to invoke the use of higher intellectual processes is unnecessary. However, before we can proceed much further in an analysis of language learning in animals we have to define more closely the characteristics of language.

## What is language?

Language has often been considered a faculty which is exclusively human. This is a tradition which was started by Descartes and has continued to be upheld by present day psycholinguists. But this having been said, begs a question: should we rate an animal's ability to transmit information as language-like purely on the basis of its similarity to our own communication system? Attempts have been made to teach language to apes and the success or failure of this venture is often assessed on the basis of whether the end product reflects those features of language that we consider uniquely human. That is, we often tend to define the important characteristics of 'language' as those which are exclusive to human language. The point to be emphasised here is that different species might go about learning to communicate in totally different ways to human beings. This being the case one would expect their resultant 'language' to have quite different features and structure from human language, yet to suit their needs admirably. To put this in its most radical form, it has been proposed that biologically determined mechanisms which are uniquely human determine the structure of human language (Geschwind, 1970; Lenneberg, 1967) – this could also be the case with non-human organisms. So what we mean, when we say we are trying to teach language to chimpanzees, is that we are trying to teach *human* language to chimpanzees. Thus whether we succeed or not might depend just as much on the structure of the communication system the chimp already possesses as on the higher intellectual processes that might be necessary for an organism to acquire human language. It might be as difficult for a chimp to learn the rules of honeybee language as it is for it to learn human language, even though the former need not require any 'sophisticated' mental processes.

  The implication from this discussion is that failure to teach chimps human-like language does not imply that they do not have the higher mental faculties necessary to master human language (for example, such faculties as symbolisation, displacement, abstraction, mastery of syntax, etc.). But it might be that they have evolved brain structures which necessitate learning a language in a particular fashion; this means their own language system may be structurally very different to ours, and our methods of teaching them language may be totally inappropriate.

Thus, without knowing more about the ontogeny and structure of animal communication systems, it is difficult to know what should be the important defining features of a language. Nevertheless, since human language is seen as the most versatile system in the animal world, mastery of this by an infrahuman organism would suggest that man does not possess a biological monopoly for the learning of his language system, and might also tell us something about the dynamics of human language acquisition itself. It has been this aspect of language learning in apes that has motivated most of the studies we shall discuss but before we can assess the degree of success that these studies achieved we have to outline the defining characteristics of human language. This again, is a matter of degree: one can be very stringent about defining criteria, or one can select out just one or two prominent characteristics of human language and evaluate the chimp's behaviour against these. Thorpe (1974) has probably been the most diligent in tabulating the characteristics of the human language system, in fact he catalogues 16 features which he considers to be important. Many of these features are possessed by animal communication systems but a number of important ones are not so obviously identifiable in infrahuman animals. These include: (1) *Discreteness*: the language consists of small elements, for example, words which do not functionally grade into each other; (2) *Tradition*: the meaning of symbols is transmitted by learning; (3) *Learnability*: learners of the communication system learn it from one another; (4) *Duality*: elements, such as words, are often meaningless on their own, but have meaning when combined together; (5) *Displacement*: the language is able to refer to things which are remote in either space or time; (6) *Productivity*: New messages can be communicated by previously unused combinations of the elements; (7) *Reflectiveness*: ability to communicate about the communication system itself; (8) *Prevarication*: the ability to communicate information other than that contained in the immediate internal or external environment – in fact this feature is caricatured in the ability to tell a lie. These, then are the eight criteria to bear in mind when assessing the results of language learning in animals.

## Language learning in apes

There is a striking dissimilarity in the results of early attempts to teach language to chimpanzees, all of which unequivocally failed, and the results of more recent studies whose success has been quite dramatic. To assess some of the reasons for this discrepancy it is probably instructive to consider features of the chimpanzee's natural communication system.

## Natural communication in chimpanzees

In wild chimpanzees vocal signalling appears to be the least utilised mode of communication, and visual gesturing seems to predominate (Marler, 1965). When vocal signals are used they usually serve as adjuncts to communication in the visual mode. Other important facts are: (1) the majority of communications seem to be involved in maintaining social organisation and less related to the external environmental space (Smith, 1973); (2) the effect of a single signal element appears to be affected by its context in a matrix of other signals (that is, there seems to be some element of what we have called Duality – Marler, 1965); (3) Apart from gross displays of fear and aggression, the communication system seems specific to chimpanzees and has little effect on other primates; (4) There is still a lot we do not know about the communication system of chimpanzees. This includes our inability to understand the significance of certain signalling displays such as the highly ritualised 'rain dance' (Van Lawick-Goodall, 1968) and also our failure to decipher how chimpanzees are able to impart fairly complex information very quickly. For instance, Menzel (1974) and Menzel and Halperin (1975) confined a group of chimpanzees in small cages at one end of a large enclosure. One of the group was led to a piece of food which was not visible from the isolation cages, shown the food, and then returned to his own isolation cage. When subsequently the entire group was released the 'leader' was rapidly and efficiently able to lead the rest of the group to the food. On some occasions it even appeared that the 'leader' was purposely withholding the information so that he could claim the food himself. Thus the chimpanzee seemed able both to convey and withhold information from conspecifics – so that their communication system might even allow for prevarication!

## Vicki

A number of early attempts tried to teach vocal language to chimpanzees and failed (Jacobsen, Jacobsen and Yoshioka, 1932; Kellogg and Kellogg, 1933). A further attempt was made in the 1950's with a young female chimpanzee called Vicki (Hayes and Hayes, 1955). They first increased base rate of vocalisations by reinforcing any vocal emission, and then aimed to shape these vocalisations into meaningful words. Some usage of vocalisation developed – such as 'ch' as a request for a drink, 'tsk' a request for a cigarette and clicking the teeth was a request for a car ride. However, Vicki continually showed a preference for other forms of communication, and much of it was imitative and consisted of 'pantomiming' the required need (such as assuming a prone position when tired). Nevertheless, the attempt to develop vocal language failed, and the sum total of her achievements were recognisable vocalisations of 'cup' 'mama', 'papa' and the request 'up'.

## Washoe

Since chimpanzees show remarkable manual dexterity, and since their natural communications system is primarily visual, Gardner and Gardner (1969) believed that language learning in chimpanzees would be more successful if these two factors were taken into account. Bearing this in mind they acquired Washoe, a 5- to 8-month-old female chimpanzee, and attempted to teach her American Sign Language (ASL), a form of communication which primarily uses hand gestures as symbols. The training mainly consisted of moulding her hands into signs and then gradually withdrawing physical prompting. Once the sign had been acquired it was always demanded in requests. After the first 21 months of training Washoe had acquired something like 40 signs which included nouns, verbs, adjectives, adverbs and prepositions, and she was able to transfer these signs to new and appropriate contexts without much trouble or error. Sentence construction, consisting of the connecting of two or more signs also occurred spontaneously early in the training. To obtain some objective criterion of Washoe's success, deaf individuals who already used ASL were asked to asses the legibility of Washoe's language. They agreed with experienced observers on what Washoe was 'saying' in nearly 98 per cent of the cases. So in comparison with earlier studies, Washoe's training was remarkably successful; she showed vocabulary, appropriate usage and rudimentary syntactic ordering, as well as making reference to objects not directly observable (displacement). However, structural evaluation of the language is difficult because the structure of ASL is quite different from the structure and characteristics of spoken English.

## Sarah

Premack (1970, 1971a) adopted a slightly different approach to language learning in chimpanzees. He proposed a functional description of language in which he hoped a more universal semantic and logical structure could be found than existed in human language. This entailed providing the animal with a fairly limited number of symbolic objects and adopting a sequence of training procedures which would enable the chimp to interrelate the symbols. The symbols were arbitrarily shaped pieces of coloured plastic which could be arranged in a linear sequence on a magnetised slate. The symbol, although being markedly different from its reference object, meant that the chimp did not have the problems of learning to *make* the sign before learning to *use* it.

Premack used a 5-year-old chimp called Sarah in his study and in order to develop the use of a symbol, it was required that a correct coloured chip should be placed on the slate in order to obtain a piece of banana. Next she had to precede the 'food' chip by one identifying one of the two trainers, Mary or Jim. After this the word 'give' was introduced and it was requested of Sarah that she insert this word between the name of the donor and the required food. After a short period of training her communicative repertoire consisted of the following

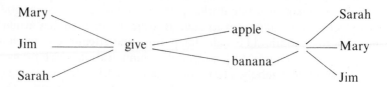

After this, further discrimination training established appropriate usage of the symbols for 'same-different'. After learning the usage of 'same-different' with just cups and spoons she was later able to transfer this concept to usage with completely new objects. Later, the interrogative '?' and 'yes' 'no' answers to it were added to her repertoire, and eventually she was able to learn to use the symbol for 'name of'. This meant she could now use the language to learn the language simply by asking her handlers the 'name of' something when she did not know it. Finally she was able to learn certain fundamentals of syntax and logic. She was quickly able to pick up the necessity of temporal ordering necessary in such phrases as 'the spoon is in the cup' and 'the apple is in the bucket'. She also mastered the use of the contingency phrase 'if-then'. For example, 'if Mary takes red then Sarah takes apple'. Sarah would observe Mary's action and then execute the consequent (Premack, 1970, 1971a, 1971b).

The extent of Sarah's achievements are quite dramatic – especially when they are compared with the early failures to teach chimpanzees a vocal language. She obviously shows some use of syntax and also exhibits the ability to master logical relations. However, she does not readily initiate communication and to be able to compare her acquisition of language with that of human children is difficult because of her strictly regimented training procedure. Nevertheless, it does demonstrate that given an appropriate medium through which to communicate, and given a logically structured training programme, chimpanzees can acquire language abilities that were once considered to be exclusively human.

### Lana

Lana is a young chimpanzee who has been taught an artificial language called 'Yerkish' (Rumbaugh, Gill and Von Glaserfeld, 1973; Rumbaugh and Gill, 1976). She has been provided in her home environment with a computer console which consists of a number of softly lit keys. On each key is a symbol which has a specific object or action referent. When keys are depressed in sequence, the sequence is displayed in order on a board (see figure 11.4). The computer controlled her food, water, access to visual stimuli through a

Figure 11.4   Lana at the computer keyboard. The overhead bar must be depressed for the keyboard to be activated. Each key has on its surface a geometric pattern (lexigram) that designates its function or meaning. Depression of a key results in a facsimile of the key's lexigram produced on a projector above the keyboard. The locations of the keys are changed frequently to ensure that keys are selected on the basis of their lexigrams, not positions. (From Rumbaugh and Gill, 1976, copyright © 1976, by the New York Academy of Sciences. Reprinted by permission).

window, and access to her handler – but only if she requested them in 'Yerkish'. In a similar way to Sarah, she was taught to request what she wanted in this artifically constructed language. Each sentence had to begin with a 'please' and terminate with a 'period' key. To evaluate her grasp of syntax she was given either incomplete sentences to finish, or she was given syntactically incorrect sentences. Her performance was extremely competent; she would quickly complete the partially presented sentences and readily erase syntactically incorrect ones. She also responded appropriately to the subtleties of word ordering; for example, her handler would punch in either the sentence 'Tim groom Lana?' or 'Lana groom Tim?'. If it were the former, Lana would immediately present herself for grooming; if the latter, she would commence to groom Tim's eyelashes.

### Conclusions

So, how do the linguistic achievements of *Pan troglodytes*, the chimpanzee, compare with the original list of characteristics we considered to be exclusive to human language? Whatever we may say, some rudimentary features of human language have been mastered. As Smith comments:

> In common with humans, these primates have shown semantic competence
> and syntax in the sense used by Chomsky (1967) and to have mastered
> language functions as described by Pribram (1971), within a system as
> required by Lenneberg (1969). Certainly a behaviourist would agree that a
> class of verbal (read 'manual') behaviours can be discriminated and
> measured. The transfer capacities of Sarah and Washoe would warm the
> heart of an investigator in cognition and concept learning. Sarah's
> performance in mastering class concepts does not compare that un-
> favourably with that of college sophomores in concept learning experi-
> ments, and the material conditional is not an easily learned language
> function for children . . . Questions unanswerable in one case are in the
> other. What Sarah lacks in spontaneity is provided by Washoe, what
> Washoe lacks in precision and control is provided by Sarah.
>
> (Smith, 1973, p. 39)

Considered together, Washoe, Sarah and Lana appear in their communications to have exhibited aspects of discreteness, duality, displacement,
productivity and reflectiveness; and although they have yet to show the ability
to map temporal space and express states of internally perceived feeling, one
feels that this must only be a matter of time and training. More recent studies
suggest that the characteristics of tradition, learnability and prevarication
may also become features of an artificial chimpanzee language. For instance,
Fouts (1973, 1975) and Fouts and Rigby (1976) have demonstrated that
chimpanzees can communicate to one another using a sign language they have
been taught by human handlers. If this is the case the time might not be far
away when mother chimps may be teaching their offspring ASL (thus fulfilling
the criteria of tradition and learnability).

## DO ANIMALS THINK?

If chimpanzees can overtly use language surely they can 'talk' to themselves?
Thinking has essentially two basic features: (1) the ability to symbolise, reason
and – at the risk of caricaturing the process – covertly 'talk to oneself' about a
problem, and (2) the ability to be 'consciously aware' of one's behaviour and
the reasons for one's behaviour (intentions). Do animals possess these
abilities? Early theorists dismissed the question on a number of different
grounds. The behaviourists insisted that thinking was not a suitable subject
matter for psychologists – unless it could be measured as, for example,
inaudible muscle movements in the larynx (Watson, 1925). However, Pavlov
(1933) believed that the ability to symbolise what was provided by language
was a necessary prerequisite for thinking; since animals did not possess any

obvious signs of language it was therefore assumed that they could not think. This type of approach is reflected in the writings of Razran:

(1) Concepts obviously cannot be communicated without the aid of behavioural symbols. (2) While higher animals communicate in some way with each other, the communication can hardly be expected to include concepts which they learn only with difficulty. (Animal experimenters would surely be sore-stymied if their cats, monkeys, or chimpanzees passed on their solutions of oddity problems – and "learning sets" – to kith and kin). (3) Animal thinking in images as symbols is not likely because of its tenuity – even human exclusive nonverbal sensory thinking is not, contrary to Freudian vocabulary, established – nor is it significant, because of its obvious abortiveness: lodged in the individual and dying with him, with no group interaction or "social heritage".

(Razran, 1971, p. 273)

The tradition in animal behaviour has always been to ignore or deny the existence of thinking in animals. This has not only been because of technical reasons – how do we measure it? – but also on philosophical grounds. Lloyd Morgan's canon continually implores us to accept as the best explanation of a behaviour that which stands lowest on the psychological scale. Yet, as Adams (1928) points out, there are logical weaknesses even with this fundamental tenet. What definitive or objective criteria do we use for assigning a psychical faculty to a place which is high or low on this psychological scale? The allocation is done purely on intuitive grounds (Griffin, 1976). And, indeed, our readiness to accept higher mental faculties, such as thinking, in animals depends upon the current climate of opinion in the behavioural sciences. We can intuitively classify types of 'mental experiences' on a scale from acceptable to taboo – where we select our cut-off line will depend on whether we are a behaviourist, a positivist, a phenomenologist, a cognitivist, and so on.

OK      PATTERN RECOGNITION
        NEURAL TEMPLATE
        SOLLWERT
        SEARCH IMAGE
        AFFECT
        SPONTANEITY
        EXPECTANCY
        COVERT VERBAL BEHAVIOUR
        INTERNAL IMAGE
        CONCEPT
        UNDERSTANDING
        INTENTION
        FEELING

AWARENESS
MENTAL EXPERIENCE
MIND (MENTAL)
THOUGHT
CHOICE
FREE WILL
TABOO    CONSCIOUSNESS

(after Griffin, 1976, p. 58)

Of course, *believing* that animals 'think' is not the same as *knowing* that animals think; this is as much a philosophical problem as a practical one, and I do not intend to get embroiled in the former here. Nevertheless, there are some points we can make about the possibility of thinking in animals. First, if we assume that symbolisation is necessary for efficient thinking, then a number of primate species appear to have the skills necessary for thinking. If chimpanzees can use and combine symbols and gestures that serve them in much the same way as words serve humans, then should we assume that they can manipulate covertly the communication systems they use overtly? Secondly, if thinking or 'awareness' is possessed by animals one might assume that it has evolved to serve some adaptive purpose. Certainly, the ability to abstract the important qualities from an event or object and recognise it, despite various distortions, is useful (Griffin, 1976); as would be the ability to 'reason out' the solution to a problem before taking action – to find a way of getting from tree A to tree B whilst still sitting safely in tree A is far better than tackling the problem in a trial-and-error fashion on the ground, with the risk of being attacked by predators.

Nevertheless, we started this chapter with a caution and perhaps we should end it with one. What looks to us like 'intelligent' behaviour may often have far simpler origins, and the same wariness should be applied to a discussion of thinking in animals: behavioural complexity by no means implies conscious awareness. However, the ability to communicate on a fundamental level with animals such as chimpanzees may only be the thin end of the wedge; this break through may well provide the technical key to a fuller analysis of the cognitive and mental abilities of animals.

# Section Three
# Conditioning in Humans

# 12 Developmental Aspects of Conditioning in Man

The human neonate comes into the world as a very vulnerable organism with a few basic reflexes in his behavioural repertoire. But from these meagre beginnings develops the human adult with all the behavioural and psychological sophistication that he will eventually possess. This process must obviously be complex but does reflect the importance of learning during development and maturation – it is certainly the case that psychological processes are primarily dependent on learning for their development and maintenance. This being the case, what mechanisms does the neonate have at his disposal in order to mould his rag-bag of reflexes into integrated patterns of behaviour, eventually being able to interact meaningfully with his environment? A lot of research has been addressed to this problem, but until recently, the picture of neonatal learning processes was far from clear. We shall see in the ensuing section that study of neonatal learning has had more than its fair share of technical and methodological difficulties, and these are now being overcome with the result that we are beginning to know more not only about learning in early childhood, but also about the nature of the basic learning mechanisms themselves.

## CLASSICAL CONDITIONING IN NEONATES

We discussed classical conditioning at some length in chapter 2 and suggested that it reflected a mechanism of learning which could be found in the widest possible variety of animals – from neurologically unsophisticated organisms to embryos in the early stages of development. Since the human neonate arrives in the world with only a handful of simple reflexes at his disposal, classical conditioning seems intuitively to be an ideal mechanism by which the child can begin to interact with his environment. But before we look for evidence of classical conditioning processes in human neonates we need to look at a number of methodological problems in studies of this kind: problems which continually plagued early investigations into neonatal learning processes.

## Methodological considerations

First, in order to be able to say anything about differences in learning capacity with differences in age it is important that performance differences cannot be attributed to age differences in such factors as motivation, potency of the reinforcer, discriminability of conditioned stimuli (sensory capacities), etc. For example, the human neonate may well have the capacity to form the associations necessary for classical conditioning to occur, but individual studies may not show this because they have used CSs which cannot be well differentiated by the child's underdeveloped sensory systems, or they may have used a reinforcer (UCS) which produces overemotionality which interferes with conditioning (Brackbill and Koltsova, 1967). It has certainly been the case that features of conditioning which were considered to reflect maturational changes in learning capacity have in fact been found to depend more on factors such as the response measure chosen (Green, 1962) and the type of CS adopted (Cornwell and Fuller, 1961). These are problems we shall return to later when we look with a more theoretical eye at classical conditioning in neonates.

A second problem is one of control procedures. Studies of classical conditioning should normally include controls for non-associative factors such as pseudoconditioning and sensitisation. Rescorla (1967) has suggested that the only real control procedure which is adequate in isolating such non-associative factors in classical conditioning is a control procedure with truly random presentation of CS and UCS. The only way it differs from the experimental procedure is that it lacks the explicit contingency between CS and UCS; thus, any unconditioned effects on behaviour that might result solely from either the CS and UCS can be detected fairly well. This type of control is important for two reasons: (1) the presentation of a particular UCS may merely increase the overall baseline rate of the associated UCR such that the UCR may even occur in the absence of the UCS; similarly, many CSs used in neonate classical conditioning studies appear to elicit defensive reactions which very often resemble the UCR in the situation (this is especially true in aversive classical conditioning; see Sameroff, 1971); (2) some of the neonatal responses which have been studied in classical conditioning experiments have a very high baseline level of spontaneous occurrence. One such example is sucking. Sucking has been studied extensively in classical and operant conditioning studies with neonates, but as Reese and Lipsitt point out, it is certainly far from ideal because 'sucking responses occur in the newborn spontaneously, in the absence of any experimentally introduced eliciting stimulus' (Reese and Lipsitt, 1973, p. 78). All of this implies that it can be very difficult to state categorically that the occurrence of a conditioned response is explicitly due to the contingency between CS and UCS. Unfortunately, a majority of studies, especially the early studies which purported to show classical conditoning in the human neonate (Marquis, 1931; Wenger, 1936), failed to utilise the necessary control procedures and so their results must be considered with some caution.

A third and final methodological problem is a more fundamental procedural one. Since many classical conditioning studies involve the sequence CS→CR→UCS, it may be that, although the contingency set up by the experimenter is between CS and UCS, the CR might be acquired and maintained through it being constantly followed by an appetitive UCS (that is, by implicit operant conditioning). Bijou and Baer (1965) have emphasised this point as a criticism of many neonate classical conditioning studies. In a study where a tone is being used as the CS, milk as the UCS and sucking is the UCR and resultant CR, the CS may well be simply a discriminative stimulus that sets the occasion for sucking to be operantly reinforced by milk. Thus, it is difficult to specify whether the response is controlled by eliciting stimuli (the antecedent CS), or by its consequences (operant reinforcement). This has been a problem with the interpretation of a large number of neonate classical conditioning studies, but it must be remembered that it is not a problem particular to neonate experiments – it is also a pertinent theoretical problem in animal studies (see chapter 7, pp. 221–222).

## Classical appetitive conditioning

Perhaps the earliest, and most cited, study which purports to demonstrate classical conditioning in the human neonate is one carried out by Marquis (1931). In her study of anticipatory sucking she used every feeding time over the first 9 days of life as her experimental sessions. Sucking was recorded by a balloon fastened under the child's chin and connected to recording apparatus. She sounded a buzzer (CS) for 5 s before presenting a nursing bottle (UCS) and found that 7 out of her 8 subjects showed evidence of conditioned sucking to the buzzer. Although these results are quite clear, and Marquis did include controls for pseudoconditioning, it is not obvious whether the conditioned response resulted from the CS–UCS contingency (classical conditioning) or whether conditioned sucking was being operantly reinforced as a result of being followed by milk. One might argue that sucking in the presence of the CS must have occurred for some reason before it could be operantly reinforced, but we have already mentioned that sucking is not a very suitable response for classical conditioning studies because of its high level of spontaneous occurrence (Reese and Lipsitt, 1973; Sameroff, 1968). Other responses which have been reported to be successfully classically conditioned are the Babkin reflex (an unconditioned mouth opening or gaping response which is elicited by pressure on the child's palm – for example, Kaye, 1965, using arm-flexion as the CS), and head-turning and eye-movement responses have been classically conditioned with children over 1 month of age (Kasatkin, Mirzoiants and Khokhitva, 1953; Koch, 1965). Classical conditioning has been reported with appetitive UCSs that have included a nipple eliciting sucking (Lipsitt and Kaye, 1964; Kaye, 1967; Marquis, 1931); vestibulation or rocking eliciting changes in respiration, heart-rate and motility (Lipsitt and Ambrose, 1967); flashing coloured lights eliciting head movements (Kasatkin et al., 1953); and adult faces which elicit head-turning (Koch, 1965).

Despite the claims of many investigators there has been much argument over the existence of classical conditioning processes in the human neonate. For instance, early Russian studies constantly implied that classical appetitive conditioning could not be established in the newborn during the first 3 weeks of life (Kasatkin and Levikova, 1935a, b; Mirzoiants, 1954), whereas American studies have suggested that classical conditioning is possible even during the first few days of life (Kaye, 1967; Lipsitt and Kaye, 1964; Lipsitt, Kaye and Bosack, 1966). Although the source of these discrepancies can often be traced to very different methodological and procedural factors, another view of these results can be taken. Recent evidence has suggested that there may well be certain biological or maturational constraints on classical conditioning in the human neonate. Before 1970 it was assumed that Pavlov's law of equipotentiality (all stimuli are potential CSs) still held. That is, if classical conditioning could not be demonstrated with one particular CS or one particular UCS, then this casts doubt upon the ability of *any* response to be classically conditioned – it casts doubt on the actual existence of a classical conditioning process. However, as we saw in chapters 5 and 6, not all stimuli appear to have equal access to individual learning mechanisms, and indeed there often appears to be a predetermination for certain kinds of stimuli to be more readily associated (see for example chapter 6, p. 204ff.). This also appears to be the case in human neonate classical conditioning. We shall discuss the theoretical implications of this a little later, but first let us look at some of the facts.

First, the CS which appears to be most potent in eliciting conditioned responding in the early neonatal period is *time*. Temporal organisation of autonomic responses (for example, body temperature, heart-rate, sleep-wake cycles) seems to be an important feature of early postnatal life (Hellbrügge, 1960) and so it seems reasonable to suggest that the neonate will be relatively sensitive to time as a CS. However, studies of temporal conditioning (see chapter 2, p. 25) in the human neonate have revealed an interesting constraint during appetitive classical conditioning: conditioned responses of the autonomic nervous system are readily acquired through temporal conditioning (Brackbill, Fitzgerald and Lintz, 1967; Fitzgerald, Lintz, Brackbill and Adams, 1967; Lipsitt and Ambrose, 1967), but motor responses have shown a firm reluctance to be conditioned in this manner (Abrahamson, Brackbill, Carpenter and Fitzgerald, 1970; Brackbill et al., 1968). Secondly, in contrast to this finding, tactile, auditory and visual CSs are more readily conditioned to somatic or motor responses than to autonomic responses (Clifton, 1974a, b;

Fitzgerald and Brackbill, 1971, 1976, pp. 357–360). What appears to be the case here is a 'preparedness' of certain classes of stimuli and responses to become associated (Seligman, 1970), and this associative selectivity may explain many of the discrepancies in the neonatal conditioning literature. Biological constraints on classical conditioning have been observed in many other animals (Shettleworth, 1972a; and chapters 5 and 6), and so it does not seem unreasonable that we should also expect to find them in the human neonate. However, having found that there is a tendency towards selective associations, the next question that theorists must ask is 'what are the possible reasons for these predeterminations?'. We shall touch on this briefly in the theoretical accounts of classical conditioning in neonates, but next let us look at the facts concerning classical aversive conditioning.

## Classical aversive conditioning

The early literature on classical aversive conditioning in human neonates is littered with unsuccessful attempts to demonstrate conditioning; this occurred either because their procedures had not included the necessary control groups, or simply because pairing a CS with an aversive UCS failed to produce a conditioned response, (Marum, 1962; Rendle–Short, 1961; Wickens and Wickens, 1940; cf. Lipsitt, 1963). In fact, the most famous example of classical aversive conditioning is one which can be interpreted in operant reinforcement terms. This is the example of 'Little Albert' (Watson and Rayner, 1920). Little Albert was a 9-month-old infant who through an explicit classical aversive conditioning procedure acquired a 'conditioned fear' of white rats, rabbits and other furry objects. Initially Little Albert would approach and pet tame white rats and rabbits (the CS); however, after the experimenters had paired his approach to these animals with a loud noise (the UCS), the rats and rabbits when subsequently shown to Little Albert also elicited crying and withdrawal (the CR) – the latter being the normal UCR to the loud noise. However, although this looks like a *prima facie* case of classical conditioning, Watson and Rayner only presented the loud noise as Little Albert went to touch the animals; thus, it is probably more like a study in operant punishment than in classical aversive conditioning.

Now, what of the bulk of studies which have failed to demonstrate classical aversive conditioning? Again, we might look towards biological factors which constrain the formation of particular associations. For instance, Brackbill and Koltsova (1967) have noted that strong aversive UCSs (1) often increase the probability of sensitisation and pseudoconditioning, and thus make the interpretation of results equivocable, and (2) they can increase levels of arousal to the point where this interferes with conditioning. In fact, it seems that aversive UCSs such as electric shock elicit a general defence reaction which competes with the conditioned response. In this sense this is a constraint of the sort mentioned by Bolles (1970; see chapter 5, pp. 192–194) in which species-specific defence reactions are readily elicited in aversive learning situations, and if they are incompatible with the response to be conditioned, then conditioning will at best be weak. A similar factor which may interfere with conditioning in young neonates is mentioned by Sameroff (1971). He suggests that in the immediate postnatal period the perceptual organisation of the child is extremely unsophisticated and is such that all of those environmental stimuli which are 'novel' or do not relate to reflex 'schemas' will elicit defence reactions. Since a CS must, before conditioning, be a relatively 'novel' stimulus, it is therefore quite likely to elicit these withdrawal or defence reactions. These defensive reactions may well interfere with the acquisition of the conditioned response – a factor which could account not only for failures of classical aversive conditioning but also classical appetitive conditioning during the first 2–3 weeks of life.

## Age differences in classical conditioning

There is little evidence in the literature for a relationship between speed of conditioned response formation (*learning capacity*) and chronological age (Fitzgerald and Brackbill, 1976, pp. 369–371; Reese and Lipsitt, 1973, pp. 94–95), but even so, it does appear that some types of response are more readily conditionable during certain age periods. For example, autonomic responses appear to be more readily conditionable at early ages than do responses of the somatic nervous system, and very young neonates (that is, under 3 weeks of age) do possess response predispositions which under many circumstances interfere with the conditioning process. This has led Fitzgerald and Brackbill (1976) to suggest that although conditionability is not directly related to chronological age, it may be related to *neurological maturation* which eventually overcomes the associative predispositions exhibited by the very young neonate.

## Theories of neonate classical conditioning

Far from attempting to construct fundamental theories of infant classical conditioning, workers in this field have primarily been preoccupied simply with demonstrating unequivocally that classical conditioning does occur in human neonates during the first few days of life. However, in recent years it has become accepted that although classical conditioning can be demonstrated in the first few days of life there are constraints on what the neonate can learn. So, as the case with animal conditioning has been, the realisation that there are constraints on learning processes has also caught up with theories of human learning.

## Cognitive 'Schemas'

Many studies show that although classical conditioning does not seem to occur readily during the first 2 weeks of life, this same classical conditioning can be fairly reliably demonstrated during the third week of life (Lipsitt, 1963; Papousek, 1967a; Polikanina, 1961; Siqueland and Lipsitt, 1966). What happens during weeks 2 and 3 to change this state of affairs? Sameroff (1971) has diligently argued through the possibilities and arrived at some tentative conclusions. One hypothesis is that the neonate is initially unable to detect changes in his environment. That is, he lacks a co-ordinated orienting reaction to environmental stimulation, and until this reaction has developed, conditioning cannot occur. Sameroff suggests that the available evidence goes against this hypothesis; neonates as young as 1 day old exhibit behavioural and cardiovascular changes to both auditory and visual stimuli (Papousek, 1967 a, b; Salapatek and Kessen, 1966; Sameroff, 1970; Wolff, 1966). So, even though such stimulation early in life may elicit defensive reactions, these studies suggest that an orienting reaction (that is, a reaction designed to prepare the organism to deal with novel stimulation – Lynn, 1966) is present in the newborn.

A second hypothesis can be constructed by outlining the steps that must occur for successful classical conditioning. First, the organism must be able to react to new stimulation and new contingencies in the environment. That the newborn does possess an orienting reaction suggests that this step is not too great a problem. The second step is for the neonate, having reacted to the new situation, to *respond differentially* to it. Sameroff suggests that

> classical conditioning is defined as the association of a previously neutral CS with the non-neutral UCS. An additional problem in newborn conditioning is that a neutral stimulus is also a new stimulus. How many newborns have had previous experience with electric shock or acetic acid vapours or even bells and buzzers? Since . . . it seems that the newborn can respond to general changes in stimulation, the first hypothesis related to his inability to be classically conditioned (i.e. that the child may lack orienting reactions) seems disconfirmed. The next hypothesis to explain his inability could be that the newborn is unable to respond differentially to the specific stimuli that have been used in studies of early classical conditioning.
>
> (Sameroff, 1971)

Sameroff points out that the newborn comes into the world with a set of built-in reflex schemas which have an optimal input. For example, the sucking schema is optimally fitted to tactual inputs such as the nipple, but can be readily adapted to different inputs in the same or related modality (for example, sucking a finger, tube, tongue, blanket etc.). Difficulties in conditioning appear to arise when one begins to depart from these schemas, such as when there is an attempt to relate two previously unrelated stimuli in different sensory modalities (for example, conditioning sucking – tactile – to a buzzer – auditory). Sameroff further suggests that failures such as these could occur for two possible reasons. First, the child may not be able to co-ordinate two separate schemas; that is, an experiment often co-ordinates artificially two different stimuli (the CS and UCS) which rarely occur together naturally in the new organism's world – it takes time for the infant to learn that stimuli from one schema (for example, auditory stimulation) can become predictive of a UCS from a very different schema (for example, tactile stimulation and sucking). Classical conditioning in neonates might be more successful if the CS and UCS are chosen from more similar modalities, thus making schematic co-ordination more easy for the infant. The second possible reason for failure to demonstrate classical conditioning is that the child may lack an independent schema for the CS. For instance, it is not clear that the newborn child initially can differentiate inputs from the auditory, visual or tactual systems other than at the peripheral level of the perceiving organ. If conditioning is to be possible, then the organism must just be able to differentiate the CS from the background stimuli in its environment, and the only way a neonate is going to be able to build up a varied repertoire of schemas is by repeated exposure to novel stimuli. This should not only increase the ability of environmental events to become effective CSs but also, as the likelihood of encountering a stimulus for which there is no schema declines, to reduce the likelihood of the CS eliciting a general defensive reaction that will interfere with conditioning.

Intuitively the logic of this account seems to make sense and it accords with much of the experimental evidence. In order to co-ordinate his differentiated perceptual response systems with other sensory-motor schemas such as sucking, the neonate must be able to differentiate the schema systems related to both the CS and UCS. For distance receptors Sameroff suggests that this development seems to take about 3 weeks – exactly the time of life when newborns begin to show reliable signs of classical conditioning.

### Biological constraints

Kasatkin was the first worker to offer any structured account of neonate classical conditioning (Brackbill, 1962; Kasatkin, 1972). He suggested that there is an inbuilt invariant order in which sensory modalities contribute effective conditioned stimuli for establishing conditioned responses; in a later paper he also affirmed that there is an inbuilt developmental sequence of effectiveness for the UCS also. In short, he suggests that the phylogenetically older sensory systems (that is vestibular, cutaneous, gustatory and olfactory modalities) are more important initial determinants of conditioning than are those sensory systems which are phylogenetically younger (visual and auditory); in similar fashion the phylogenetically older, vegetative responses can be conditioned earlier than motor responses which are phylogenetically younger. A third developmental feature of neonate classical conditioning has been added by Brackbill and Fitzgerald (1969); apart from differential rates of maturation in response and sensory systems, they suggest that stimulus–response specificity also follows a developmental pattern.

What evidence is available does support some kind of developmental account. In line with Kasatkin's analysis, responses of the autonomic nervous system do classically condition more readily in the immediate postnatal period (Fitzgerald and Brackbill, 1976) and stimuli from modalities other than the visual and auditory systems do seem to make the better CSs. Moreover there is a great deal of initial associative specificity as Brackbill and Fitzgerald point out. The predisposition of the infant to associate only certain types of stimuli with certain types of responses (see p. 326) also undergoes development until, eventually, this specificity gives way to a more eclectic process. However, these accounts are difficult to integrate into a unified theory as yet. Nevertheless, Fitzgerald and Brackbill (1976) do provide a catalogue of facts which should demand the close attention of any theorist of neonate classical conditioning: (1) Learning is constrained by stimulus–response specificity: autonomic responses are more readily linked to temporal CSs than to tactile, auditory, or visual CSs, while the opposite is true for somatic responses. Conditionability is not determined *solely* by the nature of the CS. (2) The role of the UCS is unclear; but electroshock is a particularly ineffective UCS, perhaps because it elicits defensive reactions which interfere with conditioning. (3) Younger organisms seem to require longer inter-stimulus intervals than do mature organisms for classical conditioning. (4) Elicitation of an orienting reaction facilitates conditionability in infant discrimination studies. (5) The success of conditioning can be affected by the 'biobehavioural' state of the neonate (that is, a continuum that varies from 'sleep to awake'). It is interesting to note that newborns are often tested when they are asleep or drowsy while older children are tested while they are awake (Graham and Jackson, 1970). This could in some degree have contributed to differential findings between newborns and infants over 3–4 weeks of age, especially where heart-rate has been used as the dependent variable. (6) Neither chronological age nor the nature of the UCS directly determine the outcome of conditioning. Neurological maturation, however, does appear to be correlated with conditionability.

**Summary**

So, the answer to the question 'Do human neonates classically condition?' seems to be both yes, and no. Up to about 3 weeks of age there appear to be constraints on what stimuli make effective CSs, what stimuli can be used as UCSs, and what CSs can become associated with what UCRs. At the moment these constraints appear to have been interpreted either in terms of cognitive development – conditionability is constrained during the period when the infant is organising his perceptions of the world; or in terms of neurological development – these constraints reflect the neurological immaturity at birth of the phylogenetically younger sensory and response systems. However, one fact is quite clear; apart from telling us something about the way the newborn infant adapts to his world, this research has also thrown much light on the classical conditioning process itself. Far from being an adaptive process possessing constant and universal features as Pavlov originally envisaged, it appears to be a much more complicated phenomenon subject to both developmental and biological constraints.

# OPERANT CONDITIONING IN CHILDREN

In contrast to infant studies of classical conditioning, it has proven to be much easier to demonstrate operant conditioning in human neonates. At first sight this seems to be paradoxical in the sense that one tends to consider classical conditioning as being the more fundamental adaptive mechanism and would thus expect to observe its operation more frequently in relatively underdeveloped organisms. These differences may reflect methodological differences in studying the two types of learning; but one thing the following studies do illustrate is that operant conditioning does provide the newborn with a method of modifying the simple reflexes that he is born with. It provides a way of increasing behavioural complexity that classical conditioning does not offer.

## Methodological considerations

There are two main problems for researchers into neonate operant conditioning. First is the problem of choosing a response: operant conditioning requires that a response first be emitted 'for other reasons' before it can be reinforced, and the newborn's behavioural repertoire is extremely limited. Historically it had always been considered the case that 'reflexive' responses were not readily amenable to operant conditioning (see chapter 2, pp. 53–55), and this made workers reticent to use these responses as operants in the newborn. However, since technical problems related to the recording of the infant's reflexive behaviours have been overcome, operant conditioning has been demonstrated with a number of responses, involving both responses of the autonomic nervous system and more integrated motor co-ordination. More specifically, the most popular operants have been sucking (modification of frequency and amplitude), and different aspects of head-turning or more general bodily orientation. The second problem involves the use of reinforcers. The operant conditioner cannot employ with newborn babies the strict deprivation schedules that control motivational levels in animals. However, despite this lack of true control over deprivation state, appetitive reinforcers have been used successfully in neonate operant studies. These include milk (Papousek, 1967a; Siqueland, 1964), sugar solution (Siqueland and Lipsitt, 1966), sweets (Bijou, 1957), biscuits (Weisberg and Fink, 1966), fruit (Gellermann, 1933) and sugar, jam and honey (Ling, 1941; Myers, 1908; Valentine, 1914). With slightly older infants, what are labelled as 'social reinforcers' and 'audio-visual' reinforcers are also effective. Operant responding can be maintained by such consequences as smiling (Etzel and Gewirtz, 1967), jostling and patting (Brackbill, 1958; Wahler, 1967), door-chimes (Simmons, 1964; Simmons and Lipsitt, 1961) and novel visual stimulation (Siqueland, 1966).

## Demonstrations of operant conditioning

Perhaps the most comprehensive series of studies in neonate operant conditioning comes from Papousek (1959, 1967a, b) who, using differential head-turning as the response, looked at conditioning, extinction and discrimination learning in groups of newborns, 3-month-old infants and 5-month-old infants. The procedure consisted of presenting a bell ($S^D$) for 10 s and if the infant responded with a left head turn during this time it was reinforced with milk from a nursing bottle. The results of Papousek's studies can be summarised in the following points: (1) Conditioning was reliably demonstrated in all age groups and, more specifically, was demonstrated in most subjects during the first 28 days of life; (2) changes in response frequency did appear to be due to the reinforcement contingency since the head-turning response was predictably influenced in turn by conditioning, extinction and reconditioning procedures; (3) A significant increase in speed of conditioning with increasing age was interpreted by Papousek as reflecting age differences in learning ability. However, this interpretation has since been criticised (Reese and Lipsitt, 1973, chapter 4) on the grounds that Papousek's study did not reliably control for age differences in maturation, sensory capacities and response capacities. More specifically, it seems from Papousek's observations that not all age groups were equally able to execute the required head-turning operant – the newborns rarely spontaneously emitted a left head-turn of more than 30 degrees, whereas the older age groups did so fairly readily.

Having established that operant conditioning can occur in the first few days of life, of what value is it to the newborn? A number of studies which have attempted to modify characteristics of elicited head-turning and the sucking response may throw some light on this. Siqueland and Lipsitt (1966) used a modified operant procedure in which tactile stimulation of the cheek served as an $S^D$ for head-turning which was reinforced by a 5 per cent dextrose solution. This study is interesting in that the stimulation of the cheek is in fact a UCS which elicits head-turning in the newborns; thus, it is – in Skinner's original terminology – a respondent. They found that a group of newborns who

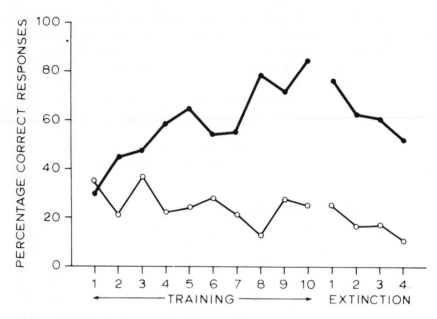

Figure 12.1   Percentage number of correct head-turns during training and extinction for the contingent reinforcement group (filled circles) and the non-contingent reinforcement group (open circles). (After Siqueland and Lipsitt, 1966).

received contingent dextrose reinforcement emitted a significantly higher number of correct head-turns than a control group who received the same number of non-contingent dextrose deliveries (see figure 12.1). These results, then, specifically suggest that components of unconditioned responses may be influenced or modified by *operant* reinforcement contingencies; this is an outcome which would not have been predicted by those theorists who stressed the independence of respondent–operant response systems (Kimble, 1961; Skinner, 1938; see chapter 2, pp. 53–55). Since this study was carried out, modification of more discrete aspects of unconditioned responses has been demonstrated using operant reinforcement. Both Sameroff (1968) and Seltzer (1968) have shown that characteristics of the sucking response in newborns can be altered using operant reinforcement. Sameroff, for instance, found that when nutritive consequences were made contingent on the 'suction' (negative pressure) and 'expression' (positive pressure) components of the sucking response, these features could be modified in accordance with the contingency. Similarly, Seltzer (1968) found that when 0.1 cm$^3$ of milk was delivered contingent upon high amplitude sucking, the frequency of these high amplitude sucks during subsequent extinction depended on the preceding schedule of reinforcement. If the infant was reinforced on an FR schedule which was greater than one, subsequent extinction of high amplitude sucking was retarded in comparison with subjects reinforced on CRF. This reflects the partial reinforcement extinction effect found in animals during operant conditioning (see chapter 2, pp. 43–45). So, using operant reinforcement, experimenters have been able to modify not only frequency, but also form, duration, amplitude and patterning of responses. These results not only cast more doubt on the traditional operant–respondent distinction, but also suggest how the new born can increase the complexity of his behavioural repertoire. As Reese and Lipsitt point out:

> Careful experimental analysis of the responses of the newborn, with the kinds of reinforcement procedures used in these studies, may allow the developmental psychologist to specify which response systems are rather strictly reflexive and resistant to modification, and which are subject to multiple control by both eliciting and reinforcing stimuli. The question of multiple control of response systems in the newborn may be closely relevant to the question of how the infant progresses from being a highly reflexive organism to an adaptive organism subject to control by reinforcing events in the environment.
>
> (Reese and Lipsitt, 1973, p. 115)

## Schedules of reinforcement

If newborns can be operantly conditioned, do they exhibit typical patterns of responding on the basic schedules of reinforcement? More precisely, do human neonates respond to contingencies of reinforcement in much the same way as other animals?

### Response shaping

Reese and Lipsitt (1973) point out that operant reinforcement seems to be an ideal way for the newborn to increase the variety of his behaviours. If this is the case it should also be fairly easy to use successive approximations to shape-up more complex or less readily emitted behaviours from simple responses. A study by Etzel and Gewirtz (1967) provides an example of this adaptability. Using a shaping procedure they attempted to change a 6-week-old infant's predominant behaviour from crying to smiling. First, they elicited smiling by presenting the child with a shiny metal saucer; following this, step by step, the elicited smiling response (which was incompatible with crying) was shaped until the eliciting stimulus could be discontinued and it met the criterion of a 'broad, full smile'. However, human infants differ somewhat from lower animals in the ease with which desired behaviours can be established. First, they are extremely imitative organisms, and a response can be established fairly rapidly using a combination of positive reinforcement accompanied by a demonstration (Corsini, 1969). Secondly, after a certain period in their development they cease to be a non-linguistic organism and begin to respond to verbal cues. Thus, response shaping by successive approximations becomes redundant and the child can simply be instructed as to the nature of the response required by the task (Spiker, 1959). However, as we shall see in the next chapter, instructions can have some unpredictable effects with both infant and adult human subjects on operant tasks (see pp. 357–359); we shall discuss these effects more fully there.

**Interval schedules**

For studies which look at schedule performance in children we have to look at studies which have used relatively older children (4–8 years of age) and originally to a study by Long, Hammack, May and Campbell (1958). Using trinkets as reinforcers, they found that on FI the typical FI *scallop* found in non-human subjects was difficult to obtain (see also chapter 13, pp. 347–351). The scallop when it was obtained was often transitory, with the most frequent type of behaviour being a steady constant rate of responding more characteristic of animal VI than FI performance. Long *et al.* (1958) concluded that in order to obtain typical FI patterning a number of procedures were often required: (1) starting the subject on VI and progressing to FI; (2) shifting from a small to a large FI value; (3) beginning an FI without prior 'shaping-up' or FR schedules – the latter seemed to establish high-rate responding which often became insensitive to changes in reinforcement contingencies. However, under VI schedules, behaviour begins to show a patterning (a steady, stable rate of responding) which is also typical of non-human animals. However, whether this reflects a sensitivity to the schedule contingencies, or whether it simply reflects a tendency in subjects to emit steady continuous rates of responding regardless of scheduled contingencies is another matter.

**Ratio schedules**

On fixed-ratio schedules Long *et al.* (1958) found that infant performance resembled the break-and-run pattern of responding usually exhibited by animals on such schedules. However, this was only obtained: (1) if the initial FR value was not too small (20 or less); (2) if the initial FR value was not too large (over 60), in which case typical FR behaviour rarely developed; (3) if the shift to larger ratios (for example, ratios of 90 or 100) was accomplished only slowly, and (4) the subjects did not become 'satiated' with the reinforcing trinkets.

**Escape and avoidance schedules**

Because of the violation of ethical standards that the use of aversive stimuli such as electric shock would necessitate, avoidance and escape learning in human subjects has mainly been studied using time-out from positive reinforcement (that is, negative reinforcement, see p. 36) as the aversive event (Baer, 1960, 1962). For example, Baer (1960) seated 4- to 6-year-old children at a table; the table contained a lever which the child could play with while he watched a cartoon film. On the first day, Baer simply recorded the operant-level of lever-pressing. On the second day, however, he arranged it so that after the first minute the cartoon film was switched off and remained off until the subject pressed the lever. Thus, if the subject pressed the lever when the film had been stopped it was an escape response, if he pressed the lever when the film was still running, it delayed the next interruption for $n$ seconds (that is, it was an avoidance response). Under contingencies such as these, Baer found that avoidance responses gradually decreased and after a number of sessions the child was only making escape responses. However, in a later study (Baer, 1962) it was found that avoidance responses would develop if an $S^D$ was programmed to indicate the onset of time-out periods (discriminated avoidance).

As an alternative to time-out from positive reinforcement as an aversive event, some experimenters have used loud tones (Penny and Croskery, 1962; Robinson and Robinson, 1961). In discriminated avoidance procedures using aversive loud tones, children of pre-school age readily learnt to acquire an avoidance response. Moreover, studies using tones as the aversive stimuli have found: (1) that more avoidance responses are made by the low anxious than high anxious rated subjects (Penney and Croskery, 1962) and (2) escape responding is significantly faster in high anxious than low anxious subjects (Penney and McCann, 1962). There are a number of possible explanations for these effects, one being that low anxiety children adapt more quickly to the aversive tone and thus settle more quickly to the avoidance contingency; high anxiety may invoke competing response tendencies which interfere with avoidance responding but sensitises the child to react to the presentation of the aversive tone (escape responding; see Penney and Kirwin, 1965).

# DISCRIMINATION LEARNING

Learning, of course, is not just simply about the learning of responses, it is also about learning to respond to specific stimuli; that is, it is also about discrimination learning. The new-born infant has to learn not only to increase his behavioural repertoire but also to distinguish different features of his environment.

## Perceptual versus dimensional discriminations

Primarily, there are two stages to the study of discrimination learning in human infants. The first involves *perceptual* discrimination – discovering whether the infant actually has the capacity to discriminate in the perceptual sense. The second involves *dimensional* discrimination – this assumes that the stimuli are perceptually discriminable, and attempts to assess how far conditioning procedures can teach the child to identify particular dimensions of the stimuli which have been chosen arbitrarily by the experimenter.

Using both operant and classical conditioning procedures human neonates of only a few weeks of age have been shown to exhibit several different types of discriminations, including left-right response discriminations, GO–NOGO discriminations and operant spatial discriminations. Using classical discrimination training procedures, Kasatkin and Levikova (1935a) were able to establish differential sucking to a tone in infants between 35 and 45 days of age, and established differentiation between a tone and a bell by 2–3 months of age. Using similar procedures, auditory and visual discriminations have been established in the first 4–5 months of life (Janos, 1959; Rendle-Short, 1961; Vakhrameeva, 1964). The series of studies mentioned earlier by Papousek (1959, 1967a, b) also provide some interesting information on neonate discrimination learning. These studies were the first to demonstrate auditory discrimination in newborns as young as 3 months of age. On presentation of a tone the infant was required to make a left head-turn to receive milk and on the sound of a buzzer, to make a right head-turn to receive milk (a left–right discrimination). When an arbitrary discrimination criterion had been achieved the discrimination was reversed. Papousek found that the first criterion was reached at around 3 months of age and the reversal criterion was also met a month later. Figure 12.2 shows not only that learning occurred more quickly the older the child but also that subsequent reversals produced 'savings' effects; that is, it looked as though infants of all age groups were beginning to form 'learning sets' (see chapter 11, p. 309) after successfully completing one or two reversals. Although children in the 53 months age group complete the discriminations more quickly than children in other age groups, they do take slightly longer to complete the two reversals. This could represent a 'floor' effect. Perhaps even more dramatically, Siqueland and Lipsitt (1966) demonstrated discrimination learning in 2- to 4-day-old infants. These newborns learned to turn their head ipsilaterally to tactile stimulation of the appropriate cheek: tactile stimulation to one side was paired with a buzzer and stimulation to the other cheek was paired with a tone – tactile stimulation overlapped the last 3 s of the 5 s auditory stimulus. Correct responses were reinforced with a dextrose solution. This procedure produced reliable and appropriate head-turning responses to the two CSs. In a subsequent similar study they looked at GO–NOGO discriminations in which a buzzer signalled no reinforcement for a head-turn ($S^A$), and a tone functioned as an $S^D$ for head-turning. Sure enough, more responses occurred to the $S^D$ than to $S^A$, and when the signalling functions of the tone and buzzer were reversed, so did the infant's tendency to respond to the two stimuli – thus demonstrating true discriminative control by tone and buzzer. Still more studies have been able to

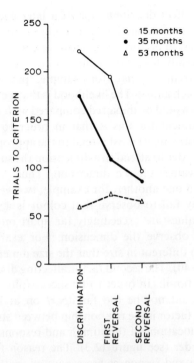

Figure 12.2   A comparison of discrimination learning and subsequent reversals of the discrimination in children of three different age groups. (After Papousek, 1967b).

demonstrate spatial discrimination in 3- to 4-month-old newborns using an operant reinforcement procedure (Caron, 1967; Siqueland, 1964).

A slightly different approach to the study of discrimination learning is to test for discrimination using a generalisation test (see chapter 2, p. 50). For example, Bower (1964) attempted to reinforce head-turns in new-borns using the presence of a large white cube as the $S^D$, and the absence of the cube as the $S^\Delta$. After a discrimination had been established, Bower obtained generalisation gradients by presenting stimuli that differed from the $S^D$ along the dimensions of size and distance. This is certainly a useful technique for studying discrimination in the newborn but unfortunately this study by Bower confounds two dimensions during generalisation testing and did not really use enough testing stimuli during the generalisation phase to produce convincing generalisation gradients. Nevertheless, the study did suggest that the infants were discriminating – at least to some degree – the features of size and distance of the $S^D$.

As the child grows older the ability to discriminate between different aspects of the environment becomes easier, and finer and even more 'abstract' discriminations can be made. Reese (1976) lists 4 characteristics of the learning task which affect discrimination at a later age. First, as might be assumed intuitively, the more similar the two objects to be discriminated, the poorer the discrimination performance. However, Reese does point out that few studies have actually tested the child's ability to discriminate between objects which are extremely *unlike* (for example, 'pinheads and cartwheels; crickets and sixteen-inch guns'). The likelihood is that a child's discrimination between such objects would be impaired compared with slightly more similar stimuli. Reese's argument for this is that in order to identify differences between two stimuli the subject must attend to the appropriate dimensions on which they differ; in order to attend to a dimension the subject must be able to perceive at least 2 values on the dimension. Thus . . .'if two values are exceedingly similar to one another, for example, two very similar shades of blue, the subject may fail to observe that colour is a possible dimension. Conversely, if the values are exceedingly far apart on the dimension, the subject may fail to observe the dimension. For example, pinheads and cartwheels may be so different in size that the size dimension is not noticed' (Reese, 1976, pp. 39–40). The second factor affecting discriminative ability is separation between stimuli. In order to be successfully 'compared' it seems that two stimuli should not be too far *apart* on at least one particular dimension. The third factor is the relationship between stimulus and response locations. When the locations of the stimuli and responses are separated the discrimination is harder (see figure 12.3). The reason for this seems to be because when stimuli and responses are separated, *position* of the stimulus becomes a possible salient one. However, when there is no separation between stimulus and response (that is, the subject responds directly to the stimulus), the last thing the subject sees are the features of the stimulus itself, rather than the position of the stimulus (Shepp, 1962). Fourthly, if the two stimuli to be discriminated differ on a number of dimensions they are more likely to be discriminated than if they vary only on one dimension. For example, *stereometric* stimuli vary in width, depth, and height as well as dimension such as colour, shape, etc. However, *planometric* stimuli vary only in width and depth. Both monkeys and human children have been found to learn discriminations between stereometric stimuli more readily than between planometric stimuli (Reese, 1963, 1964).

## 'Preference' effects in discrimination tasks

In the literature on infant discrimination learning there are a number of studies which, on the face of it, appear to provide paradoxical results. For example, children of one age group may show a difficulty in mastering a particular discrimination while children of a *lower* age group may master the task fairly quickly. Apart from some obvious procedural differences there is another explanation for those results which appear to contest the validity of a

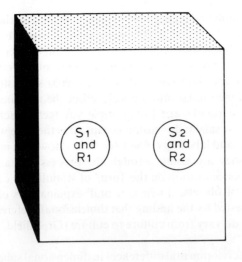

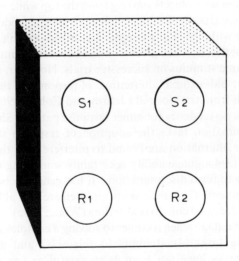

Figure 12.3 Examples of a simple apparatus in which the locations of the stimuli and response buttons are identical (top) and where stimulus and response locations are separated (bottom).

simple developmental progression in child discrimination learning: children from different age groups appear to have specific preferences either as to the dimensions of a stimulus they will primarily attend to, or to the kinds of reinforcers which provide the best incentive to learn.

First, let us consider the dimensions that a child will more readily attend to. In pre-school children the preference appears to be to attempt to discriminate between two stimuli first on the basis of colour, second on the basis of number, third on the basis of size and finally on the basis of form (Lee, 1965). However, kindergarten children show a difference in their 'preference' order. For them, form seems to be the most salient dimension, with colour coming second (Lee, 1965; Suchman and Trabasso, 1966). Lee (1965) suggested that these age differences in dimensional salience may reflect the teaching that the child is undergoing at a particular age. For example, younger pre-school children may find colour a most salient dimension because it is the time when their parents are pointing out and labelling colours for them. However, in their first year or so at school, they are going through the process of having to learn the alphabet, so a concentration on the 'form' of stimuli may come to emphasise this dimension of objects. 'Environmental' explanations of such preference effects are suggested by the finding that dimensional preferences exhibited by young children do vary from culture to culture (Greenfield, Reich and Olver, 1966).

Apart from developmental differences in dimensional salience other factors which affect discrimination learning have been related to age. For instance, (1) 3-year-olds tend to scan objects starting from the top while 5-year-olds use the bottom as their focal point (Braine, 1965); (2) response tendencies or response sets also change with age. For example, 3-year-olds exhibit strong perserveration tendencies; that is, when given a choice between 2 stimuli, they will tend to choose the same stimulus on successive trials. However, more than 75 per cent of four-year-olds exhibit 'alternation' of responding: they alternate their choice of stimuli from trial to trial (Jeffrey and Cohen, 1965). Ten year olds apparently show no tendency for either response pattern (Shusterman, 1963). In most discrimination tasks the adoption of response strategies such as perseveration or alternation are bound to interfere with the learning of the discrimination; (3) language ability is certainly something which is likely to affect the ability to form discriminations. It has generally been assumed that with the development of linguistic skills, children are now able to utilise words to serve as self-generated cues to responding (White, 1965) – something which is an obvious advantage when it comes to solving a complex discrimination or concept learning task; (4) attempts to relate IQ and ability to master discrimination tasks have not been as successful as one might intuitively imagine they would be. However, there is some evidence for a weak correlation between IQ and performance in visual discrimination tasks (Zeaman and House, 1967). This poor correlation may reflect the fact that brighter children may attempt to solve discrimination problems in a totally different way from less able children, and the strategy they adopt may well pay dividends on most discrimination tasks but on the rest it may prove to be a hindrance (Osler and Trautman, 1961). This being the case, the discovery of a correlation between IQ and ability to solve discriminative tasks may well depend on the nature of the task.

Finally, the incentive value of the reinforcer does seem to affect how well a discrimination will be learnt. In older children less preferred incentives produce poorer discrimination learning than highly preferred incentives (Bisett and Rieber, 1966; Miller and Estes, 1961). But paradoxically, less preferred incentives have been found to work better with younger children in a probability learning task (Stevenson and Hoving, 1964; Stevenson and Weir, 1959). One explanation for this is perhaps that different age groups seem to value reinforcers differently and the above two studies assumed that what was a strong reinforcer for an older child would be the same for the younger age group. However, a study by Bisett and Rieber (1966) showed that there were not only age-group differences in the ranking of reinforcers but also sex differences as well (see table 12.1) – a fact which is highly relevant to the study of discrimination learning.

**Table 12.1   Reinforcer Preferences and Age Differences in Children (Ranked in order of preference, 1 = most frequently chosen, 8 = least frequently chosen).**

| | All | Boys | | Girls | |
| --- | --- | --- | --- | --- | --- |
| Reinforcer | Children | 6–7 | 10–11 | 6–7 | 10–11 |
| Jewelry | 1 | 2 | 3 | 1 | 1 |
| Pennies | 2 | 4 | 2 | 2 | 2 |
| Cars | 3 | 1 | 1 | 5 | 4 |
| Beatle Cards | 4 | 5 | 5 | 3 | 3 |
| Trinkets | 5 | 3 | 4 | 4 | 5 |
| Marble Chips | 6 | 7 | 7 | 6 | 6 |
| Washers | 7 | 6 | 6 | 7 | 8 |
| Paper Clips | 8 | 8 | 8 | 8 | 7 |

(After Bisett and Rieber, 1966)

# CONCLUSIONS

A study of the fundamental principles of learning in the human neonate provides an enlightening insight into the adaptive roles of basic conditioning procedures. Classical conditioning provides the newborn with a mechanism which can increase the number of stimuli in the environment which will elicit important reflexive responses. Similarly, operant conditioning provides a process for expanding the child's behavioural repertoire beyond the collection of reflexes he brings with him into the world. Both processes provide a means for learning more about the new environment in which the newborn infant finds himself; this can be thought of in terms of finding out what responses are beneficial and should be retained, and secondly as a means of sharpening perceptual abilities to cope with necessary discriminations. Finally, although our knowledge of the developmental aspects of neonatal learning is still in the formative stages, the research into neonatal conditioning processes has also provided an insight into the nature of the conditioning processes themselves – into the possible constraints that the biology of the organism places both on these processes and the study of them, and also into the pervasiveness of the operant conditioning process and its ability to modify subtle aspects of even the most 'reflexive' behaviours in an environmentally inexperienced organism.

# 13 Human Operant Performance

It may seem somewhat paradoxical to include a whole chapter devoted almost solely to the performance of human subjects on schedules of reinforcement, especially after we have stated in chapter 3 that it is best to consider schedules of reinforcement not as phenomena *per se* but as one particular tool for teasing out the characteristics of the fundamental mechanisms which underlie learning. However, justification for this chapter is based not on the grounds of inflating the importance of schedules of reinforcement, but on the fact that human schedule performance exhibits many peculiarities, and these peculiarities tell us something about the different factors which can come to control human and animal behaviour respectively.

The Skinnerian tradition has emphasised that basic principles of learning are common to all species, including man, and primarily that assumption has been based on the face similarity of the performances generated on schedules of reinforcement by a wide variety of species. Morse, for example, was one of those who had been impressed by the apparent universality of basic schedule performance. He writes:

> Schedules of reinforcement . . . are important because they represent the most intensively studied and best understood body of information on the generation and maintenance of operant behaviour . . . (In schedules of reinforcement) radically different patterns of responding and associated general demeanor can be made to appear, change and disappear in the same subject over brief periods of time. Furthermore, any member of most species will give a similar performance on the same schedules.
>
> (Morse, 1966, p. 57)

Now, although the gross patterning of responding exhibited on cumulative records by different species does have face similarity, we have noted – especially in chapter 5 – that the implications of this are not as theoretically important as some workers have made out. First, when analysed closely, the patterns of responding emitted by different species do exhibit subtle differences which in turn appear to reflect species-differences in the variables which come to control behaviour on a particular schedule (Lowe *et al.*, 1974; Lowe and Harzem, 1977; Staddon, 1965, 1970, 1977).

345

Secondly, if 2 different species do exhibit similar behaviour patterns this by no means implies that the mechanisms underlying the behaviours are similar in the 2 species – they may well have evolved different mechanisms to cope with similar problems (White *et al.*, 1973). So, bearing these points in mind, what do we make of human operant performance? Do similar variables operate to control human schedule behaviour as control animal schedule behaviour? The enthusiasm for general principles of learning was certainly encouraged by early studies of human performance on schedules of reinforcement: simple responses could be shaped and acquired in much the same way as lever-pressing in the rat and key-pecking in the pigeon, and simple schedule performance showed much the same patterning as that of animals (Ayllon and Azrin, 1964; Holland, 1958; Laties and Weiss, 1963). However, since this time it has come to be recognised that human schedule performance is characterised primarily by its wide intra- and inter-subject variability. It is often difficult to match these performances with those of animals on similar schedules, and the behaviour of many human subjects frequently entails low rates of responding often interspersed with long and unpredictable pauses (Barrett and Lindsley, 1962; Lindsley, 1960; Orlando and Bijou, 1960; Sidman, 1962; Spradlin and Girardeau, 1966). Why, then, do we seem to get this messy hotch-potch of behaviours with humans when animals are so pleasingly predictable in their performances? Is human behaviour subject to different controlling variables? Do human operant experiments really use a faithful analogue of animal procedures? Or, quite simply, is the behaviour of human beings simply not controlled by environmental contingencies of reinforcement? Before we start to search for the source of any answers to these questions, first let us look more closely at some of the facts of human performance on schedules of reinforcement.

## HUMAN BEHAVIOUR ON SCHEDULES OF REINFORCEMENT

In order to replicate an experimental environment similar to the rat's Skinner-box, most human studies utilise a fairly simple response – such as pressing a button or telegraph key – and seat the subject in a relatively bare room facing a consol on which there are various stimulus lights and counters (see figure 13.1 as an example). The reinforcer is usually points, which are accumulated on a counter in front of the subject – these points act as tokens which may be exchanged for money or other desired items (see p. 355ff. for a fuller account of the reinforcers used in these situations). Finally, rather than expend the effort of 'shaping' the subjects to make the appropriate operant response, subjects are usually issued with minimal instructions regarding the response manipulandum and the consequences of the response (for example, 'Your task is to earn as many points as you can. Points are shown on the counter. Points are

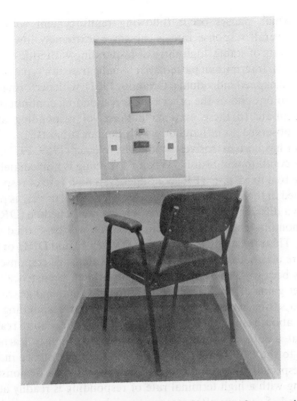

Figure 13.1 A typical cubicle in which human operant performance is tested. The subject is seated in front of a panel containing a counter (used to tally the number of reinforcers obtained) and stimulus keys – similar to pigeon pecking-keys – which are used as the operant manipulanda.

available for pressing the telegraph key', etc.) We shall discuss later the effects that different instructions and different types of operant response can have on the resulting performance.

## Fixed-interval

The steady-state performance of animals on an FI schedule is characterised by a pause after reinforcement followed by a positively accelerating response rate up to the next reinforcement (the FI 'scallop' – see chapter 3, pp. 63–64). Post-reinforcement pause duration is usually directly related to the FI value, indicating a temporal discrimination on the part of the animal. However, the pattern of responding engendered by human subjects on FI schedules varies remarkably both within and between subjects. Most typically, human FI performance seems to fall into two distinct categories. One is a low rate pattern in which pauses occur following reinforcement with occasional scalloping – overall response rate is, however, extremely low. The second category consists of stable, high rates of responding with little or no pausing (see figure 13.2). Intermittent patterns are usually transitory, particularly with mentally handicapped individuals (Orlando, 1961), with performance eventually shifting to one end or the other of this responding continuum. So, what factors contribute to these extreme patterns of responding and why is 'scalloping' observed so infrequently with human subjects?

One factor has been extensively studied by Weiner (1969, 1972), and this is the effect of conditioning history. When FI training is immediately preceded by exposure to a fixed-ratio (FR) contingency, a high rate of responding will be maintained in the subsequent FI performance; if FI training is preceded by exposure to a differential reinforcement of low rate schedule (DRL) the low rate of responding engendered by this schedule will be carried over to FI responding. This rate interaction also occurs when FI and DRL, or FI and FR schedules are alternated. Weiner has suggested that this response induction effect arises because an FI schedule reinforces such persistence – it fails to make all net gains of reinforcement contingent upon change. Since the FI schedule does not make any specifications about responding or, more specifically, about response rate between reinforcers, it is quite reasonable to suppose that a subject will simply transfer his previously learned rate of responding to the FI schedule. Nevertheless, such persistence in humans is surprising, especially since Ferster and Skinner (1957) have demonstrated that FI scalloping with a high terminal rate of responding is readily acquired by pigeons only 4–5 sessions after extensive training on a low rate schedule (for example, DRL). This being so, some factors other than the scheduled contingencies must be influencing the human performance.

Conditioning histories, however, do not seem a sufficient explanation of aberrant human FI performance; most subjects do not have pretraining before exposure to the FI schedule nor have they ever been in a remotely similar experimental environment before – yet they still produce either high or low rate responding on FI. One very powerful determinant of these rate extremes is the type of instruction administered to the subject before training. We shall discuss the effects of instructions in more detail later but suffice it to say here that, as a general rule, high rate responding on FI seems to result when information is given about the response requirement, whilst low rate responding results from information in the instructions concerning the scheduling of reinforcement (Kaufman, Baron and Kopp, 1966).

Weiner (1962, 1964, 1965) has also shown that cost (point loss per response) can be an aversive event which suppresses the continuous responding of some subjects between reinforcement on FI. Both real and 'imagined' cost (that is, the subject is told that responses will be penalised but in fact are not) act to attenuate responding; but real cost produces a more marked and consistent suppressive effect than imagined cost. The introduction of response cost may

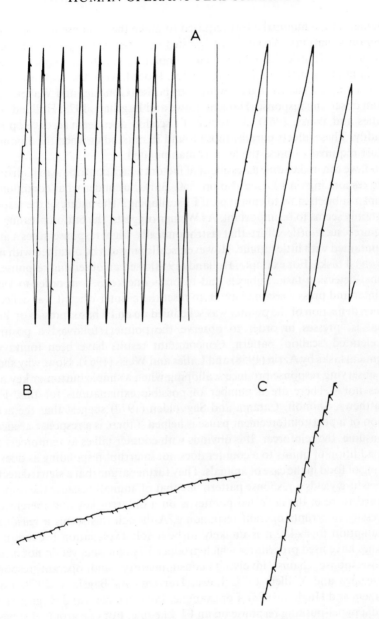

Figure 13.2 Typical cumulative records of human fixed-interval performance. (A) These two records show the maintenance of extremely high rates of responding between reinforcers, a common feature of human FI performance. (B) Low-rate FI performance; responses are emitted relatively slowly and with no obvious relationship to the temporal parameters of the schedule. (C) When responding does come under the control of the FI schedule it usually resembles the *break-and-run* patterning found in animals on FR schedules.

counteract the minimal effort required to make the response in most human operant situations. Weiner has suggested that the button pressing response, because of the little effort required to complete it, may be an important factor in maintaining high rate FI performance. Indeed, when the response is made more complex, such as a handwriting response (Gonzalez and Waller, 1974), or an observing response (Harzem, Lowe and Bagshaw, 1978; Holland, 1958; Laties and Weiss, 1963), the typical FI scallop does seem to develop more readily. These effects parallel those found in rat studies when the amount of effort required to press the lever is manipulated.

So, we can make some assessment of factors which contribute to high or low rate responding on FI, (see also pp. 360–361), but under what conditions do human subjects start to emit typical FI scalloping? The nature of the response certainly seems to be important. As Weiner suggests, the button pressing task requires such little effort that astronomically high response rates can be maintained with little fatigue. However, scalloping can be obtained with more involved tasks. For example, Holland (1958) used an observing response in a signal-detection task. Subjects had to press one key, A, in order to view a pointer and press a second key, B, to reset the pointer when it was deflected. When deflection of the pointer was scheduled on an FI basis, presses on key A (that is, presses in order to observe the pointer) followed a positively accelerated 'scallop' pattern. Concomitant results have been found with identical tasks by Azrin (1958) and Laties and Weiss (1963). Now, why should an observing response produce scalloping when a simple button or key press does not ? There are a number of possible explanations for this. First, Matthews, Shimoff, Catania and Sagvolden (1977) suggest that the acquisition of a post-reinforcement pause is helped if there is a response needed to 'consume' the reinforcer. It is obvious with counter tallies as reinforcers that the addition of points to a counter does not interrupt responding as does the eating of food in the case of animals. They further argue that a signal detection procedure yields a response pattern like that of animals because the response needed to reset the deflected pointer is on a different key and therefore by necessity interrupts operant responding. Although this seems a reasonable assumption to make, it is unlikely to be a sole explanation because other studies have used procedures which produce FI scalloping yet do not involve 'two-response chains' involving 'consummatory' and operant responses (Gonzalez and Waller, 1974; Lowe, Harzem and Bagshaw, 1978; Lowe, Harzem and Hughes, 1978). For example, Lowe, Harzem and Hughes (1978) used a panel-pushing response on an FI schedule. For one group of subjects, responses on a panel were reinforced by points on various FI schedules. For a second group, responses on the panel still produced point reinforcers on FI schedules, but also produced brief illuminations of a digital clock. Responding in the first group varied considerably both within and between subjects – response patterns falling either in the high or low rate categories typical of many FI performances with a simple key-press operant. However, subjects in the second group all produced characteristic FI scalloping. Furthermore, post-reinforcement pause duration was an increasing function of FI value, and running rate (response rate calculated after exclusion of the post-reinforcement pause) was a decreasing function of the FI value – results identical to those found with animals. Lowe, Harzem and Hughes (1978) propose that scalloping was obtained in their study because responding to produce the digital clock transferred the control of behaviour from internal 'self-produced' cues to external cues related to the scheduled contingencies. Subjects often formulate their own 'hypotheses' as to the necessary conditions to produce reinforcers, then respond only according to these internal formulations and ignore the external contingencies scheduled by the experimenter. These 'self-produced' cues often involve counting of responses and also counting as a means of estimating temporal parameters: a strategy which often produces great intra- and inter-subject variability in responding. Lowe, Harzem and Hughes (1978) point out that, before typical FI schedule performance is to be attained, studies must implement procedures which (1) minimise subject-produced cues such as counting, and (2) bring the behaviour under the control of stimuli produced by the experimenter (see also pp. 360–361).

## Fixed-ratio

Responding by animals on an FR schedule typically consists of a 'break-and-run' pattern – a pause after reinforcement followed by an abrupt transition to a high rate of responding which is maintained up to the next reinforcer (see chapter 3, pp. 70–71). Generally, the topography of responding by human subjects on FR schedules is tolerably comparable with the break-and-run pattern exhibited by non-human organisms. The occurrence of post-reinforcement pauses has been reported with children (Long et al., 1958; Weisberg and Fink, 1966), mentally handicapped children (Bijou and Orlando, 1961; Orlando and Bijou, 1960), adults (Holland, 1958) and mentally handicapped adults (Ellis, Barnett and Pryer, 1960). A study by Wallace and Mulder (1973) also replicated animal findings by demonstrating in mentally handicapped adults that the length of the post-reinforcement pause was directly related to the FR value. FR size had no effect on response rate when responding resumed following the post-reinforcement pause. However, these results conflict with the findings of other human studies which have reported the absence of increased post-reinforcement pausing with human subjects of high FR values (Ellis et al., 1960; Holland, 1958; Hutchinson and Azrin, 1961).

Similarly, as we mentioned in the last chapter, Long et al. (1958) reported deteriorative changes in performance when the ratio size was increased too rapidly; an effect which appears analogous to the 'ratio strain' phenomenon reported in pigeons by Ferster and Skinner (1957). In contrast to this, Ellis et *al.* (1960) found that FR responding was maintained for most of their subjects on FR schedules as high as 1024!

So unlike human FI performance, FR performance by human subjects does exhibit some similarity to that shown by animals. The 'break-and-run' pattern of responding is frequently observed, although some studies have reported the absence of post-reinforcement pausing. Nevertheless, perhaps a caution should be added here. Even FR responding in animals does appear to be quite variable and does not always correspond to the 'break-and-run' pattern – especially at high FR values. Indeed, a proportion of animal subjects even fail to acquire responding on FR schedules at all. Bearing this in mind, it is as well to consider if there is any truly stereotyped FR animal performance with which human FR performance can be compared.

## Variable-interval and variable-ratio

Behaviour on both VI and VR schedules is characterised by a short pause after reinforcement followed by a transition to a steady response rate which is maintained until the next reinforcer. Given equivalent reinforcement frequencies, response rate on VR is usually higher than on VI (see chapter 3, pp. 74–80).

The performance of human subjects on VI has been shown to resemble closely that of other organisms (Long *et al.*, 1958; Lindsley, 1960). Long *et al.* noted, however, that the construction of the VI programme was an important factor in determining the regularity of the subjects' performance. If there were too few short intervals in the programme, there was considerable pausing and the cumulative records were generally irregular. In another study, Lindsley (1960) found that although there were individual differences in response rate on a VI 1-min schedule, each individual's rate was steady and relatively consistent during many successive experimental periods. Similarly, an experiment by Holland (1958), using a signal-detection task, found that response rates of his subjects decreased as the average interval of the VI schedule was increased from 15 s to 2 min, data that are consistent with those found in animal studies (Catania and Reynolds, 1968).

VR schedules, however, have been used only infrequently with human subjects. Orlando and Bijou (1960) studied the VR performance of developmentally retarded children and found that like non-human subjects, these children responded at high steady rates on the VR schedule, with pauses being infrequent and not obviously related to the receipt of the reinforcer; VR yielded fewer pauses than FR. In a more recent study, Matthews *et al.* (1977) compared the response rates of subjects on yoked VI and VR schedules. This experiment replicated with humans a procedure similar to that used by Thomas and Switalski (1966) with pigeons. One subject responded to obtain points on a VR schedule while a second subject acted as a 'yoked' partner, and received points for key-pressing only when one was obtained by his yoked partner. Essentially then, when the VR subject obtained a point this made available a reinforcer for his partner, whose next key-press obtained a reward. Both subjects received the same frequency and distribution of rewards, but one subject was responding on a VR schedule and one on what was essentially a VI schedule. In line with Thomas and Switalski's results with pigeons, Matthews *et al.* found that the VR subjects in general responded at a higher rate than their yoked VI partners, even though frequency of reward was the same for the two conditions.

## Differential reinforcement of low rate

Under DRL contingencies a response is reinforced only if a specified minimum interval has elapsed since the preceding response. In most animals this schedule generates a constant low rate of responding with the modal IRT usually occurring around the DRL minimum IRT criterion (see chapter 3, pp. 79–82). In human subjects, DRL schedules have proved to be quite effective in maintaining stable and efficient low rate behaviour. Kane (cited by Spradlin and Girardeau, 1966), studied the behaviour of 2 moderately retarded girls on DRL. Response rates were low, as expected. One subject on DRL20 s exhibited short 'bursts' of responding during the first 5 sec after reinforcement. The other subject overshot DRL20 s most of the time. When the degree of spring tension on the manipulandum was increased, both subjects showed a general increase in efficiency. However, when tension was low, the subjects undershot more often. One subject was observed repeatedly to exhibit certain chains of behaviour such as pulling the manipulandum, looking in the goal box, walking to the opposite side of the room, tapping the wall and then slowly walking back to the panel and pulling the manipulandum. This behaviour seemed to act as a chain of superstitious behaviour which mediated the temporal requirements of the DRL schedule (see chapter 3, pp. 83–84).

Holland (1958), studying observing responses on DRL30 s, found that when his subjects responded just a little sooner than the required interval they often emitted a short 'burst' of responses at a rapid rate. 'Bursting' was also found in the DRL performance of a child studied by Goddard (cited by Sidman, 1962a). He also found that older children were more precise in timing their responses than younger children, producing IRT distributions with much less variability than is usually found with rats, but which were comparable with the performance of monkeys. Many of the children were observed to go through a 'superstitious' chain of behaviours between each lever pull.

Mediating behaviour having been observed during human DRL performance, Bruner and Revusky (1961) attempted to make a systematic study of this aspect of human DRL responding. Subjects were required to space their responses on a telegraph key by at least 8.2 s; also present were 3 other keys, none of which had any scheduled consequences. They found that a given subject would repeat the same sequence of responses but with slight variations from one minute to the next. The systematic response pattern of all subjects on collateral keys 'filled up' the necessary temporal delay between responses on the 'DRL' key. The operant level and extinction phase produced very erratic response patterns and post-experimental interviews revealed that all subjects were convinced that reinforcers could be obtained only by a pattern of responses on at least one collateral key in order to 'set up' the reinforcer on the 'DRL' key.

So perhaps not surprisingly, human subjects appear to be able to adapt their behaviour quite efficiently to DRL contingencies; their resultant behaviour has also been shown to exhibit both instances of response 'bursts' and collateral chains of 'mediating' behaviour – phenomena which are common features of non-human DRL performance.

## Summary of human schedule performance

All this evidence indicates that although a claim can be made for similarity between human and animal performance under some task conditions, the behaviour of human subjects does exhibit enormous intra- and inter-subject variability. In some cases subjects cease responding at unpredictable times, and for long periods; in other cases, they incessantly respond at inexplicably high rates with their behaviour showing a total lack of sensitivity to schedule contingencies. This is nowhere more apparent than on fixed-interval schedules of reinforcement and suggests that we must often look elsewhere than the experimenter programmed contingencies for those factors which are controlling behaviour. It is important to understand what these other factors might be if we are to evolve a comprehensive experimental analysis of human behaviour. We shall discuss what these other factors might be shortly, but suffice it to say here that the study of operant conditioning in human beings differs from that of animals in some important respects. First, man has linguistic capabilities that other organisms do not possess, and verbal cues – if only in the form of 'talking to oneself' – can come to control behaviour. Secondly, the psychology experiment with human subjects is a social situation where 'social' variables can come to influence the subjects' behaviour. The experimenter and subject usually have to interact in some way, however briefly, and this can lead to the subject not only forming expectations of what is required of him in the experimental task, but also – depending on the impression he has formed of the experimenter – affect his level of motivation to perform in the task. Neither of these factors need be considered seriously in animal studies but their potency in controlling human behaviour is often underrated. However, before we look at these factors, let us look at the role of the reinforcer in human operant performance. Unlike rats and pigeons, it is usually impractical and, what is more, considered unethical, to manipulate deprivation states in human subjects. The choice of reinforcer in these studies is thus often arbitrary and as a consequence many studies have reported deterioration in the orderliness of their subjects' performances which reflect a deterioration similar to that found in animals which have become satiated after insufficient deprivation. Can some of the unpredictability and variability in human schedule performance be accounted for by the type of reinforcer used?

# THE NATURE OF THE REINFORCER

Researchers investigating human operant performance have usually adopted generalised conditioned reinforcers such as tokens or points. These reinforcers are easily administered and do not involve disruptive consummatory activities. An advantage of generalised conditioned reinforcers is that they may be redeemable for a whole variety of merchandise such as money, trinkets, toys, candy, fruit, articles of clothing, etc. The experimenter is not limited to using one item, which may be a reinforcer for one subject but not for others. Money has been used successfully as a reinforcer by, for example, Lindsley (1964), Schwitzgebel and Schwitzgebel (1961), and Slack (1960). Tokens have been used by researchers dealing with mentally handicapped subjects (Ayllon and Azrin, 1965; Birnbrauer and Lawler, 1964; Girardeau and Spradlin, 1964; Watson, Lawson and Sanders, 1965), and are established rather quickly as reinforcers, even with moderately and severely mentally handicapped children. Studies by Watson *et al.* (1965) have looked at both edible (candy and food) and 'manipulable' (movies, sound, mechanical toys) reinforcers with severely mentally handicapped children. Initially they found there was a preference for the manipulable, but over 13 sessions there was no evidence for a difference between the 2 classes of reinforcers. Long *et al.* (1958), using trinkets as reinforcers with children, found a deterioration in the orderliness of their subjects' behaviour after several sessions of experimentation. Since the altered performance was similar to that of lower animals which had become satiated after insufficient deprivation, they substituted new trinkets which ameliorated the deterioration in performance. Stoddard (cited by Sidman, 1962a), using candy reinforcers with children, observed similar changes, and was able to reverse these deteriorating performances by varying the nature of the reinforcer within each experimental session; the children received candy, pennies, any of a wide variety of trinkets, or tokens that could be exchanged later for more valuable toys.

Counter tallies are the most commonly used reinforcer for normal human subjects on schedules of reinforcement. Little indication has been found that providing money as payment for counter tallies is an essential feature of the reinforcing event or that it leads to behaviours that differ from behaviour when payment is not provided (Kaufman *et al.*, 1966; Verplanck, 1956; Weiner, 1962, 1964). However, the effectiveness of counter tallies alone as reinforcing agents is dependent upon the use of other reinforcers to ensure the subject's participation in the experiment. In the experiments of Kaufman *et al.* (1966) subjects who were reinforced only with counter tallies either fulfilled a course requirement or earned an hourly wage through their participation.

Social consequences have also been used in operant and instrumental studies with mentally handicapped individuals (Barnett, Pryer and Ellis, 1959). Spradlin and Girardeau (1966) suggest that the effectiveness of social approval as a reinforcer for mentally handicapped individuals will depend upon the unique history of the child. If an adult's comments have, in the past, set the occasion for a response to be reinforced, it is quite likely that comments by adults now will be reinforcing. However, social consequences may also exert a covert effect on behaviour, especially where experimenter–subject interactions are necessary. Bijou and Sturges (1959) have pointed out that interaction between subject and experimenter can markedly alter the degree of control exercised by an ostensible reinforcer. The individual brings with him into the laboratory a history of social interaction, and socially mediated factors such as signs of approval or expectation from the experimenter may override the effects of reinforcers like candy or trinkets. This is nowhere better expressed than in a quote from Bijou and Baer on operant studies with children.

> Since children represent a species highly sensitive to social reinforcement, and indeed represent an age in which their parents and other members of society typically are striving to implant and develop this sensitivity, it is clear that unrecognised social reinforcement contingencies may abound in almost any experimental situation. For example, instructions that tell the subject what he *may* do are sometimes responded to as if he had been told that he *must* do. Children are observed to give thousands of extinction responses after only the modest reinforcement programs (one child sat and responded with tears rolling down his cheeks. When asked what the trouble was he replied, "I don't want to do this anymore." When asked why he didn't stop, he said, "You didn't tell me I could stop"). While it is not wise to give complete credence to the verbal behaviour of children as indicators of controlling variables, it is likely that a process like this has played a part in many experimental contingencies.

(Bijou and Baer, 1966, p. 722)

Bijou and Sturges (1959) suggest that human subjects be given minimal instructions, that all reinforcers be presented by mechanical or electronic means, and that the experimenter have as little contact with the subject as possible.

So, as a general rule, human schedule performance largely seems to be unaffected by type of reinforcer. Generalised conditioned reinforcers such as tokens, points or money appear to be the most effective in maintaining behaviour; the controlling potency of more specific reinforcers, such as trinkets or candy, often weakens over time in a fashion which resembles satiation in non-human organisms on appetitive reinforcers. One important factor to emerge is the necessity of controlling spurious sources of reinforcement provided through experimenter-subject interaction. However, many studies of human operant performance attempt to control this factor by delivering the reinforcer mechanically and housing the subject in a separate sound-proofed room.

## THE INFLUENCE OF INSTRUCTIONS

Since experimenters working with human subjects do not often have the unlimited access to their subjects that animal workers do, it is often imperative to induce rapid and stable baseline performance in order that the appropriate variables can be manipulated after as few sessions as possible. Verbal instructions given to human subjects have been considered to provide a means for rapidly establishing this baseline. However, this leaves the experimenter with the problem of deciding how much information about the task he should convey to the subject and how this information should be worded – instructions which have minimal differences in emphasis can produce maximal differences in performance. It is often impossible to verbalise the contingencies accurately and one should not expect different subjects to interpret them similarly. Before discussing the important aspects of control of human behaviour by instructions, let us look at some of the work that has been carried out on the role of instructions in influencing human schedule performance.

In many studies, substantial numbers of subjects have failed to acquire the desired response strategy despite the scheduling of reinforcement contingencies deemed favourable for acquisition (Ader and Tatum, 1961; Ayllon and Azrin, 1964; Turner and Solomon, 1962). By comparison, addition of instructions about the desired response results in rapid adoption of the response (Ayllon and Azrin, 1964; Baron and Kaufman, 1966), but may also induce inappropriately high rates, particularly on temporally-based schedules of reinforcement (Kaufman et al., 1966; Weiner, 1962). More detailed instructions about reinforcing contingencies, as well as the response itself, typically produce response rates approximating the requirements of the reinforcement schedules (Dews and Morse, 1958; Kaufman et al., 1966; Weiner, 1962). Furthermore, several studies have shown that instructions about the reinforcement schedule may have effects overriding those of the reinforcement schedule itself. Thus, instructions can induce behaviours in the absence of scheduled reinforcers (Ayllon and Azrin, 1964; Kaufman *et al.*, 1966), and can produce behaviours more in accord with instructions than with actually programmed reinforcement (Kaufman *et al.*, 1966; Lippman and Meyer, 1967).

Instructions, therefore, can often be the most powerful single determinant of human schedule performance. For instance, Kaufman *et al.* (1966) found that subjects on a VI schedule paused for a considerable period of time after reinforcement when instructed that reinforcement was scheduled on an FI basis. This prevented reinforcement of the shorter inter-reinforcement intervals of the actually programmed VI schedule. However, some indication was obtained that instructional control could be weakened by introducing extreme discrepancies between instructions about the schedule and the schedule actually programmed – normally when the latter was extinction. It has already been noted that cost (point loss per response) can suppress the continuous no-cost responding of some human subjects under FI, but Weiner (1962, 1964, 1965) has found that both real and imagined cost (that is, subjects were instructed that each response would lose points, but this did not in fact happen) suppressed responding.

Skinner (1966) has raised the question of whether instructions about contingencies have the same behavioural effects as actual exposure to those contingencies. Skinner suggested that since subjects usually cannot verbalise accurately the contingencies to which they have been exposed, they cannot be expected to react appropriately to descriptions of contingencies provided by experimenters. On these grounds he contends that verbal instructions should not be used as a substitute for the actual arrangement and manipulation of contingencies, although he does concede that instructions may be of value as an alternative to shaping when concern is more with the eventual performance of a response pattern than with its acquisition. Kaufman *et al.* (1966) also suggest that instructions are no substitute for real exposure to the actual contingencies since, although the schedule instructions can produce behaviours similar to the terminal behaviours usually observed when such schedules are used, they do not indicate to the subject what rate of responding to assume, only under what circumstances the reinforcer will appear. However, Lippman and Meyer (1967) have suggested that instructions which place some emphasis (however small) on the response required to obtain reinforcement may resemble the 'setting operation for a *vigilance task*', thus inducing the high rates of responding characteristic of so many human subjects on FI.

Sidman (1962a) has pointed out that instructions, as well as providing information about response–reinforcer relationships, can also affect the potency of the reinforcer. He suggested that the signal-detection procedure utilised by Holland (1958) has proved to be an extremely effective method for using a generalised reinforcer that is based upon instructions to the subject; the instructions in turn derive their effectiveness from an unspecified, but evidently general, set of cultural reinforcing practices.

On the basis of recent studies, Baron, Kaufman and Stauber (1969) make some broad conclusions about the effects of instructions. First, instructions represent an external, observable determinant of behaviour whose influences, although complex, can be investigated in a straightforward, objective manner; and secondly, instructions given to humans provide a means of evoking and controlling operant behaviours whose establishment in other ways would be impractical, if not impossible. Once behaviour has been established, various experimental contingencies become accessible to study. They conclude by stating that

> The use of instructional manipulations in the study of human behaviour may be viewed as playing a role parallel to such manipulations as deprivation and drug administration in work with subhuman subjects; by increasing the probability of desired behaviours in this way a means is provided whereby the controlling influences of reinforcement contingencies may be studied effectively.
>
> (Baron, Kaufman and Stauber, 1969)

This discussion leads on to two further considerations arising from human operant performance. First, given that instructions do tend to contaminate the control of human performance by schedule contingencies, what kinds of procedures can we evolve that will allow us to make a meaningful analysis of the effects of schedule variables on human responding – and thus compare these effects with the results of animal studies? Secondly, during our daily lives a large proportion of our behaviour is changed not by actual contact with reinforcement contingencies, but by verbal communication of contingencies. We live our lives not by 'trial and error' learning but by following 'rules' – written-down laws, culturally transmitted norms and taboos, and more casual communication of experiences between individuals. For example, a large proportion of us do not find out that Shakespeare's plays are entertaining by making random visits to the theatre and by being reinforced for this activity moreso by Shakespeare's plays than, say, any other author's plays; we initially go to see Shakespearian plays because others, teachers or friends, for example, have recommended them – they suggest that if we emit this differential behaviour, it will be reinforced. Similarly, when ice-skating, we do not learn to avoid thin ice by falling through it into the water, we take note of warning signs that say 'Danger – Thin Ice'. Furthermore, verbal communication of contingencies can influence our everyday behaviour to the extent that we even ignore the testimony of our own experience and, to this extent, the behavioural effects of instructions do not simply represent a vexing experimental problem, they truly reflect the way our behaviour is controlled in day-to-day situations.

## SCHEDULE CONTINGENCIES VERSUS SELF-PRODUCED CUES

Let us just recap over the last two points in the previous section and state them in slightly different ways: First, many types of instructions seem to produce 'cognitive sets' which override external contingencies – that is, they generate subject-produced cues which control behaviour. Second, if we want to compare schedule control in humans with that in animals we have to evolve experimental preparations which minimise control by self-produced cues and maximise control by programmed contingencies. The two points are linked in the sense that before we can evolve the latter preparations we have to understand what kinds of self-produced cues can come to control behaviour on schedules of reinforcement. Two examples of self-produced controlling cues are evident from many of the experiments on human schedule perform-ance. The first comes from using instructions which indicate the required response. This situation resembles what Lippman and Meyer (1967) call the 'setting operations for a vigilance task': that is, it primes the subject simply to respond, and respond, and respond – the self-produced cue is of the kind 'This is what the experimenter said I must do so I will do it!' Typically, this type of strategy produces inappropriately high rates of responding, especially on FI schedules, and more importantly a total insensitivity to any of the schedule contingencies. A weaker, but similar, effect of indicating or demonstrating the required response is that it induces a belief in the subject that reinforcers are dependent on specific *numbers* of responses. This is especially prevalent on FI schedules: after many hours of training on FI a great many subjects will still maintain that the reinforcer is dependent on making a specific *number* of responses, (Leander, Lippman and Meyer, 1968; Lowe, Harzem and Hughes, 1978). (Even in situations where the subject cannot equate a specific number of responses with the receipt of the reinforcer, he will often still maintain that the experimenter is continually changing a ratio requirement rather than believe that reinforcers are primarily scheduled on a temporal basis!) In other situations, where the subject has reported that there is a temporal con-tingency, counting often serves as a mediating behaviour which spans the temporal gap between reinforcers (Laties and Weiss, 1963). Even so, continuous high-rate responding can still be maintained under such circum-stances if the subject uses lever- or button-presses as an aid to counting (Lowe, Harzem and Hughes, 1978), and on other occasions it can lead to unusually low response rates when the subject is especially skilled at counting out time intervals. In both situations responding is not directly under the control of the schedule contingencies, but in fact under the control of self-produced cues. Do we have to conclude from this that human schedule behaviour neither shows the regularity exhibited by animal behaviour nor is in any way capable of comparison? Not necessarily; since we understand some of the self-produced cues which control human schedule behaviour we can take some steps towards eliminating them or producing a 'shaping-up' procedure which is 'purer' in the sense that it reflects control of behaviour only by schedule variables. There are a number of ways of going about this.

## Response shaping

The wording of instructions is an important source for generating 'internal' cues which control behaviour. The radical solution is thus to eliminate them altogether and shape-up the subjects' response using successive approximations (see chapter 2, pp. 40–41). This type of approach has been used successfully by Matthews et al. (1977), and the ensuing behaviour of subjects on VI and VR schedules showed rate and patterning characteristics similar to that found in non-human animals.

## The use of external cues

If the experimenter is reasonably aware of the nature of internal cues (for example, counting on temporally-based schedules) then he can perhaps attenuate them by providing comparable cues externally. For example, counting out temporal intervals is a common strategy of human subjects on FI schedules, their counting is often not accurate and leads to behaviour which resembles 'break-and-run' patterning rather than typical FI scalloping (Lowe, Harzem and Bagshaw, 1978). That is, they count out what they believe to be the interval duration and then respond rapidly at a steady rate until they obtain their reinforcer. However, covert counting can be replaced by instituting a second response which briefly illuminates a digital clock (Lowe, Harzem and Hughes, 1978; Lowe, Harzem and Bagshaw, 1978; see pp. 350–351 for details of the experiment). The studies of Lowe, Harzem, Hughes and Bagshaw found that, with interference from counting eliminated, responding to produce brief illuminations of the digital clock was open to the direct effect of the temporal contingencies. Sure enough, they found that responding to illuminate the digital clock exhibited a 'scalloped' pattern similar to that found in other animals on FI schedules.

## The use of concurrent tasks

A further method of eliminating self-produced cues is to provide the subject with a concurrent task of some kind. Laties and Weiss (1963) have utilised such a procedure and found that when the subject had to engage in a fairly simple problem solving task at the same time as responding on a telegraph key, inter-reinforcement counting was effectively eliminated, and typical FI scalloping established in subjects whose behaviour had previously been characterised by very high or low rate responding.

### Conditioning without awareness

Those self-produced cues which are most resistant to elimination seem to be mainly generated in circumstances where the subject is completely aware of that aspect of his behaviour which produces the reinforcer. This allows him to generate hypotheses about the relationship between response and reinforcer. However, procedures which enable an experimenter to reinforce very discrete aspects of a subject's behaviour – such that the subject cannot verbalise the reinforcement contingencies – do produce performances which can be favourably compared with those of non-human animals. Hefferline, Keenan and Harford (1959) and Hefferline and Parera (1963) have pioneered such studies in 'covert reinforcement'. In order to avoid aversive noise stimulation, human subjects had to produce a very small thumb-twitch (detected electromyographically); the subjects were not told beforehand that this was the response which would avoid the aversive noise, but even so the frequency of criterion thumb-twitches did increase with training on this procedure. After the experiment, none of these subjects could verablise what response was needed to avoid the noise stimulation, nor – when told that it was a discrete thumb-twitch – could they produce this thumb-twitch voluntarily. As Hefferline *et al.* (1959) point out, being 'aware' of a contingency of reinforcement is by no means a prerequisite for a human subject to come under the control of that contingency; and secondly, the behaviour produced by such a procedure can allow an assessment of human operant behaviour in the absence of internally generated cues. Similar results in an 'appetitive' procedure are reported by Keehn (1969). First he reinforced college students with money for pressing a telegraph key on a VR200 schedule; however, during the session the reinforcement of key-presses was discontinued and reinforcers were delivered instead after every eighth eye-blink (FR8). Under these conditions, key-pressing still continued but the frequency of eye-blinks rose from about 13 to 30 blinks per minute. The subject still believed that key-pressing produced reinforcers and had no idea that blinking was involved. In other similar studies, the rate of such 'covert' operants has been observed to rise and fall in accordance with the prevailing schedule of reinforcement (Keehn, 1967).

These four methods, then, are examples of the way in which self-produced cues can be eliminated to allow an assessment of the schedule factors which control human operant performance. These kinds of studies suggest that when internal control of responding can be eliminated, subsequent operant performance closely resembles that of other organisms. Similarly they imply that although self-produced cues are extremely important in controlling human behaviour, reinforcement contingencies can also exert extremely delicate control over behaviour, even in the absence of an awareness of either the reinforcement contingency or the nature of the reinforced responses.

# CONCLUSIONS

From a reading of this chapter it should be apparent that a study of human operant performance is more involved than a study of animal operant performance; schedule contingencies are not the only factors which can come to control behaviour – indeed they are quite often the least effective variables in controlling operant performance! The most prominent fact to emerge from this discussion is that verbally mediated internal cues regularly gain control over behaviour and override the effect of any scheduled reinforcement contingencies. It is thus important for a full understanding of human operant performance to know, (1) the conditions under which self-produced cues are generated, what these cues consist of, and how they influence simple operant performance; and (2) whether schedule performance which is 'cleansed' of these internal controlling factors resembles that of other animals – in other words, can we still maintain the Skinnerian position that there are principles of learning and behavioural control which are universal to all animals? The answer to the latter question is probably yes, but we must remember that procedures are still being devised which eliminate self-produced cues and maximise control by schedule contingencies.

# 14 Behaviour Modification

Although the analysis of human behaviour is more involved than that of animal behaviour, the evidence in chapters 12 and 13 suggests that human behaviour does still seem to comply with the basic principles of conditioning and learning. There are certain to be subtle, but sometimes important, species differences in susceptibility to conditioning contingencies, and human behaviour is no exception, but by and large the behaviour of human beings in simple learning situations can be predicted and controlled if one has knowledge of the functional relationships between behaviour and environment. This knowledge consists of knowing such things as what contingencies exist between behaviour and its consequences, what stimuli are present when behaviour is emitted, what consistent relationships exist between different events in a person's life, and so on. Now it is at this point where we can make the jump from *passive prediction* of behaviour to *active control* of it. If we can predict an individual's behaviour from knowing the controlling variables that exist in the environment, then we can intervene to change those controlling variables and thus alter the individual's behaviour. This, of course, is a thesis which stems from the work of Skinner and we shall discuss this approach to human behaviour more fully in the final chapter. However, this present chapter serves as a bridge between the facts of conditioning in humans and the more broader theoretical speculations on the causes and control of human behaviour made by Skinner and his fellow radical behaviourists. In this context, then, there are a series of statements which should precede a discussion of behaviour modification techniques, and these serve as the assumptions upon which this branch of psychology is based: (1) human behaviour is governed by the fundamentals of learning which are embodied in operant and classical conditioning principles; (2) human behaviours (even the maladaptive and 'bizarre' behaviours clinicians commonly describe as pathological) are acquired, maintained and modified by these principles of learning; (3) since this approach emphasises that the 'causes' of behaviour are located in the environment (in the form of contingencies) and not 'within' the individual (in the form of 'motives', 'intentions', *etcetera*), these causes are directly accessible and can be manipulated; (4) if the causes of behaviour can be manipulated, then the behaviour of the individual can also be manipulated in a controlled and predictable fashion.

So in these assumptions we have the beginnings of a technology of behaviour, and this technology was first applied in the 1950s and 1960s to behavioural problems in the clinical setting. Again the assumption here is that many manifestations of what are labelled clinically as 'pathological' behaviours can be reasonably interpreted in terms of 'faulty learning'; more simply, it is assumed that an individual has come into contact with environmental contingencies which have developed inappropriate responses in certain situations (for example, phobias). If the maladaptive behaviour has been learnt, then it is reasonable to assume that it can be 'unlearnt' using therapies based on learning principles. The first such techniques were collectively labelled behaviour therapy (Eysenck, 1960); they consisted of procedures primarily based on classical conditioning principles and were designed to alleviate neurotic and phobic symptoms. Wolpe expresses the reasoning behind these therapies:

> Because human neurotic habits of reaction can often be dated from particular experiences that involve stimuli to which the patient has come to react with anxiety, and because these habits can be altered through the techniques of psychotherapeutic interviews, there is a *prima facie* presumption that neurotic reactions owe their existence to the learning process.
>
> (Wolpe, 1962)

From these beginnings developed the broader field of behaviour modification, and techniques then evolved from single-client therapies based on classical conditioning principles to group therapy and management procedures based on operant conditioning. The introduction of operant principles into the field of therapy was an important step because operant psychology offered a wider range of possibilities. First, it provided a method for changing complex and integrated behaviour patterns in a way that classical conditioning procedures could not; secondly, it allowed the introduction of group management procedures where therapy could be conducted in large groups and in settings which resembled those found in everyday life (for example, token economy systems, see p. 378ff.); and thirdly, it provided a framework in which those variables which controlled behaviour could be analysed, and as a consequence of this analysis, more appropriate therapy could be devised. Sandler and Davidson emphasise the importance of a *functional analysis* of this kind:

> The views expressed by Skinner and other operant theorists suggest that a better understanding of pathological conditioning can be accomplished by analysing the interactions between (a) the variables involved in an individual's behaviour history, and (b) those determinants currently impinging upon the organism. With the knowledge of the former we can better predict how the latter will influence behaviour.
>
> (Sandler and Davidson, 1973, pp. 63–64)

However, as well as introducing new therapeutic techniques, the advent of behaviour modification also marked a radical theoretical change in thinking about pathological behaviour. The traditional approach to pathological behaviour had been based on the 'medical' model of mental illness and this latter approach claimed that the behavioural abnormalities exhibited by phobics, schizophrenics, neurotics, and so on were merely the manifest symptoms of some underlying 'cause'. This cause might either be medical – in the sense that it reflected abnormalities in the central nervous system, brain biochemistry, or general physiological functioning of the individual; or it might be a deep-rooted personal problem – such as an 'inner conflict' or another similar psychodynamic process. Thus, this approach assumes that a 'cure' is not implemented unless an underlying cause has been eradicated – mere removal of the symptoms of this cause does not constitute a cure.

The assumptions of behaviour therapists and behaviour modifiers demand that they conceptualise the problem differently. Since they assume that maladaptive behaviours are acquired via normal learning processes, there is thus nothing pathological about them, and the removal of the behaviour therefore removes the problem. The behaviour *is* the problem, and once it is removed or changed then the individual's problem is removed. Similarly, because there is no internal or root cause of the behaviour, the behaviour is not a 'symptom' of anything, it simply exists by being maintained by environmental contingencies. The crucial test of these two opposed conceptualisations rests on the existence of 'symptom substitution' effects: if there is a root cause to a behavioural problem then elimination of the 'symptoms' without treatment of the cause should result in new, and possibly equally maladaptive symptoms, taking the place of the original ones. An unqualified answer to this test cannot readily be given and there is evidence in the literature to support either view; however, we shall discuss more fully the problem of symptom substitution later in this chapter (p. 388ff.).

At this point, before we progress to more detailed discussion, it would be wise to expound a few definitions and to assess the current scope of therapeutic techniques based on conditioning principles. *Behaviour therapy* is a term that is usually reserved for those techniques which are based on classical conditioning principles, while *behaviour modification* refers to therapies based on operant conditioning. However, this distinction is becoming more and more eroded as techniques combine both operant and classical principles in a single therapy, and currently it is more usual to think of behaviour therapy as a special form of behaviour modification (Brown, Wienckowski and Stolz, 1975). In general, behaviour modification typically attempts to influence behaviour by changing the environment and the way that individuals interact, rather than by intervening in a more direct way using drugs or surgery. Behaviour modification does *not* include psychosurgery, electroconvulsive therapy (ECT), and the non-contingent administration of drugs (that is, the administration of drugs independently of the individual's on-going behaviour).

From their beginnings in the 1950s the users of behaviour modification techniques have expanded into many clinical and management settings. For example, (1) autistic children, who might otherwise have to be continually restrained because of their self-destructive behaviour, have been helped by properly designed behaviour modification schemes (Lovaas, Koegel and Simmons, 1973); (2) severely mentally handicapped children, previously considered incapable of acquiring anything but the most fundamental responses, have been shown capable of acquiring relatively advanced linguistic and self-help skills (Baer and Guess, 1971); (3) behaviour modification techniques have helped to improve teaching methods and classroom management; (4) finally, apart from their now accepted use in clinical settings, such techniques have been extended to social problems such as 'the facilitation of cooperative living in a public housing project, decreasing littering, encouraging the use of public transportation, and enabling unemployed persons to find jobs' (Brown, Wienckowski and Stolz, 1975, pp. 3–4).

Perhaps the single most important factor contributing to the widespread popularity of behaviour modification techniques is that theories of psychopathy based on learning principles readily suggest appropriate therapeutic action. As Bandura points out, most other theories of psychopathy do not have this benefit:

> The major deficiencies of theories that explain behaviour primarily in terms of conjectural inner causes would have been readily demonstrated had they been judged, not in terms of their facility in interpreting behavioural phenomena that have already occurred, but rather on the basis of their efficiency in predicting or modifying them. Because the internal determinants propounded by these theories (such as mental structures, Oedipal complexes, collective unconscious) could not be experimentally induced, and rarely possessed unequivocal consequences, psychodynamic formulations enjoyed an immunity to genuine empirical verification. If progress in the understanding of human behaviour is to be accelerated, psychological theories must be judged by their predictive power, and by the efficacy of the behavioural modification procedures that they produce.
>
> (Bandura, 1969, p. 16)

But before we look specifically at the techniques of behaviour modification that this approach engenders, let us look in more detail at the way in which maladaptive behaviours can be considered to result from 'faulty learning'.

## LEARNING PROCESSES AND 'PATHOLOGICAL' BEHAVIOUR

Perhaps the single most important premise of the behavioural approach to therapy is the assumption that much of the abnormal behaviour commonly labelled as pathological can be considered as resulting from perfectly *normal* learning processes. It is not important to consider if there is anything physiologically wrong with the organism who exhibits these behaviours, it is simply claimed that any organism who is subjected to certain environmental contingencies will exhibit predictable behavioural abnormalities. Well, how true is this claim? There appears to be quite a reasonable amount of evidence to suggest that certain conditioning procedures do induce behaviours which resemble some types of maladaptive behaviours found in clinical patients. These procedures reliably and predictably generate such behavioural tendencies in both human and animal subjects.

## Experimental neuroses

Perhaps the oldest of these experimental procedures stems from the works of Pavlov. In studies of classical discrimination learning he and his colleagues found that certain kinds of experimental procedures induced signs of anxiety and irritability in previously friendly and co-operative dogs (Pavlov, 1927, pp. 291ff). In fact, the symptoms resembled those of behaviour found in human beings labelled as 'neurotic'. One particular procedure involved training the dog to salivate when a circle was presented on a screen. Following this, the subject was taught a discrimination between the circle (S+) and an ellipse (S−), with an initial ratio between the semi-axes of 2:1. This discrimination was learnt quite rapidly and as training progressed the shape of the ellipse was changed until it was almost circle-like (a ratio between the semi-axes of 9:8). Kimble describes the subsequent change in the dog's behaviour at this point:

> The hitherto quiet dog began to squeal in its stand, kept wriggling about, tore off with its teeth the apparatus for mechanical stimulation of the skin and bit through the tubes leading from the animal's room to the observer's. . . . On being taken into the experimental room the dog now barked violently . . . In short, it presented symptoms of a condition which, in human beings, we would call neurosis. Among the additional symptoms reported by later investigators have been (1) signs of anxiety such as whining and trembling: (2) a breakdown of the precision of the CR . . . (3) refusal of the dog to eat in the experimental apparatus or room . . . and (4) signs of strong 'inhibition' such as yawning, drowsiness and sleep.
>
> (Kimble, 1961, p. 441)

Numerous other studies (Masserman, 1943, 1950; Schneiderman, Pearl, Wilson, Metcalf, Moore and Swadlow 1971), have described similar behavioural effects under conditions which require a very difficult discrimination. These effects are usually quite predictable and the neurosis is often extremely persistent (Anderson and Parmenter, 1941).

Masserman (1943) has also reported 'neurotic-like' behaviour in cats in an approach-avoidance conflict situation. He trained cats to operate a switch which produced a stimulus signalling that food was available. The food was delivered automatically into a receptacle, but at the moment of feeding the cat would receive either an unpleasant airblast across its face or an electric shock delivered to its feet.

Both the 'approach-avoidance' procedure and the 'difficult discrimination' procedure generate a similar behavioural syndrome. The normal manifestations of this syndrome include (1) hyper-irritability, (2) resistance to entering the experimental chamber, (3) a tendency to avoid eating in the experimental chamber, (4) abnormalities in heart-rate and respiration in the experimental chamber, and (5) an incidence of aggression being induced in previously passive animals.

## Learned helplessness

In chapter 4 we discussed the phenomenon known as 'learned helplessness' (pp. 150–153) and mentioned that many of the subjects who failed to learn an avoidance response after exposure to unavoidable shock did exhibit neurotic-like behaviour that was similar in many respects to that of experimental neurosis. But apart from this, learned helplessness has other implications for an analysis of pathological behaviour. Seligman (1976) has suggested that the paradigm which produces learned helplessness in animals may be analogous to those processes which produce certain kinds of pathological depression in human beings. For example, a person may go through a period in their life when events around them seem totally out of their control; they experience a number of 'unavoidable aversive events', such as close friends or relatives dying unexpectedly, they may lose their job through no fault of their own, and so on. This being the case, says Seligman, they 'learn to be helpless', they become convinced that their behaviour has no effect on the environment and so they cease to emit behaviours which would enable them to cope with day to day living. In fact, they exhibit all the symptoms of what we call 'depression'. Of course, not all manifestations of what we call depression can be explained in this way, but it does suggest a way in which depression can be considered as a *learnt behaviour* rather than an internal mental or physiological state. If we can conceptualise ways in which depression might be learnt, this should also suggest means by which it might be successfully 'unlearnt'.

## Classical conditioning and the development of phobias

Again we can go back to a previous chapter to find a conditioning procedure which produces maladaptive or pathological behaviour. In chapter 12 we mentioned the early study of Watson and Rayner (1920). By utilising a classical conditioning procedure (but see p. 327) they were able to condition fear and avoidance reactions to a white rat in a 9-month-old child ('Little Albert'). Although Watson and Rayner appeared to be successful in their intention to demonstrate how phobias can be acquired through conditioning, their procedure is particularly unclear and not well-documented. Nevertheless, it seems to suggest that 'many of the phobias in psychopathology are true conditioned emotional reactions either of the direct or transferred type' (Watson and Rayner, 1920, p. 14), and, certainly, classical conditioning does seem to provide an ideal process for the transfer of emotional responses from an aversive stimulus (UCS) to an originally neutral event (CS). However, there is a note of caution that should be introduced here. Certain objects or events tend to be more popular as the focus for phobias than others, and some of these phobic stimuli are not the kinds of stimuli that an individual is likely to encounter regularly enough to acquire conditioned responses to them. For example, a common phobia is a dislike of snakes, yet it is highly unlikely that we have come into contact frequently with pairings of snakes and aversive UCSs. If all avoidance phobias resulted from classically conditioned fear then there should be more people walking around with phobias about lambs and kittens than snakes and spiders (Rachlin, 1976)! However, in spite of this there does seem to be some case for attributing certain kinds of phobias and fetishes to classical conditioning processes. An experimental study conducted by Rachman (1966a) provides a compelling example of this. In this study he attempted to demonstrate the acquisition of a sexual fetish through classical conditioning. The subjects (all male) were individually shown slides of attractive female nudes (the UCS), with each slide being preceded by a coloured slide of a pair of black, knee-length women's boots (the CS). The conditioned response was defined as changes in penis volume as measured by a phallo-plethysmograph. The CS was presented for 15 s followed by 30 s of the UCS. Three subjects acquired the conditioned response within 30–40 trials and one even exhibited generalisation of this response to other types of footwear. A subsequent study by Rachman and Hodgson (1968) further demonstrated that the acquisition of this 'fetish' was attributable to the contingency between CS and UCS, and not to any uncontrolled pseudoconditioning effects.

Other experimental studies have implicated the classical conditioning process in psychosomatic reactions. For example, Noelpp and Noelpp-Eschenhagen (1951, 1952) demonstrated that if an auditory stimulus (CS) was reliably followed by induced asthmatic attacks (UCS) in guinea pigs, then the guinea pigs came to exhibit respiratory dysfunctions characteristic of bronchial asthma even during the auditory stimulus alone. Dekker, Pelser and Groen (1957) have since extended this analysis to asthmatic attacks in humans, showing that the mere presentation of a mouthpiece through which asthma inducing nebulised allergens were inhaled also had the power to provoke asthmatic attacks.

It is this kind of evidence which strongly implicates classical conditioning processes in the development of many phobic reactions and psychosomatic disorders. However, despite the compelling simplicity of this analysis it must be emphasised that it cannot provide a full account of such behavioural phenomena, and indeed it seems to be the case that people appear to have 'preparedness' for acquiring phobic reactions to some stimuli (for example, snakes) rather than others (for example, lambs).

## Conditioned anxiety

The quasi-fear reaction we commonly label 'anxiety' is frequently manifest as a symptom of behavioural disorders. It is sometimes cited as a cause of the behavioural abnormality, in other cases it is considered merely as a correlate of it. We have already discussed the problems of defining anxiety and of envisaging its role in psychopathology (chapter 4, pp. 148–150), but it is clear that certain conditioning procedures produce behavioural changes which resemble the behavioural changes we normally attribute to anxiety. This is particularly true of the conditioned suppression procedure. This is important because it again suggests that there are normal causes for some pathological behaviours (see p. 150, Sidman, 1960a); that is, these behaviours are produced by *normal* learning processes and do not represent any kind of dysfunctioning of the organism's psychological or physiological processes. The conditioned suppression procedure provides a real example of the way in which some pathological anxiety states can be attributed to learning.

## Operant reinforcement of maladaptive behaviours

Of all the applications of conditioning principles to an explanation of pathological behaviour, perhaps the most obvious and simple one is to suggest that maladaptive or bizarre behaviours are developed and maintained because they have reinforcing consequences. They may either reduce anxiety (that is, they function as an avoidance response), or they may help the individual to acquire certain things which he considers valuable or important (a particularly potent example of such a reinforcer is attention and approval given by other members of one's family or society, etc.; praise and attention consequences are usually termed *social reinforcers*).

The way in which attention can act as a reinforcer – especially for the mentally handicapped and institutionalised psychiatric patients – is well illustrated in a study by Ayllon and Michael (1959). They report the behaviour of a mentally handicapped patient called Lucille, who persistently interrupted the work of the nurses in the hospital by visiting their office up to 16 times a day. After some initial complaint the nurses began to tolerate the visits, they talked to Lucille and led her back to the ward. They hoped the visits would eventually cease – but they did not. It seemed in this situation that the attention of the nurses was reinforcing Lucille's 'visiting' behaviour and Ayllon and Michael reasoned that if she were to be ignored by the nurses when she visited them, this behaviour would eventually extinguish. After about 50 days of this extinction procedure, Lucille's visits had fallen to only around 2 a day, indicating that her behaviour had in fact been controlled by the attentive consequences supplied by the nurses. A similar study is reported by Williams (1959) in the modification of tantrum behaviour in a 21-month-old infant. The child had been seriously ill in very early life and one parent had always spent time at his bedside waiting for him to fall asleep. When a parent was not present the child cried and had a tantrum. This tantrum behaviour was eventually extinguished by allowing the child to cry without reinforcing it with attention – whenever the child cried the parent left the room and did not re-enter until crying had stopped. The child cried for 45 min on the first night of treatment, but by the tenth night, the child no longer screamed and 'was observed to smile as the parents left the room'. To underline the fact that this tantrum behaviour was controlled by attentive consequences, Williams also reports that the tantrum behaviour was reinstated by an aunt who unwittingly reinforced the behaviour by re-entering the bedroom when the child cried.

Now, 'visiting' behaviour and tantrums in young children are fairly common responses, and it is not unreasonable to expect that they might be acquired and maintained by operant reinforcement. But what about the more bizarre behaviour patterns that are frequently exhibited by psychiatric patients? Can we attribute the ontogeny of these behaviours to operant reinforcement? It is true that many clinicians commonly characterise such behaviours as manifestations of underlying psychodynamic processes, or that they simply result from 'thought disorders'. However, a revealing study by Ayllon, Haughton and Hughes (1965) provides some insight into processes that might generate apparently bizarre behaviour in institutionalised patients. They reinforced a female schizophrenic simply for carrying a broom: whenever she was observed holding the broom a nurse would approach her, offer her a cigarette, or give her a token which could be exchanged for a cigarette. Eventually, when this behaviour was well established, it was transferred from a continuous to an intermittent reinforcement schedule until the patient was carrying the broom around for a considerable part of the day. It was at this point that Ayllon *et al.* called in two psychiatrists (who were unaware of the reinforcement schedule) to give their opinions on the nature of this behaviour. One of them gave the following reply:

Her constant and compulsive pacing, holding a broom in the manner she does, could be seen as a ritualistic procedure, a magical action . . . Her broom would be then: (1) a child that gives her love and she gives him in return her devotion, (2) a phallic symbol, (3) the sceptre of an omnipotent queen . . . this is a magical procedure in which the patient carries out her wishes, expressed in a way that is far beyond our solid, rational and conventional way of thinking and acting.

(Ayllon, Haughton and Hughes, 1965, p. 3)

There are two points to be emphasised here. First, the description given by this psychiatrist may well represent what goes on in the patient's head – we do not know for sure – but it does not reflect in any way the process by which the behaviour was acquired. Secondly, although Ayllon *et al.* have systematised the acquisition process in this instance, it seems reasonable to suppose that contingencies of this kind could be unwittingly set up quite frequently – especially in the confined environments of an institution such as a psychiatric hospital. Although this study does caricature the process of acquisition it certainly implies that operant reinforcement can play a powerful role in establishing even bizarre behaviour patterns.

In conclusion then, these conditioning procedures do produce behavioural effects which resemble many kinds of pathological behaviour. However, having said this, it is best to be cautious about what this means. As Sidman (1960b) points out, just because we can experimentally produce behaviour which *looks like* certain kinds of pathological behaviour in humans, it does not mean that the processes which produced the behaviour in the natural and experimental cases are identical. To be able to claim that the processes are identical in the two cases requires a much fuller functional analysis of the variables which shape and maintain these behaviours in man. While this task is a difficult one – since it often requires painstaking analysis into the detail of individual cases – what these procedures do provide is a factual framework around which possible therapeutic techniques can be moulded.

# TECHNIQUES OF BEHAVIOURAL CHANGE

## Classical conditioning procedures

### Extinction procedures

If maladaptive phobias or fetishes can be considered as maintained by pairing of the phobic stimulus (CS) with an unconditioned aversive stimulus (UCS) then perhaps the most obvious therapeutic approach is to extinguish the conditioned response. Classical conditioning principles tell us that this can be achieved in a number of ways; by presenting the CS in the absence of the UCS, by reconditioning the CS to an acceptable UCS, or simply by presenting the UCS in the context of new CSs (but this is likely to create many new phobias at the expense of eradicating the old one!). Bearing these operations in mind, there are a number of therapies which are derived from classical conditioning extinction procedures. However, as we shall see, they are often embedded uneasily within a psychodynamic framework – a fact which is perhaps not surprising since many of these therapies evolved in the 1950s, when a psychodynamic approach to psychopathology was still prevalent.

*Massing and flooding*

Pavlovian extinction has been integrated into one particular procedure called *massing*. This involves presenting the CS (or phobic stimulus) at repeated, regular intervals so that more extinction trials occur than do acquisition trials in a specified period of time. The extreme of this technique is known as *flooding*, where the subject is continually exposed to the CS for long periods. It is assumed that the CR (for example high anxiety levels and emotional reactions) will eventually extinguish because the CS ceases to be paired with an aversive UCS. The first study with human subjects was carried out by Malleson (1959). He instructed his client to feel more frightened as he related characteristics of the phobic stimulus to him. After an initial increase in distress, the client showed rapid recovery. Similarly, Miller and Levis (1971) demonstrated an increased tendency to approach and contact snakes in a group of snake-phobic girls who underwent a flooding procedure. They exposed the girls to snakes for periods of 0, 15, 30 or 45 min, and found that a significant number of subjects in each group showed improvement. There is some evidence that the longer the exposure period to the phobic stimulus the greater the likelihood of improvement (Rachman, 1966b; Wolpin and Raines, 1966, but see Morganstern, 1973, pp. 328–329).

In a slightly different study, Polin (1959) found that extended exposure to the CS with opportunity to emit the CR produced more rapid extinction than exposure when the emission of the CR was physically prevented; and with one or two exceptions, this seems to be generally true of human subjects.

Studies of massing and flooding which have used animal analogues of the human therapeutic procedure have testified to the efficacy of these extinction procedures (Baum, 1969, 1970; Baum and Oler, 1968), and also suggested further procedural refinements. But in many cases extrapolation from the animal results to the human therapeutic procedure has been difficult (Morganstern, 1973).

*Implosion therapy*

This is an extinction procedure first devised by Stampfl and Levis (1967). They assumed that many phobic reactions are not simply responses maintained by classical conditioning, but that they are usually classically conditioned fear responses which become maintained by the anxiety-reducing effect of avoiding the phobic stimulus. (Thus it is initially developed by classical conditioning but is eventually maintained by an operant avoidance response.) Morganstern describes the procedure in the following way:

> The procedure of implosion therapy begins with an initial assessment of the crucial stimuli that are associated with the patients' anxiety. From this assessment an Avoidance Serial Cue Hierarchy is constructed, with conditioned stimuli ordered according to the extent to which they are associated with the original primary reinforcement (UCS). Symptom-contingent cues . . . low on the Avoidance Serial Cue Hierarchy are the ones presented and extinguished first . . . Cues . . . higher on the Avoidance Serial Cue Hierarchy . . . because they are more threatening and more highly avoided . . . are extinguished last . . . the emphasis (is) simply on the extinction of all the anxiety-evoking cues which provide both motivational and reinforcing properties for perpetuating the patient's avoidance responses.
>
> (Morganstern, 1973, p. 319)

Thus a person afraid of snakes would be asked to imagine himself picking up a snake, after a while he would be asked to imagine how slimy the snake was, and then imagine the snake biting him, and so on. The implication here is that presenting each stimulus in an imagined form should extinguish the classical conditioned fear response that it normally elicits. The efficacy of this technique has been verified in a number of studies. For example, Hogan and Kirchner (1967) compared implosion therapy and relaxation (that is, the subject had to visualise a neutral image such as a quiet walk) as different therapies for rat-phobia in female students. After one session, a majority of the implosion group picked up a white rat whereas only 2 out of 22 in the relaxation group did so. Similarly, Levis and Carrera (1967) assigned subjects to one of three therapy groups, (1) implosion therapy, (2) 'conventional' therapy, consisting of insight and 'supportive' therapy, and (3) no treatment. Results indicated that subjects in the implosion group exhibited the greatest improvement on post-therapeutic assessment tests.

Certainly, implosion therapy does appear to be superior to a number of conventional therapeutic techniques in the treatment of phobias. However, one must be cautious about drawing definite conclusions on the matter, since many of the psychodynamic speculations considered to be a necessary part of implosion therapy (for example, introducing psychodynamic themes during the implosion process – Stampfl, 1967) have been omitted, and yet therapy has still been successful. Secondly, when implosion therapy has been successful it can be argued that success was not due *in toto* to the extinction procedure, but at least in part to 'the demand characteristics of the situation, the expectancy of the subjects, or other extra-therapy variables' (Morganstern, 1973, p. 332).

## Counter-conditioning

Counter-conditioning is also a procedure loosely based on Pavlovian extinction principles, but as well as extinguishing the conditioned response it, at the same time, develops a new, acceptable response. It is based on the principle that an organism cannot do two incompatible things at once. The procedure is illustrated well in an animal analogue study carried out by Klein (1969). Initially rats were trained on a discriminated avoidance task to run between compartments in a shuttle-box (see chapter 4, p. 132). This response was subsequently extinguished, but during the extinction procedure the rats were divided into three groups; one group was confined to the compartment where shock had been delivered and was also given food in this compartment (the counter-conditioning group); animals in a second group were simply confined in the compartment without food (flooding); and members of the third group were not confined at all (extinction). The results suggested that the counter-conditioning procedure was the most effective in eliminating the conditioned response.

A variation of counter-conditioning is known as systematic desensitisation by reciprocal inhibition (Wolpe, 1958). In this procedure the anxiety-escape responses to the phobic stimulus are extinguished and, while this process is occurring, these responses are gradually replaced by the 'fading-in' of responses which are incompatible with them.

## Systematic desensitisation

By carefully analysing those stimuli which evoke emotional or anxiety responses in the client, the therapist constructs a ranked list of events or stimuli to which the client reacts with increasing anxiety or avoidance. Having done this, the client is trained to relax while at the same time being in the presence of, or thinking about, the stimulus at the bottom of the fear-inducing hierarchy. When the client claims he is able to relax calmly while in this situation, he then progresses on to the next most fear-inducing event on the hierarchy (Wolpe and Lazarus, 1966). The assumption here is that relaxation is incompatible with fear and anxiety, and the two cannot occur together (the principle of reciprocal inhibition). Thus, very gradually, the fear reaction to the phobic stimulus is extinguished and replaced by relaxation as the conditioned response.

Systematic desensitisation has been by far the most widely utilised Pavlovian-based therapy, and arguably is one of the most effective (Morganstern, 1973; Paul, 1966; Wilkins, 1972). However, whether its efficacy is dependent on the strict procedural guidelines originally laid down by Wolpe is a different matter. Wolpe's procedure seems to derive logically from conditioning principles yet there appear to be a number of aspects of the procedure which are redundant when it comes to successful therapy. For instance, it does not appear to be necessary to structure the stimulus events into a systematic fear hierarchy in order to alleviate phobic reactions (Wilson and Smith, 1968; Wolpin and Raines, 1966). Secondly, a number of studies have found that relaxation is not strictly necessary to reduce anxiety (Lazarus, 1965; Paul, 1969; Rachman, 1968). Finally, it also seems that there is a good deal of social reinforcement involved in systematic desensitisation. For instance, reduction in phobic responding will occur if the therapist simply praises the client for approach, contacting or thinking about aspects of the phobic stimulus (Leitenberg, Agras, Thompson and Wright, 1969; Wagner and Cauthen, 1968). These problems of theoretical interpretations are ones we shall talk about again later (pp. 386–388).

## Classical aversive conditioning

The therapies we have talked about so far have followed a logical progression from (1) simple extinction procedures (massing and flooding), through (2) graduated extinction where the CR is slowly 'faded-out' (implosion therapy), (3) graduated extinction plus conditioning of an incompatible response (counter-conditioning), to (4) graduated extinction combined with graduated conditioning of an incompatible response (systematic desensitisation). An alternative approach is usually necessary when attempting to teach an individual conditioned fear to a formally attractive stimulus. Rather than extinguish the CRs to the attractive stimulus, the most direct approach is simply to pair that stimulus with an aversive UCS. This is known as classical aversive conditioning. Studies by Voegtlin and Lemere (1942) and Lemere and Voegtlin (1950) serve as examples of this technique. In their procedure alcoholic patients were given injections of emetine or apomorphine, which quickly elicit both nausea and vomiting (UCS). Immediately prior to vomiting the patient is given a drink of his favourite alcoholic beverage (CS). The procedure is successful with many patients in that the CS (alcohol) eventually comes to elicit a conditioned response of nausea. It is sometimes difficult to introduce into the therapeutic situation the stimulus for the undesired behaviour and as a consequence, verbal, pictorial or imaginal representations of the actual stimulus object are used in the therapy situation, with the hope that the subsequent aversive reaction will generalise to their real-life counterparts. For example – a typical treatment for homosexuality would be to show the client nude male and female slides while presenting an aversive electric shock when the male slides appear (Costello, 1963; James, 1962; Thorpe, Schmidt and Castell, 1963). (Such clients usually refer themselves for therapy and willingly undergo the therapeutic process. Classical aversion therapy is often unsuccessful for those who are coerced into treatment (Freund, 1960).)

Although this therapy appears to be a fairly straightforward extrapolation from classical conditioning principles, there are a number of questions we need to ask about its efficacy and theoretical 'purity'. For instance, this procedure can often implicitly contain operant components; in the aversive conditioning of alcoholic behaviour for example, is the nausea-producing drug acting as a classical UCS or as an operant punisher (Rachman, 1965; Rachman and Teasdale, 1969)? Secondly, is the aversive reaction that is established in the therapeutic setting readily generalised to real-life?

## Operant conditioning procedures

### The use of positive reinforcement

*Token economies*

One great advantage of operant conditioning techniques is that they are more readily adaptable to group therapy and group management situations, and perhaps the first group management procedure of this kind was called the *token economy* (Ayllon and Azrin, 1968). In this type of programme the participants receive tokens when they have engaged in an appropriate behaviour and they can, at some later time, exchange these tokens for a variety of desired (and hopefully positively reinforcing) items. The token acts as a generalised conditioned reinforcer and can be delivered with the minimum of delay after the required behaviour has occurred. This procedure is used both as a therapeutic and management procedure (see pp. 405–406). Its primary use as a therapeutic technique is with institutionalised persons such as psychiatric patients, and a typical token economy scheme will require that in order to obtain tokens the individual performs various self-care behaviours such as combing hair, bathing, brushing teeth, making their bed, etc. and also help in the general day-to-day running of the ward by running errands, cleaning the ward, etc. (Ayllon and Azrin, 1965). With more withdrawn patients tokens may be given simply to encourage socialisation. In general, such a programme helps to increase the sociability of institutionalised individuals and also strengthens behaviour that is compatible with that needed in society at large (for example, regular performance on a job, self-care, exchange of currency for desired items). Token economies are also used in half-way houses – not only for psychiatric patients, but also for social offenders, delinquents and drug addicts. They are also used in more every-day settings as a useful management procedure. In this context their adoption is most commonly been to aid classroom management in schools; the intention here is to replace punitive control methods, such as corporal punishment, with a positive counterpart that allows learning to occur in a more benevolent and less fearful atmosphere.

As a technique for behaviour modification, the token economy has a number of advantages: (1) it does not need professional therapists to dole out tokens; for example, once the nurse on the token economy ward has been informed of the target behaviour for each patient, she can contingently deliver the tokens; (2) such programmes prevent the deterioration of normal social behaviour on the admittance of an individual into an institution; and (3) they can be used in any group institution (Kazdin, 1975; Kazdin and Bootzin, 1972, for fuller evaluative reviews of token economy schemes).

Certainly the token economy is probably the most widely utilised technique of behaviour modification based on operant conditioning principles, primarily because it is soundly based on conditioning theory and also because the client–therapist ratio can normally be greater than 1:1! Nevertheless, operant conditioning principles have been utilised in behaviour modification to more sophisticated ends and in more sophisticated ways.

## Response shaping

Operant conditioning is effective only if the target response is emitted in the first place; it can then be reinforced. However, what happens when the target response occurs infrequently or has *never* occurred in the known history of the subject? This is a problem we encountered in chapter 2 (pp. 40–41), and the solution was to attempt to reinforce successive approximations to the behaviour. This is an especially useful strategy when attempting to develop relatively skilful behaviours in mentally handicapped individuals and severely withdrawn psychotic patients. An early study by Isaacs, Thomas and Goldiamond (1960) serves to illustrate this method. They attempted to reinstate verbal behaviour in a psychotic patient who had been mute since commitment 19 years earlier. Since he was withdrawn and exhibited little psychomotor activity they might well have waited for another 19 years before the subject emitted a verbal response eligible for reinforcement! They tackled this task in two steps; (1) it was necessary to find an appropriate reinforcer for this individual, and (2) it was necessary to break down the target behaviour so that it could be reached by reinforcing a series of approximations to it. They solved the first problem by discovering that the patient moved his eyes when chewing-gum was waved before his face; this looked as though it might function as an effective reinforcer for this individual. The second problem was solved by constructing a hierarchy of responses which eventually would lead to verbal behaviour; the first responses in the hierarchy were to be fairly simple, discrete responses whose operant levels were high enough for them to occur within a training session. The shaping programme went as follows: (1) when the patient moved his eyes towards the chewing-gum, the experimenter gave him the gum. After 2 weeks the probability of this response to the gum was quite high; (2) the experimenter now only gave the patient the gum when he moved both eyes and lips. By the end of the third week, those behaviours were well under the control of the reinforcer; (3) the experimenter then withheld giving the gum until the patient made a vocalisation of some sort. By the end of the fourth week the patient was moving his lips and eyes and making an audible 'croak'; (4) during weeks 4 and 5, the experimenter asked the patient to 'say gum', repeating this each time the patient vocalised. At the end of week 6 the patient spontaneously said 'gum please'; (5) in later sessions the patient verbally responded to questions by the experimenter, but the behaviour seemed to be stimulus-bound in the sense that he would reply verbally to only the experimenter and usually only in the experimental room; (6) to counteract the specificity of this behaviour the patient was placed back in the ward environment and the nurses were instructed to tend to his needs – but only if he verbalised them. This not only acted to reduce the stimulus-bound nature of the behaviour but also transferred the control from a chewing-gum reinforcer to a wider range of more 'natural' reinforcers.

*Modelling*

In order to speed up the acquisition of a response the process of shaping can be supplemented by *modelling*. Instead of waiting for the desired response to be emitted, a person who already knows how to engage in the response (usually the therapist or experimenter) demonstrates it for the individual who is learning. This combines processes of operant reinforcement and observational learning (see chapter 9), and is especially effective in teaching skills to the mentally handicapped – a category of individuals who can be particularly imitative. For example, in the teaching of language to non-verbal children, modelling is particularly useful. The child is first given a period of imitation training – basically he just learns to imitate the therapist – and this involves the child imitating a series of gross motor movements such as clapping, standing, etc. (Bricker and Bricker, 1970; Buddenhagen, 1971). After this has been established, the therapist attempts to focus imitation onto finer movements round the mouth; the therapist will generally model mouth shapes necessary for sounds such as 'eeee' or 'o' (Sloane, Johnston and Harris, 1968; Stark, Giddan and Meisel, 1968). Finally the therapist will now introduce sounds to accompany the mouth movements, and eventually attempt to string these sounds together to produce meaningful words (see Harris, 1975, for a review of the role of modelling in language training with children). Modelling is a technique which is used frequently in behaviour modification, not only to develop basic skills in the mentally handicapped but also to teach quite sophisticated skills to individuals with psychological and social problems. For example, if a person were learning socially appropriate ways to greet members of the opposite sex, the therapist would find it advantageous to model this behaviour for the client. For a fuller account of modelling and its role in behaviour modification, the interested reader is referred to Bandura (1969, chapter 3).

*Contingency contracting*

One of the quickest ways to circumvent the need for shaping and modelling procedures is to inform the client verbally of the contingencies whenever this is possible. This is essentially the purpose in contingency contracting and involves the 'striking of a bargain' between two or more people. For example, a therapist and his client may decide together on the behavioural goals and on the consequences (either reinforcing or punishing) that the client will receive if he does or does not achieve these goals. An instance of this type of procedure is illustrated in a study by Henry Boudin with drug abusers:

> The contracts made between the drug abusers and the therapists cover a large number of aspects of the addicts' lives. For example, an addict might agree to set up a joint bank account with his therapist, to which the addict deposits his own money. If a urine test indicates that he has broken his promise not to use illegal drugs, funds are taken from that account by the therapist and sent to some organization that the addict strongly dislikes. Contracts work both ways: If the therapist is late for an appointment with the addict or misses a therapy session, he can be required to deposit money to the addict's account. A contract involving positive reinforcement might specify that if the addict completes some amount of time or a job, he would receive a few movie passes or discounts on some number of phonograph records.
>
> (from Brown, Wienckowski and Stolz, 1975)

Perhaps one of the areas in which contingency contracting is most frequently used is in marriage guidance counselling. In this situation, a husband and wife will state the behaviours they would like to see exhibited by their spouse, and also the kinds of things they themselves find reinforcing. Once these have been defined, the contingencies can be set up: for example, the husband may like to see the house kept tidy, and the wife may like to be given chocolates or flowers more regularly – so, if the wife keeps the house tidy, the husband gives her chocolates and vice versa.

In reality all that contingency contracting is attempting to do is to systematise relationships in line with operant conditioning principles. Each person entering into the contract agrees both to change his own behaviour and to provide reinforcement for the changes that the other person makes.

## Punishment procedures

*Painful consequences*

Studies of aversive control in animals have tended to use electric shock as the noxious stimulus primarily because, as Azrin and Holz (1966) point out, electric shock can be presented immediately on execution of the selected response, and its intensity and duration can be closely controlled. However, the ethical and moral issues involved in presenting such noxious stimuli to human subjects are more complex. Nevertheless, as a general rule of thumb, electric shock is used as a punisher only when it is necessary to implement *immediately* the suppression or elimination of a particular behaviour. An example of such a situation is when individuals exhibit self-injurious behaviours; autistic children and severely mentally handicapped individuals are particularly prone to indulge in behaviours which are self-mutilating. These can include face-slapping, head-banging (against walls or any other hard object), hair-pulling, face-scratching, finger-biting, etc. all of which need to be eliminated fairly rapidly in the interests of the individuals themselves. As we saw in chapter 4 (p. 121ff.), the most effective method for suppressing a behaviour rapidly is punishment, and a number of studies have utilised electric shock as a punisher for self-injurious behaviours (Corte, Wolf and Locke, 1971; Lovaas and Simmons, 1969; Risley, 1968). Corte *et al.* compared the effectiveness of 3 techniques for eliminating self-injurious behaviours in severely mentally handicapped adolescents: (1) elimination of the social consequences of self-injurious behaviour (that is, they were removed from all social interaction so that this could not act as a reward for self-mutilation); (2) reinforcement (using food reward) of behaviour other than that which was self-injurious (a differential reinforcement of other behaviour schedule); and (3) punishment of self-injurious behaviour using electric shock. In terms of the subsequent reduction in frequency of self-mutilating behaviours the electric shock punishment procedure was the most effective. Although this result testifies to the usefulness of electric shock punishment in this kind of context, it is at first sight a paradoxical finding: the self-mutilating behaviour is presumably painful yet painful consequences still eliminate it. The paradox will not be resolved until we know a little more about the factors which normally maintain self-mutilating behaviour.

*Time-out from positive reinforcement (TO)*

A different kind of punishment procedure is that which uses removal of positive reinforcement as the noxious event (Coughlin, 1972; Leitenberg, 1965). In general, this consists of a loss of privileges following a particular behaviour, or more simply the withdrawal of attention by the therapist when an undesired behaviour is exhibited by the client. In animal studies, bar-pressing and key-pecking responses can be quickly eliminated if they produce a stimulus which is correlated with extinction (Coughlin, 1972). However, its use in behaviour modification programmes has been refined so as to satisfy the needs of the individual programme. For example, in certain kinds of institutions (for example, schools, half-way houses for social offenders) the removal of privileges and presentation of stimulus changes correlated with this punishment can best be achieved by placing the individual in what is known as a 'time-out room'. This is usually a small, bare room where the offender must spend a specified period of time immediately after the undesired piece of behaviour. This procedure has been successful in reducing the frequency of disruptive and aggressive behaviour in psychiatric patients (Bostow and Bailey, 1969), institutionalised delinquents (Tyler and Brown, 1967), and school children (Wasik, Senn, Welch and Cooper, 1969) as well as improving the eating habits (Barton, Guess, Garcia and Baer, 1970) and toilet behaviour (Azrin and Foxx, 1971) of mentally handicapped children.

## Procedures using drugs

We mentioned near the beginning of this chapter that the non-contingent administration of drugs did not constitute a behaviour modification technique. However, drugs can be used in conjunction with operant principles in two important ways. First, drugs can be used specifically as punishing stimuli when made contingent upon particular behaviours. The type of drug will often depend on the behaviour being eliminated. For example, in the treatment of alcoholics, nausea-inducing drugs can be administered after the client has been given an alcoholic drink – the after-effects are aversive, and related to the character of the behavioural problem (Lemere and Voegtlin, 1950; Rachman and Teasdale, 1969). Similarly, succinylcholine chloride (Anectine) is a drug which induces unpleasant sensations of drowning and suffocation, and has been used successfully to reduce an individual's dependency on sniffing various substances such as model airplane glue (Blanchard, Libet and Young, 1973). Again, the drug is aversive, and because its physiological effects seem appropriate to the particular behavioural problem its use is successful.

Secondly, drugs can be used to induce deprivation states so that certain reinforcers can be more effective. For example, drugs can be used to induce a state of thirst accompanied by an extremely dry mouth: these conditions make iced lemon water a particularly potent reinforcer. Similarly, the relaxation necessary in certain behaviour therapy techniques (notably systematic desensitisation) can be induced more readily by administering drugs which enhance relaxation (Brady, 1966).

## Biofeedback and behavioural self-control

Nearly all of the therapies we have talked about so far have been based on fundamental conditioning principles such as reinforcement and punishment, and have required a professional or paraprofessional agent to set up the conditioning contingencies and administer the consequences of the client's behaviour. This tends to characterise the therapy situation as 'mechanistic', the therapist as a controller, and the client as a passive, manipulated organism. This, of course, is the least glamorous perspective of behaviour modification and we shall discuss the implications of such a conceptualisation later. However, many individuals do not need nor desire constant management of their behaviour by outside agents; they want to change their own behaviour but just do not know how to go about it. In this situation, behaviour modification can offer techniques whereby the individual can control and manipulate *his own behaviour*. Two specific techniques are important in this respect; one helps the individual to become aware of aspects of his own behaviour so that he can bring them under 'voluntary' control (biofeedback), and the other provides the individual with a framework within which he can analyse his behaviour, identify the variables controlling it, and hence systematically modify it (behavioural self-control).

## Biofeedback

In chapter 2 (pp. 55–57) we discussed experiments by Miller and his colleagues which suggested that responses of the autonomic nervous system could be controlled by contingencies of operant conditioning. In Skinner's original 1938 terminology this implied that autonomic responses could be brought under 'voluntary' control. This research, combined with that of Kamiya (1968) in the development of voluntary control over complex psychological states, suggested that if (1) the sensory consequences of visceral or autonomic responses could be enhanced (that is, they could be made more discriminative), and (2) operant reinforcers could be made contingent upon specific changes in these responses, then individual control of autonomic and visceral responses might be achieved. This is usually implemented in practice by monitoring a particular visceral response, for example, blood-pressure, and linking changes in this response to a visual or auditory signal. Increases in blood-pressure can then be registered as increases in tone frequency or as the movement of a marker along a scale on a visual display unit. The subject is simply asked to try and increase or decrease his blood pressure by observing the exteroceptive feedback stimulus. Using this methodology, a great variety of responses which were formally considered to be involuntary have been brought under the voluntary control of the subject. These include the regulation of blood-pressure and heart-rate, especially in individuals for whom this is clinically desirable (Benson, Shapiro, Tursky and Schwartz, 1971; Engel and Bleecker, 1974; Kristt and Engel, 1975; Shapiro, Tursky and Schwartz, 1970; see Miller, 1975 and Blanchard *et al.*, 1974, for reviews of clinical applications of biofeedback techniques); the regulation of brain-waves in order to control particular psychological states and aid relaxation (Brown, 1970, 1971; Nowlis and Kamiya, 1970; Plotkin, 1976; Shapiro, 1976); the regulation of gastric secretions in ulcer patients (Welgan, 1974); and the control of less well-defined psychological phenomena such as headaches (Budzynski, Stoyva and Adler, 1970), and pain (Melzack and Perry, 1975).

## Behavioural self-control

Biofeedback teaches the client to become aware of quite discrete aspects of his behaviour, a different technique enables the subject to become more aware of the factors controlling his behaviour. Basically, what behavioural self-control aims to do is to teach the individual the principles of behaviour control (that is, the principles of conditioning) so that he can use these principles as a framework for understanding why he does what he does; it also enables the individual to set up his own contingencies of reinforcement so that in effect he can himself reinforce those behaviours he considers desirable. As a method of behaviour modification this technique has a number of advantages over those we have discussed so far. As Thoresen and Mahoney point out:

> If a person can be helped to manage his own behaviour, less professional time may be required for the desired change. Moreover, the person may be the best possible agent to change his own behaviour – he certainly has much more frequent access to it than anyone else, particularly when the behaviour is covert. Self-control strategies may also avoid some of the generalization and maintenance problems that often plague therapist-centred strategies. . . . Finally, training in self-control may provide an individual with technical and analytic talents that will facilitate subsequent attempts at self-control with different behaviours.
>
> (Thoresen and Mahoney, 1974, p. 7)

Behavioural self-control involves two important subsections (1) Environmental planning, and (2) Behavioural programming.

## Environmental planning

This essentially involves fading out undesirable responses by decreasing the number of stimuli which will elicit them, and increasing the frequency of desirable responses by increasing the number of stimuli that will evoke them. For example, eating responses occur in the presence of very many environmental cues that eventually gain control over the response. To regulate over-eating therefore it is important to *reduce* the number of these cues. This can be done by eating only in specific places so that the number of cues eliciting eating or 'nibbling' can be reduced, and secondly by never engaging in reinforcing activities while eating (for example, watching television, reading) (Ferster, Nurnberger and Levitt, 1962; Stuart, 1967).

## Behavioural programming

In contrast to manipulating the antecedent stimuli which can control a behaviour, the individual can also manipulate the consequence of his behaviour by laying down a set of rules which he must follow. Thoresen and Mahoney list some of the possible strategies.

1. *Self-observation* the recording, charting, and/or display of information relevant to a controlled response (e.g. charting one's weight).

2. *Positive self-reward*   the self-administration or consumption of a freely available reinforcer only after performance of a specific, positive response (e.g. treating one's self to a special event for having lost weight).

3. *Negative self-reward*   the avoidance of or escape from a freely avoidable aversive stimulus only after performance of a specific, positive response (e.g. removing an uncomplimentary pig poster from one's dining room whenever a diet is adhered to for a full day).

4. *Positive self-punishment*   the removal of a freely available reinforcer after the performance of a specific, negative response (e.g. tearing up a dollar bill for every 100 calories in excess of one's daily limit).

5. *Negative self-punishment*   the presentation of a freely available aversive stimulus, after the performance of a specific, negative response (e.g. presenting one's self with a noxious odour after each occurrence of snacking).

(Thoresen and Mahoney, 1974, p. 21–22)

So, behavioural self-control does seem to be the logical goal for behaviour modification; it hands over contingency management and response selection to the individual himself, it can enhance the individual's perception of the causes of his own behaviour, and also removes some of the stigma of control and manipulation from therapies based on conditioning principles.

## Relationship of behaviour modification techniques to conditioning theory

Since their innovation in the late 1950s and early 1960s, behaviour modification and behaviour therapy techniques have multiplied until there are literally hundreds of different procedures either closely or loosely based on conditioning principles. So it is probably instructive at this point to see how closely these techniques actually do adhere to learning theory or how much they are simply means to an end. To put this in another way, the goal for psychotherapy is primarily to alleviate behavioural or personal problems, and conditioning initially offered one valuable means to this end. Since then, however, techniques have been refined and amalgamated so that although the goal is still the same, the means may be losing sight of theory. There are two facets of this problem. First, many techniques have successfully combined either (1) different theoretical approaches to therapy, for example, implosive therapy, which combines principles of classical conditioning with a psychodynamic interpretation of anxiety hierarchies (see pp. 374–375); or (2) combined principles from classical *and* operant conditioning; for example, Feldman and

MacCulloch (1965) developed a technique for treatment of homosexuality which combines principles of classical aversive conditioning with operant escape and avoidance responding. Each client viewed a projected slide of a nude male. After 8 s a painful electric shock was delivered and maintained until the client responded by closing a switch which terminated shock, removed the male slide and replaced it with a female nude slide; (3) or more recently, techniques have tended to combine various principles of operant conditioning into a single therapy. One such recent example is overcorrection (Foxx and Azrin, 1972, 1973). Overcorrection is a behaviour modification method combining elements of positive reinforcement, punishment and extinction (although in some cases it is difficult to identify which aspect of the procedure corresponds to which process!). For example, a patient in a psychiatric hospital who overturns a bed in a dormitory might be required not only to remake the bed, but also to straighten the bedclothes on all the other beds in the dormitory. Remaking the bed is praised (reinforcement), the physical effort of straightening all the other beds is aversive (punishment for overturning the bed in the first place) and since the attention gained by overturning beds may have reinforced this behaviour in the past, the overcorrection technique effectively extinguishes this source of reinforcement.

Secondly, although techniques may superficially seem adequately tied to theory, a closer analysis can often reveal striking anomalies. For example, although the classical aversive conditioning technique used by Lemere and Voegtlin (1950) is often quite effective in eliminating alcoholic behaviour, it has been argued that this procedure is actually *backward conditioning* in which the UCS (aversive drug) is presented before the CS (alcoholic drink), and it is generally considered that backward conditioning can more readily produce inhibition of the CR than enhancement of it (Mackintosh, 1974, pp. 58–60; Rackham and Teasdale, 1969). Similarly, if one looks more closely at the systematic desensitisation procedure, the assumption is that the therapist is attempting to form a classical association between a UCS (relaxation) and the phobic stimulus (CS). But for this to be the case the CS must *predict* the UCS; however, since the therapist attempts to maintain the state of relaxation *continually* during therapy there is really no contingency between CS and UCS at all, and under these conditions classical conditioning should not occur. A further anomaly concerns the status of self-reinforcement in behavioural self-control techniques. Goldiamond (1976) has argued that self-reinforcement cannot be considered as true operant reinforcement in the same way that we reinforce a rat in a Skinner-box. He stresses that it is not the delivery of the reinforcer which is the important operation in operant conditioning but the setting up of the contingency between response and reinforcer. So for operant reinforcement to occur there must be a specified contingency between response and reinforcer and some *independent* evaluation of when the response has occurred. In a sense all operant reinforcers are self-administered because it is the conditionees response that 'produces' the reinforcer; it is an independent evaluation of the contingency which defines operant reinforcement and it is almost impossible for an individual truly to set up his own contingencies because he can readily change them, or simply 'cheat'!

There is no doubt that techniques of behavioural change have tended to outstrip theory (London, 1972) with the result that when specific procedures fail it is more often the technique that is questioned rather than the theoretical underpinnings. This is an extremely important problem dependent to a large extent on how we evaluate the efficacy of behaviour modification techniques. Are failures simply a result of using an inappropriate technique, or do they question the assumptions about pathological behaviour on which behaviour therapies are based?

# EVALUATION OF TECHNIQUES OF BEHAVIOURAL CHANGE

When it comes to evaluating the worth of behaviour therapy and behaviour modification techniques there are two relevant questions that need to be asked: (1) when a therapy does result in behavioural improvement, how sure can we be that the improvement results directly from the therapeutic procedure and not from 'placebo' effects? and (2) what criterion does the behaviour analyst actually adopt to indicate improvement, and is this adequate to avoid the charge that behaviour modification merely substitutes one maladaptive behaviour for another?

We have already mentioned a number of experimental studies which have suggested that therapies based on conditioning principles do have a significantly greater effect on the target behaviour than 'non-treatment' or conventional 'supportive' therapies (Hogan and Kirchner, 1967; Levis and Carrera, 1967; Miller and Levis, 1971). However, there are still studies which report 'placebo' or 'Hawthorne-like' effects. For instance, a client may well show improvement simply because a new technique is being tried out on him, or simply because he 'believes' the treatment will work (placebo effect). Placebo effects may well be operating in flooding and systematic desensitisation studies (Morganstern, 1973; Paul, 1966; Sloane, 1975; Wagner and Cauthen, 1968); and it is not unconceivable that Hawthorne effects might appear when, for example, token economy programmes are introduced into institutional or group settings. This places great importance not only on the comparison of therapeutic treatment with non-treatment control groups, but also on the rigour of 'improvement' measures. It requires that the therapist monitor behaviours other than the particular target behaviour in order to interpret any likely symptom substitution effects, and secondly, it requires that the therapist continues monitoring these behaviours after his criterion for elimination or establishment of the target behaviour has initially been met.

It is this latter problem of what behaviours to monitor and how long to monitor them for that has aroused controversy concerning the appearance of symptom substitution following behaviour therapy/modification. Some theorists have rejected the suggestion that symptom substitution does occur with behavioural procedures (Cahoon, 1968; Ullman and Krasner, 1965), while others have suggested that because evaluative measures are based on the theoretical assumptions of behaviour modification, we are bound to overlook symptom substitution should it occur. For example, Willems (1974) suggests that because behaviour therapists believe that there are no underlying causes to a behaviour problem – that is, that the behaviour itself is all that needs to be changed – there is thus a tendency to monitor only the target behaviour. He further adds that even if one denies that there is an underlying cause to the behavioural problem, symptom substitution could still occur because: (1) we may have failed to evaluate how response systems are interlinked; for example, changing the frequency of one response may affect the frequency of a related one – there may be a hierarchy of responses, each inter-related with one another (see also Bandura, 1969); (2) changing the behaviour of the client may radically alter the way that other people react to him, and since the other people in an individual's environment are contingency arrangers and reinforcement dispensers, this may reflect back on the client's behaviour in an undesirable way. This implies that symptom substitution can be expected even within a behavioural analysis of pathological behaviour; in this case it is not because some hypothetical underlying cause has not been eradicated, but because we have failed to recognise the systems-like nature of behaviour in which responses are interlinked, and changing the frequency of one particular behaviour can also affect the contingencies which may be maintaining or shaping others. Willems gives two interesting examples within this context:

One subject-mother was observed to nag (emit commands) at rates of up to 100 or more per hour, and the child complied at a very low rate. The rate of the mother's commands was reduced to an average of 15 per hour, and, correspondingly, the proportion of compliance on the part of the child went up. This was the outcome that had been designated as successful. However, the investigator went on to report difficulty in dealing with this case. As the study progressed and as the shaping of nagging succeeded, the mother's rate of eating went up, she gained weight, and she reported frequent anxiety and tension. Finally, she abandoned the child and left town. These events were seen by the investigator as only an unfortunate and vexing interruption of the treatment programme.

In the Probation Department of Los Angeles, some explicit use has been made of token systems and other behaviour modification techniques in dealing with deviant behaviour among adolescent boys. The probation officers were successful in reducing the rate of petty vandalism, such as stealing hubcaps and items from stores. However, as the petty vandalism went down, the rates of more serious offences, such as stealing cars and destroying property, went up.

(Willems, 1974, p. 157)

The psychodynamic therapist would claim that these undesirable behavioural substitutions came about because the behaviour modifier only tackled a 'symptom' of some more important underlying cause; the behaviour modifier, however, could reasonably claim they resulted from the fact that changing one aspect of behaviour indirectly changed other contingencies in the environment, and until we understand *how* this trellis-work of responses and contingencies is structured we are unlikely to be able to predict whether 'symptom substitution' will result or not. One implication of the behavioural account of 'symptom substitution' is that it does not presuppose the quality of any substituted responses – merely because we have changed the frequency of one behaviour does not mean that we may indirectly affect other behaviours *for the worse*. Indeed, there are very many accounts in the behaviour modification literature where behaviour modification programmes directed at quite specific behaviours, have also had beneficial effects on other behaviours – they may have increased the frequency of socially desirable responses in formally withdrawn psychiatric patients, or the frequency of self-help skills in the mentally handicapped (Ayllon, 1963; Ayllon and Haughton, 1962).

## CONCLUSIONS

In this chapter we have discussed learning interpretations of pathological behaviour, the theoretical underpinnings of therapies based on conditioning principles, and the efficacy and evaluation of these techniques. It must be clear to the reader that whether one considers behaviour modification techniques to be successful or not depends on the evaluation criteria one adopts, and, where there is a paradigm clash—as there is in approaches to psychotherapy—the criteria for success will not be obvious. However, it is clear that behaviour modification techniques have been successful in ameliorating behaviour problems in a wide variety of settings: clinics, psychiatric hospitals, half-way houses, school-rooms, and every-day environmental settings. But the problem in many cases still remains to link this success unequivocally to the conditioning principles explicit in the techniques, and this is not a necessary worry for clinicians whose primary aim is to alleviate behavioural problems – the ends often dictate the means. However, in chapters 12 and 13 we noted that human behaviour is affected by basic laws of conditioning, so, at this stage in our knowledge, it is not unreasonable to assume that certain pathological behaviours can be accounted for by 'faulty' learning and that many of the hard-nosed techniques of behaviour modification are successful because they incorporate basic principles of conditioning. Nevertheless, behaviour modification is just one facet of a technology of behaviour change. We mentioned at the beginning of this chapter that the passive prediction of behaviour naturally invited active control of it, and this raises important ethical issues–there is, for example, only a thin line between therapy and more perfidious management of behaviour. These issues have been side-stepped in this chapter so that they can be discussed in the following chapter within the broader framework of conditioning models of man.

# 15 Radical Behaviourism and Behavioural Engineering

The previous three chapters have illustrated either the phenomenon of conditioning in human beings or the way in which conditioning principles have been utilised in attempts to change human behaviour. Yet, in the view of the majority of laymen, 'conditioning' probably represents only one way in which human behaviour can be influenced – conditioning in the vernacular English implies an unconscious, almost insidious, process of behaviour modification which is less important than more 'cognitive' processes such as 'choosing', 'thinking', 'reasoning', 'free-will', etc. In fact, to most non-psychologists the term 'conditioning' implies the conditioned reflex, and the establishment of 'reflexive' or 'instinctive' habit-like behaviours. Indeed, many of the important theoretical 'schools' within psychology attempt to explain human behaviour within conceptual frameworks which do not allude to principles of conditioning at all. Now, with these points in mind, how pervasive are the basic principles of conditioning in an analysis of human behaviour? Is all human behaviour the result of conditioning processes, or must one conclude that such processes supply only a small fraction of the 'causes' of our behaviour? The first point to make is to stress that accounts of human behaviour which do not allude to principles of conditioning are not necessarily incompatible with conditioning accounts. In many cases the behavioural phenomena that we are dealing with are extremely complex and for an adequate analysis they therefore require a framework conceptually more elaborate than conditioning terminology. For example, although in theory we might be able to conceive of the disruptive behaviour of a teenage delinquent in terms of a vast multiplicity of conditioning contingencies that have shaped his behaviour during his lifetime, in practice this is not really possible: first, because we don't have a history of his life in anything like the detail necessary, and secondly, simply relating these contingencies to the adolescent's present behaviours would be a totally unenviable task. So, what does happen in these cases is that an analysis on a more molar level is attempted – an analysis which couches the 'causes' of the individual's present behaviour in terms which are not so obviously related to conditioning. For instance, his disruptive behaviour may be related to the 'lack of a mother-figure during early infancy', or his 'treatment by peers during early adolescence', etc. These accounts point to the crucial factors in the individual's life but do not explicitly spell out how they have the effects they do. Staunch advocates of conditioning theory would probably argue that these factors have the effects they do because they set up a particular sequence or trellis of reinforcement contingencies which have a profound and predictable 'shaping-up' effect on that individual's later behaviour. But even so, the details of these contingencies are still not readily specifiable. However, one direct benefit of a conditioning approach to complex human behaviour is that it does tend to help the psychologist to eliminate explanatory 'deadwood'. For example, conditioning theory emphasises that the causes of behaviour are all located in the external environment in the form of contingencies. Thus, to say that the delinquent's behaviour is caused by his 'personality' is no explanation at all, because we then have to account for how all those behaviours which make up the individual's personality are themselves shaped-up – the answer, says the conditioning theorist, must eventually lie in environmental contingencies. Skinner originally argued the behaviourist's mode of interpretation in the following way:

> The *enthusiastic* person is, as the etymology of the word implies, energized by a "god within". It is only a modest refinement to attribute every feature of the behaviour of the physical organism to a corresponding feature of the "mind" or of some inner "personality" . . . The inner man wills an action, the outer executes it. The inner loses his appetite, the outer stops eating. The inner man wants and the outer gets. The inner has the impulse which the outer obeys . . .
>
> The practice of looking inside the organism for an explanation of behaviour has tended to obscure the variables which are immediately available for a scientific analysis. These variables lie outside the organism, in its immediate environment and in its environmental history. They have a physical status to which the usual techniques of science are adapted, and they make it possible to explain behaviour as other subjects are explained in science. These independent variables are of many sorts, and their relations to behaviour are often subtle and complex, but we cannot hope to give an adequate account of behaviour without analyzing them.
>
> (Skinner, 1953, pp. 29 and 31)

This chapter has so far been angled from the viewpoint of the committed conditioning theorist who upholds that: (1) human behaviour, however complex, can in principle be understood in terms of the basic laws of conditioning, and (2) the causes of behaviour are located in the environment in the form of contingencies, and not inside the individual in the form of mental dispositions or personality traits, etc. But there are various objections to these premises: just because the behaviour of non-human animals obeys many of the principles of conditioning does not mean that complex human behaviour necessarily does the same; because *in theory* we can explain complex human behaviour in conditioning terms does not mean that this is how it is in *practice*; indeed, the whole spirit of an explanation of human behaviour in the reductionist, mechanistic and animalistic terminology that conditioning theory implies to some critics has also been condemned either as 'dehumanising' or just simply trivial (Chomsky, 1972; Koch, 1964; Koestler, 1967). These are arguments that we shall discuss later, but suffice it to say here that the conditioning model is just one paradigm for interpreting human behaviour, and its eventual validity will depend on how efficiently and faithfully it can account for the processes which mould and maintain human behaviour. The remainder of this chapter deals with, first of all, a *strong* version of the conditioning approach to human behaviour – radical behaviourism – and how this approach might account for aspects of human behaviour which do not appear to be immediately interpretable in conditioning terms, and finally, the ethical and practical problems involved in the behavioural technology that conditioning theory has created.

# RADICAL BEHAVIOURISM

In chapter 2 we talked about the basic tenets of Behaviourism as they were originally formulated by Watson (p. 17). These tenets outlined a strict scientific methodology for the study of behaviour, and because of this became known as *methodological behaviourism*. Since these early days, however, methodological behaviourism has largely been superseded by a more 'liberal' behaviourism commonly called *radical behaviourism*. (This approach is also commonly called simply *the experimental analysis of behaviour* (Skinner, 1966, 1969).) This approach to the study of behaviour is primarily based on the writings of Skinner and, although it still maintains that behaviour is the only legitimate subject matter of psychology and that the 'causes' of behaviour are to be found in the environment, it is first of all less dismissive in its dealings with concepts such as 'consciousness', 'private events', 'thinking', etc., and secondly, less ready to reduce complex behaviour patterns down to simple 'reflexes' or 'stimulus–response' chains in order to explain them (that is, it is more functional than reductionist in its analysis). There is no single complete consensus as to the principles which are important in radical behaviourism but the following propositions probably convey both the bones and the flavour of this approach:

(1) Explanations of behaviour should be couched *not* in terms of 'events taking place somewhere else, at some other level of observation, described in different terms and measured if at all in different dimensions' (Skinner, 1950), but in terms of the *functional relationships* that exist between behaviour and the environment. For instance, explanations of behaviour which stress

unobservable processes within the organism may divert attention away from those environmental variables which control the behaviour, and since these environmental variables are open to direct scrutiny and manipulation it means that more direct experimental control can be exerted over the behaviour itself.

(2) A science of behaviour should be developed inductively, first by studying individual organisms whose behaviour is brought under strict experimental control (this renders unnecessary group designs using statistical comparisons), and secondly, by making inductive generalisations about the facts gleaned from this process and not by setting up hypothetical postulates which can then be tested, the latter being a deductive approach.

(3) Explanations of behaviour should attempt to relate the behaviour of an organism to the observable consequences of that behaviour (the principle of operant reinforcement). This serves two purposes: first, it acts as a general guiding principle by which the variables controlling behaviour can be pinpointed; and secondly, operant reinforcement provides a procedure for controlling behaviour so that the process of learning can be studied in more detail (in this latter sense it can be considered as a technique which can be used by learning theorists of any theoretical persuasion).

(4) Certain fundamental laws of learning are common to a large number of quite different species of animals. More precisely, the principle of operant reinforcement can be seen to apply to human beings, rats, pigeons, fish, etc., and, although the behaviour of these species differs in its complexity and achievement, they still share some basic psychological mechanisms.

(5) Internal psychological phenomena such as 'thinking', 'hallucinating', 'dreaming', 'consciousness', etc. can often be thought of as behaviours which are both shaped-up and maintained in the same way that external, observable behaviours are controlled. They have previously been neglected in a behaviourist approach because of the technical problems involved in recording and controlling them, and secondly because the traditional conceptions of such phenomena as *causes* of overt behaviour were anathema to behaviourism. A reworking of these concepts in behavioural terms ratifies their existence, and, perhaps more importantly, suggests processes by which these phenomena are developed in the individual.

These five points are, then, a fairly rough sketch of the radical behaviourist philosophy (for further accounts see Blackman, 1974, pp. 217—222; Carpenter, 1974, pp. 1–61; Skinner, 1953, 1969, 1974), and they represent a framework loosely adapted from conditioning theory in which all behaviour – human and animal – can be analysed. Criticisms of the radical behaviourist philosophy are abundant, and quite frequently simply reflect the fact that different theorists choose to work in different paradigms (Katahn and Koplin, 1968), and each views his own paradigm as the most efficient and consistent in explaining the facts of behaviour. However, other criticisms need more serious attention because they stress either that radical behaviourism is incapable of explaining certain important behavioural phenomena or that a fundamental principle such as operant reinforcement – while applicable to the behaviour of non-human animals – is not a consistent controller of the behaviour of man. There are certainly a number of psychological phenomena which are considered to be uniquely human and which at first sight seem to defy an analysis in operant terms. Two important members of this class are thinking and consciousness. How does an approach which is primarily based on fairly simple conditioning principles attempt to account for these?

## Thinking and the reporting of 'Private Events'

To account for thinking was probably the biggest problem that the methodological behaviourists originally encountered. Thinking certainly existed, we are all witness to that fact, yet it could not be readily observed nor could it be measured; but perhaps more importantly, was it to be considered as an initiator of behaviour, or just simply as a behaviour? In tackling this problem Watson attempted to equate thinking with subvocal speech. Talking in a loud voice is overt behaviour, and as the voice diminishes to an inaudible whisper the whole process becomes covert. Watson (1914, 1920) found support for his assessment in the fact that experiments demonstrated that thinking – without movement of the vocal chord – is accompanied by small microcurrents from the nerves which project to the speech organs. In a similar way, a study by Max (1937) showed that such currents could also be obtained from the finger-muscles of congenital deaf mutes when they were thinking. However, this assessment is not particularly satisfying because it seems to be more obsessed with actually objectively detecting the existence of thinking rather than providing a structured account of how, as a behaviour, it is acquired. Nevertheless, Watson did conceive of thinking as a behaviour and this is how the radical behaviourist also conceives of it:

> Thinking is behaving. It does not explain overt behaviour but is itself simply more behaviour to be explained. The mistake of the mentalists is to allocate this form of behaviour to the mind. Thinking has the dimensions of behaviour, not of some fanciful inner process which finds expression in behaviour.
>
> (Martin and Crawford, 1976, p. 37)

If thinking is behaviour, this immediately raises the question of how the contingencies of reinforcement which develop and maintain thinking are set up. First, how is the appropriate 'thinking response' detected and defined by an external reinforcing agent, and secondly, how are the contingencies for its reinforcement arranged?

The assumptions of a radical behaviourist approach to covert behaviours are quite simple: 'We need not suppose that events which take place within an organism's skin have any special properties for that reason' (Skinner, 1953, p. 257), and a functional equivalence is assumed to exist between overt and covert events (Cautela and Baron, 1977; Day, 1968; Homme, 1965). Nevertheless, this leaves unanswered the question of how certain covert behaviours are shaped-up. For example, how do I come to recognise that a toothache is in fact a toothache, and convey my experience of the toothache to others? More specifically, how can we come to recognise in ourselves an internal state such as 'I was on the point of going home at one o'clock'? Skinner suggests that private events such as these examples do have public events which accompany them. For instance, a toothache is commonly accompanied by characteristic overt behaviours – these are not well defined but may include wincing and holding one's jaw for example. Individuals in the external world can recognise these symptoms as toothache and provide the contingencies by which the sufferer can come to *discriminate* the pain as a toothache. They may ask 'Do you have a toothache?' or 'Is your tooth painful?'. Skinner (1945, 1953) suggests that it is in this way that our language of private events is developed. Control which is initially exerted by external stimuli becomes transferred to the internal correlates of these overt be- haviours. In the case of reporting that 'I was on the point of going home at one o'clock', such behaviour probably became reinforced when private stimuli which are generated in addition to the public act of 'going home' gain discriminative control over the response. When the private stimuli occur alone, then so does the behaviour of reporting that one was 'on the point of going home'.

Now all this is very well in theory, but what about in practice? Do covert behaviours follow the rules of discriminative control and operant reinforce- ment? First of all, some indications of covert conditioning can be pointed to if we search through the details of the previous chapter. A number of therapeutic techniques based on conditioning principles utilise 'covert' contingencies to treat behaviour disorders. For example, therapies such as flooding, massing and systematic desensitisation often involve the condition- ing of a 'covert' response during treatment (see p. 373ff.) – the client is asked to *imagine* his behaviour in certain phobic situations, indeed in many examples of counterconditioning and systematic desensitisation he is even asked to imagine the consequences of this behaviour. Although the success of these techniques does not necessarily directly attest to the validity of covert conditioning (see p. 388ff.), it certainly adds momentum to an argument in favour of covert conditioning. A second indication that conditioning is indeed a viable method of shaping and altering private events comes from physiolog- ical studies. It is now well-known that certain physiological changes occur in conjunction with the reporting of certain non-specific psychological states: changes in EEG wave forms accompany states of relaxation and pain for example (see p. 383ff.). More specifically, even viewing or thinking about different geometric patterns such as a square or a circle influence the EEG-evoked potential wave form in different ways (John, 1967). The fact that physiological changes are tied to internal psychological events, and that changing the physiological variable also alters the psychological variable (see the evidence on biofeedback techniques, p. 383) strengthens the belief that covert behaviour does respond to the laws of conditioning. Finally, however, more recent developments in this field have concerned what is explicitly known as *covert conditioning*. In this procedure the subject is asked to imagine a response to be modified and then is asked to immediately imagine a consequence. If he wants the response to be established he imagines a pleasant consequence, if he wants to eliminate the response he imagines an aversive consequence (Cautela, 1976). Normally this has been used as a therapy procedure where the client generates the response covertly but intends its overt counterpart to be modified. However, responses which are uniquely covert can be manipulated using this method: these include increasing the frequency of mental imagery (Mahoney, Thoresen and Danaher, 1972), changing mood states (Jacobs, 1971), producing hallucinations (Hefferline, Bruno and Davidowitz, 1971), and even manipulating changes in an individual's self-concept (Krop, Calhoun and Verrier, 1971).

This evidence does suggest quite strongly that 'private' events can be governed by the laws of conditioning – it does not, however, imply that these are the *only* psychological laws that govern private events. The importance of these studies is that they do provide evidence of theory in practice.

## Consciousness

Dualistic accounts of man and his behaviour stem from the writings of the early classical scholars, and were given further impetus in the teachings of Descartes. In essence this philosophy maintains that there are two different kinds of subject matter for psychology–body and mind. The body obeys the material laws of physics, while the mind is a less tangible entity which directs the behaviour of the body (that is, it provides motives and intentions, and makes 'choices' and 'decisions', etc.). This view of man and his behaviour is extremely pervasive, and it is not difficult to see why: when the external causes of behaviour are not readily identifiable–which is often the case with much human behaviour – we tend to locate the causes of the behaviour within the organism. Hence, the disruptive behaviour of the delinquent is said to be caused by his personality, rather than the environment that may have shaped this behaviour. The dualistic approach to behaviour is very difficult to shake off–even by psychologists, many of whom still refer to psychology as the study of behaviour *and* mental life!

In its simplest and most caricatured form the dualistic approach can be represented as a little man or homunculus residing in the organism who receives incoming information and directs outward behaviour. In fact he represents what we would normally call 'consciousness'. However, what remains a mystery in these accounts of behaviour is an explanation of the behaviour of the homunculus! How does the homunculus perceive incoming pictures of the world? How does he 'choose' which response to make in which situation? Skinner (1963, 1969) has addressed himself to a discussion of this type of explanation:

What *are* the private events to which, at least in a limited way, a man may come to respond in ways we call "perceiving" or "knowing"?
. . . The search for copies of the world within the body particularly in the nervous system, still goes on, but with discouraging results. If the retina could suddenly be developed like a photographic plate it would yield a poor picture. The nerve impulses in the optic tract must have an even more tenuous resemblence to "what is seen". The patterns of vibrations which strike our ear when we listen to music are quickly lost in transmission. The bodily reactions to substances tasted, smelled and touched would scarcely qualify as faithful reproductions. These facts are discouraging for those who are looking for copies of the real world within the body, but they are fortunate for psychophysiology as a whole. At some point the organism must do more than create duplicates. It must see, hear, smell, and so on, as forms of *action* rather than of *reproduction. It must do some of the things it is differentially reinforced for doing when it learns to respond discriminatively.*
(Skinner, 1969, pp. 230–232)

If a reproduction of the real world were suddenly discovered in the visual cortex we would still have to explain how the organism *perceived* this reproduction. Therefore, behaviours that we normally consider in a passive way, must at some point require action or behaviour. In this sense it must surely be more profitable to consider the experience of 'seeing' or 'perceiving' as behaviours. If this is so we do not require a homunculus to observe the 'reproductions' of the real world that are screened in our brains. Now, once private events such as 'seeing' are considered as actions or behaviours, they become amenable to an analysis within a framework which is applicable to overt behaviour.

This only gets us partially to the nub of the question concerning 'consciousness'. It is true that we often consider our 'experiences of perceiving' as evidence of consciousness, but the concept of consciousness really goes beyond this. There is a second level of consciousness which extends beyond our first-order experiences of seeing and perceiving the outside world. For instance, 'people are conceived of as not only capable of monitoring and controlling their performance, but of monitoring the control they exercise in the first order performance' (Harré, 1971, p. 116). They not only monitor the world, they also monitor their monitoring of the world: in simple terms they are *aware* that they are seeing. Psychologists are often very unsure about what awareness implies: to some it is a 'state' of the organism, to others it is a phenomenon which is not accountable within traditional psychological or even scientific terms (see for example, Burt, 1968; Wann, 1964). However, the radical behaviourist again conceptualises this kind of awareness as a behaviour; it is 'seeing that you are seeing' (Skinner, 1963). The problem here, however, is to describe how behaviours such as 'seeing that you are seeing' are developed because, as Skinner points out:

> It is not, however, seeing our friend which raises the question of conscious content but 'seeing that we are seeing him'. There are no natural contingencies for such behaviour. We learn to see that we are seeing only because a verbal community arranges for us to do so. We usually acquire the behaviour when we are under appropriate visual stimulation, but it does not follow that the thing seen must be present when we see that we are seeing it.
>
> (Skinner, 1969, p. 233)

Thus, the behaviour of 'seeing' can be developed directly by natural environmental contingencies, but 'seeing you are seeing' or 'monitoring one's monitoring' is developed indirectly by contingencies arranged by the verbal community. This might occur in the following way: (1) When asked 'What are you doing?' this shapes up the response of observing our own external behaviour and reporting it; (2) there is no reason why this should not also apply to behaviours that are 'private', such as seeing; (3) the behaviour of seeing can also occur in the absence of the external stimuli which have shaped it up (such as the actual external object), just as in the rat the behaviour of lever-pressing can occur in the absence of the external $S^D$s and reinforcers which originally controlled it; this enables us to experience hallucinations, evoke mental images, and dream, etc.; (4) just as the behaviour of seeing can occur in the absence of its external controlling stimuli, so can the behaviour of 'seeing that you are seeing'. In practice this would involve monitoring our mental imagery, or in simple terms 'seeing that we were thinking'.

Consciousness is not an easy concept to come to terms with, primarily because it means different things to different people, and in some cases can be a word used simply to convey the whole richness and achievement of mental life. However, the important points to glean from this section are: (1) that radical behaviourism – unlike methodological behaviourism – does not attempt to ignore the facts of consciousness and conscious experience; (2) the experiences of 'seeing' and 'being aware that one is seeing' can be considered as behaviours which can be differentially shaped-up, either by natural contingencies, or by contingencies arranged by the verbal community. The final note, however, must again be a cautionary one: just because in principle the phenomenon of conscious awareness can be explained within a conditioning framework, does not mean this is how it is in practice. Indeed, perhaps the greatest criticism of this analysis is that simply by attempting to reduce the concept of conscious awareness to a behaviour we may have already lost the unique flavour of what most people understand by the term consciousness. Nevertheless, it seems to me that in the last analysis this argument must rest on a matter of belief about the status of human mental life and faith in the validity of one's theoretical framework. What the radical behaviourist has done is simply to show that the existence of conscious experience need not be as mystical as tradition and solipsism would have us believe.

## CRITICAL ISSUES IN A TECHNOLOGY OF BEHAVIOUR

One of the advantages of conditioning theory that is continually stressed by its supporters is that it readily suggests a technology of behavioural intervention (see p. 366ff.; Bandura, 1969, p. 16; Sandler and Davidson, 1973). For instance, it provides therapeutic techniques that are soundly based on theory, easily and quickly deduced from the theory, and, most importantly, seem to work in practice. This is behaviour modification, and is just one aspect of the behavioural technology derived from conditioning theory. However, other aspects of this technology are not as easy to justify as is behaviour modification; it is one thing to apply these techniques to the alleviation of distress, but is behavioural intervention of this kind justified when dealing with behaviours which are part and parcel of our normal, every-day life? We are here talking about the construction of contingency management programmes for the control and manipulation of the behaviour of whole communities. The bulk of the arguments in favour of adopting such an approach can be found in Skinner's controversial book *Beyond Freedom and Dignity* (1971) and the reader is urged to consult this book before making up his own mind on the ethical and moral issues of such an approach. Skinner argues that since we have the technology to manipulate the behaviour of whole populations, we should not let our societies, and more importantly, our culture, evolve haphazardly to the stage where the existence of mankind itself is threatened – factors such as overpopulation, pollution, and nuclear power need urgent attention if disaster is to be averted. He suggests that factors such as these can only be reliably controlled if we set up contingencies of reinforcement which are favourable for changing the behaviour of those people in society who are responsible for creating these problems; for example, in the case of overpopulation this would require setting up contingencies which reinforce people for having small families – at present most societies have welfare systems which set up the converse. Now, the reader who is encountering these ideas for the first time may feel obvious and justified alarm about some of the consequences of instigating such a technology – we shall discuss some of the causes for this alarm shortly. However, the ardent critics of this approach fall into two broad categories: (1) those who see the adoption of such a technology resulting in the erosion of personal freedom, a greater degree of bureaucratic control of individual behaviour, and the evolution of totalitarian states, and (2) those who claim that introduction of such a technology is pointless because operant conditioning just does not work with complex human behaviour or on such a large scale. (Note that on logical grounds one cannot assert *both* of these points, although some critics do so!). Before discussing these issues in detail, a look at the basic features of Skinner's manifesto is required.

## Beyond freedom and dignity

In order to check the drift of western societies towards self-destruction, Skinner argues that we must radically alter our conception of 'freedom'. Traditionally, the notion of freedom has been inextricably bound up with the concept of autonomous man – an organism who possesses free will, has total jurisdiction over his own behaviour, and is responsible for his own actions. The radical behaviourist view is of course quite different; man's behaviour is shaped by his environment and in this sense man is not autonomous, he does not possess free-will, and – if the argument is taken to its logical end – he is not responsible for his own behaviour (in the sense that it is not a simple product of causal states within him). So, by this view, even in our present society man is not free – his behaviour is shaped by whatever contingencies of reinforcement exist in his environment. Skinner's main argument progresses thus: the behaviour of individuals in our societies is at present generated by 'sloppy', inconsistent and badly controlled contingencies of reinforcement, many of which are contrary to the best long-term interests of the society; the illusion of freedom (that is, free-will, self-determination, etc.) is encouraged by the fact that these contingencies are uncontrolled and ill-specified. Now, if people can be shown evidence of the fact that their behaviour is controlled by environmental factors, and that their traditional concept of free-will is illusory, then they will be more readily inclined to accept the need for a technology of behaviour control which replaces inconsistent contingencies with consistent ones – consistent in the sense that they take into account the best interests of the individual and the society. This is what Skinner calls 'replacing autonomous man with scientific man'. To summarise, the crux of the manifesto is that our behaviour is already controlled by external contingencies, so why not regularise these contingencies to our own benefit?

Two further characteristics of the behavioural control in our present societies are also mentioned by Skinner as in need of change: (1) the more common adoption of punishment rather than reinforcement procedures for changing behaviour, and (2) the sensitivity of behaviour to control by 'short-term', immediate reinforcers, rather than long-term reinforcers, with the former not usually being in the best long-term interests of the society as a whole.

In the case of the first factor – adoption of punishment over reinforcement procedures – Skinner advocates a systematic replacement of punishment procedures with reinforcement procedures. In many senses, Skinner characterises 'freedom' in his designed environment as freedom from aversive control. It is not difficult to see why aversive procedures are more often adopted – punishment stops undesired behaviour almost immediately, thus reinforcing the punisher for using this method! Skinner suggests that to rid a society of aversive control will 'make life less punishing and in doing so (will) release for more reinforcing activities the time and energy consumed in the avoidance of punishment' (Skinner, 1971, p. 81). Although many people agree with Skinner's views about punishment, they replace aversive procedures with 'permissiveness' rather than positively reinforcing contingencies; when 'permissiveness' often fails to have the desired behavioural results, the punishment procedures are simply reintroduced. The second factor – the potency of short-term reinforcers over long-term reinforcers – is an important one because it concerns the survival of our species. The problem for a technology of behaviour is to convert long-term large-scale reinforcers – such as the stemming of population growth – into immediate personal reinforcers. Platt emphasises that this integration has important ethical consequences:

> . . . it could be said that happiness is having short-run reinforcers congruent with medium-run and long-run ones, and wisdom is knowing how to achieve this. And ethical behaviour results when short-run personal reinforcers are congruent with long-run group reinforcers. This makes it easy to "be good", or more exactly, to "behave well".
>
> (Platt, 1972, p. 48)

This has been a very cursory summary of the notions put forward by Skinner in *Beyond Freedom and Dignity*, and the reader is again urged to seek Skinner's views in the original. However, before we discuss the important criticisms the main points are as follows: (1) the notion of autonomous man needs to be replaced by the notion of 'scientific' man, (2) 'freedom' should imply freedom from aversive control, and (3) long-term reinforcers which control the survival of cultures, societies and even mankind itself should be converted into more tangible short-term reinforcers; short-term reinforcers which are incongruent with these more important long-term reinforcers should be eliminated.

## Fear of control

One of the commonest reasons put forward for condemning a technology of behaviour is that it could lead to wholesale control of almost every aspect of our daily lives. Now, it is certainly true that the word 'control' has been bandied around frequently in this chapter, and it is a term constantly used by radical behaviourists when they refer to behaviour. But first, what kind of control are the critics objecting to, and secondly, how is the term 'control' used by protagonists on different sides of the argument?

The more virulent of its critics claim that behavioural engineering is not only 'a flat-earth view of the mind' or a 'crude slot-machine model', but it is also as 'nasty as it is naïve' (Koestler, 1967, pp. 3–18), and as 'congenial to an anarchist as to a Nazi' (Chomsky, 1972, p. 32). The arguments surrounding these criticisms are involved ones because they consist of a 'cocktail' of inter-related points concerning philosophical, ideological and psychological views of man – it is difficult to extract one of these factors and discuss it without losing the flavour of the whole argument. The concern here is not that behavioural engineering should be outlawed because it *might* be used by a tyrant, a Hitler and so on – this is a problem which surrounds the introduction of any new radical technology, whether based on physics or biology – but that the radical behaviourist's conception of man as being manipulated by his environment and not truly 'responsible' for his own actions *invites* tyrannical control (see Chomsky, 1972, as an example of a fuller account of this view). However, if we can extract the problem of control from this account without detracting from it too greatly, there are three questions which stand out as being relevant: (1) Who will be the controller? (2) Can we adequately differentiate between those behaviour programmes which benefit the con-troller and those that benefit the controlled? and (3) Does the rejection of 'autonomous man' imply that behaviour is now totally out of the individual's control?

**The controller**

To deal with the first problem, who then is going to be the controller? Who is it that is going to define what the goals of a designed culture should be, and also prescribe and impose on us the ethics of such a society?

The reply given by most advocates of Skinner's theory is that it is usually not the aims that are in question, it is the means. Within limits we all probably agree – 'controllers' and 'controllees' – about problems such as overpopulation, pollution, the nuclear arms race and the need to come to terms with such factors for the sake of our own survival. In short, the controllers of most democratic societies are usually teachers or officials whose methods and values are accepted by the members of the society. What Skinner's theory therefore attempts to do is not to respecify the ethics or aspirations of the society but to provide a more efficient means for fulfilling these aspirations. However, this does not come to terms with the fact that a technology of behavioural engineering could still be misused by a malevolent controlling elite. In fact, as Holland (1973) points out, the dangers here need to be doubly

stressed because tyrannical control based on positive reinforcement techniques looks more humane than tyranny based on the traditional methods of
punishment and aversive control – it is therefore easier to deceive a society
into accepting it and, of course, less easy to detect it as 'tyrannical'. Advocates
of Skinner's theory claim that what this requires is countercontrol, that is,
behavioural contracting between members of a society and the leaders of the
society. What 'contracting' implies is that the controllers specify the
behaviours required of the members of the society, while the members of the
society also specify the behaviours they will reinforce in the controllers (it is in
essence an expanded form of contingency contracting, see chapter 14,
p. 381). As Platt outlines this argument:

> The symmetry of the reinforcement method in which all those involved are
> reinforcing each other for behavioural modification, means that any new
> methods of control are matched by correspondingly powerful new methods
> of countercontrol. This is what the better integration of a society means –
> like the integration of an organism. If the stomach does not get enough
> blood from the heart it demands more, and gets it, or we become sick and
> the heart may be forced to stop. So leaders or officials managing a society
> are endangering themselves as well as everyone else if the countercontrol
> demands for equity, or rights, or variety are not satisfied and the society
> tears itself apart.
>
> (Platt, 1972, p. 50)

Skinner argues that tyrannical injustice often arises – even in 'free'
societies – when the authority or controller is detached from or simply
unaware of the consequences of his actions (Skinner 1971). A society which
brings important consequences to bear on the controlling behaviour of the
controller is less likely to find itself abused by such an elite.

## Who benefits from behavioural engineering?

The question of controlling the controller is obviously an important one and
behavioural engineering is open to the same risks as any new technology when
it comes to abuse by malevolent groups or individuals. Unfortunately,
advocates of Skinner's technology can only broadly hint at, rather than
specify in more practical terms, how such abuses can be avoided. Further
discussion of this question can be found in Perelman (1973), Ritchie-Calder
(1973) and Carpenter (1974). However, the kinds of problems entailed here
can be seen on a smaller scale when we look closely at behaviour modification
programmes. For instance, behaviour modification schemes are not always
instigated to benefit the targets of the programme. Holland (1973) quotes a
number of examples where behaviour modification techniques are used
explicitly as management procedures to benefit the modifier himself. They

may be used to control irritating social problems or they may be used to supplement disciplinary rules in situations where well-ordered behaviour is required (for example, to keep recruits in order in army establishments). On the other hand, schemes which start off explicitly designed to benefit the target may drift insidiously in structure and spirit until they eventually benefit the modifier. This can sometimes be found in token economy schemes which are left to 'run themselves' (Cullen, Hattersley and Tennant, 1977). It may be intended to use a token economy programme as a therapeutic device in a psychiatric ward, and it may be initially designed for such ends, but what starts as a therapeutic procedure on the ward can often drift into a management procedure, where the programme is used simply to keep the ward clean and tidy. This reduces the burden of work on the nursing staff and does not necessarily constitute therapy.

## Autonomous man, scientific man and the problem of responsibility

The final question on control revolves around the concept of 'responsibility'. If man's behaviour is controlled by his environment can he be held responsible for his actions? If the answer were to be no, what moral implications would this have for the way we behave? The logical repercussion of replacing autonomous man with scientific man is the elimination of the belief that men are solely responsible for their acts. Although Skinner believes that the 'freedom' and responsibility accorded to individuals by the idea of autonomous man has been useful in helping guard against tyranny, it has also had its negative side. For instance, it has also been used as a justification for punishment – if an individual was 'free' then this behaviour had no outside causes and any corrective measures needed to be taken directly against the individual and his body. However, it is difficult to estimate the value of according man with responsibility for his own actions, primarily because the notion of 'responsibility' is a culture-bound one. In some societies responsibility for one's own actions is indeed a cherished and enobling feature of the individual and essential to the fabric of the society; in other societies – modern China is an example – the notions of individual 'freedom' which are allied to responsibility are less valued than the freedom and dignity of the society as a whole. As mentioned earlier, Skinner argues that the notion of autonomous man and all that that implies is mythical, an impediment to the implementation of a technology of behaviour, and in many cases may actually be detrimental to the survival of the society and the individuals in it. Perhaps the only way in which such 'enobling myths' might be retained in a designed society based on operant conditioning principles are as 'necessary positive reinforcers for a humane and decent society' (Wheeler, 1973). Nevertheless, perhaps the crux of this question really revolves around how one uses the term 'freedom', and Skinner does indeed use it in a rather idiosyncratic way by confusing the notions of free-will, self-determination and 'individualism'. His thesis thus draws critics from the full spectrum of academia, each one defending his own notion of freedom. Rather than perpetuate the melée here, the reader is referred to Carpenter (1974) for an expansive evaluation of Skinner's use of the term 'freedom'.

## Dehumanising man

One pervasive criticism of Skinner's whole approach to human behaviour is that it simply 'dehumanises' man – it makes out man to be either machine-like or just another animal by stripping him of those qualities which are uniquely human. These qualities are primarily those that characterise human mental life and social achievement. There are two broad aspects of Skinner's system that these critics particularly object to: (1) that radical behaviourism does not convey the significance of emotions, feelings, purposes, etc. and especially the important role played by expression of these feelings during human interaction; and (2) they claim that an explanation of human behaviour based on principles derived primarily from rat behaviour can in principle *never* come to terms with those facets of behaviour which are uniquely human.

In the case of the first objection, it is quite true that Skinner has tried to play down the importance of internal factors such as emotions and motives to an explanation of behaviour, simply because he believes they do not play a causal role in determining behaviour. In fact, he suggests that attempting to explain behaviour in terms of internal feelings is as unprofitable as was animism in the explanation of the behaviour of inanimate objects. For instance, primitive physics consisted primarily of attributing to inanimate objects feelings and intentions – for example, meteorites glowed in the sky because they were 'jubilant' at eventually reaching the mother earth. Skinner claims that human feelings and intentions are no more an explanation of human behaviour than 'jubilance' is an explanation of why meteorites glow. The objection to this is that although inanimate objects may not in reality have feelings and intentions, human beings do; and because they do, no account of human behaviour can be complete without conveying the importance of these feelings and intentions to the individual and to his interactions with others. This is emphasised by Chomsky

. . . Skinner notes that physics advanced only when it 'stopped personifying things' and attributing to them 'wills, impulses, feelings, purposes', and so on. Therefore, he concludes, the science of behaviour will progress only when it stops personifying people and avoids reference to 'internal states'. No doubt physics advanced by rejecting the view that a rock's wish to fall is a factor in its 'behaviour', because in fact a rock has no such wish. For Skinner's argument to have any force, he must show that people have wills, impulses, feelings, purposes, and the like, no more than rocks do. If people differ from rocks in this respect, then a science of human behaviour will have to take account of this fact.

(Chomsky, 1972, pp. 14–15).

Basically, the argument claims that although Skinner's system may be most efficient at predicting and controlling behaviour, it does not indicate the important and subtle role played in human interactions by the expression and reception of feelings and intent. Supporters of this argument claim that this is nowhere more evident than in certain behaviour modification techniques. For instance, they claim that contingency contracting can foster a manipulative, exchange orientation to social interactions, and token economies place an emphasis on the materialistic evaluation of human efforts. Similarly, Rappoport illustrates this point concisely in the following passage discussing the use of behaviour modification techniques with children.

... adult practitioners must train themselves to inhibit spontaneous expressions of emotion in favour of expressions arranged to produce some desired behaviour in the child. And assuming that the child "learns" to produce that which may never be openly demanded of him, he may also eventually learn that smiles and frowns are not genuine signs of reflexive emotions, but rather are signals indicating how well he is conforming to authority . . . the legitimate emotional meaning of facial expressions may become corrupted for both adults and children.

(Rappoport, 1972, preface)

This is just one of the many ways in which a technology of behaviour is considered by some critics to 'dehumanise' man: an analysis of human interactions in terms of contingencies of reinforcement seems to devalue the importance of feelings and intentions in human relationships. The behaviour modifier would probably reply by asserting that the techniques which have been criticised in this respect are by no means ideal; they often entail setting up response-reinforcer relationships that rarely occur naturally (such techniques include rewarding verbalisations in a child with food, or expressing 'insincere' delight when a psychotic adult makes his bed). They do not replicate the natural contingencies which set up such behaviours in the majority of us and so we should not think of them as artificially arranged duplicates of the natural contingencies which control verbal and social behaviours (Ferster, 1967).

The second prong of the 'dehumanising' criticism concerns the extrapolation from rats to man, and the claim of the radical behaviourist that the principles of learning derived from non-human animals have important and central relevance to human behaviour. In the words of Koestler, this leads the radical behaviourists to be 'driven by an almost fanatical urge to deny, at all costs, the existence of properties which account for the humanity of man and the rattiness of the rat' (Koestler , 1967, p. 15). Skinner, of course, would disagree; he admits that there are very stark and important differences between the behaviour of the rat and the behaviour of man. However, he claims that they do share similar psychological mechanisms, and *in principle* it is possible to account for those behaviours which are uniquely human – such as the reporting of private events (Skinner, 1945), conscious awareness (Skinner, 1963), language acquisition (Skinner, 1957), and creativity (Skinner, 1972) – in terms of the psychological mechanisms that man shares with the rat. However, even if what in principle is also true in practice – and this is by no means clear – taking the laboratory rat as the basis for our psychological model can have an important influence on our conception not only of psychological mechanisms but also our views of freedom and dignity. To say that the behaviours of organisms as different as rat and man, are controlled by basically similar mechanisms must surely miss a lot – for instance, the achievements of man are far greater and more significant than those of the rat and this difference must logically have some basis in differences in behavioural-psychological mechanisms. It is certainly true that the behaviours of both man and rat do obey the laws of conditioning, but if one presses this point too far it can be viewed as a denial of the existence of 'human' behaviour, and also as a denial of the existence of more flexible and sophisticated psychological mechanisms in man.

One further important implication of the 'rat model' of human behaviour is its influence on our conceptions of freedom and dignity. Morgan makes the point accordingly

> . . . perhaps it should be pointed out that we who work in animal behaviour laboratories are likely to develop eccentric views of freedom and dignity. To experiment on animals, that is, to starve, to shock, or poison them, to remove parts of their brains or sense organs, to paralyse them; to do this in very many cases merely to see what will happen (our ignorance of animals is so profound as to illustrate perfectly Kierkegaard's guess that 'most of what passes for science . . . is mere curiosity'), and to do this on the explicit assumption that the minds of animals have enough in common with our own to make the results of these experiments interesting . . . Well, to do all of this requires a dissociation from feeling so profound as to put the judgements of the experimenters in serious danger.
>
> (Morgan, 1971, p. 61)

Animal psychologists in general do treat their experimental subjects with care and compassion, but it can still be argued that the experimenter is using them to his own ends and – apart from his concern about their general health and well-being – he does treat them as scientific objects. In a sense, their freedom and dignity is denied before even a single psychological experiment is performed on them. This being the case, to extrapolate directly from rat to man also implicitly invites the extrapolation of our notions of the freedom and dignity of our laboratory animals. As Morgan (1971) puts it 'How long before we start treating people as objects?'

The issues involved here are difficult ones, as this chapter in general will probably have indicated. However, if the important questions can be outlined briefly, they are probably the following: (1) Can psychological principles evolved from the behaviour of non-human animals account for the achievement, sophistication and uniqueness of human behaviour? (2) Perhaps more importantly, are the techniques of behaviour modification developed from these principles able to produce individuals who do not simply just behave in a mechanical, unfeeling, 'manipulated' way? and (3) Does a psychological-model of man which is based on principles derived from non-human animals invite us to conceive of the freedom and dignity of man in the same way that we conceive of the freedom and dignity of the laboratory animal? Like most of the questions posed in this chapter, the answers to these questions are likely to depend more on ideological and epistomological argument than on empirical psychological verification.

## CONCLUDING REMARKS: CONDITIONING IN ANIMALS AND MAN

Sections one and three of this book have been concerned with conditioning in animals and man respectively, and it is important to reiterate the differing research strategies that these two topics employ. While the primary aim of theorists in animal learning has been to elaborate the details of the *associative mechanisms* which underlie conditioning (chapter 7), conditioning theory in human psychology has to date been primarily concerned with the control, prediction and manipulation of behaviour – a goal which can be achieved at the level of *controlling variables*. For instance, operant and classical conditioning work quite efficiently as procedures for changing the behaviour of both animals and man – the rat in the Skinner-box and the patient in the psychiatric hospital. Furthermore, the therapist using behaviour modification techniques is not usually interested in the mechanisms which underlie operant reinforcement in a psychiatric patient; he needs to know what the variables are which control the individual's behaviour and so can be manipulated in order to change that behaviour.

The distinction between description of mechanism and description of controlling variables as different kinds of explanation of behaviour is quite important. This can be explained more clearly with an example. I can explain how a car moves (in effect, how it behaves) by telling you that turning the ignition key starts the car, the accelerator pedal controls the speed at which the car will go, and so on; that is, in the latter case, I am saying there is a functional relationship between pressing the accelerator pedal and the speed of the car. That is an explanation of the car's behaviour in terms of controlling variables, it tells me how to alter the behaviour of the car without knowing how this works in practice. The second level of explanation is analogous to elaborating details of the mechanism mediating these functional relationships. By describing to you the workings of the carburettor, pistons, gear-box, etc. I am also explaining how the car behaves. If you simply want to know how to drive the car you need only to have knowledge of the controlling variables, it does not matter to you how the engine mediates these relationships; and indeed, the same controlling variables apply to many different kinds of motor car (that is, all go faster when the accelerator pedal is pressed and stop when the brake is applied) but the details of the underlying mechanism (the engine) can be quite different. Although this is a much simplified, and in many ways a naïve example, it does illustrate a number of points about the different emphases in human and animal conditioning theory, and also illustrates a number of misconceptions concerning extrapolation from animals to man. For instance, because operant reinforcement is a variable which controls behaviour in both the laboratory rat and the psychiatric patient, it does not imply that it has its effects *in the same way* in the two organisms. It is one thing to say that operant reinforcement as a principle of behaviour change is common to a very large number of animals, it is quite another to say that a *mechanism* of operant reinforcement is also common. For example, what we discover about the associative mechanisms underlying operant conditioning in the laboratory rat may bear little resemblance to the mechanisms underlying operant conditioning in, say, the pigeon, let alone man (see chapter 7, pp. 241–242). The survival value of learning about the consequences of one's behaviour is enormous, and so living creatures which respond to operant conditioning procedures may do so because their ancestors independently evolved mechanisms to deal with response-consequence contingencies, or alternatively they refined existing mechanisms to cope more efficiently with these kinds of contingencies. The moral of this argument for extrapolations from animal to man is that, although it is quite reasonable to assert that the behaviours of both man and the laboratory rat are susceptible to contingencies of operant conditioning, to assert also that the mechanisms underlying this control are identical is grossly misleading and highly unlikely. Indeed, one of the major complaints of those writers who condemn conditioning interpretations of human behaviour is that they fail to see how complex human behaviours can be established by associative mechanisms elaborated to account for conditioning in the rat (Chomsky, 1959; Koestler, 1967). Such accounts, they claim, are either naïve or impractical. Of course they are, but at present many conditioning theorists – especially radical behavicurists – are not concerned with elaborating the mechanisms underlying conditioning accounts of human behaviour, they are primarily concerned with specifying the controlling relationships between environment and behaviour.

The final point of emphasis is that, although radical behaviourism and its associated theories and technologies owe much to the study of conditioning in non-human animals, the debt is primarily for providing a conceptual framework within which to describe human behaviour. To be concerned with associative learning in the rat does not make a research worker a radical behaviourist, and indeed many animal psychologists would deny that their work had any relevence to human behaviour at all. Similarly, the radical behaviourist – especially the behaviour modifier – would find little of interest in a description of the mechanisms of associative learning in non-human animals. The one important connection between the two disciplines is now probably only historical. At the present time the study of animal conditioning has reached the stage where it is attempting to elaborate the mechanisms underlying conditioning (and at the same time realising that its dependance on a few species, such as the rat and pigeon, has perhaps been misleading – see chapters 5 and 6), while the study of conditioning in humans has developed in a number of directions: it is concerned with the kinds of controlling variables that can influence human behaviour in conditioning situations (see chapters 12 and 13), it has developed as a technology of behaviour change (see chapter 14), and it has developed a broad ranging theory of human behaviour which in principle claims to be able to explain, and at the same time demystify, many of the phenomena of human mental life.

# References

Abrahamson, D., Brackbill, Y., Carpenter, R. & Fitzgerald, H. E. (1970). Interaction of stimulus and response in infant conditioning. *Psychosom. Med.*, **32**, 319–325.

Adams, D. K. (1928). The inference of mind. *Psychol. Rev.*, **35**, 235–252.

Adams, D. L. & Allen, J. D. (1971). Compound stimulus control by discriminative stimuli associated with high and moderate response rates. *J. exp. Anal. Behav.*, **16**, 201–205.

Ader, R. & Tatum, R. (1961). Free-operant avoidance conditioning in humans. *J. exp. Anal. Behav.*, **3**, 275–276.

Agranoff, B. W. (1967). Agents that block memory. In G. C. Quarton, T. Melnechuk & F. O. Schmitt (eds), *The Neurosciences: A study program*, pp. 756–764. New York, The Rockefeller Univ. Press.

Agranoff, B. W., Davies, R. E. & Brink, J. J. (1966). Chemical studies on memory fixation in goldfish. *Brain Res.*, **1**, 303–309.

Alloway, T. M. (1969). Effects of low temperature upon acquisition and retention in the grain beetle (*Tenebrio molitor*). *J. comp. physiol. Psychol.*, **69**, 1–8.

Alloway, T. M., Riedesel, M. L. & McNamara, M. C. (1973). Hibernation: effects on memory or performance? *Science*, **181**, 86–87.

Amsel, A. (1958). The role of frustrative nonreward in noncontinuous reward situations. *Psychol. Bull.*, **55**, 102–119.

Amsel, A. (1962). Frustrative nonreward in partial reinforcement and discrimination learning. *Psychol. Rev.*, **69**, 301–328.

Amsel, A. & Roussel, J. (1952). Motivational properties of frustration: I. Effect on a running response of the addition of frustration to the motivational complex. *J. exp. Psychol.*, **43**, 363–368.

Anderson, D. O. & Parmenter, R. (1941). A long-term study of the experimental neurosis in sheep and in the dog. *Psychosom. Med. Monogr.*, **2**, No. 3 and 4.

Anger, D. (1956). The dependence of interresponse times upon the relative reinforcement of different interresponse times. *J. exp. Psychol.*, **52**, 145–161.

Anger, D. (1963). The role of temporal discriminations in the reinforcement of Sidman avoidance behaviour. *J. exp. Anal. Behav.*, **6**, 477–506.

Angermeier, W. F. (1960). Some basic aspects of social reinforcements in albino rats. *J. comp. physiol. Psychol.*, **53**, 364–367.

Angermeier, W. F. (1962). The effect of a novel and novel noxious stimulus upon the social operant behaviour in the rat. *J. genet. Psychol.*, **100**, 151–154.

Angermeier, W. F., Schaul, L. T. & James, W. T. (1959). Social conditioning in rats. *J. comp. physiol. Psychol.*, **52**, 370–372.

Appel, J. B. (1963). Punishment and shock intensity. *Science*, **141**, 528–529.

Armstrong, E. A. (1950). The nature and function of displacement activities. *Symp. Soc. exp. Biol.*, **4**, 361–384.

Atnip, G. W. (1977). Stimulus – and response–reinforcer contingencies in autoshaping, operant, classical and omission training procedures in rats. *J. exp. Anal. Behav.*, **28**, 59–69.

Autor, S. M. (1960). The strength of conditioned reinforcers as a function of frequency and probability of reinforcement. Doctoral dissertation, Harvard Univ.

Ayllon, T. (1963). Intensive treatment of psychotic behaviour by stimulus satiation and food reinforcement. *Behav. Res. Ther.*, **1**, 53–61.

Ayllon, T. & Azrin, N. H. (1964). Reinforcement and instructions with mental patients. *J. exp. Anal. Behav.*, **6**, 327–331.

Ayllon, T. & Azrin, N. H. (1965). The measurement and reinforcement of behaviour in psychotics. *J. exp. Anal. Behav.*, **8**, 351–383.

Ayllon, T. & Azrin, N. H. (1968). *The token economy: A motivational system for therapy and rehabilitation.* New York, Appleton-Century-Crofts.

Ayllon, T. & Haughton, E. (1962). Control of the behaviour of schizophrenic patients by food. *J. exp. Anal. Behav.*, **5**, 343–352.

Ayllon, T., Haughton, E. & Hughes, H. B. (1965). Interpretation of symptoms: Fact or fiction. *Behav. Res. Ther.*, **3**, 1–7.

Ayllon, T. & Michael, J. (1959). The psychiatric nurse as a behavioural engineer. *J. exp. Anal. Behav.*, **2**, 323–334.

Azrin, N. H. (1956). Some effects of two intermittent schedules of immediate and nonimmediate punishment. *J. Psychol.*, **42**, 3–21.

Azrin, N. H. (1958). Some effects of noise on human behaviour. *J. exp. Anal. Behav.*, **1**, 183–200.

Azrin, N. H. (1959a). Some notes on punishment and avoidance. *J. exp. Anal. Behav.*, **2**, 260.

Azrin, N. H. (1959b). A technique for delivering shock to pigeons. *J. exp. Anal. Behav.*, **2**, 161–163.

Azrin, N. H. (1960). Effects of punishment intensity during variable-interval reinforcement. *J. exp. Anal. Behav.*, **3**, 123–142.

Azrin, N. H. (1970). Punishment of elicited aggression. *J. exp. Anal. Behav.*, **14**, 7–10.

Azrin, N. H. & Foxx, R. M. (1971). A rapid method of toilet training the institutionalized retarded. *J. appl. Behav. Anal.*, **4**, 89–99.

Azrin, N. H. & Hake, D. F. (1969). Positive conditioned suppression: Conditioned suppression using positive reinforcers as the unconditioned stimuli. *J. exp. Anal. Behav.*, **12**, 117–173.

Azrin, N. H. & Holz, W. C. (1966). Punishment. In W. K. Honig (ed.) *Operant behaviour: Areas of research and application*, pp. 380–447. New York, Appleton–Century–Crofts.

Azrin, N. H., Holz, W. C. & Hake, D. F. (1963). Fixed-ratio punishment. *J. exp. Anal. Behav.*, **6**, 141–148.

Azrin, N. H. & Hutchinson, R. R. (1967). Conditioning of the aggressive behaviour of pigeons by a fixed-interval schedule of reinforcement. *J. exp. Anal. Behav.*, **10**, 395–402.

Azrin, N. H., Hutchinson, R. R. & Hake, D. F. (1963). Pain-induced fighting in the squirrel monkey. *J. exp. Anal. Behav.*, **6**, 620.

Azrin, N. H., Hutchinson, R. R. & Hake, D. F. (1966). Extinction-induced aggression. *J. exp. Anal. Behav.*, **9**, 191–204.

Baer, D. M. (1960). Escape and avoidance responses of preschool children to two schedules of reinforcement withdrawal. *J. exp. Anal. Behav.*, **3**, 155–159.

Baer, D. M. (1962). A technique of social reinforcement for the study of child behaviour: Behaviour avoiding reinforcement withdrawal. *Child Dev.*, **33**, 847–858.

Baer, D. M. & Guess, D. (1971). Receptive training of adjectival inflections in mental retardates. *J. appl. Behav. Anal.*, **4**, 129–139.

Bandura, A. (1969). *Principles of Behaviour Modification.* New York, Holt, Rinehart & Winston.

Bankart, C. P., Bankart, B. M. & Burkett, M. (1974). Social factors in acquisition of bar-pressing by rats. *Psychol. Rep.*, **34**, 1051–1054.

Barlow, E. D. & Dewardener, H. E. (1959). Compulsive water drinking. *Q. Jl. Med.*, **28**, 235–258.

Barlow, H. B. (1975). Visual experience and cortical development. *Nature*, **258**, 199–204.

Barlow, H. B. & Pettigrew, J. D. (1971). Lack of specificity of neurones in the visual cortex of young kittens. *J. Physiol.*, **218**, 98–100.

Baron, A. & Antonitis, J. J. (1961). Punishment and preshock as determinants of bar-pressing behaviour. *J. comp. physiol. Psychol.*, **54**, 716–720.

Baron, A. & Kaufman, A. (1966). Human free-operant avoidance of "time-out" from monetary reinforcement. *J. exp. Anal. Behav.*, **9**, 557–565.

Baron, A., Kaufman, A. & Stauber, K. A. (1969). Effects of instructions and reinforcement feedback on human operant behaviour maintained by fixed-interval reinforcement. *J. exp. Anal. Behav.*, **12**, 701–712.

Barrera, F. J. (1974). Centrifugal selection of signal directed pecking. *J. exp. Anal. Behav.*, **22**, 341–355.

Barrett, B. H. & Lindsley, O. R. (1962). Deficits in acquisition of operant discrimination and differentiation shown by institutionalized retarded children. *Am. J. ment. Defic.*, **67**, 424–436.

Barnett, C. D., Pryer, M. W. & Ellis, N. R. (1959). Experimental manipulation of verbal behaviour in defectives. *Psychol. Rep.*, **5**, 393–396.

Barton, E. S., Guess, D., Garcia, E. & Baer, D. M. (1970). Improvement of retardates' mealtime behaviours by timeout procedures using multiple baseline techniques. *J. appl. Behav. Anal.*, **3**, 77–84.

Bateson, P. P. G. (1964). Effect of similarity between rearing and testing conditions on chick's following and avoidance responses. *J. comp. physiol. Psychol.*, **57**, 100–103.

Bateson, P. P. G. (1966). The characteristics and context of imprinting. *Biol. Rev.*, **41**, 177–220.

Bateson, P. P. G. (1969). The development of social attachments in birds and man. *Adv. Sci.*, **25**, 279–288.

Bateson, P. P. G. (1973). Internal influences on early learning in birds. In Hinde, R. A. & Stevenson-Hinde, J. (eds), *Constraints on Learning*, pp. 101–116. London, Academic Press.

Bateson, P. P. G. & Chantrey, D. F. (1972). Retardation of discrimination learning in monkeys and chicks previously exposed to both stimuli. *Nature*, **237**, 173–174.

Bateson, P. P. G. & Reese, E. P. (1969). Reinforcing properties of conspicuous stimuli in the imprinting situation. *Anim. Behav.*, **17**, 692–699.

Baum, M. (1969). Extinction of an avoidance response motivated by intense fear: Social facilitation of the action of response prevention (flooding) in rats. *Behav. Res. Ther.*, **7**, 57–62.

Baum, M. (1970). Extinction of avoidance responding through response prevention (flooding). *Psychol. Bull.*, **74**, 276–284.

Baum, M. (1974). On two types of deviation from the matching law: bias and undermatching. *J. exp. Anal. Behav.*, **22**, 231–242.

Baum, M. & Oler, I. D. (1968). Comparison of two techniques for hastening extinction of avoidance-responding in rats. *Psychol. Rep.*, **23**, 807–813.

Baum, W. M. (1973). The correlation-based law of effect. *J. exp. Anal. Behav.*, **20**, 137–153.

Bedford, J. & Anger, D. (1968). Flight as an avoidance response in pigeons. Paper presented at the meeting of the Psychonomic Soc., St. Louis.

Behrend, E. R. & Bitterman, M. E. (1961). Probability-matching in the fish. *Am. J. Psychol.*, **74**, 542–551.

Behrend, E. R. & Bitterman, M. E. (1966). Probability-matching in the goldfish. *Psychon. Sci.*, **6**, 327–328.

Benel, R. A. (1975). Intra- and inter-specific observational learning in rats. *Psychol. Rep.*, **37**, 241–242.

Beninger, R. J. (1972). Positive behavioural contrast with qualitatively different reinforcing stimuli. *Psychon. Sci.*, **29**, 307–308.

Beninger, R. J., Kendall, S. B. & Vanderwolf, C. H. (1974). The ability of rats to discriminate their own behaviours. *Can. J. Psychol.*, **28**, 79–91.

Benson, H., Shapiro, D., Tursky, B. & Schwartz, G. E. (1971). Decreased systolic blood pressure through operant conditioning techniques in patients with essential hypertension. *Science*, **173**, 740–742.

Bermant, G. (1961). Response latencies of female rats during sexual intercourse. *Science*, **133**, 1771–1773.

Bernheim, J. W. & Williams, D. R. (1967). Time-dependent contrast effects in a multiple schedule of food reinforcement. *J. exp. Anal. Behav.*, **10**, 243–249.

Biederman, G. B., D'Amato, M. R. & Keller, D. M. (1964). Facilitation of discriminated avoidance learning by dissociation of CS and manipulandum. *Psychon. Sci.*, **1**, 229–230.

Bijou, S. W. (1957). Patterns of reinforcement and resistance to extinction in young children. *Child Dev.*, **28**, 47–55.

Bijou, S. W. & Baer, D. M. (1965). *Child development*, Vol. 2. *Universal Stages of Infancy.* New York, Appleton–Century–Crofts.

Bijou, S. W. & Baer, D. M. (1966). Operant methods in child behaviour and development. In W. K. Honig (ed.), *Operant Behaviour: Areas of Research and Application*, pp. 718–789. New York, Appleton–Century–Crofts.

Bijou, S. W. & Orlando, R. (1961). Rapid development of multiple schedule performance with retarded children. *J. exp. Anal. Behav.*, **4**, 7–16.

Bijou, S. W. & Sturges, R. T. (1959). Positive reinforcers for experimental studies with children – consumables and manipulables. *Child Dev.*, **30**, 151–170.

Bindra, D. (1972). A unified account of classical conditioning and operant training. In A. H. Black & W. F. Prokasy (eds), *Classical Conditioning II: Current Research and Theory*, pp. 453–481. New York, Appleton–Century–Crofts.

Bindra, D. (1974). A motivational view of learning, performance and behaviour modification. *Psychol. Rev.*, **81**, 199–213.

Birch, H. G. (1945). The relation of previous experience to insightful problem solving. *J. comp. Psychol.*, **38**, 367–383.

Birkimer, J. C. & Drane, D. L. (1968). Stimulus intensity dynamism with stimuli equal decibel distances above and below background. *Psychon. Sci.*, **12**, 213–214.

Birnbrauer, J. S. & Lawler, J. (1964). Token reinforcement for learning. *Ment. Retard.*, **2**, 275–279.

Bisett, B. M. & Rieber, M. (1966). The effects of age and incentive value on discrimination learning. *J. exp. Child Psychol.*, **3**, 199–206.

Bitterman, M. E. (1965a). Phyletic differences in learning. *Am. Psychol.*, **20**, 396–410.

Bitterman, M. E. (1965b). The evolution of intelligence. *Scient. Am.*, Jan.

Bitterman, M. E. (1971). Visual probability learning in the rat. *Psychon. Sci.*, **22**, 191–192.

Bitterman, M. E. (1975). The comparative analysis of learning. *Science*, **188**, 699–709.

Bitterman, M. E., Wodinsky, J. & Candland, D. K. (1958). Some comparative psychology. *Am. J. Psychol.*, **71**, 94–110.

Black, A. H. (1958). The extinction of avoidance responses under curare. *J. comp. physiol. Psychol.*, **51**, 519–524.

Black, A. H. (1971). Autonomic aversive conditioning in infrahuman subjects. In F. R.

Brush (ed.) *Aversive Conditioning and Learning*, pp. 3–104. New York, Academic Press.

Black, A. H. & Young, G. A. (1972). Constraints on the operant conditioning of drinking. In R. M. Gilbert & J. R. Millenson (eds), *Reinforcement: Behavioural Analyses*, pp. 35–50. New York, Academic Press.

Black, M. (1968). *The Labyrinth of Language*. New York, Praeger.

Black, R. W. (1969). Incentive motivation and the parameters of reward in instrumental conditioning. In W. J. Arnold & D. Levine (eds), *Nebraska Symposium on Motivation*, Vol. 17, pp. 85–137. Lincoln, University of Nebraska Press.

Blackman, D. E. (1968a). Conditioned suppression or facilitation as a function of the behavioural baseline. *J. exp. Anal. Behav.*, **11**, 53–61.

Blackman, D. E. (1968b). Response rate, reinforcement frequency and conditioned suppression. *J. exp. Anal. Behav.*, **11**, 503–516.

Blackman, D. E. (1970). Conditioned suppression of avoidance behaviour in rats. *Q. Jl exp. Psychol.*, **22**, 547–553.

Blackman, D. E. (1972). Conditioned anxiety and operant behaviour. In R. M. Gilbert & J. D. Keehn (eds), *Schedule Effects: Drugs, Drinking and Aggression*, pp. 26–49. Toronto, Univ. of Toronto Press.

Blackman, D. E. (1974). *Operant Conditioning: an Experimental Analysis of Behaviour*. London, Methuen.

Blakemore, C. (1973). Environmental constraints on development in the visual system. In R. A. Hinde & J. Stevenson-Hinde (eds), *Constraints on Learning*, pp. 51–73. London, Academic Press.

Blakemore, C. & Cooper, C. F. (1970). Development of the brain depends on the visual environment. *Nature*, **228**, 477–478.

Blanchard, E. B., Libet, J. M. & Young, L. D. (1973). Apneic aversion and covert sensitization in the treatment of a hydrocarbon inhalation addiction: A case study. *J. Behav. Ther. exp. Psychiat.*, **4**, 383–387.

Blanchard, E. B., Young, L. D. & Jackson, M. S. (1974). Clinical applications of biofeedback training. *Arch. gen. Psychiat.*, **30**, 573.

Bloomfield, T. M. (1967a). Behavioural contrast and relative reinforcement frequency in two multiple schedules. *J. exp. Anal. Behav.*, **10**, 151–158.

Bloomfield, T. M. (1967b). Some temporal properties of behavioural contrast. *J. exp. Anal. Behav.*, **10**, 159–164.

Bloomfield, T. M. (1969). Behavioural contrast and the peak shift. In R. M. Gilbert & N. S. Sutherland (eds), *Animal Discrimination Learning*, pp. 215–241. London, Academic Press.

Blough, D. S. (1958). A method for obtaining psychophysical thresholds from the pigeon. *Science*, **126**, 304–305.

Blough, D. S. (1966). The reinforcement of least-frequent interresponse times. *J. exp. Anal. Behav.*, **9**, 581–591.

Blough, D. S. (1967). Stimulus generalization as signal detection in pigeons. *Science*, **158**, 940–941.

Blough, D. S. (1969). Generalization gradient shape and summation in steady-state tests. *J. exp. Anal. Behav.*, **12**, 91–104.

Blue, S., Sherman, J. G. & Pierrel, R. (1971). Differential responding as a function of auditory stimulus intensity without differential reinforcement. *J. exp. Anal. Behav.*, **15**, 371–377.

Boe, E. E. & Church, R. M. (1967). Permanent effects of punishment during extinction. *J. comp. physiol. Psychol.*, **63**, 486–492.

Bolles, R. C. (1970). Species-specific defense reactions and avoidance learning. *Psychol. Rev.*, **77**, 32–48.

Bolles, R. C. (1971). Species-specific defense reactions. In F. R. Brush (ed.), *Aversive*

*Conditioning and Learning*, pp. 183–233. New York, Academic Press.

Bolles, R. C. (1972). Reinforcement, expectancy and learning. *Psychol. Rev.*, **79**, 394–409.

Bolles, R. C. (1975). *Learning Theory*. New York, Holt, Rinehart & Winston.

Bolles, R. C. & Grossen, N. E. (1969). Effects of an informational stimulus on the acquisition of avoidance behaviour in rats. *J. comp. physiol. Psychol.*, **68**, 90–99.

Bolles, R. C. & Grossen, N. E. (1970). Function of the CS in shuttle-box avoidance learning by rats. *J. comp. physiol. Psychol.*, **70**, 165–169.

Bolles, R. C., Moot, S. A. & Grossen, N. E. (1971). The extinction of shuttlebox avoidance. *Learn. Motiv.*, **2**, 324–333.

Bolles, R. C. & Popp, R. J. (1964). Parameters affecting the acquisition of Sidman avoidance. *J. exp. Anal. Behav.*, **7**, 315–321.

Bolles, R. C., Stokes, L. W. & Younger, M. S. (1966). Does CS termination reinforce avoidance behaviour? *J. comp. physiol. Psychol.*, **62**, 201–207.

Bond, N. (1973). Schedule-induced polydipsia as a function of the consummatory rate. *Psychol. Rec.*, **23**, 377–382.

Boneau, C. A. & Axelrod, S. (1962). Work decrement and reminiscence in pigeon operant responding. *J. exp. Psychol.*, **64**, 352–354.

Borgealt, A. J., Donahoe, J. W. & Weinstein, A. (1972). Effects of delayed and trace components of a compound CS on conditioned suppression and heart rate. *Psychon. Sci.*, **26**, 13–15.

Bostow, D. E. & Bailey, J. B. (1969). Modification of severe disruptive and aggressive behaviour using brief timeout and reinforcement procedures. *J. appl. Behav. Anal.*, **2**, 31–37.

Bower, G., McLean, J. & Meacham, J. (1966). Value of knowing when reinforcement is due. *J. comp. physiol. Psychol.*, **62**, 184–192.

Bower, G., Starr, R. & Lazarovitz, L. (1965). Amount of response-produced change in the CS and avoidance learning. *J. comp. physiol. Psychol.*, **52**, 727–729.

Bower, T. G. R. (1964). Discrimination of depth in premature infants. *Psychon. Sci.*, **1**, 368.

Bowlby, J. A. (1958). The nature of the child's tie to his mother. *Int. J. Psychoanal.*, **39**, 350–373.

Boyd, H. & Fabricius, E. (1965). Observations on the incidence of following of visual and auditory stimuli in naive mallard ducklings (*Anas platyrhynchos*). *Behaviour*, **25**, 1–15.

Brackbill, Y. (1958). Extinction of the smiling response in infants as a function of reinforcement schedule. *Child Dev.*, **29**, 115–124.

Brackbill, Y. (1962). Research and clinical work with children. In R. A. Bower (ed.), *Some Views on Soviet Psychology*, pp. 99–164. Washington D.C., American Psychological Association.

Brackbill, Y. & Fitzgerald, H. E. (1969). Development of the sensory analyzers during infancy. In L. P. Lipsitt & H. W. Reese (eds), *Advances in Child Development and Behaviour*, Vol. 4. New York, Academic Press.

Brackbill, Y., Fitzgerald, H. E. & Lintz, L. M. (1967). A developmental study of classical conditioning. *Monogr. Soc. Res. Child Dev.*, **32**, No. 8.

Brackbill, Y. & Koltsova, M. M. (1967). Conditioning and learning. In Y. Brackbill (ed.), *Infancy and Early Childhood*, pp. 207–286. New York, Free Press.

Brackbill, Y., Lintz, L. M. & Fitzgerald, H. E. (1968). Differences in the autonomic and somatic conditioning of infants. *Psychosom. Med.*, **30**, 193–201.

Bradley, R. J. (1971). Some serial properties of burst responding on DRL. *Psychon. Sci.*, **24**, 28–30.

Brady, J. P. (1966). Brevital-relaxation treatment of frigidity. *Behav. Res. Ther.*, **4**, 71–77.

Brady, J. V., Kelly, D. & Plumlee, L. (1969). Autonomic and behavioural responses of the rhesus monkey to emotional conditioning. *Ann. N.Y. Acad. Sci.*, **159**, 959–975.

Braine, L. G. (1965). Age changes in the mode of perceiving geometric forms. *Psychon. Sci.*, **2**, 155–156.

Braveman, N. & Capretta, P. J. (1965). The relative effectiveness of two experimental techniques for the modification of food preferences in rats. *Proc. 73rd Annual Convention Am. Psychol. Ass.*, pp. 129–130. Washington D.C., Am. Psychol. Ass.

Breland, K. & Breland, M. (1961). The misbehaviour of organisms. *Am. Psychol.*, **16**, 661–664.

Breland, K. & Breland, M. (1966). *Animal Behaviour.* New York, The Macmillan Company.

Brener, J. & Hothersall, D. (1966). Heart rate control under conditions of augmented sensory feedback. *Psychophysiology*, **3**, 23–28.

Brethower, D. M. & Reynolds, G. S. (1962). A facilitative effect of punishment on unpunished behaviour. *J. exp. Anal. Behav.*, **5**, 191–199.

Bricker, W. A. & Bricker, D. D. (1970). A program of language training for the severely handicapped child. *Exceptional Children*, **37**, 101–111.

Broadbent, D. E. (1958). *Perception and Communication.* Oxford, Pergamon Press.

Brogden, W. J. (1939a). Unconditioned stimulus-substitution in the conditioning process. *Am. J. Psychol.*, **52**, 46–55.

Brogden, W. J. (1939b). Sensory pre-conditioning. *J. exp. Psychol.*, **25**, 323–332.

Brogden, W. J. (1939c). The effect of frequency of reinforcement upon the level of conditioning. *J. exp. Psychol.*, **24**, 419–431.

Brogden, W. J., Lipman, E. A. & Culler, E. (1938). The role of incentive in conditioning and extinction. *Am. J. Psychol.*, **51**, 109–117.

Brown, B. (1970). Recognition of aspects of consciousness through association with EEG alpha activity represented by a light signal. *Psychophysiology*, **6**, 442–452.

Brown, B. (1971). Awareness of EEG-subjective activity relationships detected within a closed feedback system. *Psychophysiology*, **7**, 451–464.

Brown, B. S., Wienckowski, L. A. & Stolz, S. B. (1975). *Behaviour Modification: Perspective on a Current Issue.* U.S. Dept. of Health, Education and Welfare, Publication No. (ADM) 75–202.

Brown, J. (1958). Some tests of the decay theory of immediate memory. *Q. Jl exp. Psychol.*, **10**, 12–21.

Brown, P. L. & Jenkins, H. M. (1968). Auto-shaping of the pigeon's key-peck. *J. exp. Anal. Behav.*, **60**, 64–69.

Brown, R. T. (1975). Following and visual imprinting in ducklings across a wide age range. *Dev. Psychobiol.*, **8**, 27–33.

Brown, R. T. & Hamilton, A. S. (1977). Imprinting: effects of discrepancy from rearing conditions on approach to a familiar imprinting object in a novel situation. *J. comp. physiol. Psychol.*, **91**, 784–793.

Browne, M. (1974). Autoshaping and the role of primary reinforcement and overt movements in the acquisition of stimulus-stimulus relations. Unpublished Doctoral dissertation, Indiana Univ.

Bruner, A. & Revusky, S. H. (1961). Collateral behaviour in humans. *J. exp. Anal. Behav.*, **4**, 349–350.

Brush, F. R. (1962). The effects of intertrial interval on avoidance learning in the rat. *J. comp. physiol. Psychol.*, **55**, 888–892.

Brush, F. R. (1966). On the differences between animals that learn and do not learn to avoid electric shock. *Psychon. Sci.*, **5**, 123–124.

Brush, F. R. (1971). *Aversive Conditioning and Learning.* New York, Academic Press.

Buddenhagen, R. (1971). *Establishing Vocalizations in Mute Mongoloid Children.* Champaign, Ill., Research Press.

Budzynski, T. H., Stoyva, J. M. & Adler, C. S. (1970). Feedback-induced relaxation: Application to tension headache. *J. Behav. Theor. exp. Psychiat.*, **1**, 205.

Bull, J. A. (1970). An interaction between appetitive Pavlovian CSs and instrumental avoidance responding. *Learn. Motiv.*, **1**, 18–26.

Bullock, D. H. & Bitterman, M. E. (1962). Food imprinting in the snapping turtle, *Chelydra serpentina. Science*, **151**, 108–109.

Burt, C. (1968). Brain and consciousness. *Br. J. Psychol.*, **59**, 55–69.

Butterfield, P. (1970). An analysis of the pair bond in the zebra finch. In J. H. Crook (ed.), *Social Behaviour in Animals and Man*. New York, Academic Press.

Cahoon, D. D. (1968). Symptom substitution and the behaviour therapies. *Psychol. Bull.*, **69**, 149–156.

Camp, D. S., Raymond, G. A. & Church, R. M. (1967). Temporal relationship between response and punishment. *J. exp. Psychol.*, **74**, 114–123.

Campbell, C. S. (1969). The development of specific preferences in thiame-deficient rats: Evidence against mediation by after-tastes. Unpublished Master's thesis: University of Illinois at Chicago Circle.

Campbell, S. L. (1962). Lever holding and behaviour sequences in shock escape. *J. comp. physiol. Psychol.*, **55**, 1047–1053.

Capaldi, E. J. (1967). A sequential hypothesis of instrumental learning. In K. W. Spence & J. T. Spence (eds), *The Psychology of Learning and Motivation*, Vol. 1, pp. 67–156. New York, Academic Press.

Caplan, M. (1970). Effects of withheld reinforcement on timing behaviour of rats with limbic lesions. *J. comp. physiol. Psychol.*, **71**, 119–135.

Capretta, P. J. (1961). An experimental modification of food preferences in chickens. *J. comp. physiol. Psychol.*, **54**, 238–242.

Carlson, J. G. (1968). Frustrative non-reinforcement of operant responding: magnitude of reinforcement and response force effects. *Psychon. Sci.*, **11**, 307–308.

Carlson, N. J. & Black, A. H. (1960). Traumatic avoidance learning. The effect of preventing escape responses. *Can. J. Psychol.*, **14**, 21–28.

Carmona, A. (1967). Unpublished Doctoral dissertation, Yale University.

Caron, R. F. (1967). Visual reinforcement in young infants. *J. exp. child Psychol.*, **5**, 489–511.

Carpenter, F. (1974). *The Skinner Primer*. London, The Free Press.

Catania, A. C. (1966). Concurrent operants. In W. K. Honig (ed.), *Operant Behaviour: Areas of Research and Application*, pp. 213–270. New York, Appleton–Century–Crofts.

Catania, A. C. & Reynolds, G. S. (1968). A quantitative analysis of the responding maintained by interval schedules of reinforcement. *J. exp. Anal. Behav.*, **11**, 327–383.

Cautela, J. R. (1976). Covert conditioning. *Mental Imagery*,

Cautela, J. R. & Baron, M. G. (1977). Covert conditioning: a theoretical analysis. *Behav. Mod.*, **1**, 351–368.

Chance, M. R. A. (1960). Köhler's chimpanzees – how did they perform? *Man*, **60**, 130–135.

Chapman, H. W. (1969). Oropharyngeal determinants of non regulatory drinking in the rat. Unpublished Ph.D. dissertation, Univ. of Pennsylvania.

Cherek, D. R., Thompson, T. & Heistad, G. T. (1973). Responding maintained by the opportunity to attack during an interval food reinforcement schedule. *J. exp. Anal. Behav.*, **19**, 113–123.

Chiang, H. & Wilson, W. A. (1963). Some tests of the diluted water hypothesis of saline consumption in rats. *J. comp. physiol. Psychol.*, **63**, 24–27.

Chillag, D. & Mendelson, J. (1971). Schedule-induced airlicking as a function of body-weight deficit in rats. *Physiol. Behav.*, **6**, 603–605.

Chiszar, D. A. & Spear, N. E. (1968). Proactive interference in retention of nondiscriminative learning. *Psychon. Sci.*, **12**, 87–88.

Chomsky, N. (1959). Verbal behaviour, by B. F. Skinner. *Language*, **35**, 26–58.

Chomsky, N. (1967). The formal nature of language. In E. H. Lenneberg (ed.), *Biological Foundations of Language.* New York, John Wiley and Sons.

Chomsky, N. (1972). Psychology and ideology. *Cognition*, **1**, 11–46.

Church, R. M. (1969). Response suppression. In B. A. Campbell & R. M. Church (eds), *Punishment and Aversive Behaviour*, pp. 111–156. New York, Appleton–Century–Crofts.

Clark, F. C. (1962). Some observations on the adventitious reinforcement of drinking under food reinforcement. *J. exp. Anal. Behav.*, **5**, 61–63.

Clifton, R. K. (1974a). Heart rate conditioning in the new-born infant. *J. exp. child Psychol.*, **18**, 9–21.

Clifton, R. K. (1974b). Cardiac conditioning and orienting in the infant. In P. Obrist, J. Brener, L. DiCara & A. H. Black (eds), *Cardiovascular Psychophysiology – Current Issues.* New York, Aldine.

Cohen, I. L. & Mendelson, J. (1974). Schedule-induced drinking with food, but not ICS, reinforcement. *Behav. Biol.*, **12**, 21–29.

Cohen, J. S., Stetner, L. J. & Michael, D. J. (1969). Effect of deprivation level on span of attention in a multi-dimension discrimination task. *Psychon. Sci.*, **15**, 31–32.

Cohen, P. S. (1968). Punishment: the interactive effects of delay and intensity of shock. *J. exp. Anal. Behav.*, **11**, 789–799.

Cohen, P. S. & Looney, T. A. (1973). Schedule-induced mirror responding in the pigeon. *J. exp. Anal. Behav.*, **19**, 395–408.

Collias, N. E. (1953). Some factors in maternal rejection by sheep and goats. *Bull. ecol. Soc. Am.*, **34**, 78.

Collias, N. E. & Collias, E. C. (1956). Some mechanisms of family integration in ducks. *Auk*, **73**, 378–400.

Corfield-Sumner, P. K., Blackman, D. E. & Stainer, G. (1977). Polydipsia induced in rats by second-order schedules of reinforcement. *J. exp. Anal. Behav.*, **27**, 265–273.

Cornwell, A. C. & Fuller, J. L. (1961). Conditioned responses in young puppies. *J. comp. physiol. Psychol.*, **54**, 13–15.

Corsini, D. A. (1969). Developmental changes in the effect of nonverbal cues on retention. *Dev. Psychol.*, **1**, 425–435.

Corte, H. E., Wolf, M. M. & Locke, D. B. (1971). A comparison of procedures for eliminating self-injurious behaviour of retarded adolescents. *J. appl. Behav. Anal.*, **4**, 201–203.

Costello, C. G. (1963). The essentials of behaviour therapy. *Can. Psychiat. Ass. J.*, **8**, 162–166.

Couch, J. V. (1974). Reinforcement magnitude and schedule-induced polydipsia: a re-examination. *Psychol. Rec.*, **24**, 559–262.

Coughlin, R. C. (1972). The aversive properties of withdrawing positive reinforcement: A review of the recent literature. *Psychol. Rec.*, **22**, 333–354.

Craggs, B. G. (1972). The development of synapses in cat visual cortex. *Invest. Ophthal.*, **11**, 377–385.

Crespi, L. L. (1942). Quantitative variation of incentive and performance in the white rat. *Am. J. Psychol.*, **55**, 467–517.

Crisler, G. (1930). Salivation is unnecessary for the establishment of the salivary conditioned reflex induced by morphine. *Am. J. Psychol.*, **94**, 553–556.

Crowder, R. G. (1967). Proactive and retroactive inhibition in the retention of a T-maze habit in rats. *J. exp. Psychol.*, **74**, 167–171.

Cullen, C. N., Hattersley, J. & Tennant, L. (1977). Behaviour modification – some implications of a radical behaviourist view. *Bull. Br. Psychol. Soc.*, **30**, 65–69.

Daly, H. B. (1972). Learning to escape cues paired with reward reductions following single-or-multiple-pellet rewards. *Psychon. Sci.*, **26**, 49–52.

D'Amato, M. R. (1973). Delayed matching and short-term memory in monkeys. In G. H. Bower (ed.), *The Psychology of Learning and Motivation: Advances in Research and Theory*, Vol. 7. New York, Academic Press.

D'Amato, M. R. & Cox, J. K. (1976). Delay of consequences and short-term memory in monkeys. In D. L. Medin, W. A. Roberts & R. T. Davis (eds), *Processes of Animal Memory*, pp. 49–78. New York, Lawrence Erlbaum Ass.

D'Amato, M. R. & Fazzaro, J. (1966). Discriminated lever-press avoidance learning as a function of type and intensity of shock. *J. comp. physiol. Psychol.*, **61**, 313–315.

D'Amato, M. R., Fazzaro, J. & Etkin, M. (1967). Discriminated bar-press avoidance maintenance and extinction in rats as a function of shock intensity. *J. comp. physiol. Psychol.*, **63**, 351–354.

D'Amato, M. R., Fazzaro, J. & Etkin, M. (1968). Anticipatory responding and avoidance discrimination as factors in avoidance conditioning. *J. exp. Psychol.*, **77**, 41–47.

D'Amato, M. R. & O'Neill, W. (1971). Effect of delay-interval illuminations on matching behaviour in the capuchin monkey. *J. exp. Anal. Behav.*, **15**, 327–333.

D'Amato, M. R. & Schiff, D. (1964). Long-term discriminated avoidance performance in the rat. *J. comp. physiol. Psychol.*, **57**, 123–126.

D'Amato, M. R. & Worsham, R. W. (1972). Delayed matching in the capuchin monkey with brief sample durations. *Learn. Motiv.*, **3**, 304–312.

D'Amato, M. R. & Worsham, R. W. (1974). Retrieval cues and short-term memory in capuchin monkeys. *J. comp. physiol. Psychol.*, **86**, 274–282.

Darby, C. L. & Riopelle, A. J. (1959). Observational learning in the rhesus monkey. *J. comp. physiol. Psychol.*, **52**, 94–98.

Davis, J. M. (1974). Socially induced flight reactions in pigeons. Unpublished doctoral dissertation, Duke University.

Davitz, J. R. & Mason, D. J. (1955). Socially facilitated reduction of a fear response in rats. *J. comp. physiol. Psychol.*, **48**, 149–51.

Day, W. F. (1969). Radical behaviourism in reconciliation with phenomenology. *J. exp. Anal. Behav.*, **12**, 315–328.

Dekker, E., Pelser, H. E. & Groen, J. (1957). Conditioning as a cause of asthmatic attacks. *J. Psychosom. Res.*, **2**, 97–108.

Denny, M. R. (1971). Relaxation theory and experiments. In F. R. Brush (ed.), *Aversive Conditioning and Learning*, pp. 235–295. New York, Academic Press.

Denti, A. & Epstein, A. (1972). Sex differences in the acquisition of two kinds of avoidance behaviour in rats. *Physiol. Behav.*, **8**, 611–615.

Del Russo, J. E. (1975). Observational learning of discriminative avoidance in hooded rats. *Anim. Learn. Behav.*, **3**, 76–80.

Delwesse, J. (1977). Schedule-induced biting under fixed-interval schedules of food or electric-shock presentation. *J. exp. Anal. Behav.*, **27**, 419–431.

Dews, P. B. (1958). Studies on behaviour: IV stimulant actions of metamphetamine. *J. Pharmacol. exp. Ther.*, **122**, 137–147.

Dews, P. B. (1962). The effect of multiple S$^\Delta$ periods on responding on a fixed-interval schedule. *J. exp. Anal. Behav.*, **5**, 319–374.

Dews, P. B. (1965). The effects of multiple S$^\Delta$ periods on responding on a fixed-interval schedule. III. Effects of changes in pattern of interruptions, parameters and stimuli. *J. exp. Anal. Behav.*, **8**, 427–433.

Dews, P. B. (1966). The effect of multiple S$^\Delta$ periods on responding on a fixed-interval schedule. V. Effect of periods of complete darkness and of occasional omissions of food presentations. *J. exp. Anal. Behav.*, **9**, 573–578.

Dews, P. B. (1969). Studies on responding under fixed-interval schedule of reinforcement: the effects on the pattern of responding of changes in requirement at reinforcement. *J. exp. Anal. Behav.*, **12**, 191–199.

Dews, P. B. (1970). The theory of fixed-interval responding. In W. N. Schoenfeld (ed.), *The Theory of Reinforcement Schedules*, pp. 43–61. New York, Appleton–Century–Crofts.

Dews, P. B. & Morse, W. H. (1958). Some observations on an operant in human subjects and its modification by dextro-amphetamine. *J. exp. Anal. Behav.*, **1**, 359–364.

Dicara, L. V. & Miller, N. E. (1968a). Changes in heart-rate instrumentally learned by curarized rats as avoidance responses. *J. comp. physiol. Psychol.*, **65**, 8–12.

Dicara, L. V. & Miller, N. E. (1968b). Instrumental learning of vasomotor responses by rats: learning to respond differentially in the two ears. *Science*, **159**, 1485–1486.

Dicara, L. V. & Miller, N. E. (1968c). Instrumental learning of systolic blood pressure by curarized rats: Dissociation of cardiac and vascular changes. *Psychosom. Med.*, **30**, 489–494.

Dickinson, A. (1972). Septal damage and response output under frustrative nonreward. In R. A. Boakes & M. S. Halliday (eds), *Inhibition and Learning*, pp. 461–466. London, Academic Press.

Dickinson, A. & Scull, J. (1975). Transient effects of reward presentation and omission on subsequent operant responding. *J. comp. physiol. Psychol.*, **88**, 447–458.

Dietz, M. N. & Capretta, P. J. (1967). Modification of sugar and sugar-saccharin preference in rats as a function of electrical shock to the mouth. *Proc. 75th Annual Convention Am. Psychol. Ass.*, pp. 161–162. Washington D.C., American Psychological Association.

Dinsmoor, J. A. & Winograd, E. (1958). Shock intensity in variable interval escape schedules. *J. exp. Anal. Behav.*, **1**, 145–148.

Dobrzeska, C., Szweijkowska, G. & Konorski, J. (1966). Qualitative versus directional cues in two forms of differentiation. *Science*, **153**, 87–89.

Domjan, M. & Wilson, N. E. (1972). Contribution of ingestive behaviours to taste-aversion learning in the rat. *J. comp. physiol. Psychol.*, **80**, 403–412.

Dücker, G. & Rensch, B. (1968). Verzogerung des Vergessens erlernber visuellen Aufgaben bei Fischen durch Dunkelhaltung. *Pfluegers Arch. für Gesamte Physiol. Mensch. Tiere*, **301**, 1–6.

Dunham, P. J. (1968). Contrasted conditions of reinforcement: A selective critique. *Psychol. Bull.*, **69**, 295–315.

Dunham, P. J. (1971). Punishment: method and theory. *Psychol. Rev.*, **78**, 58–70.

Dunham, P. J. (1972). Some effects of punishment upon unpunished responding. *J. exp. Anal. Behav.*, **17**, 443–450.

Dunlap, W. P., Hughes, L. F., Dachowski, L. & O'Brien, T. H. (1974). The temporal course of the frustration effect. *Learn. Motiv.*, **5**, 484–497.

Egger, M. D. & Miller, N. E. (1962). Secondary reinforcement in rats as a function of information value and reliability of the stimulus. *J. exp. Psychol.*, **64**, 97–104.

Egger, M. D. & Miller, N. E. (1963). When is a reward reinforcing? An experimental study of the information hypothesis. *J. comp. physiol. Psychol.*, **56**, 132–137.

Eckerman, D. A. (1967). Stimulus control by part of a complex $S^\Delta$. *Psychon Sci.*, 7, 299–300.

Ellis, N. R., Barnett, C. D. & Pryer, M. W. (1960). Operant behaviour in mental defectives: exploratory studies. *J. exp. Anal. Behav.*, **3**, 63–69.

Ellison, G. D. (1964). Differential salivary conditioning to traces. *J. comp. physiol. Psychol.*, **57**, 373–380.

Emurian, H. H. & Weiss, S. J. (1972). Compounding discriminative stimuli controlling free-operant avoidance. *J. exp. Anal. Behav.*, **17**, 249–256.

Engel, B. T. & Bleecker, E. R. (1974). Application of operant conditioning techniques to the control of the cardiac arrhythmias. In P. A. Obrist, A. H. Black, J. Brener & L. V. Dicara (eds), *Cardiovascular Psychophysiology*. Chicago, Aldine.

Estes, W. K. (1944). An experimental study of punishment. *Psychol. Monogr.*, **57**, (3, whole No. 263).

Estes, W. K. (1969). Outline of a theory cf punishment. In B. A. Campbell & R. M. Church (eds), *Punishment and Aversive Behaviour*, pp. 57–82. New York, Appleton–Century–Crofts.

Estes, W. K. & Skinner, B. F. (1941). Some quantitative properties of anxiety. *J. exp. Psychol.*, **29**, 390–400.

Etkin, M. W. (1972). Light produced interference in a delayed matching task with capuchin monkeys. *Learn. Motiv.*, **3**, 313–324.

Etzel, B. C. & Gewirtz, J. L. (1967). Experimental modification of caretaker-maintained high-rate operant crying in a 6- and a 20-week old infant (Infans tyramotearus): Extinction of crying with reinforcement of eye contact and smiling. *J. exp. child Psychol.*, **5**, 303–317.

Eysenck, H. J. (ed.) (1960). *Behaviour Therapy and the Neuroses*. London, Pergamon Press.

Fabricius, E. (1951). Zur ethologie junger Anatiden. *Acta Zool. Fenn.*, **68**, 1–175.

Fabricius, E. & Boyd, H. (1954). Experiments on the following reactions of ducklings. *Wildfowl Trust Ann. Rep.*, *1952–53*, 84–89.

Falk, J. L. (1961). Production of polydipsia in normal rats by an intermittent food schedule. *Science*, **133**, 195–196.

Falk, J. L. (1964). Studies on schedule-induced polydipsia. In W. J. Wagner (ed.), *Thirst: First International Symposium on Thirst in the Regulation of Body Water*, pp. 95–116. New York, Pergamon Press.

Falk, J. L. (1966a). Schedule-induced polydipsia as a function of fixed-interval length. *J. exp. Anal. Behav.*, **9**, 37–39.

Falk, J. L. (1966b). Analysis of water and NaCl solution acceptance by schedule-induced polydipsia. *J. exp. Anal. Behav.*, **9**, 111–118.

Falk, J. L. (1967). Control of schedule-induced polydipsia: type, size, and spacing of meals. *J. exp. Anal. Behav.*, **10**, 199–206.

Falk, J. L. (1969). Conditions producing psychogenic polydipsia in animals. *Ann. N. Y. Acad. Sci.*, **157**, 569–593.

Falk, J. L. (1971). The nature and determinants of adjunctive behaviour. *Physiol. Behav.*, **6**, 577–588.

Fantino, E. (1969). Conditioned reinforcement, choice, and the psychological distance to reward. In D. P. Hendry (ed.), *Conditioned Reinforcement*, pp. 163–191. Homewood, Illinois, The Dorsey Press.

Fantino, E. & Cole, M. (1968). Sand-digging in mice: functional autonomy? *Psychon. Sci.*, **10**, 20–30.

Fantino, E. & Duncan, B. (1972). Some effects of interreinforcement time upon choice. *J. exp. Anal. Behav.*, **17**, 3–14.

Fantino, E., Weigele, S. & Lancy, D. (1972). Aggressive display in the siamese fighting fish (*Betta splendens*). *Learn. Motiv.* **3**, 457–468.

Fantz, R. L. (1965). Ontogeny of perception. In A. M. Schrier, H. F. Harlow & F. Stollnitz (eds), *Behaviour of Nonhuman Primates*, Vol. II, pp. 365–403.

Fantz, R. L. (1967). Visual perception and experience in early infancy. In H. W. Stevenson, E. H. Hess & H. L. Rheingold (eds), *Early Behaviour*. New York, Wiley.

Feldman, M. P. & MacCulloch, M. J. (1965). The application of anticipatory avoidance learning to the treatment of homosexuality. I. Theory, technique and preliminary results. *Behav. Res. Ther.*, **2**, 165–183.

Feldman, R. S. & Bremner, F. J. (1963). A method for rapid conditioning of stable, avoidance bar pressing behaviour. *J. exp. Anal. Behav.*, **6**, 393–394.

Felton, M. & Lyon, D. O. (1966). The post-reinforcement pause. *J. exp. Anal. Behav.*, **9**, 131–134.

Ferraro, D. P., Schoenfeld, W. N. & Snapper, A. G. (1965). Sequential response effects in the white rat during conditioning and extinction on a DRL schedule. *J. exp. Anal. Behav.*, **8**, 255–260.

Ferster, C. B. (1964). Arithmetic behaviour in chimpanzees. *Scient. Am.*, **210**, 98–106.

Ferster, C. B. (1967). Arbitrary and natural reinforcement. *Psychol. Rec.*, **17**, 341–347.

Ferster, C. B., Nurnberger, J. I. & Levitt, E. B. (1962). The control of eating. *J. Mathetics*, **1**, 87–109.

Ferster, C. B. & Skinner, B. F. (1957). *Schedules of Reinforcement*. New York, Appleton–Century–Crofts.

Fields, P. E. (1932). Studies in concept formation. *Comp. Psychol. Monogr.*, **9**, 70–

Finch, G. (1938). Salivary conditioning in atropinized dogs. *Am. J. Psychol.*, **124**, 136–141.

Fischer, G. J. (1966). Discrimination and successive spatial reversal learning in chicks that fail to imprint vs. ones that imprint strongly. *Percept. Mot. Skills*, **23**, 579–584.

Fischer, G. J. (1967). Comparisons between chicks that fail to imprint and ones that imprint strongly. *Behaviour*, **29**, 262–267.

Fisher, J. & Hinde, C. A. (1949). The opening of milk bottles by birds. *Br. Birds*, **42**, 347–357.

Fitzgerald, H. E. & Brackbill, Y. (1971). Tactile conditioning of an autonomic and somatic response in young infants. *Conditional Reflex*, **6**, 41–51.

Fitzgerald, H. E. & Brackbill, Y. (1976). Classical conditioning in infancy: development and constraints. *Psychol. Bull.*, **83**, 353–376.

Fitzgerald, H. E., Lintz, L. M., Brackbill, Y. & Adams, G. (1967). Time perception and conditioning of an autonomic response in human infants. *Percept. Mot. Skills*, **24**, 479–486.

Fitzgerald, R. D. (1963). Effects of partial reinforcement with acid on the classically conditioned salivary response in dogs. *J. comp. physiol. Psychol.*, **56**, 1056–1060.

Flagg, S. F. (1975). Transformation of positions and short-term memory in rhesus monkeys. Unpublished Doctoral dissertation. Washington State University, Pullman.

Flory, R. K. (1969). Attack behaviour as a function of minimum inter-food interval. *J. exp. Anal. Behav.*, **12**, 825–828.

Flory, R. K. & Ellis, B. B. (1973). Schedule-induced aggression against a slide-image target. *Bull. Psychon. Soc.*, **2**, 287–290.

Flory, R. K. & O'Boyle, M. K. (1972). The effect of limited water availability on schedule-induced polydipsia. *Physiol. Behav.*, **8**, 147–149.

Foree, D. D. & Lolordo, V. M. (1973). Attention in the pigeon: the differential effects of food-getting vs shock-avoidance procedures. *J. comp. physiol. Psychol.*, **85**, 551–558.

Foree, D. D. & Lolordo, V. M. (1975). Stimulus-reinforcer interactions in the pigeon. The role of electric shock and the avoidance contingency. *J. exp. Psychol. anim. Behav. Process*, **104**, 39–46.

Fouts, R. S. (1973). Capacities for language in the Great Apes. *Proc. IXth Int. Congr. Anthropol. Ethnol. Sci.* The Hague, Mouton and Co.

Fouts, R. S. (1975). Communication with chimpanzees. In I. Eibl-Eiblesfeldt & G. Kurth (eds), *Hominisation and Verhalten*. Stuttgart, Gustav Fischer.

Fouts, R. S. & Rigby, R. L. (1976). Man-chimpanzee communication. In T. A. Seberk (ed.), *How Animals Communicate*. Bloomington, Indiana Univ. Press.

Fowler, H. (1971). Suppression and facilitation by response contingent shock: In F. R. Brush (ed.), *Aversive Conditioning and Learning*, pp. 537–604. New York, Academic Press.

Fowler, H., Goldman, L. & Wischner, G. J. (1968). Sodium amytal and the shock-right intensity function for visual discrimination learning. *J. comp. physiol. Psychol.*, **65**, 155–159.

Foxx, R. M. & Azrin, N. H. (1972). Restitution: A method of eliminating aggressive – disruptive behaviour of retarded and brain damaged patients. *Behav. Res. Ther.*, **10**, 15–27.

Foxx, R. M. & Azrin, N. H. (1973). The elimination of autistic self-stimulatory behaviour by over-correction. *J. appl. Behav. Anal.*, **6**, 1–14.

Frank, J. & Staddon, J. E. R. (1974). Effects of restraint on temporal discrimination behaviour. *Psychol. Rec.*, **24**, 123–130.

French, J. W. (1942). The effect of temperature on the retention of a maze habit in fish. *J. exp. Psychol.*, **31**, 79–87.

Freund, K. (1960). Some problems in the treatment of homosexuality. In H. J. Eysenck (ed.), *Behaviour Therapy and the Neuroses*, pp. 312–325. New York, Pergamon.

Frey, P. W. (1969). Within-and between-sessions CS intensity performance effects in rabbit eyelid conditioning. *Psychon. Sci.*, **17**, 1–2.

Fromer, R. (1963). Conditioned vasomotor responses in the rabbit. *J. comp. physiol. Psychol.*, **56**, 1050–1055.

Gallup, G. G. (1965). Aggression in rats as a function of frustrative nonreward in a straight alley. *Psychon. Sci.*, **3**, 99–100.

Gallup, G. G. (1966). Mirror image reinforcement in monkeys. *Psychon. Sci.*, **5**, 39.

Gallup, G. G. & Hare, G. K. (1969). Activity following partially reinforced trials. Evidence for a residual frustration effect due to conditioned frustration. *Psychon. Sci.*, **16**, 41–42.

Galvani, P. F., Riddell, W. I. & Foster, K. M. (1975). Passive avoidance in rats and gerbils as a function of species-specific exploratory tendencies. *Behav. Biol.*, **13**, 277–290.

Gamzu, E. & Schwam, E. (1974). Autoshaping and automaintenance of a key-press response in squirrel monkeys. *J. exp. Anal. Behav.*, **21**, 361–371.

Gamzu, E. R. & Williams, D. R. (1973). Associative factors underlying the pigeon's key pecking in autoshaping procedures. *J. exp. Anal. Behav.*, **19**, 225–232.

Gamzu, E. R., Williams, D. R. & Schwartz, B. (1973). Pitfalls of organismic concepts: "Learned Laziness"? *Science*, **181**, 367–368.

Garcia, J., Hawkins, W. G. & Rusiniak, K. W. (1974). Behavioural regulation of the milieu interne in man and rat. *Science*, **185**, 824–831.

Garcia, J. & Koelling, R. A. (1966). Relation of cue to consequence in avoidance learning. *Psychon. Sci.*, **4**, 123–124.

Garcia, J., McGowan, B. K. & Green, K. F. (1972). Biological constraints on conditioning. In M. E. P. Seligman & J. L. Hager (eds) *Biological Boundaries of Learning*, pp. 21–43. New York, Appleton–Century–Crofts.

Gardner, B. T. & Gardner, R. A. (1969). Teaching sign language to a chimpanzee. *Science*, **165**, 664–672.

Gardner, W. M. (1969). Autoshaping in bobwhite quail. *J. exp. Anal. Behav.*, **12**, 279–281.

Geller, A., Robustelli, F., Barondes, S. H., Cohen, H. D. & Jarvik, M. E. (1969). Impaired performance by post-trial injections of cycloheximide in a passive avoidance task. *Psychopharmacologia*, **14**, 371–376.

Gellerman, L. W. (1933). Form discrimination in chimpanzees and two-year old children: I. Form (Triangularity) *per se*. *Pedagog. Semin.*, **42**, 3–27.

Gentry, W. D. (1968). Fixed-ratio schedule-induced aggression. *J. exp. Anal. Behav.*, **11**, 813–817.

Gentry, W. D. & Schaeffer, R. W. (1969). The effect of FR response requirement on aggressive behaviour in rats. *Psychon. Sci.*, **14**, 236–238.

Geschwind, N. (1970). The organization of language and the brain. *Science*, **170**, 940–944.

Gibson, E. J., Walk, R. D. & Tighe, T. J. (1959). Enhancement and deprivation of visual stimulation during rearing as factors in visual discrimination learning. *J. comp. physiol. Psychol.*, **52**, 74–81.

Gilbert, R. & Beaton, J. (1967). Imitation and cooperation by hooded rats. *Psychon. Sci.*, **8**, 43.

Gilbert, R. M. (1974a). Schedule-induced ethanol polydipsia in rats with restricted fluid availability. *Psychopharmac.*, **38**, 151–157.

Gilbert, R. M. (1974b). Ubiquity of schedule-induced polydipsia. *J. exp. Anal. Behav.*, **21**, 277–284.

Girardeau, F. L. & Spradlin, J. E. (1964). Token rewards on a cottage program. *Ment. Retard.*, **2**, 245–351.

Gleitman, H. (1971). Forgetting of long-term memories in animals. In W. K. Honig & P. H. R. James (eds), *Animal Memory*. New York, Academic Press.

Gleitman, H. & Jung, L. (1963). Retention in rats: The effect of proactive interference. *Science*, **142**, 1683–1684.

Gleitman, H. & Steinman, F. (1963). Retention of runway performance as a function of proactive interference. *J. comp. physiol. Psychol.*, **56**, 834–838.

Glickman, S. E. (1973). Responses and reinforcement. In R. A. Hinde & J. Stevenson-Hinde, *Constraints on Learning*, pp. 207–241. London, Academic Press.

Goldiamond, I. (1976). Self-reinforcement. *J. appl. Behav. Anal.*, **9**, 509–514.

Gonzalez, F. A. & Waller, M. B. (1974). Handwriting as an operant. *J. exp. Anal. Behav.*, **21**, 165–176.

Gonzalez, R. C., Fernhoff, D. & David, F. G. (1973). Contrast, resistance to extinction and forgetting in rats. *J. comp. physiol. Psychol.*, **84**, 562–571.

Goodrich, K. P. (1960). Running speed and drinking rate as functions of sucrose concentration and amount of consummatory activity. *J. comp. physiol. Psychol.*, **53**, 245–250.

Gormezano, I. (1965). Yoked comparisons of classical and instrumental conditioning of the eyelid response; and an addendum on "voluntary responders". In W. F. Prokasy (ed.), *Classical Conditioning*, pp. 48–70. New York, Appleton–Century–Crofts.

Gormezano, I. (1966). Classical conditioning. In J. B. Sidowski (ed.), *Experimental Methods and Instrumentation in Psychology*, pp. 385–420. New York, McGraw-Hill.

Gormezano, I. & Hiller, G. W. (1972). Omission training of the jaw-movement response of the rabbit to a water US. *Psychon. Sci.*, **29**, 276–278.

Gorry, T. H. & Ober, S. E. (1970). Stimulus characteristics of learning over long delays in monkeys. Paper delivered at the 10th Ann. meeting of the Psychon. Soc., San Antonio.

Gottlieb, G. (1965). Imprinting in relation to parental and species identification by avian neonates. *J. comp. physiol. Psychol.*, **59**, 345–356.

Gottlieb, G. (1966). Species identification by avian neonates: contributory effect of perinatal auditory stimulation. *Anim. Behav.*, **14**, 282–290.

Gottlieb, G. & Klopfer, P. H. (1962). The relation of developmental age to auditory and visual imprinting. *J. comp. physiol. Psychol.*, **55**, 821–826.

Gottwald, P. (1967). The role of punishment in the development of conditioned suppression. *Physiol. Behav.*, **2**, 283–286.

Gould, J. L. (1974). Honey bee communication: Misdirection of recruits by foragers with covered ocelli. *Nature*, **252**, 300–301.

Gould, J. L. (1975). Honey bee communication: the dance-language controversy. *Science*, **189**, 685–693.

Gould, J. L. (1976). The dance-language controversy. *Q. Rev. Biol.*, **51**, 211.

Graham, F. K. & Jackson, J. C. (1970). Arousal systems and infant heart rate responses. In H. W. Reese & L. P. Lipsitt (eds) *Advances in Child Development and Behaviour*, Vol. 5. New York, Academic Press.

Grant, D. S. & Roberts, W. A. (1973). Trace interaction in pigeon short-term memory. *J. exp. Psychol.*, **101**, 21–29.

Gray, J. A. (1965a). Relation between stimulus intensity and operant response rate as a function of discrimination training and drive. *J. exp. Psychol.*, **69**, 9–24.

Gray, J. A. (1965b). Stimulus intensity dynamism. *Psychol. Bull.*, **63**, 180–196.

Gray, T. & Appignanesi, A. A. (1973). Compound conditioning: elimination of blocking effect. *Learn. Motiv.*, **4**, 374–380.

Gray, V. A. (1976). Stimulus control of differential reinforcement of low-rate responding. *J. exp. Anal. Behav.*, **25**, 199–207.

Green, P. C. (1962). Learning, estimation and generalization of conditioned responses by young monkeys. *Psychol. Rep.*, **10**, 731.

Greenfield, P. M., Reich, L. C. & Olver, R. R. (1966). On culture and equivalence: II. In J. S. Bruner, R. R. Olver, P. M. Greenfield *et al.*, *Studies in Cognitive Growth*, pp. 270–318. New York, Wiley.

Grether, W. F. (1938). Pseudo-conditioning without paired stimulation encountered in attempted backward conditioning. *J. comp. Psychol.*, **25**, 91–96.

Grice, G. R. (1948). The relation of secondary reinforcement to delayed reward in visual discrimination learning. *J. exp. Psychol.*, **38**, 1–16.

Grice, G. R. & Hunter, J. J. (1964). Stimulus intensity effects depend upon the type of experimental design. *Psychol. Rev.*, **71**, 247–256.

Griffin, D. R. (1976). *The Question of Animal Awareness*. New York, The Rockefeller Univ. Press.

Gross, C. G., Rocha-Miranda, C. E. & Bender, D. B. (1972). Visual properties of neurons in inferotemporal cortex of the macaque. *J. Neurophysiol.*, **35**, 96–111.

Grossen, N. E., Kostansek, D. J. & Bolles, R. C. (1969). Effects of appetitive discriminative stimuli on avoidance behaviour. *J. exp. Psychol.*, **81**, 340–343.

Grusec, T. (1968). The peak-shift in stimulus generalization: equivalent effects of errors and non-contingent shock. *J. exp. Anal. Behav.*, **11**, 39–49.

Guthrie, E. R. (1935). *The Psychology of Learning*. New York, Harper.

Guttman, N. (1953). Operant conditioning, extinction and periodic reinforcement in relation to concentration of sucrose used as reinforcing agent. *J. exp. Psychol.*, **46**, 213–224.

Guttman, N. & Kalish, H. I. (1956). Discriminability and stimulus generalization. *J. exp. Psychol.*, **51**, 79–88.

Haber, A. & Kalish, H. I. (1963). Prediction of discrimination from generalization after variations on schedule of reinforcement. *Science*, **142**, 412–413.

Hake, D. F. & Azrin, N. H. (1965). Conditioned punishment. *J. exp. Anal. Behav.*, **8**, 279–293.

Halliday, M. S. & Boakes, R. A. (1971). Behavioural contrast and response independent reinforcement. *J. exp. Anal. Behav.*, **16**, 429–434.

Halliday, M. S. & Boakes, R. A. (1972). Discrimination involving response-independent reinforcement: implications for behavioural contrast. In R. A. Boakes & M. S. Halliday (eds), *Inhibition and Learning*, pp. 73–97. London, Academic Press.

Halliday, M. S. & Boakes, R. A. (1974). Behavioural contrast without response rate reduction. *J. exp. Anal. Behav.*, **22**, 453–462.

Hanson, H. M. (1959). Effects of discrimination training on stimulus generalization. *J. exp. Psychol.*, **58**, 321–334.

Harlow, H. F. (1937). Experimental analysis of the role of the original stimulus in conditioned responses in monkeys. *Psychol. Rec.*, **1**, 62–68.

Harlow, H. F. (1943). Solution by rhesus monkeys of a problem involving the Weigh Principle using the matching-form-sample method. *J. comp. Psychol.*, **36**, 217–227.

Harlow, H. F. (1949). The formation of learning sets. *Psychol. Rev.*, **56**, 51–56.

Harlow, H. F. & Harlow, M. K. (1962). Principles of primate learning. In *Little Club Clinics in Development Medicine*, No. 7. London, Heinemann.

Harré, R. (1971). Joynson's dilemma. *Bull. Br. Psychol. Soc.*, **24;** 115–119.

Harris, J. D. (1943). Studies of nonassociative factors inherent in conditioning. *Comp. Psychol. Monogr.*, **18**, (1, serial No. 93).

Harris, S. L. (1975). Teaching language to nonverbal children – with emphasis on problems of generalization. *Psychol. Bull.*, **82**, 565–580.

Harzem, P. (1969). Temporal discrimination. In R. M. Gilbert & N. S. Sutherland (eds), *Animal Discrimination Learning*, pp. 299–334. London, Academic Press.

Harzem, P., Lowe, C. F. & Bagshaw, M. (1978). Verbal control in human operant behaviour. *Psychol. Rec.*, in press.

Harzem, P., Lowe, C. F. & Davey, G. C. L. (1975a). After-effects of reinforcement magnitude: dependence upon context. *Q. Jl. exp. Psychol.*, **27**, 579–584.

Harzem, P., Lowe, C. F. & Davey, G. C. L. (1975b). Two-component schedules of differential-reinforcement-of-low-rate. *J. exp. Anal. Behav.*, **24**, 33–42.

Hawkins, T. D., Schrot, S. H., Githens, S. H. & Everett, P. B. (1972). Schedule-induced polydipsia: An analysis of water and alcohol ingestion. In R. M. Gilbert & J. D. Keehn (eds), *Schedule Effects: Drugs, Drinking and Aggression*. Toronto, Toronto Univ. Press.

Hayes, K. J. & Hayes, C. (1955). The cultural capacity of chimpanzees. In J. A. Garan (ed.), *The Non-human Primates and Human Evolution*, pp. 256–372. Detroit, Wayne Univ. Press.

Hearst, E. (1965). Approach, avoidance and stimulus generalization. In D. I. Mostofsky (ed.), *Stimulus Generalization*, pp. 331–355. Stanford, Stanford Univ. Press.

Hearst, E. (1969). Stimulus intensity dynamism and auditory generalization for approach and avoidance behaviour in rats. *J. comp. physiol. Psychol.*, **68**, 111–117.

Hearst, E., Besley, S. & Farthing, G. W. (1970). Inhibition and the stimulus control of operant behaviour. *J. exp. Anal. Behav.*, **14**, 373–409.

Hearst, E. & Jenkins, H. M. (1974). *Sign Tracking: the Stimulus-Reinforcer Relation and Directed Action*. Austin, Texas, Psychon. Soc.

Hearst, E., Koresko, M. B. and Poppen, R. (1964). Stimulus generalization and the response-reinforcement contingency. *J. exp. Anal. Behav.*, **7**, 369–380.

Hefferline, R. F., Bruno, L. J. J. & Davidowitz, J. (1971). Feedback control of covert behaviour. In K. J. Connolly (ed.), *Mechanisms of Motor Skill Development*. New York, Academic Press.

Hefferline, R. F., Keenan, B. & Harford, R. A. (1959). Escape and avoidance conditioning in human subjects without their observation of the response. *Science*, **130**, 1338–1339.

Hefferline, R. F. & Parera, T. B. (1963). Proprioceptive discrimination of a covert operant without its observation by the subject. *Science*, **139**, 834–835.

Held, R. & Hein, A. (1967). On the modifiability of form perception. In W. Wathen-Dunn (ed.), *Models for Perception of Speech and Visual Forms*. Cambridge, M.I.T. Press.

Hellbrügge, J. (1960). The development of circadian rhythms in infants. In *Cold Spring*

*Harbor Symposium on Quantitative Biology* (Vol. 25). Cold Spring Harbor, New York, Biological laboratory.

Heller, D. P. (1968). Absence of size constancy in visually deprived rats. *J. comp. physiol. Psychol.*, **65**, 336–339.

Hemmes, N. S. (1973). Behavioural contrast in the pigeon depends upon the operant. *J. comp. physiol. Psychol.*, **85**, 171–178.

Hemmes, N. S. (1975). Pigeon's performance under differential reinforcement of low rate schedules depends upon the operant. *Learn. Motiv.*, **6**, 344–357.

Hemmes, N. S. & Eckerman, D. A. (1972). Positive interaction (induction) in multiple variable-interval, differential-reinforcement-of-high-rate schedules. *J. exp. Anal. Behav.*, **17**, 51–57.

Hendry, D. P. (1969). *Conditioned Reinforcement*. Homewood, Illinois, The Dorsey Press.

Hendry, D. P. & Rasche, R. H. (1961). Analysis of a new nonnutritive positive reinforcer based on thirst. *J. comp. physiol. Psychol.*, **54**, 477–483.

Herman, L. M. (1975). Interference and auditory short-term memory in the bottlenosed dolphin. *Anim. Learn. Behav.*, **3**, 43–48.

Herman, R. L. & Azrin, N. H. (1964). Punishment by noise in an alternative response situation. *J. exp. Anal. Behav.*, **7**, 185–188.

Herrnstein, R. J. (1955). Behavioural consequences of the removal of a discriminative stimulus associated with variable-interval reinforcement. Unpublished doctoral dissertation, Harvard Univ.

Herrnstein, R. J. (1958). Some factors influencing behaviour in a two-response situation. *Trans. N. Y. Acad. Sci.*, **21**, 35–45.

Herrnstein, R. J. (1961). Relative and absolute strength of response as a function of frequency of reinforcement. *J. exp. Anal. Behav.*, **4**, 267–272.

Herrnstein, R. J. (1969). Method and theory in the study of avoidance. *Psychol. Rev.*, **76**, 49–69.

Herrnstein, R. J. (1970). On the law of effect. *J. exp. Anal. Behav.*, **13**, 243–266.

Herrnstein, R. J. & Hineline, P. N. (1966). Negative reinforcement as shock-frequency reduction. *J. exp. Anal. Behav.*, **9**, 421–430.

Herrnstein, R. J., Loveland, D. H. & Cable, C. (1976). Natural concepts in pigeons. *J. exp. Psychol. Anim. Behav. Process*, **2**, 285–302.

Herrnstein, R. J. & Morse, W. H. (1958). A conjunctive schedule of reinforcement. *J. exp. Anal. Behav.*, **1**, 15–24.

Herrnstein, R. J. & Sidman, M. (1958). Avoidance conditioning as a factor in the effects of unavoidable shocks as food-reinforced behaviour. *J. comp. physiol. Psychol.*, **51**, 380–385.

Hess, E. H. (1962). Imprinting and the critical period concept. In E. L. Bliss (ed.), *Roots of Behaviour*, pp. 254–263. New York, Hoeber.

Hess, E. H. (1964). Imprinting in birds. *Science*, **146**, 1128–1139.

Hess, E. H. (1972). The natural history of imprinting. *Ann. N.Y. Acad. Sci.*, **193**, 124–136.

Hess, E. H. (1973). *Imprinting*. New York, Van Nostrand Reinhold.

Hess, E. H. & Hess, D. B. (1969). Innate factors in imprinting. *Psychon. Sci.*, **14**, 129–130.

Hilgard, E. R. & Bower, G. H. (1966). *Theories of Learning*. New York, Appleton–Century–Crofts.

Hilgard, E. R. & Campbell, A. A. (1936). The course of acquisition and retention of conditioned eyelid responses in man. *J. exp. Psychol.*, **19**, 227–247.

Hill, W. F., Cotton, J. W., Spear, N. E. & Duncan, C. P. (1969). Retention of T-maze learning after varying intervals following partial and continuous reinforcement. *J. exp. Psychol.*, **79**, 584–585.

Hill, W. F. & Wallace, W. P. (1967). Reward magnitude and number of training trials as joint factors in extinction. *Psychon. Sci.*, **7**, 267–268.

Hinde, R. A. (1966). *Animal Behaviour.* New York, McGraw-Hill, first edition.

Hinde, R. A. (1970). *Animal Behaviour: A Synthesis of Ethology and Comparative Psychology.* New York, McGraw-Hill.

Hinde, R. A. (1972). *Non-verbal Communication.* London, Cambridge University Press.

Hineline, P. N. (1970). Negative reinforcement without shock reduction. *J. exp. Anal. Behav.*, **14**, 259–268.

Hineline, P. N. (1973). Varied approaches to aversion: A review of *Aversive Conditioning and Learning*, edited by F. Robert Brush. *J. exp. Anal. Behav.*, **19**, 531–540.

Hineline, P. N. & Herrnstein, R. J. (1970). Timing in free-operant and discrete-trial avoidance. *J. exp. Anal. Behav.*, **13**, 113–126.

Hineline, P. N. & Rachlin, H. (1969). Escape and avoidance of shock by pigeons pecking a key. *J. exp. Anal. Behav.*, **12**, 533–538.

Hiroto, D. A. (1974). Locus of control and learned helplessness. *J. exp. Psychol.*, **102**, 187–193.

Hiroto, D. S. & Seligman, M. E. P. (1975). Generality of learned helplessness in man. *J. Personal. soc. Psychol.*, **31**, 311–327.

Hirsch, H. V. B., & Spinelli, D. N. (1971). Modification of the distribution of receptive field orientation in cats by selective visual exposure during development. *Expl. Brain Res.*, **12**, 509–527.

Hoffman, H. S. (1966). The analysis of discriminated avoidance. In W. K. Honig (ed.), *Operant Behaviour: Areas of Research and Application*, pp. 499–530. New York, Appleton–Century–Crofts.

Hoffman, H. S. (1968). The control of distress vocalization by an imprinted stimulus. *Behaviour*, **30**, 175–191.

Hoffman, H. S. (1969). Stimulus factors in conditioned suppression. In B. A. Campbell & R. M. Church (eds), *Punishment and Aversive Behaviour*, pp. 185–234. New York, Appleton–Century–Crofts.

Hoffman, H. S. & Fleshler, M. (1959). Aversive control with the pigeon. *J. exp. Anal. Behav.*, **2**, 213–218.

Hoffman, H. S. & Fleshler, M. (1962). The course of emotionality in the development of avoidance. *J. exp. Psychol.*, **64**, 288–294.

Hoffman, H. S., Fleshler, M. & Chorny, H. (1961). Discriminated bar-press avoidance. *J. exp. Anal. Behav.*, **4**, 309–316.

Hogan, J. A. (1964). Operant Control of Preening in Pigeons. *J. exp. Anal. Behav.*, **7**, 351–354.

Hogan, J. A. (1967). Fighting and reinforcement in the Siamese fighting fish (*Betta splendens*). *J. comp. physiol. Psychol.*, **64**, 356–359.

Hogan, J. A., Kleist, S. & Hutchings, C. S. L. (1970). Display and food as reinforcers in the Siamese fighting fish (*Betta splendens*). *J. comp. physiol. Psychol.*, **70**, 351–357.

Hogan, R. A. & Kirchner, J. H. (1967). Preliminary report of the extinction of learned fears via short-term implosive therapy. *J. abnorm. Psychol.*, **72**, 106–109.

Holland, J. G. (1958). Human Vigilance. *Science*, **128**, 61–67.

Holland, J. (1973). Political implications of applying behavioural psychology. In R. Ulrich, T. Stachnik & J. Mabry (eds), *Control of Human Behaviour*, Vol. 3, pp. 413–419. New York, Scott, Foresman.

Holland, P. C. (1977). Conditioned stimulus as a determinant of the form of the Pavlovian conditioned response. *J. exp. Psychol. Anim. Behav. Process*, **3**, 77–104.

Holland, P. C. & Rescorla, R. A. (1975). Second-order conditioning with food unconditioned stimulus. *J. comp. physiol. Psychol.*, **88**, 459–467.

Holman, G. L. (1969). The intragastric reinforcement effect. *J. comp. Physiol. Psychol.*, **69**, 432–441.

Holz, W. C. & Azrin, N. H. (1962). Interactions between the discriminative and aversive properties of punishment. *J. exp. Anal. Behav.*, **5**, 229–234.

Homme, L. E. (1965). Perspectives in psychology: XXIV. Control of coverants, the operants of the mind. *Psychol. Rec.*, **15**, 501–511.

Honig, W. K., Boneau, C. A., Burstein, K. R. & Pennypacker, H. S. (1963). Positive and negative generalization gradients obtained after equivalent training conditions. *J. comp. physiol. Psychol.*, **56**, 111–116.

Honig, W. K. & Slivka, R. M. (1964). Stimulus generalization of the effects of punishment. *J. exp. Anal. Behav.*, **7**, 21–25.

Hubel, D. H. & Wiesel, T. N. (1963). Receptive fields of cells in striate cortex of very young, visually inexperienced kittens. *J. Neurophysiol.*, **26**, 994.

Hubel, D. H. & Wiesel, T. N. (1970). The period of susceptibility to the physiological effects of unilateral eye closure in kittens. *J. Physiol.*, **206**, 419–436.

Hughes, R. N. (1969). Social facilitation of locomotion and exploration in rats. *Br. J. Psychol.*, **60**, 385–388.

Hull, C. L. (1943). *Principles of Behaviour.* New York, Appleton–Century–Crofts.

Hull, C. L. (1947). The problem of primary stimulus generalization. *Psychol. Rev.*, **54**, 120–134.

Hull, C. L. (1949). Stimulus intensity dynamism (V) and stimulus generalization. *Psychol. Rev.*, **56**, 67–76.

Hull, C. L. (1952). *A Behaviour System.* New Haven, Yale University Press.

Humphreys, L. G. (1939). The effect of random alternation of reinforcement and extinction of conditioned eyelid reactions. *J. exp. Psychol.*, **25**, 141–158.

Hunt, E. L. (1949). Establishment of conditioned responses in chick embryos. *J. comp. physiol. Psychol.*, **42**, 107–117.

Hunt, H. F. & Brady, J. V. (1955). Some effects of punishment and inter current "anxiety" on a simple operant. *J. comp. physiol. Psychol.*, **48**, 305–310.

Hutchinson, R. R. & Azrin, N. H. (1961). Conditioning of mental hospital patients to fixed-ratio schedules of reinforcement. *J. exp. Anal. Behav.*, **4**, 87–95.

Hutchinson, R. R., Azrin, N. H. & Hunt, G. M. (1968). Attack produced by intermittent reinforcement of a concurrent operant response. *J. exp. Anal. Behav.*, **11**, 485–495.

Hutt, P. J. (1954). Rate of bar pressing as a function of quality and quantity of food reward. *J. comp. physiol. Psychol.*, **47**, 235–239.

Imanishi, K. (1957). Social behaviour in Japanese monkeys, *Macaca fuscata, Psychologia*, **1**, 47–54.

Imbert, M. & Buisseret, P. (1975). Receptive field characteristics and plastic properties of visual cortical cells in kittens reared with or without visual experience. *Exp. Brain Res.*, **22**, 25–36.

Isaacs, W., Thomas, J. & Goldiamond, I. (1960). Application of operant conditioning to reinstate verbal behaviour in psychotics. *J. Speech Hear. Disorders*, **25**, 8–12.

Iversen, L. L. & Iversen, S. D. (1975). *Behavioural Pharmacology.* New York, O.U.P.

Jacobs, A. (1971). Mood-emotion-effect: the nature of and manipulation of affective states with particular reference to positive affective states and emotional illness. In A. Jacobs & L. B. Sachs (eds), *The Psychology of Private Events.* New York, Academic Press.

Jacobsen, C. F., Jacobsen, M. M. & Yoshioka, J. G. (1932). Developments of an infant chimpanzee during her first year. *Comp. Psychol. Monogr.*, **9**, 1–94.

Jacquet, Y. F. (1972). Schedule-induced licking during multiple schedules. *J. exp. Anal. Behav.*, **17**, 413–423.

James, B. (1962). Case of homosexuality treated by aversion therapy. *Br. Med. J.*, **1**, 768–770.

James, J. P. & Mostoway, W. W. (1968). Stimulus intensity dynamism: effect of nonreinforcement of the highest intensity stimulus. *Can. J. Psychol.*, **23**, 49–55.

Janos, O. (1959). Development of higher nervous activity in premature infants. *Pavlov J. Higher Nervous Activity.*, **9**, 760–767.

Jansen, P. E., Goodman, E. D., Jowaisas, D. & Bunnell, B. N. (1969). Paper as a positive reinforcer for acquisition of a bar-press response by the golden hamster. *Psychonom. Sci.*, **16**, 113–114.

Jarrad, L. E. & Moise, S. L. (1970). Short-term memory in the stumptail macaque: effect of physical restraint of behaviour on performance. *Learn. Motiv.*, **1**, 267–275.

Jarvik, M. E., Goldfarb, T. L. & Carley, J. L. (1969). Influence of interference on delayed matching in monkeys. *J. exp. Psychol.*, **81**, 1–6.

Jaynes, J. (1956). Imprinting: the interaction of learned and innate behaviour: I. development and generalization. *J. comp. physiol. Psychol.*, **49**, 201–206.

Jeffrey, W. E. & Cohen, L. B. (1965). Response tendencies of children in a two choice situation. *J. exp. Child Psychol.*, **2**, 248–254.

Jenkins, H. M. (1965). Generalization gradients and the concept of inhibition. In D. I. Mostofsky (ed.), *Stimulus Generalization*, pp. 55–61. Stanford, Stanford University Press.

Jenkins, H. M. (1973). Effects of the stimulus-reinforcer relation on selected and unselected responses. In R. A. Hinde & J. Stevenson-Hinde (eds), *Constraints on Learning*. New York, Academic Press.

Jenkins, H. M. (1977). Sensitivity of different response systems to stimulus-reinforcer and response-reinforcer relations. In H. Davis & H. M. B. Hurwitz (eds), *Operant-Pavlovian Interactions*, pp. 47–62. Hillsdate N.J., Lawrence Erlbaum.

Jenkins, H. M. & Harrison, R. H. (1960). Effect of discrimination training on auditory generalization. *J. exp. Psychol.*, **59**, 246–253.

Jenkins, H. M. & Moore, B. R. (1973). The form of the autoshaped response with food or water reinforcers. *J. exp. Anal. Behav.*, **20**, 163–181.

Jensen, C. & Fallon, D. (1973). Behavioural after-effects of reinforcement and its omission as a function of reinforcement magnitude. *J. exp. Anal. Behav.*, **19**, 459–468.

John, E. R. (1967). *Mechanisms of Memory*. New York, Academic Press.

John, E. R., Chesler, P., Bartlett, F. & Victor, I. (1968). Observation learning in cats. *Science*, **159**, 1489–1491.

Johnson, R. N. (1970). Spatial probability learning and brain stimulation in rats. *Psychon. Sci.*, **18**, 33.

Kagan, J. (1955). Differential reward value of incomplete and complete sexual behaviour. *J. comp. physiol. Psychol.*, **48**, 59–64.

Kalat, J. W. & Rozin, P. (1973). "Learned Safety" as a mechanism in long-delay taste-aversion learning in rats. *J. comp. physiol. Psychol.*, **83**, 198–207.

Kalish, H. I. & Guttman, N. (1957). Stimulus generalization after equal training on two stimuli. *J. exp. Psychol.*, **53**, 139–144.

Kalish, H. I. & Guttman, N. (1959). Stimulus generalization after training on three stimuli: a test of the summation hypothesis. *J. exp. Psychol.*, **57**, 268–272.

Kamano, D. K. (1970). Types of Pavlovian conditioning procedures used in establishing CS + and their effect upon avoidance behaviour. *Psychon. Sci.*, **18**, 63–64.

Kamin, L. J. (1954). Traumatic avoidance learning: the effects of CS–US interval with a trace-conditioning procedure. *J. comp. physiol. Psychol.*, **47**, 65–72.

Kamin, L. J. (1956). The effects of termination of the CS and avoidance of the US on avoidance learning. *J. comp. physiol. Psychol.*, **49**, 420–424.

Kamin, L. J. (1957a). The effects of termination of the CS and avoidance of the US on avoidance learning: an extension. *Can. J. Psychol.*, **11**, 48–56.

Kamin, L. J. (1957b). The retention of an incompletely learned avoidance response. *J. comp. physiol. Psychol.*, **50**, 457–460.

Kamin, L. J. (1959). The delay-of-punishment gradient. *J. comp. physiol. Psychol.*, **52**, 434–437.

Kamin, L. J. (1968). 'Attention-like' processes in classical conditioning. In M. R. Jones (ed.), *Miami Symposium on the Prediction of Behaviour: Aversive Stimulation*, pp. 9–33. Miami, University of Miami Press.

Kamin, L. J. (1969). Predictability, surprise, attention and conditioning. In B. A. Campbell & R. M. Church (eds), *Punishment and Aversive Behaviour*, pp. 279–296. New York, Appleton–Century–Crofts.

Kamin, L. J., Brimer, C. J. & Black, A. H. (1963). Conditioned suppression as a monitor of fear of the CS in the course of avoidance training. *J. comp. physiol. Psychol.*, **56**, 497–501.

Kamiya, J. (1968). Conscious control of brain waves. *Psychol. Today*, **1**, 57–60.

Karpicke, J., Christoph, G., Peterson, G. & Hearst, E. (1977). Signal location and positive versus negative conditioned suppression in the rat. *J. exp. Psychol. Anim. Behav. Process.*, **3**, 105–118.

Kasatkin, N. I. (1972). First conditioned responses and the beginning of the learning process in the human infant. In G. Newton & A. H. Riesen (eds), *Advances in Psychobiology*, Vol. 1. New York, Wiley.

Kasatkin, N. I. & Levikova, A. M. (1935a). On the development of early conditioned reflexes and differentiations of auditory stimuli in infants. *J. exp. Psychol.*, **18**, 1–19.

Kasatkin, N. I. & Levikova, A. M. (1935b). The formation of visual conditioned reflexes and their differentiation in infants. *J. essen. Psychol.*, **12**, 416–435.

Kasatkin, N. I., Mirzoiants, N. S. & Khokhitva, A. (1953). Conditioned orienting responses in children in the first year of life. *Zhurnal Vysshei Nervnoi Deyatel'nosti imeni I. P. Pavlova*, **3**, 192–202.

Katahn, M. & Koplin, J. H. (1968). Paradigm clash: comment on "Some recent criticisms of behaviourism and learning theory with special reference to Breger and McGaugh and to Chomsky". *Psychol. Bull.*, **69**, 147–148.

Katzev, R. (1967). Extinguishing avoidance responses as a function of delayed warning signal termination. *J. exp. Psychol.*, **75**, 339–344.

Kaufman, A., Baron, A. & Kopp, R. M. (1966). Some effects of instructions on human operant behaviour. *Psychon. Monogr. Suppl.*, **1**, No. 11, 243–250.

Kaye, H. (1965). The conditioned Babkin reflex in human newborns. *Psychon. Sci.*, **2**, 287–288.

Kaye, H. (1967). Infant sucking behaviour and its modification. In L. P. Lipsitt & C. C. Spiker (eds), *Advances in Child Development and Behaviour*, Vol. 3, pp. 1–52. New York, Academic Press.

Kazdin, A. E. (1975). *Behaviour Modification in Applied Settings*. Homewood, Illinois, Dorsey Press.

Kazdin, A. E. & Bootzin, R. R. (1972). The token economy: an evaluative review. *J. appl. Behav. Anal.*, **5**, 343–372.

Keehn, J. D. (1967). Experimental studies of "the unconscious": Operant conditioning of unconscious eyeblinking. *Behav. Ther.*, **5**, 95–102.

Keehn, J. D. (1969). Consciousness, discrimination and the stimulus control of behaviour. In R. M. Gilbert & N. S. Sutherland (eds), *Animal Discrimination Learning*, pp. 273–298. London, Academic Press.

Keesey, R. E. & Kling, J. W. (1961). Amount of reinforcement and free-operant responding, *J. exp. Anal. Behav.*, **4**, 125–132.

Kehoe, J. (1963). Effects of prior and interpolated learning on retention in pigeons. *J. exp. Psychol.*, **65**, 537–545.

Kelleher, R. T. (1966). Chaining and conditioned reinforcement. In W. K. Honig (ed), *Operant Behaviour: Areas of Research and Application*, pp. 160–212. New York: Appleton–Century–Crofts.

Kelleher, R. T., Fry, W. & Cook, L. (1959). Interresponse time distribution as a function of differential reinforcement of temporally spaced responses. *J. exp. Anal. Behav.*, **2**, 91–106.

Keller, K. (1974). The role of elicited responding in behavioural contrast. *J. exp. Anal. Behav.*, **21**, 249–257.

Kello, J. E. (1972). The reinforcement-omission effect on fixed-interval schedules: frustration or inhibition? *Learn. Motiv.*, **3**, 138–147.

Kellogg, W. N. & Kellogg, L. A. (1933). *The Ape and the Child: A Study of Environmental Influence upon Early Behaviour.* New York, McGraw-Hill.

Kessen, W. (1953). Response strength and conditioned stimulus intensity. *J. exp. Psychol.*, **45**, 82–86.

Khavari, K. A. & Eisman, E. H. (1971). Some parameters of latent learning and generalized drives. *J. comp. physiol. Psychol.*, **77**, 463–469.

Killeen, P. (1969). Reinforcement frequency and contingency as factors in fixed-ratio behaviour. *J. exp. Anal. Behav.*, **12**, 391–395.

Killeen, P. (1975). On the temporal control of behaviour. *Psychol. Rev.*, **82**, 89–115.

Kimble, G. A. (1961). *Hilgard and Marquis' Conditioning and Learning.* New York, Appleton–Century–Crofts.

King, G. D. (1974). Wheel-running in the rat induced by a fixed time presentation of water. *Anim. Learn. Behav.*, **2**, 325–328.

King, J. A. & Weisman, R. G. (1964). Sand digging contingent upon bar pressing in deermice (*Peromyscus*). *Anim. Behav.*, **12**, 446–450.

Kirby, A. J. (1968). Explorations of the Brown-Jenkins auto-shaping phenomenon. Unpublished thesis, Dalhousie University.

Kissileff, H. R. (1969). Food-associated drinking in the rat. *J. comp. physiol. Psychol.*, **67**, 284–300.

Klein, S. B. (1969). Counterconditioning and fear reduction in the rat. *Psychon. Sci.*, **77**, 150–151.

Klein, S. B. (1972). Adrenal-pituitary influence system in reactivation of avoidance-learning memory in the rat after intermediate intervals. *J. comp. physiol. Psychol.*, **79**, 341–359.

Klein, S. B. & Spear, N. E. (1970). Reactivation of avoidance-learning memory in the rat after intermediate retention intervals. *J. comp. physiol. Psychol.*, **72**, 498–504.

Klopfer, P. H. (1957). Empathetic learning in ducks. *Am. Nat.*, **91**, 61–63.

Klopfer P. H. (1959). Social interactions in discrimination learning with special reference to feeding behavior in birds. *Behaviour*, **14**, 282–299.

Klopfer, P. H. (1961). Observational learning in birds: the establishment of behavioural modes. *Behaviour*, **17**, 71–80.

Klopfer, P. H. (1971). Imprinting: determining its perceptual basis in ducklings. *J. comp. physiol. Psychol.*, **75**, 378–385.

Klopfer, P. H., Adams, D. K. & Klopfer, M. S. (1964). Maternal "imprinting in goats". *Proc. Natn Acad. Sci.*, **52**, 911–914.

Knutson, J. F. (1970). Aggression during the fixed-ratio and extinction components of a multiple schedule of reinforcement. *J. exp. Anal. Behav.*, **13**, 221–231.

Knutson, J. F. & Kleinknecht, R. A. (1970). Attack during differential reinforcement of low rate of responding. *Psychon. Sci.*, **19**, 289–290.

Knutson, J. F. & Schrader, S. P. (1975). A concurrent assessment of schedule-induced aggression and schedule-induced polydipsia in the rat. *Anim. Learn. Behav.*, **3**, 16–20.

Koch, J. (1965). The development of the conditioned orienting reaction to humans in 2–3 month old infants. *Activ. Nerv. Sup.*, **7**, No. 2, 141–142.

Koch, S. (1964). Psychology and emerging conceptions of knowledge as unitary. In T. W. Wann (ed.), *Behaviourism and Phenomenology*, pp. 1–41. Chicago, Univ. of Chicago Press.

Köhler, W. (1925). *The Mentality of Apes*. London, Routledge and Kegan Paul, third edition, 1973.

Kohn, B. (1976). Observation and discrimination learning in the rat: effects of stimulus substitution. *Learn. Motiv.*, **7**, 303–312.

Kohn, B. & Dennis, M. (1972). Observation and discrimination learning in the rat: specific and nonspecific effects. *J. comp. physiol. Psychol.*, **78**, 292–296.

Koestler, A. (1967). *The Ghost in the Machine*. London, Hutchinson.

Konorski, J. (1948). *Conditioned Reflexes and Neuron Organization*. Cambridge Univ. Press.

Konorski, J. (1967). *Integrative Activity of the Brain*. Chicago, Univ. of Chicago Press.

Konorski, J. & Miller, S. (1937). On two types of conditioned reflex. *J. gen. Psychol.*, **16**, 264–272.

Konorski, J. & Szwejkowska, G. (1952). Chronic extinction and restoration of conditioned reflexes: IV. The dependence of the course of extinction and restoration of conditioned reflexes on the "history" of the conditioned stimulus (the principle of the primacy of first training). *Acta Biol. Exp.*, **16**, 95–113.

Konorski, J. & Szwejkowska, G. (1956). Reciprocal transformations of heterogeneous conditioned reflexes. *Acta Biol. Exp.*, **17**, 141–165.

Koshtoyants, Kh. S. (1955). *Pavlov: Selected Works*. Moscow, Foreign Languages Publishing House.

Kovach, J. K. (1971). Effectiveness of different colors in the elicitation and development of approach behaviour in chicks. *Behaviour*, **38**, 154–168.

Kovach, J. K. & Hess, E. H. (1963). Imprinting: effect of painful stimulation on the following behaviour. *J. comp. physiol. Psychol.*, **56**, 461–464.

Kraeling, D. (1961). Analysis of amount of reward, as a variable in learning. *J. comp. physiol. Psychol.*, **54**, 560–565.

Kramer, T. J. & Rodriguez, M. (1971). The effect of different operants on spaced responding. *Psychon. Sci.*, **25**, 177–178.

Krane, R. V. & Wagner, A. R. (1975). Taste aversion learning with a delayed shock vs implications for the "generality of the laws of learning". *J. comp. physiol. Psychol.*, **88**, 882–889.

Krieckhaus, E. F. & Wolf, G. (1968). Acquisition of sodium by rats. Interaction of innate mechanisms and latent learning. *J. comp. physiol. Psychol.*, **65**, 197–201.

Kristt, D. A. & Engel, B. T. (1975). Learned control of blood pressure in patients with high blood pressure. *Circulation*, **51**, 370–378.

Krop, H., Calhoun, B. & Verrier, R. (1971). Modification of the "self-concept" of emotionally disturbed children by covert reinforcement. *Behav. Ther.*, **2**, 201–204.

Lashley, K. S. & Ball, J. (1929). Spinal conduction and kinesthetic sensitivity in the maze habit. *J. comp. Psychol.*, **9**, 71–105.

Laties, V. G. & Weiss, B. (1963). Effects of a concurrent task on Fixed-interval responding in humans. *J. exp. Anal. Behav.*, **3**, 431–436.

Laties, V. G., Weiss, B., Clark, R. L. & Reynolds, M. D. (1965). Overt "mediating" behaviour during temporally spaced responding. *J. exp. Anal. Behav.*, **8**, 107–116.

Laties, V. G., Weiss, B. & Weiss, A. B. (1969). Further observations on overt "mediating" behaviour and the discrimination of time. *J. exp. Anal. Behav.*, **12**, 43–57.

Lawson, R., Mathis, P. R. & Pear, J. J. (1968). Summation of response rates to discriminative stimuli associated with qualitatively different reinforcers. *J. exp. Anal. Behav.*, **11**, 561–568.

Lazarus, A. A. (1965). A preliminary report on the use of directed muscular activity in counterconditioning. *Behav. Res. Ther.*, **2**, 301–303.

Leander, J. D., Lippman, L. G. & Meyer, M. E. (1968). Fixed-interval performance as related to subjects' vocalizations of the reinforcement contingency. *Psychol. Rec.*, **18**, 469–474.

Lee, L. C. (1965). Concept utilization in preschool children. *Child Dev.*, **36**, 221–227.

Leitenberg, H. (1965). Is time-out from positive reinforcement an aversive event? A review of the experimental evidence. *Psychol. Bull.*, **64**, 428–441.

Leitenberg, H., Agras, W. S., Thompson, L. E. & Wright, D. (1968). Feedback in behaviour modification: An experimental analysis in two phobic cases. *J. appl. Behav. Anal.*, **1**, 131–137.

Lemere, F. & Voegtlin, W. L. (1950). An evaluation of aversive treatment of alcoholism. *Q. Jl. Stud. Alcohol.*, **11**, 199–204.

Lenneberg, E. H. (1967). *Biological Foundations of Language.* New York, Wiley.

Lenneberg, E. H. (1969). On explaining language. *Science*, **164**, 135–42.

Leonard, D. W. & Theios, J. (1967). Classical eyelid conditioning in rabbits under prolonged single alternation conditions of reinforcement. *J. comp. physiol. Psychol.*, **64**, 273–276.

Lett, B. T. (1973). Delayed reward learning: disproof of the traditional theory. *Learn. Motiv.*, **3**, 237–246.

Lettvin, J. Y., Maturana, H. R., McCulloch, W. W. & Pitts, W. H. (1959). What the frog's eye tells the frog's brain. *Proc. Inst. Radio Eng.*, **47**, 1940–1951.

Leventhal, A. G. & Hirsch, H. V. B. (1975). Cortical effect of early selective exposure to diagonal lines. *Science*, **190**, 902–904.

Levine, G. & Loesch, R. (1967). Generality of response intensity following nonreinforcement. *J. exp. Psychol.*, **75**, 97–102.

Levis, D. J. (1976). Learned helplessness: A reply and an alternative S-R interpretation. *J. exp. Psychol. (Gen.)*, **105**, 47–65.

Levis, D. J. & Carrera, R. (1967). Effects of ten hours of implosive therapy in the treatment of outpatients: A preliminary report. *J. abnorm. Psychol.*, **72**, 504–508.

Levitsky, D. & Collier, G. (1968). Schedule-induced wheel running. *Physiol. Behav.*, **3**, 571–573.

Likely, D. G. (1974). Autoshaping in the rhesus monkey. *Anim. Learn. Behav.*, **2**, 203–206.

Lindsley, O. R. (1960). Characteristics of the behaviour of chronic psychotics as revealed by free-operant conditioning methods. *Diseases of the Nervous System (Monograph Supplements)*, **21**, 66–78.

Lindsley, O. R. (1964). Direct measurement and prosthesis of retarded behaviour. *J. Educ.*, **147**, 62–81.

Ling, B. C. (1941). Form discrimination as a learning cue in infants. *Comp. Psychol. Monogr.*, **17**, No. 2, 66.

Lippman, L. G. & Meyer, M. E. (1967). Fixed-interval performance as related to instructions and to subjects' verbalisations of the contingency. *Psychon. Sci.*, **8**, 135–136.

Lipsitt, L. P. (1963). Learning in the first year of life. In L. P. Lipsitt & C. C. Spiker (eds), *Advances in Child Development and Behaviour*, Vol. 1, pp. 147–194. New York, Academic Press.

Lipsitt, L. P. & Ambrose, J. A. (1967). A preliminary report of temporal conditioning to three types of neonatal stimulation. Paper presented at the meeting of the Soc. Res. Child Dev., New York.

Lipsitt, L. P. & Kaye, H. (1964). Conditioned sucking in the human newborn. *Psychon. Sci.*, 1, 29–30.

Lipsitt, L. P., Kaye, H. & Bosack, T. N. (1966). Enhancement of neonatal sucking through reinforcement. *J. exp. Child Psychol.*, 4, 163–168.

Liu, S. S. (1971). Differential conditioning and stimulus generalization of the rabbits nictitating membrane response. *J. comp. physiol. Psychol.*, 77, 136–142.

Loehrl, H. (1959). Zur frage des zeitpunktes einer prägung auf die heimatregion beim halsbandschnäpper (*Ficedulla albiocollis*). *J. Orn.*, 100, 132–140.

Logan, F. A. (1954). A note on stimulus intensity dynamism (V). *Psychol. Dev.*, 61, 77–80.

Logan, F. A. & Spanier, D. (1970). Relative effect of delay of food and water reward. *J. comp. physiol. Psychol.*, 72, 102–104.

Lolordo, V. M. (1967). Similarity of conditioned fear responses based upon different aversive events. *J. comp. physiol. Psychol.*, 64, 154–158.

Lolordo, V. M. (1971). Facilitation of food-reinforced responding by a signal for response-independent food. *J. exp. Anal. Behav.*, 15, 49–55.

Lolordo, V. M., McMillan, J. C. & Riley, A. L. (1974). The effects upon food-reinforced pecking and treadle-pressing of auditory and visual signals for response-independent food. *Learn. Motiv.*, 5, 24–41.

London, P. (1972). The end of ideology in behaviour modification. *Am. Psychol.*, 27, 913–920.

Long, E. R., Hammack, J. T., May, F. & Campbell, B. J. (1958). Intermittent reinforcement of operant behaviour in children. *J. exp. Anal. Behav.*, 1, 315–340.

Lore, R., Blanc, A. & Suedfeld, P. (1971). Empathetic learning of a passive avoidance response in domesticated *Rattus norvegicus*. *Anim. Behav.*, 19, 112–114.

Lorenz, K. Z. (1935). Der kumpan in der umwelt des vogels. In K. Lorenz, *Studies in Animal and Human Behaviour*, 1970, Vol. 1, 101–258, translated by R. Martin. Cambridge, Mass., Harvard University Press.

Lorenz, K. Z. (1969). Innate bases of learning. In K. Pribram (ed.), *On the Biology of Learning*. New York, Harcourt, Brace & World.

Lorge, I. (1936). Irrelevant rewards in animal learning. *J. comp. Psychol.*, 21, 105–128.

Lotter, E. C., Woods, S. C. & Vasselli, J. R. (1973). Schedule-induced polydipsia: An artifact. *J. comp. physiol. Psychol.*, 83, 478–484.

Lovaas, O. I., Koegel, R., Simmons, J. Q. & Long, J. S. (1973). Some generalization and follow-up measures on autistic children in behaviour therapy. *J. appl. Behav. Anal.*, 6, 131–165.

Lovaas, O. I. & Simmons, J. Q. (1969). Manipulation of self-destruction in three retarded children. *J. appl. Behav. Anal.*, 2, 143–157.

Lowe, C. F., Davey, G. C. L. & Harzem, P. (1974). Effects of reinforcement magnitude on interval and ratio schedules. *J. exp. Anal. Behav.*, 22, 553–560.

Lowe, C. F., Davey, G. C. L. & Harzem, P. (1976). After-effects of reinforcement magnitude on temporally spaced responding. *Psychol. Rec.*, 26, 33–40.

Lowe, C. F. & Harzem, P. (1977). Species differences in temporal control of behaviour. *J. exp. Anal. Behav.*, 28, 189–201.

Lowe, C. F., Harzem, P. & Bagshaw, M. (1978). Species differences in temporal control of behaviour. II. Human performance. *J. exp. Anal. Behav.*, 29, 351–361.

Lowe, C. F., Harzem, P. & Hughes, S. (1978). Determinants of operant behaviour in humans: some differences from animals. *Q. Jl. exp. Psychol.*, 30, 373–386.

Lucas, G. A. (1975). The control of key pecks during automaintenance by prekeypeck omission training. *Anim. Learn. Behav.*, 3, 33–36.

Lynn, R. (1966). *Attention, Arousal and the Orientation Reaction.* Oxford, Pergamon.

Lyon, D. O. (1963). Frequency of reinforcement as a parameter of conditioned suppression. *J. exp. Anal. Behav.*, 7, 289–291.

Lyon, D. O. & Ozolins, D. (1970). Pavlovian conditioning of shock elicited aggression: A discrimination procedure. *J. exp. Anal. Behav.*, 13, 325–331.

Maatsch, J. L. (1959). Learning and fixation after a single shock trial. *J. comp. physiol. Psychol.*, 52, 408–410.

McAllister, W. R. & McAllister, D. E. (1971). Behavioral measurement of conditioned fear. In F. R. Brush (ed.) *Aversive Conditioning and Learning.* New York, Academic Press.

Macdonald, A. (1946). The effect of adaptation to the unconditioned stimulus upon the formation of conditioned avoidance responses. *J. exp. Psychol.*, 36, 1–12.

Macdonald, G. E. (1968). Imprinting: drug produced isolation and the sensitive period. *Nature*, 217, 1158–1159.

Macdonald, G. E. & Solandt, A. (1966). Imprinting: effects of drug-induced immobilization. *Psychon. Sci.*, 5, 95–96.

Macfarlane, D. A. (1930). The role of kinesthesis in maze learning. *Univ. Calif. Publ. Psychol.*, 4, 277–305.

McHose, J. H. (1970). Relative reinforcement effects: $S_1/S_2$ and $S_1/S_1$ paradigms in instrumental conditioning. *Psychol. Rev.*, 77, 135–146.

McKearney, J. W. (1969). Fixed-interval schedules of electric shock presentation: Extinction and recovery of performance under different shock intensities and fixed-interval durations. *J. exp. Anal. Behav.*, 12, 301–313.

McKearney, J. W. (1972). Schedule-dependent effects: effects of drugs, and maintenance of responding with response-produced electric shocks. In R. M. Gilbert & J. D. Keehn (eds), *Schedule Effects: Drugs, Drinking and Aggression*, pp. 3–25. Toronto, University of Toronto Press.

Mackintosh, N. J. (1969). Comparative studies of reversal and probability learning: rats, birds and fish. In R. M. Gilbert & N. S. Sutherland (eds), *Animal Discrimination Learning*, pp. 137–162, 175–185. London, Academic Press.

Mackintosh, N. J. (1970). Attention and probability learning. In D. Mostofsky (ed.), *Attention: Contemporary Theory and Analysis*, pp. 173–191. New York, Appleton–Century–Crofts.

Mackintosh, N. J. (1974). *The Psychology of Animal Learning.* London, Academic Press.

Mackintosh, N. J., Little, L. & Lord, J. (1972). Some determinants of behavioural contrast in pigeons and rats. *Learn. Motiv.*, 3, 148–161.

Mackintosh, N. J., Lord, J. & Little, L. (1971). Visual and spatial probability learning in pigeons and goldfish. *Psychon. Sci.*, 24, 221–223.

McMillan, J. C. (1971). Percentage reinforcement on fixed-ratio and variable-interval performances. *J. exp. Anal. Behav.*, 15, 297–302.

Macphail, E. M. (1968). Avoidance responding in pigeons. *J. exp. Anal. Behav.*, 11, 625–632.

Mahoney, M. J., Thoresen, C. E. & Danaher, B. G. (1972). Covert behaviour modification: An experimental analogue. *J. Behav. Ther. exp. Psychiat.*, 3, 7–14.

Maier, R. A. & Maier, B. M. (1970). *Comparative Animal Learning.* Brooks/Cole Publishing Co.

Maier, S. F. (1970). Failure to escape traumatic electric shock: incompatible skeletal-motor responses or learned helplessness. *Learn. Motiv.*, 1, 157–169.

Maier, S. F., Albin, R. W. & Testa, T. J. (1973). Failure to learn to escape in rats previously exposed to inescapable shock depends on the nature of the escape response. *J. comp. Physiol. Psychol.*, 85, 581–592.

Maier, S. F., Alloway, T. A. & Gleitman, H. (1967). Proactive inhibition in rats after prior partial reversal: A critique of the spontaneous recovery hypothesis. *Psychon. Sci.*, **9**, 63–64.

Maier, S. F., Anderson, C. & Lieberman, D. A. (1972). Influence of control of shock on subsequent shock-elicited aggression. *J. comp. physiol. Psychol.*, **81**, 94–100.

Maier, S. F. & Gleitman, H. (1967). Proactive interference in rats. *Psychon. Sci.*, **7**, 25–26.

Maier, S. F. & Seligman, M. E. P. (1976). Learned helplessness: theory and evidence. *J. exp. Psychol.* (Gen.), **105**, 3–46.

Malleson, N. (1959). Panic and phobia. *Lancet*, **1**, 225–227.

Malone, J. C. & Staddon, J. E. R. (1973). Contrast effects in maintained generalization gradients. *J. exp. Anal. Behav.*, **19**, 167–179.

Malott, R. W. & Cumming, W. W. (1964). Schedules of inter-response time reinforcement. *Psychol. Rec.*, **14**, 211–252.

Marchant, R. G., III, Mis, F. W. & Moore, J. W. (1972). Conditioned inhibition of the rabbits nictitating membrane response. *J. exp. Psychol.*, **95**, 408–411.

Maritain, J. (1957). Language and the theory of sign. In R. N. Anshen (ed.), *Language: An Inquiry into its Meaning and Function.* New York, Harper & Row.

Marler, P. (1965). Communication in monkeys and apes. In I. Devore (ed.), *Primate Behaviour: Field Studies of Monkeys and Apes*, pp. 544–584. New York, Holt, Rinehart & Winston.

Marquis, D. P. (1931). Can conditioned responses be established in the newborn infant? *J. genet. Psychol.*, **39**, 479–492.

Martin, J. & Crawford, J. (1976). Thought operants. *J. gen. Psychol.*, **95**, 33–45.

Marum, K. D. (1962). A study of classical conditioning in the human infants. Unpublished Master's thesis, Brown University,

Massermann, J. H. (1943). *Behaviour and Neurosis.* Chicago, University of Chicago Press.

Masserman, J. H. (1946). *Principles of Dynamic Psychiatry.* Philadelphia, Saunders.

Masserman, J. H. (1950). Experimental neuroses. *Scient. Am.*, **182**, 38–50.

Masterson, F. A. (1970). Is termination of a warning signal an effective reward for the rat? *J. comp. physiol. Psychol.*, **72**, 471–475.

Matthews, B. A., Shimoff, E., Catania, C. & Sagvolden, T. (1977). Uninstructed human responding: sensitivity to ratio and interval contingencies. *J. exp. Anal. Behav.*, **27**, 453–467.

Max, L. W. (1937). Experimental study of the motor theory of consciousness: IV. Action-current responses in the deaf during awakening, kinesthetic imagery and abstract thinking. *J. comp. Psychol.*, **24**, 301–344.

Meehl, P. E. (1950). On the circularity of the law of the effect. *Psychol. Bull.*, **47**, 52–75.

Mellgren, R. L. & Ost, J. W. P. (1971). Discriminative stimulus preexposure and learning of an operant discrimination in the rat. *J. comp. physiol. Psychol.*, **77**, 179–187.

Meltzer, D. & Brahlek, J. A. (1970). Conditioned suppression and conditioned enhancement with the same positive UCS: An effect of CS duration. *J. exp. Anal. Behav.*, **13**, 67–73.

Melvin, K. B. & Anson, J. E. (1969). Facilitative effect of punishment on aggressive behaviour in the Siamese fighting fish. *Psychon. Sci.*, **14**, 89–90.

Melzack, R. & Perry, C. (1975). Self-regulation of pain: the use of alpha-feedback and hypnotic training for the control of chronic pain. *Exp. Neurol.*, **46**, 452–469.

Mendelson, J. & Chillag, D. (1970). Schedule-induced air licking in rats. *Physiol. Behav.*, **5**, 535–537.

Mentzer, T. L. (1966). Comparison of three methods for obtaining psychophysical thresholds from the pigeon. *J. comp. physiol. Psychol.*, **61**, 96–101.

Menzel, E. W. (1974). A group of young chimpanzees in a one-acre field. In A. M.

Schrier & F. Stollmitz (eds), *Behaviour of Nonhuman Primates*, Vol. 5. New York, Academic Press.

Menzel, E. W. & Halperin, S. (1975). Purposive behaviour as a basis for objective communication between chimpanzees. *Science*, **189**, 652–654.

Meyer, D. R., Cho, C. & Weseman, A. F. (1960). On problems of conditioned discriminated lever-press avoidance responses. *Psychol. Rev.*, **67**, 224–228.

Michels, K. M. (1957). Response latency as a function of the amount of reinforcement. *Br. J. Anim. Behav.*, **5**, 50–52.

Miczek, K. A. & Grossman, S. P. (1971). Positive conditioned suppression: Effects of CS duration. *J. exp. Anal. Behav.*, **15**, 243–247.

Migler, B. (1964). Effects of averaging data during stimulus generalization. *J. exp. Anal. Behav.*, **7**, 303–307.

Millenson, J. R. & de Villiers, P. A. (1972). Motivational properties of conditioned anxiety. In R. M. Gilbert & J. R. Millenson (eds), *Reinforcement: Behavioural Analyses*, pp. 98–128. New York, Academic Press.

Miller, B. V. & Levis, D. J. (1971). The effects of varying short visual exposure times to a phobic test stimulus on subsequent avoidance behaviour. *Behav. Res. Ther.*, **9**, 17–21.

Miller, J. D. (1970). Audibility curve of the chinchilla. *J. acoust. Soc. Am.*, **48**, 513–523.

Miller, L. B. (1969). Compounding of pre-aversive stimuli. *J. exp. Anal. Behav.*, **12**, 293–299.

Miller, L. B. (1971). Compounding of discriminative stimuli from the same and different sensory modalities. *J. exp. Anal. Behav.*, **16**, 337–342.

Miller, L. B. & Estes, B. W. (1961). Monetary reward and motivation in discrimination learning. *J. exp. Psychol.*, **61**, 501–504.

Miller, N. E. (1948). Studies of fear as an acquirable drive. *J. exp. Psychol.*, **38**, 89–101.

Miller, N. E. (1960). Learning resistance to pain and fear: Effects of overlearning, exposure and rewarded exposure in context. *J. exp. Psychol.*, **60**, 137–145.

Miller, N. E. (1969). Learning of visceral and glandular responses. *Science*, **163**, 434–445.

Miller, N. E. (1975). Applications of learning and biofeedback to psychiatry. In A. M. Freedman, H. I. Kaplan & B. J. Sadock (eds), *Comprehensive Text-book of Psychiatry, III*. Baltimore, Williams and Wilkins.

Miller, N. E. & Banuazizi, A. (1968). Instrumental learning by curarized rats of a specific visceral response, intestinal or cardiac. *J. comp. physiol. Psychol.*, **65**, 1–7.

Miller, N. E. & Carmona, A. (1967). Modification of a visceral response, salivation in thirsty dogs, by instrumental training with water reward. *J. comp. physiol. Psychol.*, **63**, 1–6.

Miller, N. E. & Dicara, L. V. (1968). Instrumental learning of urine formation by rats; changes in renal blood flow. *Am. J. Physiol.*, **215**, 677–683.

Miller, S. & Konorski, J. (1928). Sur une forme particulierè des reflexes conditionnels. *C. R. Séanc. Soc. Biol.*, **99**, 1155–1157.

Mirzoiants, N. S. (1954). The conditioned orienting reflex and its differentiation in the child. *Zh. Vysshei Nerv. Deyat. Pavlova*, **4**, 616–619.

Miyadi, D. (1964). Social life of Japanese monkeys. *Science*, **143**, 783–786.

Moise, S. L. (1970). Short-term retention in *Maccaca speciosa* following interpolated activity during matching from sample. *J. comp. physiol. Psychol.*, **73**, 506–514.

Moltz, H. (1960). Imprinting: empirical bases and theoretical significance. *Psychol. Bull.*, **57**, 291–314.

Moltz, H. (1961). An experimental analysis of the critical period for imprinting. *Trans. N.Y. Acad. Sci.*, **23**, 452–463.

Moltz, H. (1963). Imprinting: An epigenetic approach. *Psychol. Rev.*, **70**, 123–138.

Moltz, H. & Stettner, L. J. (1961). The influence of patterned-light deprivation on the critical period for imprinting. *J. comp. physiol. Psychol.*, **54**, 279–283.

Moore, B. R. (1973). The role of directed Pavlovian reactions in simple instrumental learning in the pigeon. In R. A. Hinde & J. Stevenson-Hinde (eds), *Constraints on Learning*. New York, Academic Press.

Morgan, C. L. (1894). *An Introduction to Comparative Psychology*. London, Scott.

Morgan, C. L. (1930). *The Animal Mind*. New York, Longmans.

Morgan, J. J. B. & Morgan, S. S. (1964). Infant learning as a developmental index. *J. genet. Psychol.*, **65**, 281–289.

Morgan, M. J. (1971). Beyond freedom and dignity: Skinner's behaviourism. *Cambridge Rev.*, Nov. 19.

Morganstern, K. P. (1973). Implosive therapy and flooding procedures: A critical review. *Psychol. Bull.*, **79**, 318–334.

Morris, R. C. (1976). Behavioural contrast and autoshaping. *Q. Jl. exp. Psychol.*, **28**, 661–666.

Morse, W. H. (1966). Intermittent reinforcement. In W. K. Honig (ed.), *Operant Behaviour: Areas of Research and Application*, pp. 52–108. New York, Appleton–Century–Crofts.

Morse, W. H., Mead, R. N. & Kelleher, R. T. (1967). Modulation of elicited behaviour by a fixed-interval schedule of electric shock presentation. *Science*, **157**, 215–217.

Mowrer, O. H. (1936). "Maturation" vs "Learning" in the development of vestibular and optokinetic nystagmus. *J. genet. Psychol.*, **48**, 383–404.

Mowrer, O. H. (1939). A stimulus-response analysis of anxiety and its role as a reinforcing agent. *Psychol. Rev.*, **46**, 553–565.

Mowrer, O. H. (1947). On the dual nature of learning – a reinterpretation of "conditioning" and "problem-solving". *Harv. educ. Rev.*, **17**, 102–148.

Mowrer, O. H. (1960). *Learning Theory and Behaviour*. New York, Wiley.

Mowrer, O. H. & Viek, P. (1948). An experimental analysis of fear from a sense of helplessness. *J. abnorm. soc. Psychol.*, **43**, 193–200.

Moyer, K. E. & Korn, J. H. (1966). Effects of UCS intensity on the acquisition and extinction of a one-way avoidance response. *Psychon. Sci.*, **4**, 121–122.

Muenzinger, K. F. (1934). Motivation in learning: I. Electric shock for correct responses in the visual discrimination habit. *J. comp. Psychol.*, **17**, 267–277.

Muntz, W. R. A. (1964). Vision in frogs. *Scient. Am.*, **210**, 111–119.

Murray, H. G. (1970). Stimulus intensity and reaction time: evaluation of a decision-theory model. *J. exp. Psychol.*, **84**, 383–391.

Myer, J. S. & Hull, J. H. (1974). Autoshaping and instrumental learning in the rat. *J. comp. physiol. Psychol.*, **86**, 724–729.

Myer, J. S. & White, R. T. (1965). Aggressive motivation in the rat. *Anim. Behav.*, **13**, 430–433.

Myers, A. K. (1959). Avoidance learning as a function of several training conditions and strain differences in rats. *J. comp. physiol. Psychol.*, **52**, 381–386.

Myers, A. K. (1962). Effects of CS intensity and quality in avoidance conditioning. *J. comp. physiol. Psychol.*, **55**, 57–61.

Myers, A. K. (1964). Discriminated operant avoidance learning in Wistar and G-4 rats as a function of type of warning stimulus. *J. comp. physiol. Psychol.*, **58**, 453–455.

Myers, C. S. (1908). Some observations on the development of color sense. *Br. J. Psychol.*, **2**, 353–362.

Neuringer, A. J. (1967). Effects of reinforcement magnitude on choice and rate of responding. *J. exp. Anal. Behav.*, **10**, 417–424.

Neuringer, A. & Neuringer, M. (1974). Learning by following a food source. *Science*, **184**, 1005–1008.

Neuringer, A. & Schneider, B. A. (1968). Separating the effects of interreinforcement time and number of interreinforcement responses. *J. exp. Anal. Behav.*, **11**, 661–667.

Nevin, J. A. (1970). On differential stimulation and differential reinforcement. In W. C. Stebbins (ed.), *Animal Psychophysics: the Design and Conduct of Sensory Experiments*, pp. 401–423. New York, Appleton–Century–Crofts.

Nevin, J. A. & Berryman, R. (1963). A note on chaining and temporal discrimination. *J. exp. Anal. Behav.*, **6**, 109–113.

Nevin, J. A. & Shettleworth, S. J. (1966). An analysis of contrast effects in multiple schedules. *J. exp. Anal. Behav.*, **9**, 305–315.

Nissen. H. W. & McCulloch, T. L. (1937). Equated and non-equated situations in discrimination learning by chimpanzees. I. Comparison with unlimited response. *J. comp. Psychol.*, **23**, 165–189.

Noelpp, B. & Noelpp-Eschenhagen, I. (1951). Das experimentelle Asthma bronchiale des Meerschweinschens: Mitterlung die Rolle bedingter Reflexes in des Pathogenese des Asthma bronchiale. *Int. Archs Allergy*, **2**, 321–329.

Noelpp, B. & Noelpp-Eschenhagen, I. (1952). Das experimentelle Asthma bronchiale des Meerschweinchens. III. Studien zur Bedeutung bedingte Reflexe. Bahrungsbereitschaft und Haftfahigkeit unter "stress". *Int. Archs Allergy*, **3**, 108–136.

Nottebohm, F. (1970). Ontogeny of bird song. *Science*, **167**, 950–956.

Nowlis, D. P. & Kamiya, J. (1970). The control of electro-encephalographic alpha rhythms through auditory feedback and the associated mental activity. *Psychophysiol.*, **6**, 476–484.

Olds, J. (1969). The central nervous system and the reinforcement behaviour. *Am. Psychol.*, **24**, 114–132.

Oley, N. N. & Slotnick, B. M. (1970). Nesting material as a reinforcement for operant behaviour in the rat. *Psychonom. Sci.*, **21**, 41–43.

Olson, C. R. & Pettigrew, J. D. (1974). Single units in visual cortex of kittens reared in stroboscopic illuminations. *Brain Res.*, **70**, 189–204.

Omello, G. & Stolerman, I. P. (1977). Cocaine and amphetamine as discriminative stimuli in rats. *Br. J. Pharmacol.*, **59**, 453.

Ordy, J. M. & Samarajski, T. (1968). Visual acuity and erg-cff in relation to the morphologic organization of the retina among diurnal and nocturnal primates. *Vision Res.*, **8**, 1205–1225.

Orlando, R. (1961). The functional role of discriminative stimuli in free operant performance of developmentally retarded children. *Psychol. Rec.*, **11**, 153–161.

Orlando, R. & Bijou, S. W. (1960). Single and multiple schedules of reinforcement in developmentally retarded children. *J. exp. Anal. Behav.*, **3**, 339–348.

Osler, S. F. & Trautman, G. E. (1961). Concept attainment. II. Effect of stimulus complexity upon concept attainment at two levels of intelligence. *J. exp. Psychol.*, **62**, 9–13.

Overmier, J. B. & Seligman, M. E. P. (1967). Effects of inescapable shock upon subsequent escape and avoidance learning. *J. comp. physiol. Psychol.*, **63**, 23–33.

Overton, D. (1964). State-dependent or "dissociated" learning produced with pentobarbital. *J. comp. physiol. Psychol.*, **57**, 3–12.

Overton, D. (1966). State-dependent learning produced by depressant and atropine-like drugs. *Psychopharmacologia*, **10**, 6–31.

Padilla, A. M., Padilla, C., Ketterer, T. & Giacolone, D. (1970). Inescapable shocks and subsequent avoidance conditioning in goldfish (*Carrasius auratus*). *Psychon. Sci.*, **20**, 295–296.

Page, H. A. (1955). The facilitation of experimental extinction by response prevention as a function of the acquisition of a new response. *J. comp. physiol. Psychol.*, **48**, 14–16.

Panksepp, J., Toates, F. M. & Oatley, K. (1972). Extinction induced drinking in hungry rats. *Anim. Behav.*, **20**, 493–498.

Papousek, H. (1959). A method of studying conditioned food reflexes in young children up to the age of six months. *Pavlov J. Higher Nerv. Activ.*, **9**, 136–140.

Papousek, H. (1967a). Conditioning during postnatal development. In Y. Brackbill & G. G. Thompson (eds), *Behaviour in Infancy and Early Childhood: A Book of Readings*, pp. 259–274. New York, Free Press.

Papousek, H. (1967b). Experimental studies of appetitional behaviour in human newborns. In H. W. Stevenson, E. H. Hess & H. L. Rheingold (eds), *Early Behaviour: Comparative and Developmental Approaches*, pp. 249–277. New York, Wiley.

Pastore, N. (1954). Discrimination learning in the canary. *J. comp. Physiol. Psychol.*, **47**, 389–390.

Pastore, N. (1962). Perceptual functioning in the duckling. *J. Psychol.*, **54**, 293.

Patten, R. L. & Rudy, J. W. (1967). The Sheffield omission training procedure applied to the conditioning of the licking response in rats. *Psychon. Sci.*, **8**, 463–464.

Paul, G. L. (1966). *Insight versus Desensitization in Psychotherapy: An Experiment in Anxiety Reduction*. Stanford, Stanford Univ. Press.

Paul, G. L. (1969). Physiological effects of relaxation training and hypnotic suggestion. *J. abnorm. Psychol.*, **74**, 425–437.

Pavlov, I. P. (1927). *Conditioned Reflexes*. Oxford, Oxford University Press.

Pavlov, I. P. (1933). Essai d'une interpretation physiologique de l'hysterie. *L'encephale*, **28**, 288–295.

Pear, J. J., Moody, J. E. & Persinger, M. A. (1972). Lever attacking by rats during free-operant avoidance. *J. exp. Anal. Behav.*, **18**, 517–523.

Pear, J. J. & Wilkie, D. M. (1971). Contrast and induction in rats on multiple schedules. *J. exp. Anal. Behav.*, **15**, 289–296.

Penney, R. K. & Croskery, J. (1962). Instrumental avoidance conditioning of anxious and nonanxious children. *J. comp. physiol. Psychol.*, **55**, 847–849.

Penney, R. K. & Kirwin, P. M. (1965). Differential adaptation of anxious and nonanxious children in instrumental escape conditioning. *J. exp. Psychol.*, **70**, 539–549.

Penney, R. K. & McCann, B. (1962). The instrumental escape conditioning of anxious and nonanxious children. *J. abnorm. soc. Psychol.*, **65**, 351–354.

Perelman, C. (1973). Behaviourism's enlightened despotism. In H. Wheeler (ed.), *Beyond the Punitive Society*, pp. 121–124. London, Wildwood House.

Perin, C. T. (1943). A quantitative investigation of the delay-of-reinforcement gradient. *J. exp. Psychol.*, **32**, 37–51.

Perkins, C. C. (1947). The relation of secondary reward to gradients of reinforcement. *J. exp. Psychol.*, **37**, 377–392.

Perkins, C. C. (1953). The relation between conditioned stimulus intensity and response strength. *J. exp. Psychol.*, **46**, 225–231.

Perkins, C. C. (1968). An analysis of the concept of reinforcement. *Psychol. Rev.*, **75**, 155–172.

Perkins, C. C., Beavers, W. O., Hancock, R. A., Hemmendinger, P. C., Hemmendinger, D. & Ricci, J. A. (1975). Some variables affecting rate of key pecking during response-independent procedures (autoshaping). *J. exp. Anal. Behav.*, **24**, 59–72.

Peterson, G. B. (1975). Response selection properties of food and brain-stimulation reinforcers in rats. *Physiol. Behav.*, **14**, 681–688.

Peterson, G. B., Ackil, J. E., Frommer, G. P. & Hearst, E. S. (1972). Conditioned approach and contact behaviour toward signals for food and brain-stimulation reinforcement. *Science*, **177**, 1009–1011.

Pettigrew, J. D. (1974). The effect of visual experience on the development of stimulus specificity by kitten cortical neurones. *J. Physiol.*, **237**, 49–74.

Pettigrew, J. D. & Freeman, R. (1973). Visual experience without lines: effect on developing cortical neurones. *Science*, **182**, 599–601.

Pettigrew, J. D. & Garey, L. J. (1974). Selective modification of single neurone properties in the visual cortex of kittens. *Brain Res.*, **66**, 160–164.

Pfaffman, C. (1969). Taste preference and reinforcement. In J. T. Tapp (ed.), *Reinforcement and Behaviour*, pp. 215–241. New York, Academic Press.

Pfungst, O. (1911). *Clever Hans, the Horse of Von Osten*. New York, Holt, Rinehart and Winston.

Pierce, J. T. & Nuttall, R. L. (1961). Self-paced sexual behaviour in the female rat. *J. comp. physiol. Psychol.*, **54**, 310–313.

Pierrel, R. & Sherman, J. G. (1963). Barnabus, the rat with college training. *Brown Alumni Monthly* (February).

Pierrel, R., Sherman, J. G., Blue, S. & Hegge, F. W. (1970). Auditory discrimination: a three-variable analysis of intensity effects. *J. exp. Anal. Behav.*, **13**, 17–35.

Platt, J. (1972). Beyond Freedom and Dignity: A revolutionary manifesto. *Cent. Mag.*, **5**, 34–52.

Platt, J. R. & Senkowski, P. C. (1970). Effects of discrete trials reinforcement frequency and changes in reinforcement frequency on preceding and subsequent fixed-ratio performance. *J. exp. Psychol.*, **85**, 95–104.

Pliskoff, S. S. & Brown, T. G. (1976). Matching with a trio of concurrent variable-interval schedules of reinforcement. *J. exp. Anal. Behav.*, **25**, 69–73.

Plotkin, W. B. (1976). On the self-regulation of the occipital alpha rhythm: control strategies, states of consciousness, and the role of physiological feedback. *J. exp. Psychol. (Gen.)*, **105**, 66–99.

Poland, S. F. & Warren, J. M. (1967). Spatial probability learning by cats. *Psychon. Sci.*, **8**, 487–488.

Polikanina, R. I. (1961). The relationship between autonomic and somatic components of a defensive conditioned reflex in premature children. *Pavlov J. Higher Nerv. Activ.*, **11**, 72–82.

Polin, A. T. (1959). The effect of flooding and physical suppression as extinction techniques on an anxiety-motivated avoidance locomotor response. *J. Psychol.*, **47**, 235–245.

Pomerleau, O. F. (1970). The effects of stimuli followed by response-independent shock on shock-avoidance behaviour. *J. exp. Anal. Behav.*, **14**, 11–21.

Powell, R. W. & Peck, S. (1969). Activity and avoidance in a Mongolian gerbil. *J. exp. Anal. Behav.*, **12**, 779–787.

Premack, D. (1970). A functional analysis of language. *J. exp. Anal. Behav.*, **14**, 107–125.

Premack, D. (1971a). On the assessment of language competence and the chimpanzee. In A. M. Schrier & F. Stollnitz (eds), *Behaviour of Nonhuman Primates*, Vol. 4. New York, Academic Press.

Premack, D. (1971b). Language in chimpanzees? *Science*, **172**, 808–822.

Presley, W. J. & Riopelle, A. J. (1959). Observational learning of an avoidance response. *J. genet. Psychol.*, **95**, 251–254.

Prewitt, E. P. (1967). Number of preconditioning trials in sensory preconditioning using CER training. *J. comp. physiol. Psychol.*, **64**, 360–362.

Pribram, K. H. (1971). *Languages of the Brain*. Englewood Cliffs, New Jersey, Prentice Hall.

Priddle-Higson, P. J., Lowe, C. F. & Harzem, P. (1976). After effects of reinforcement on variable-ratio schedules. *J. exp. Anal. Behav.*, **25**, 347–354.

Prokasy, W. F. (1965). Classical eyelid conditioning: Experimenter operations, task demands, and response shaping. In W. F. Prokasy (ed.), *Classical Conditioning: A Symposium*, pp. 208–225. New York, Appleton–Century–Crofts.

Pusakulich, R. L. & Nielson, H. C. (1976). Cue use in state-dependent learning. *Physiol. Psychol.*, **4**, 421.

Quartermain, D. & McEwen, B. S. (1970). Temporal characteristics of amnesia induced by inhibition of protein synthesis. *Nature*, **228**, 677–678.

Rachlin, H. (1969). Autoshaping of key pecking in pigeons with negative reinforcement. *J. exp. Anal. Behav.*, **12**, 521–531.

Rachlin, H. (1976). *Behaviour and Learning*. San Francisco, W. H. Freeman & Co.

Rachlin, H. & Hineline, P. N. (1967). Training and maintenance of key-pecking in the pigeon by negative reinforcement. *Science*, **157**, 954–955.

Rachman, S. (1965). Aversion therapy: chemical or electrical? *Behav. Res. Ther.*, **2**, 289–299.

Rachman, S. (1966a). Sexual fetishism: An experimental analogue. *Psychol. Rec.*, **16**, 293–296.

Rachman, S. (1966b). Studies in desensitization: II. Flooding. *Behav. Res. Ther.*, **4**, 1–6.

Rachman, S. (1968). *Phobias: Their Nature and Control*. Springfield. Illinois, Charles C. Thomas.

Rachman, S. & Hodgson, R. J. (1968). Experimentally induced "sexual fetishism": Replication and development. *Psychol. Rec.*, **18**, 25–27.

Rachman, S. & Teasdale, J. (1969). *Aversion Therapy and Behaviour Disorders: An Analysis*: Coral Gables, Fla, University of Miami Press.

Rackham, D. W. (1971). Conditioning of the pigeon's courtship and aggressive display. Unpublished thesis, Dalhousie University.

Ramer, D. G. & Wilkie, D. M. (1977). Spaced food but not electrical brain stimulation induces polydipsia and air licking. *J. exp. Anal. Behav.*, **27**, 507–514.

Rappoport, L. (1972). *Personality Development*. Glenview, Scott, Foresman.

Razran, G. (1971). *Mind in Evolution: An East-West Synthesis of Learned Behaviour and Cognition*. Boston, Houghton Mifflin.

Reberg, D. & Black, A. H. (1969). Compound testing of individually conditioned stimuli as an index of excitatory and inhibitory properties. *Psychon. Sci.*, **17**, 30–31.

Reese, H. W. (1963). Discrimination learning set in children. In L. P. Lipsitt & C. C. Spiker (eds), *Advances in Child Development and Behaviour*, Vol. 1, pp. 115–145. New York, Academic Press.

Reese, H. W. (1964). Discrimination learning set in rhesus monkeys. *Psychol. Bull.*, **61**, 321–340.

Reese, H. W. (1976). *Basic Learning Processes in Childhood*. New York, Holt, Rinehart and Winston.

Reese, H. W. & Lipsitt, L. P. (1973). *Experimental Child Psychology*. New York, Academic Press.

Rendle-Short, J. (1961). The puff test. *Archs Dis. Childh.*, **36**, 50–57.

Renner, K. E. (1963). Influence of deprivation and availability of goal box cues on the temporal gradient of reinforcement. *J. comp. physiol. Psychol.*, **56**, 101–104.

Rensch, B. & Dücker, G. (1966). Verzögerung des Vergessens erlernter visirellen Aufgaben bei Tieren durch Chlorpromazin. *Pfluegers Arch. Ges. Physiol. Mensch. Tiere*, **289**, 200–214.

Rescorla, R. A. (1966). Predictability and number of pairings in Pavlovian fear conditioning. *Psychon. Sci.*, **4**, 383–384.

Rescorla, R. A. (1967). Pavlovian conditioning and its proper control procedures. *Psychol. Rev.*, **74**, 71–80.

Rescorla, R. A. (1973a). Effect of US habituation following conditioning. *J. comp. physiol. Psychol.*, **82**, 137–143.

Rescorla, R. A. (1973b). Second-order conditioning: implications for theories of learning. In F. J. McGuigan & D. B. Lumsden (eds), *Contemporary Approaches to Conditioning and Learning*. Washington D.C., V. H. Winston.

Rescorla, R. A. (1974). Effect of inflation of the unconditioned stimulus value following conditioning. *J. comp. physiol. Psychol.*, **86**, 101–106.

Rescorla, R. A. & Solomon, R. L. (1967). Two-process learning theory: Relationships between Pavlovian conditioning and instrumental learning. *Psychol. Rev.*, **74**, 151–182.

Rescorla, R. A. & Wagner, A. R. (1972). A theory of Pavlovian conditioning: Variations in the effectiveness of reinforcement and nonreinforcement. In A. H. Black & W. F. Prokasy (eds), *Classical Conditioning. II. Current Research and Theory*, pp. 64–99. New York, Appleton–Century–Crofts.

Revusky, S. H. (1971). The role of interference in association over a delay. In W. K. Honig & P. H. R. James (eds), *Animal Memory*, pp. 155–213. New York, Academic Press.

Revusky, S. H. & Bedarf, E. W. (1967). Association of illness with prior ingestion of novel foods. *Science*, **155**, 219–220.

Revusky, S. H. & Garcia, J. (1970). Learned associations over long delays. In G. H. Bower (ed.), *The Psychology of Learning and Motivation*, Vol. 4, pp. 1–84. New York, Academic Press.

Reynierse, J. H. (1966). Excessive drinking in rats as a function of number of meals. *Can. J. Psychol.*, **20**, 82–86.

Reynolds, G. S. (1961a). Behavioural Contrast. *J. exp. Anal. Behav.*, **4**, 57–71.

Reynolds, G. S. (1961b). Attention in the pigeon. *J. exp. Anal. Behav.*, **4**, 203–208.

Reynolds, G. S. (1966). Discrimination and emission of temporal intervals by pigeons. *J. exp. Anal. Behav.*, **9**, 65–68.

Reynolds, G. S. (1968). *A Primer of Operant Conditioning*. Glenville, Illinois, Scott Foresman and Co.

Reynolds, G. S., Catania, A. C. & Skinner, B. F. (1963). Conditioned and unconditioned aggression in pigeons. *J. exp. Anal. Behav.*, **6**, 73–74.

Reynolds, G. S. & McLeod, A. (1970). On the theory of inter-response-time reinforcement. In G. H. Bower (ed.), *The Psychology of Learning and Motivation*, Vol. 4, pp. 85–107. New York, Academic Press.

Ricci, J. A. (1973). Keypecking under response-independent food presentation after long simple and compound stimuli. *J. exp. Anal. Behav.*, **19**, 509–516.

Richards, R. W. & Rilling, M. (1972). Aversive aspects of a fixed-interval schedule of food reinforcement. *J. exp. Anal. Behav.*, **17**, 405–411.

Richardson, W. K. & Clark, D. B. (1976). A comparison of the key-peck and treadle-press operants in the pigeon: differential reinforcement of low-rate schedules of reinforcement. *J. exp. Anal. Behav.*, **26**, 237–256.

Richardson, W. K. & Loughead, T. E. (1974a). The effect of physical restraint on behaviour under the differential-reinforcement-of-low-rate schedule. *J. exp. Anal. Behav.*, **21**, 455–462.

Richardson, W. K. & Loughead, T. E. (1974b). Behaviour under large values of the differential-reinforcement-of-low-rate schedule. *J. exp. Anal. Behav.*, **22**, 121–129.

Riesen, A. H. & Aarons, L. (1959). Visual movement and intensity discrimination in cats after early deprivation of pattern vision. *J. comp. physiol. Psychol.*, **52**, 142–149.

Riesen, A. H., Greenberg, B., Granston, A. S. & Fantz, R. L. (1953). Solutions of patterned string problems of young gorillas. *J. comp. physiol. Psychol.*, **46**, 19–22.

Riess, D. (1971). Shuttleboxes, Skinner boxes and Sidman avoidance in rats: Acquisition and terminal performance as a function of response topography. *Psychon. Sci.*, **25**, 283–285.

Rilling, M. (1967). Number of responses as a stimulus in fixed-interval and fixed-ratio schedules. *J. comp. physiol. Psychol.*, **63**, 60–65.

Riopelle, A. J. (1960). Observational learning of a position habit by monkeys. *J. comp. physiol. Psychol.*, **53**, 426–428.

Risley, J. R. (1968). The effects and side-effects of punishing the autistic behaviours of a deviant child. *J. app. Behav. Anal.*, **1**, 21–34.

Ritchie-Calder, L. (1973). Beyond B. F. Skinner. In H. Wheeler (ed.), *Beyond the Punitive Society*, pp. 212–216. London, Wildwood House.

Roberts, W. A. (1966). Learning and motivation in the immature rat. *Am. J. Psychol.*, **29**, 3–23.

Roberts, W. A. (1969). Resistance to extinction following partial and consistent reinforcement with varying magnitudes of reward. *J. comp. physiol. Psychol.*, **67**, 395–400.

Roberts, W. A. (1972). Spatial separation and visual differentiation of cues of factors influencing short-term memory in the rat. *J. comp. physiol. Psychol.*, **78**, 284–291.

Roberts, W. A. (1974). Spaced repetition facilitates short-term retention in the rat. *J. comp. physiol. Psychol.*, **86**, 164–171.

Roberts, W. A. & Grant, D. S. (1974). Short-term memory in the pigeon with presentation time precisely controlled. *Learn. Motiv.*, **5**, 393–408.

Roberts, W. A. & Grant, D. S. (1976). Studies of short-term memory in the pigeon using the delayed matching to sample procedure. In D. L. Medin, W. A. Roberts & R. T. Davis (eds), *Processes of Animal Memory*, pp. 79–112. New York, Lawrence Erlbaum Ass.

Robinson, N. M. & Robinson, H. B. (1961). A method for the study of instrumental avoidance conditioning with children. *J. comp. physiol. Psychol.*, **54**, 20–23.

Roeder, K. (1963). *Nerve Cells and Insect Behaviour*. Cambridge, Mass., Harvard University Press.

Rohles, F. H. & Devine, J. V. (1966). Chimpanzee performance on a problem involving the concept of middleness. *Anim. Behav.*, **14**, 159–162.

Rohles, F. H. & Devine, J. V. (1967). Further studies of the middleness concept with the chimpanzee. *Anim. Behav.*, **15**, 107–112.

Rosenblith, J. Z. (1970). Polydipsia induced in the rat by a second-order schedule. *J. exp. Anal. Behav.*, **14**, 139–144.

Rozin, P. & Kalat, J. W. (1971). Specific hungers and poison avoidance as adaptive specializations of learning. *Psychol. Rev.*, **78**, 459–486.

Rozin, P. & Kalat, J. W. (1972). Learning as a situation-specific adaptation. In M. E. P. Seligman & J. L. Hager (eds), *Biological Boundaries of Learning*, pp. 66–96. New York, Appleton–Century–Crofts.

Rozin, P. & Mayer, J. (1961). Regulation of food intake in the goldfish. *Am. J. Physiol.*, **201**, 968–74.

Ruggiero, F. T. & Flagg, S. F. (1976). Do animals have memory? In D. L. Medin, W. A. Roberts & R. T. Davies (eds), *Processes of Animal Memory*, pp. 1–19. New York, Lawrence Erlbaum Ass.

Rumbaugh, D. M. & Gill, T. V. (1976). Language and the acquisition of language-type skills by a chimpanzee (*Pan*). *Ann. N.Y. Acad. Sci.*, **270**, 90–123.

Rumbaugh, D. M., Gill, T. & Von Glaserfeld, E. C. (1973). Reading and sentence completion by a chimpanzee (*Pan*). *Science*, **182**, 731–733.

Sadowsky, S. (1966). Discrimination learning as a function of stimulus location along an auditory intensity continuum. *J. exp. Anal. Behav.*, **9**, 219–225.

Salapatek, P. & Kessen, W. (1966). Visual scanning of triangles in the human newborn. *J. exp. Child Psychol.*, **3**, 155–167.

Salzen, E. A. (1970). Imprinting and environmental learning. In L. R. Aronson, E.

Tobach, D. S. Lehrman & J. S. Rosenblatt (eds), *Development and Evolution of Behaviour*. San Francisco, Freeman & Co.

Salzen, E. A. & Meyer, C. C. (1967). Imprinting: reversal of a preference established during the critical period. *Nature*, **215**, 785–786.

Salzen, E. A. & Meyer, C. C. (1968). Reversibility of imprinting. *J. comp. physiol. Psychol.*, **66**, 269–275.

Sameroff, A. J. (1968). The components of sucking in the human newborn. *J. exp. Child Psychol.*, **6**, 607–623.

Sameroff, A. J. (1970). Respiration and sucking as components of the orienting reaction in newborns. *Psychophysiology*, **7**, 213–222.

Sameroff, A. J. (1971). Can conditioned responses be established in the newborn infant: 1971? *Devl. Psychol.*, **5**, 411–442.

Sandler, S. & Davidson, R. S. (1973). *Psychopathology: Learning Theory Research, and Applications*. New York, Harper & Row.

Sanger, D. J. & Blackman, D. E. (1976). Rate-dependent effects of drugs: a review of the literature. *Pharmac. Biochem. Behav.*, **4**, 73–83.

Schick, K. (1971). Operants. *J. exp. Anal. Behav.*, **15**, 413–423.

Schiller, P. H. (1952). Innate constituents of complex responses in primates. *Psychol. Rev.*, **59**, 177–191.

Schlosberg, H. (1936). Conditioned responses in the white rat: II. Conditioned responses based upon shock to the foreleg. *J. genet. Psychol.*, **49**, 107–138.

Schneider, B. A. (1969). A two-state analysis of fixed-interval responding in the pigeon. *J. exp. Anal. Behav.*, **12**, 677–687.

Schneiderman, N. (1966). Interstimulus interval function of the nictitating membrane response of the rabbit under delay versus trace conditioning. *J. comp. physiol. Psychol.*, **62**, 397–402.

Schneiderman, N. E., Fuentes, I. & Gormezano, I. (1962). Acquisition and extinction of the classically conditioned eyelid response in the albino rabbit. *Science*, **136**, 650–652.

Schneiderman, N. E., Pearl, L., Wilson, W., Metcalf, F., Moore, J. W. & Swadlow, H. A. (1971). Stimulus control in rabbits (*Oryctolagus cuniculus*) as a function of different intensities of intracranial stimulation. *J. comp. physiol. Psychol.*, **76**, 175–186.

Schoenfeld, W. N. (1969). Avoidance in behaviour therapy. *J. exp. Anal. Behav.*, **12**, 669–674.

Schoenfeld, W. N. (1970) (ed.). *The Theory of Reinforcement Schedules*. New York, Appleton–Century–Crofts.

Schoenfeld, W. N. & Cole, B. K. (1972). *Stimulus Schedules: the t-Systems*. New York, Harper & Row.

Schulman, A. H., Hale, E. B. & Graves, H. R. (1970). Visual stimulus characteristics for initial approach responses in chicks (*Gallus domesticus*). *Anim. Behav.*, **18**, 461–466.

Schuster, C. R. & Balster, R. L. (1977). The discriminative stimulus properties of drugs. *Advances in Behavioural Pharmacology*, Vol. 1, pp. 85–138. New York, Academic Press.

Schuster, C. R. & Woods, J. H. (1966). Schedule-induced polydipsia in the monkey. *Psychol. Rep.*, **19**, 823–828.

Schuster, R. H. (1969). A functional analysis of conditioned reinforcement. In D. P. Hendry (ed.), *Conditioned Reinforcement*, pp. 192–234. Homewood, Illinois, The Dorsey Press.

Schuster, R. & Rachlin, H. (1968). Indifference between punishment and free shock: Evidence for the negative law of effect. *J. exp. Anal. Behav.*, **11**, 777–786.

Schwartz, B. (1977). Studies of operant and reflexive key pecks in the pigeon. *J. exp. Anal. Behav.*, **27**, 301–313.

Schwartz, B. & Coulter, G. (1973). A failure to transfer control of keypecking from food reinforcement to escape from and avoidance of shock. *Bull. Psychon. Soc.*, **1**, 307–309.

Schwartz, B. & Williams, D. R. (1971). Discrete-trials spaced responding in the pigeon: the dependence of efficient performance on the availability of a stimulus for collateral pecking. *J. exp. Anal. Behav.*, **16**, 155–160.

Schwartz, B. & Williams, D. R. (1972). The role of the response-reinforcer contingency in negative auto-maintenance. *J. exp. Anal. Behav.*, **21**, 351–357.

Schwitzgebel, R. & Schwitzgebel, R. (1961). Reduction of adolescent crime by a research method. *J. Soc. Ther.*, **7**, 212–215.

Scobie, S. R. (1972). Interaction of an aversive Pavlovian conditional stimulus and aversively and appetitively motivated operants in rats. *J. comp. physiol. Psychol.*, **79**, 171–188.

Scull, J. W. (1973). The Amsel frustration effect: interpretations and research. *Psychol. Bull.*, **79**, 352–361.

Sebeok, T. A. & Ramsay, A. (1969) (eds). *Approaches to Animal Communication.* The Hague, Mouton.

Segal, E. F. (1965). The development of water drinking on a dry-food free-reinforcement schedule. *Psychon. Sci.*, **2**, 29–30.

Segal, E. F. (1972). Induction and the provenance of operants. In R. M. Gilbert & J. R. Millenson (eds), *Reinforcement: Behavioural Analyses*, pp. 1–34. New York, Academic Press.

Segal, E. F. & Deadwyler, S. A. (1965). Determinants of polydipsia: VI. Taste of the drinking solution on DRL. *Psychon. Sci.*, 101–102.

Segal, E. F. & Holloway, S. M. (1963). Timing behaviour in rats with water drinking as a mediator. *Science*, **140**, 888–889.

Segal, E. F., Oden, D. L. & Deadwyler, S. A. (1965). Determinants of polydipsia: IV. Free-reinforcement schedules. *Psychon. Sci.*, **3**, 11–12.

Seligman, M. E. P. (1970). On the generality of the laws of learning. *Psychol. Rev.*, **77**, 406–418.

Seligman, M. E. P. (1976). *Helplessness.* San Francisco, Freeman & Co.

Seligman, M. E. P. & Beagley, G. (1975). Learned helplessness in the rat. *J. comp. physiol. Psychol.*, **88**, 534–541.

Seligman, M. E. P. & Campbell, B. A. (1965). Effect of intensity and duration of punishment on extinction of an avoidance response. *J. comp. physiol. Psychol.*, **59**, 295–297.

Seligman, M. E. P. & Maier, S. F. (1967). Failure to escape from traumatic shock. *J. exp. Psychol.*, **74**, 1–9.

Seltzer, R. J. (1968). Effects of reinforcement and deprivation on the development of non-nutritive sucking in monkeys and humans. Unpublished Doctoral dissertation, Brown University,

Serota, R. G., Roberts, R. B. & Flexner, L. B. (1972). Acetoxycycloheximide-induced transient amnesia: protective effects of adrenergic stimulants. *Proc. natn. Acad. Sci.*, **69**, 340–342.

Sevenster, P. (1968). Motivation and learning in sticklebacks. In D. Ingle (ed.), *The Central Nervous System and Fish Behaviour*, pp. 233–245. Chicago, University of Chicago Press.

Seward, J. P., Pereboom, A. C., Butler, B. & Jones, R. B. (1957). The role of prefeeding in an apparent frustration effect. *J. exp. Psychol.*, **54**, 445–450.

Sgro, J. A., Dyal, J. A. & Anastasio, E. J. (1967). Effects of constant delay of reinforcement on acquisition asymptote and resistance to extinction. *J. exp. Psychol.*, **73**, 634–636.

Shanab, M. E. & Peterson, J. L. (1969). Polydipsia in the pigeon. *Psychon. Sci.*, **15**, 51–52.

Shapiro, D. (1976). Biofeedback and the regulation of complex psychological processes. In *Biofeedback and Behaviour: a NATO Symposium.*

Shapiro, D., Crider, A. B. & Tursky, B. (1964). Differentiation of an autonomic response through operant reinforcement. *Psychon. Sci.*, **1**, 147–148.

Shapiro, D., Tursky, B. & Schwartz, G. (1970). Control of blood pressure in man by operant conditioning. *Circul. Res.*, **26**, and **27**, 1.

Shasoua, V. E. (1973). Seasonal changes in the learning and activity patterns of goldfish. *Science*, **181**, 572–574.

Sheffield, F. D. (1965). Relation between classical conditioning and instrumental learning. In W. F. Prokasy (ed.), *Classical Conditioning*, pp. 302–322. New York, Appleton–Century–Crofts.

Shepp, B. E. (1962). Some cue properties of anticipated rewards in discrimination learning of retardates. *J. comp. physiol. Psychol.*, **55**, 856–859.

Shettleworth, S. J. (1972a). Constraints on learning. *Advances in the study of behaviour*, **4**, 1–68. New York, Academic Press.

Shettleworth, S. J. (1972b). Stimulus relevance in the control of drinking and conditioned fear responses in domestic chicks, (*Gallus gallus*), *J. comp. physiol. Psychol.*, **80**, 175–198.

Shettleworth, S. J. (1973). Food reinforcement and the organization of behaviour in golden hamsters. In R. A. Hinde & J. Stevenson-Hinde, *Constraints on Learning*, pp. 243–263. London, Academic Press.

Shettleworth, S. J. (1975). Reinforcement and the organization of behaviour in golden hamsters: Hunger, environment and food reinforcement. *J. exp. Psychol. Anim. Behav. Proc.*, **104**, 56–87.

Shimp, C. P. (1966). Probabilistically reinforced choice behaviour in pigeons. *J. exp. Anal. Behav.*, **9**, 443–455.

Shimp, C. P. (1967). The reinforcement of short interresponse times. *J. exp. Anal. Behav.*, **10**, 425–434.

Shimp, C. P. & Moffitt, M. (1974). Short-term memory in the pigeon: stimulus-response associations. *J. exp. Anal. Behav.*, **12**, 745–757.

Shull, R. L. & Pliskoff, S. S. (1967). Changeover delay and concurrent schedules: Some effects on relative performance measures. *J. exp. Anal. Behav.*, **10**, 517–527.

Shusterman, R. J. (1963). The use of strategies in two-choice behaviour of children and chimpanzees. *J. comp. physiol. Psychol.*, **56**, 96–100.

Sidman, M. (1953). Avoidance conditioning with brief shock and no exteroceptive warning signal. *Science*, **118**, 157–158.

Sidman, M. (1954). The temporal distribution of avoidance responses. *J. comp. physiol. Psychol.*, **47**, 399–402.

Sidman, M. (1960a). Normal sources of pathological behaviour. *Science*, **132**, 61–68.

Sidman, M. (1960b). *Tactics of Scientific Research: Evaluating Experimental Data in Psychology.* New York, Basic Books.

Sidman, M. (1962a). Operant techniques. In A. J. Bachrach (ed.), *Experimental Foundations of Clinical Psychology*, pp. 170–210. New York, Basic Books.

Sidman, M. (1962b). Reduction of shock frequency as reinforcement for avoidance behaviour. *J. exp. Anal. Behav.*, **5**, 247–257.

Sidman, M. (1966). Avoidance behaviour. In W. K. Honig (ed.), *Operant Behaviour:*

*Areas of Research and Application*, pp. 448–498. New York, Appleton–Century–Crofts.

Sidman, M. & Fletcher, F. G. (1968). A demonstration of autoshaping with monkeys. *J. exp. Anal. Behav.*, **11**, 367–369.

Sidman, M. E., Herrnstein, R. J. & Conrad, D. G. (1957). Maintenance of avoidance behaviour by unavoidable shocks. *J. comp. physiol. Psychol.*, **50**, 553–557.

Siegel, A. I. (1953). Deprivation of visual form definition in the ring dove: II. Perceptual-motor transfer. *J. comp. physiol. Psychol.*, **46**, 249–252.

Siegel, S. & Domjan, M. (1971). Backward conditioning as an inhibitory procedure. *Learn. Motiv.*, **2**, 1–11.

Simmons, M. W. (1964). Operant discrimination learning in human infants. *Child Dev.*, **35**, 737–748.

Simmons, M. W. & Lipsitt, L. P. (1961). An operant-discrimination apparatus for infants. *J. exp. Anal. Behav.*, **4**, 233–235.

Siqueland, E. R. (1964). Operant conditioning of head turning in four month infants. *Psychon. Sci.*, **1**, 223–224.

Siqueland, E. R. & Lipsitt, L. P. (1966). Conditioned head-turning in human newborns. *J. exp. Child Psychol.*, **3**, 356–376.

Skinner, B. F. (1938). *The Behaviour of Organisms*. New York, Appleton–Century–Crofts.

Skinner, B. F. (1945). The operational analysis of psychological terms. *Psychol. Rev.*, **52**, 270–277.

Skinner, B. F. (1948). "Superstition" in the pigeon. *J. exp. Psychol.*, **38**, 168–172.

Skinner, B. F. (1950). Are theories of learning necessary? *Psychol. Rev.*, **57**, 193–216.

Skinner, B. F. (1953). *Science and Human Behaviour*. New York, Macmillan.

Skinner, B. F. (1957a). The experimental analysis of behaviour. *Am. Scient.*, **45**, 343–371.

Skinner, B. F. (1957b). *Verbal Behaviour*. New York, Appleton–Century–Crofts.

Skinner, B. F. (1963). Behaviourism at Fifty. *Science*, **140**, 951–958.

Skinner, B. F. (1966). Operant behaviour. In W. K. Honig (ed.), *Operant Behaviour: Areas of Research and Application*, pp. 12–32. New York, Appleton–Century–Crofts.

Skinner, B. F. (1969). *Contingencies of Reinforcement: a Theoretical Analysis*. New York, Appleton–Century–Crofts.

Skinner, B. F. (1971). *Beyond Freedom and Dignity*. New York, Alfred A. Knopf.

Skinner, B. F. (1972). *Cumulative Record (third edition)*. New York, Appleton–Century–Crofts.

Skinner, B. F. (1974). *About Behaviourism*. London, Jonathan Cape.

Skinner, B. F. & Morse, C. W. (1958). Fixed-interval reinforcement of running in a wheel. *J. exp. Anal. Behav.*, **1**, 371–379.

Slack, C. W. (1960). Experimenter–subject psychotherapy: a new method of introducing intensive office treatment for unreachable cases. *Ment. Hyg.*, **44**, 238–256.

Slivka, R. M. & Bitterman, M. E. (1966). Classical appetitive conditioning in the pigeon: partial reinforcement. *Psychon. Sci.*, **4**, 181–182.

Sloane, H. N., Johnston, M. K. & Harris, F. R. (1968). Remedial procedures for teaching verbal behaviour to speech deficient or defective young children. In H. N. Sloane & B. D. MacAulay (eds), *Operant Procedures in Remedial Speech and Language Training*. Boston, Houghton Mifflin.

Sloane, R. B. (1975). *Psychotherapy versus behaviour therapy*. New York, Harvard University Press.

Sluckin, W. (1964). *Imprinting and Early Learning*. London, Methuen.

Sluckin, W. (1973). *Imprinting and Early Learning*. Chicago, Aldine.

Sluckin, W. & Salzen, E. A. (1961). Imprinting and perceptual learning. *Q. Jl exp. Psychol.*, **13**, 65–77.

Smith, F. V. (1965). Instinct and learning in the attachment of lamb and ewe. *Anim. Behav.*, **13**, 84–86.

Smith, F. V. & Nott, K. H. (1970). The "critical period" in relation to the strength of the stimulus. *Z. Tierpsychol.*, **27**, 108–115.

Smith, J. B. (1974). Effects of response rate reinforcement frequency and the duration of a stimulus preceding response-independent food. *J. exp. Anal. Behav.*, **21**, 215–221.

Smith, J. B. & Clark, F. C. (1974). Intercurrent and reinforced behaviour under multiple spaced-responding schedules. *J. exp. Anal. Behav.*, **21**, 445–454.

Smith, L. T. (1973). Pan troglodytes: usurper or companion? *Bio. Psychol. Bull.*, **3**, 30–41.

Smith, M. C. (1968). CS-US interval and US intensity in classical conditioning of the rabbits nictitating membrane response. *J. comp. physiol. Psychol.*, **66**, 678–687.

Smith, O. A., McFarland, W. L. & Taylor, E. (1961). Performance in a shock-avoidance conditioning situation interpreted as pseudoconditioning. *J. comp. physiol. Psychol.*, **54**, 154–157.

Smith, R. F. (1967). Behavioural events other than key striking which are counted as responses during pigeon pecking. Doctoral dissertation, Indiana University.

Smith, R. F. (1974). Topography of the food-reinforced key-peck and the source of 30-millisecond interresponse times. *J. exp. Anal. Behav.*, **21**, 541–551.

Solomon, R. L. (1964). Punishment. *Am. Psychol.*, **19**, 239–253.

Solomon, R. L., Kamin, L. J & Wynne, L. C. (1953). Traumatic avoidance learning: the outcomes of several extinction procedures with dogs. *J. abnorm. soc. Psychol.*, **48**, 291–302.

Solomon, R. L. & Turner, L. H. (1962). Discriminative classical conditioning in dogs paralysed by curare can later control discriminative avoidance responses in the normal state. *Psychol. Rev.*, **69**, 202–219.

Soltysik, S. (1963). Inhibitory feedback in avoidance conditioning. *Boln Inst. Estud. Med. Biol.*, **21**, 433–449.

Soltysik, S. & Jaworska, K. (1962). Studies on the aversive classical conditioning. 2. On the reinforcing role of shock in the classical leg flexion conditioning. *Acta Biol. Exp.*, **22**, 181–191.

Spear, N. E. (1967). Retention of reinforcer magnitude. *Psychol. Rev.*, **74**, 216–234.

Spear, N. E. (1971). Forgetting as retrieval failure. In W. K. Honig & P. H. R. James (eds), *Anim. Memory*. New York, Academic Press.

Spear, N. E. (1973). Retrieval of memory in animals. *Psychol. Rev.*, **80**, 163–194.

Spear, N. E., Gordon, W. C. & Martin, P. A. (1973). Warm-up decrement as failure in memory retrieval in the rat. *J. comp. physiol. Psychol.*, **85**, 601–614.

Spear, N. E. & Parsons, P. J. (1976). Analysis of a reactivation treatment: ontogenetic determinants of alleviated forgetting. In D. L. Medin, W. A. Roberts & R. T. Davis (eds), *Processes of Animal Memory*, pp. 135–165. New York, Lawrence Erlbaum Ass.

Spelt, D. K. (1948). The conditioning of the human fetus in utero. *J. exp. Psychol.*, **38**, 375–376.

Spence, K. W. (1937). Experimental studies of learning and the higher mental processes in infra-human primates. *Psychol. Bull.*, **34**, 806–850.

Spence, K. W. (1951). Theoretical interpretations of learning. In C. P. Stone (ed.), *Comparative Psychology*, pp. 239–291. Englewood Cliffs, New Jersey, Prentice-Hall.

Spence, K. W. (1956). *Behaviour Theory and Conditioning*. New Haven, Yale University Press.

Sperling, S. E., Perkins, M. E. & Duncan, H. J. (1977). Stimulus generalization from feeder to response key in the acquisition of autoshaped pecking. *J. exp. Anal. Behav.*, **27**, 469–478.

Spiker, C. C. (1959). Performance on a difficult discrimination following pretraining with distinctive stimuli. *Child Dev.*, **30**, 513–521.

Spradlin, J. & Girardeau, F. (1966). The behaviour of moderately and severely retarded persons. In N. Ellis (ed.), *International Review of Research in Mental Retardation*, Vol. 1, pp. 132–168. New York: Academic Press.

Squier, L. H. (1969). Autoshaping key responses with fish. *Psychon. Sci.*, **17**, 177–178.

Staddon, J. E. R. (1965). Some properties of spaced responding in pigeons. *J. exp. Anal. Behav.*, **8**, 19–28.

Staddon, J. E. R. (1970). Temporal effects of reinforcement: A negative "frustration" effect. *Learn. Motiv.*, **1**, 227–247.

Staddon, J. E. R. (1972). Temporal control and the theory of reinforcement schedules. In R. M. Gilbert & J. R. Millenson (eds), *Reinforcement: Behavioural Analyses*, pp. 212–263. New York, Academic Press.

Staddon, J. E. R. (1974). Temporal control, attention and memory. *Psychol. Rev.*, **81**, 355–391.

Staddon, J. E. R. (1977). Schedule-induced behaviour. In W. K. Honig & J. E. R. Staddon (eds), *Handbook of Operant Behaviour*. Englewood Cliffs, New Jersey, Prentice-Hall.

Staddon, J. E. R. & Ayres, S. L. (1975). Sequential and temporal properties of behaviour induced by a schedule of periodic food delivery. *Behaviour*, **54**, 26–49.

Staddon, J. E. R. & Innis, N. K. (1966). An effect analogous to "frustration" as interval reinforcement schedules. *Psychon. Sci.*, **4**, 287–288.

Staddon, J. E. R. & Innis, N. K. (1969). Reinforcement omission on fixed-interval schedules. *J. exp. Anal. Behav.*, **12**, 689–700.

Staddon, J. E. R. & Simmelhag, V. L. (1971). The "superstition" experiment: A reexamination of its implications for the principles of adaptive behaviour. *Psychol. Rev.*, **78**, 3–43.

Stampfl, T. G. (1967). Implosive therapy. Part I. The theory. In S. G. Armitage (ed.), *Behaviour Modification Techniques in the Treatment of Emotional Disorders*. Battle Creek, Mich., V. A. Publication.

Stampfl, T. G. & Levis, D. J. (1967). Essentials of implosive therapy: A learning-theory-based psychodynamic behavioural therapy. *J. abnorm. Psychol.*, **72**, 496–503.

Stark, J., Giddan, J. J. & Meisel, J. (1968). Increasing verbal behaviour in the autistic child. *J. Speech and Hear. Disorders*, **33**, 42–48.

Starr, B. & Staddon, J. E. R. (1974). Temporal control of periodic schedules: signal properties of reinforcement and blackout. *J. exp. Anal. Behav.*, **22**, 535–545.

Stebbins, W. C., Mead, P. B. & Martin, J. M. (1959). The relation of amount of reinforcement to performance under a fixed-interval schedule. *J. exp. Anal. Behav.*, **2**, 351–356.

Stein, L. (1964). Excessive drinking in the rat: superstition or thirst? *J. comp. physiol. Psychol.*, **58**, 237–242.

Steiner, J. (1967). Observing responses and uncertainty reduction. *Q. Jl exp. Psychol.*, **19**, 18–29.

Steiner, J. (1968). Positive reinforcement. In L. Weiskrantz (ed.), *Analysis of Behavioural Change*, pp. 4–18. New York, Harper.

Stevenson, H. W. & Hoving, K. L. (1964). Probability learning as a function of age and incentive. *J. exp. Child Psychol.*, **1**, 64–70.

Stevenson, H. W. & Weir, M. W. (1959). Variables affecting children's performance in a probability learning task. *J. exp. Psychol.*, **57**, 403–412.

Storms, L. H., Boraczi, G. & Broen, W. E. (1962). Punishment inhibits an instrumental response in hooded rats. *Science*, **135**, 1133–1134.

Stricher, E. M. & Adair, E. R. (1966). Body fluid balance, taste and postprandial factors in schedule-induced polydipsia. *J. comp. physiol. Psychol.*, **62**, 449–454.

Stroebel, C. F. (1969). Biologic rhythm correlates of disturbed behaviour in the rhesus monkey. *Bibliotheca Primatologica*, **9**, 91–105.

Stuart, R. B. (1967). Behavioural control over eating. *Behav. Res. Ther.*, **5**, 357–365.

Stubbs, D. A. (1971). Second-order schedules and the problem of conditioned reinforcement. *J. exp. Anal. Behav.*, **16**, 289–313.

Suchman, R. G. & Trabasso, T. (1966). Color and form preferences in young children. *J. exp. Child Psychol.*, **3**, 177–187.

Sutherland, N. S. (1957). Visual discrimination of orientation by octopus. *Br. J. Psychol.*, **48**, 55–71.

Sutherland, N. S. & Holgate, V. (1966). Two cue discrimination learning in rats. *J. comp. physiol. Psychol.*, **61**, 198–207.

Tacher, R. S. & Way, J. (1968). Motivational properties of nonreward. *Psychon. Sci.*, **10**, 103–104.

Tachibana, T., Yamaguchi, M. & Haruki, Y. (1974). Some factors in imitative behaviour in cats. *A. Anim. Psychol.*, **23**, 61–68.

Tait, R. W., Marquis, H. A., Williams, R., Weinstein, L. & Suboski, M. D. (1969). Extinction of sensory preconditioning using CER training. *J. comp. physiol. Psychol.*, **69**, 170–172.

Taub, S. E. & Berman, A. J. (1969). Movement and learning in the absence of sensory feedback. In S. J. Freedman (ed.), *The Neuropsychology of Spatially Oriented Behaviour*, pp. 173–192. Homewood, Illinois, The Dorsey Press.

Taus, S. N. & Hearst, E. (1970). Effects of intertrial (blackout) duration on response rate to a positive stimulus. *Psychon. Sci.*, **19**, 265–266.

Teitelbaum, P. (1966). The use of operant methods in the assessment and control of motivational states. In W. K. Honig (ed.), *Operant Behaviours: Areas of Research and Application*, pp. 565–608. New York, Appleton–Century–Crofts.

Telegdy, G. A. & Cohen, J. S. (1971). Cue utilization and drive level in albino rats. *J. comp. physiol. Psychol.*, **75**, 248–253.

Terman, M. (1970). Discrimination of auditory intensity by rats. *J. exp. Anal. Behav.*, **13**, 145–162.

Terrace, H. S. (1963). Discrimination learning with and without "errors". *J. exp. Anal. Behav.*, **6**, 1–27.

Terrace, H. S. (1966a). Stimulus control. In W. K. Honig (ed.), *Operant Behaviour: Areas of Research and Application*, pp. 271–344. New York, Appleton–Century–Crofts.

Terrace, H. S. (1966b). Discrimination learning and inhibition. *Science*, **154**, 1677–1680.

Terrace, H. S. (1968). Discrimination learning, the peak shift and behavioural contrast. *J. exp. Anal. Behav.*, **11**, 727–741.

Terrace, H. S., Gibbon, J., Farrell, L. & Baldock, M. D. (1975). Temporal factors influencing the acquisition of an autoshaped key peck. *Anim. Learn. Behav.*, **3**, 53–62.

Thach, J. S. (1965). Comparison of Social and Nonsocial reinforcing stimuli. *Proc. 73rd Am. Psychol. Ass.*, Washington D.C., American Psychological Association.

Theios, J. (1963). Simple conditioning as two-stage all-or-none learning. *Psychol. Rev.*, **70**, 403–417.

Thielcke-Poltz, H. & Thielcke, G. (1960). Akustisches lernen vershieden alter schallisolierten amseln *Turdus merula L.*, und die entwicklung erlernter motive ohne und mit künstlichem einfluss von festosteron. *Z. Tierpsychol.*, **17**, 211–244.

Thomas, D. R. & Switalski, R. W. (1966). Comparison of stimulus generalization following variable-ratio and variable-interval training. *J. exp. Psychol.*, **71**, 236–240.

Thompson, D. M. (1965). Punishment by $S^D$ associated with fixed-ratio reinforcement. *J. exp. Anal. Behav.*, **8**, 189–194.

Thompson, R. & McConnell, J. (1955). Classical conditioning in the planarian, *Dugesia dorotocephala*. *J. comp. physiol. Psychol.*, **48**, 65–68.

Thompson, T. & Bloom, W. (1966). Aggressive behaviour and extinction-induced response-rate increase. *Psychon. Sci.*, **5**, 335–336.

Thompson, T. & Schuster, C. R. (1968). *Behavioural Pharmacology*. Englewood Cliffs, New Jersey, Prentice-Hall.

Thoresen, C. E. & Mahoney, M. J. (1974). *Behavioural Self-Control*. New York, Holt, Rinehart & Winston.

Thorndike, E. L. (1911). *Animal Intelligence: Experimental Studies*. New York, MacMillan.

Thorndike, E. L. (1931). *Human Learning*. New York, Appleton–Century–Crofts.

Thorpe, J. G., Schmidt, E. & Castell, D. A. (1963). A comparison of positive and negative (aversive) conditioning in the treatment of homosexuality. *Behav. Res. Ther.*, **1**, 357–362.

Thorpe, W. H. (1944). Type of learning in insects and other arthropods. Part III. *Br. J. Psychol.*, **34**, 66–76.

Thorpe, W. H. (1945). The evolutionary significance of habitat selection. *J. Anim. Ecol.*, **14**, 67–70.

Thorpe, W. H. (1956). *Learning and Instinct in Animals*. London, Methuen.

Thorpe, W. H. (1961). *Bird Song: the Biology of Vocal Communication and Expression in Birds*. Cambridge, England, Cambridge University Press.

Thorpe, W. H. (1974). *Animal Nature and Human Nature*. Garden City, New York, Doubleday.

Thurmond, J. B., Binford, J. R. & Loeb, M. (1970). Effects of signal-to-noise variability over repeated sessions in an auditory vigilance task. *Percept. Psychophys.*, **7**, 100–102.

Tinbergen, N. (1952). "Derived" activities: their causation, biological significance, origin and emancipation during evolution. *Q. Rev. Biol.*, **27**, 1–32.

Tinklepaugh, O. L. (1928). An experimental study of representative factors in monkeys. *J. comp. Psychol.*, **8**, 197–236.

Trapold, M. A., Carlson, J. G. & Myers, W. A. (1965). The effect of noncontingent fixed- and variable-interval reinforcement upon subsequent acquisition of the fixed-interval scallop. *Psychon. Sci.*, **2**, 261–262.

Traupmann, K. L. (1971). Acquisition and extinction of an instrumental running response with single- or multiple-pellet reward. *Psychon. Sci.*, **22**, 61–63.

Tretter, R., Cyander, M. S. & Singer, W. (1975). Modification of direction selectivity of neurones in the visual cortex of kittens. *Brain Res.*, **84**, 143–149.

Turner, E. R. A. (1964). Social feeding in birds. *Behavior*, **24**, 1–47.

Turner, L. H. & Solomon, R. L. (1962). Human traumatic avoidance learning. *Psychol. Monogr.*, **76**, (40, whole No. 559).

Tyler, V. O. & Brown, G. D. (1967). The use of swift, brief isolation as a group control device for institutionalized delinquents. *Behav. Res. Ther.*, **5**, 1–9.

Ullman, L. P. & Krasner, L. (eds) (1965). *Case Studies in Behaviour Modification*. New York, Holt, Rinehart & Winston.

Ulrich, R. E. & Azrin, N. H. (1962). Reflexive fighting in response to aversive stimulation. *J. exp. Anal. Behav.*, **5**, 511–520.

Ulrich, R. E., Johnston, M., Richardson, J. & Wolff, P. (1963). The operant conditioning of fighting behaviour in rats. *Psychol. Rec.*, **13**, 465–470.

Ulrich, R. E., Wolf, P. C. & Azrin, N. H. (1964). Shock as an elicitor of intra- and inter-species fighting behaviour. *Anim. Behav.*, **12**, 14–15.

Vakhrameeva, I. A. (1964). Characteristics of the formation and development of bilateral conditioned movement reflexes in young children. In N. I. Kasatkin (ed.), *From the Simple to the Complex*, pp. 115–126. Leningrad, Izdatel'stvo "Nauka".

Valentine, C. W. (1914). The colour perception and colour preferences of an infant during its fourth and eighth month. *Br. J. Psychol.*, **6**, 363–386.

Vanderwolf, C. H. (1969). Hippocampal electrical activity and voluntary movement in the rat. *Electroenceph. Clin. Neurophysiol.*, **26**, 407–418.

Vanderwolf, C. H. (1971). Limbic-diencephalic mechanisms of voluntary movement. *Psychol. Rev.*, **78**, 83–113.

Van Hemel, P. E. (1970). Retrieving as a reinforcer in nulliparous mice. Unpublished Doctoral dissertation. The John Hopkins University.

Van Hemel, P. E. (1972). Aggression as a reinforcer: operant behaviour in the mouse-killing rat. *J. exp. Anal. Behav.*, **17**, 237–245.

Van Houten, R., O'Leary, K. D. & Weiss, S. J. (1970). Summation of conditioned suppression. *J. exp. Anal. Behav.*, **13**, 75–81.

Van Lawick-Goodall, J. (1968). Behaviour of free-living chimpanzees of the Gombe Stream area. *Anim. Behav. Monogr.*, **1**, 165–311.

Van Sluyters, R. C. & Blakemore, C. (1973). Experimental creation of unusual neuronal properties in visual cortex of kitten. *Nature*, **246**, 506–508.

Vatsuro, E. G. (1948). *Study of the Higher Nervous Activity of Anthropoids (Chimpanzees)*. Moscow, Medgiz.

Verplanck, W. S. (1956). The operant conditioning of human motor behaviour. *Psychol. Bull.*, **53**, 70–83.

Villarreal, J. (1967). Schedule-induced pica. Paper presented to the Eastern Psychol. Ass., Boston.

Voegtlin, W. L. & Lemere, F. (1942). The treatment of alcohol addiction: A review of the literature. *Q. J. Stud. Alcohol*, **2**, 717–803.

Von Frisch, K. (1967). *The Dance Language and Orientation of Bees*. Cambridge, Harvard University Press.

Von Frisch, K. (1972). *Bees, their Vision, Chemical Senses and Language*. Ithaca, Cornell Univ. Press.

Von Frisch, K. (1974). Decoding the language of the bee. *Science*, **185**, 663–668.

Wagner, A. R. (1959). The role of reinforcement and nonreinforcement in an "apparent frustration effect". *J. exp. Psychol.*, **57**, 130–136.

Wagner, A. R. (1969). Stimulus validity and stimulus selection in associative learning. In N. J. Mackintosh & W. K. Honig (eds), *Fundamental Issues in Associative Learning*, pp. 90–122. Halifax, Dalhouse Univ. Press.

Wagner, A. R. & Rescorla, R. A. (1972). Inhibition in Pavlovian conditioning: application of a theory. In R. A. Boakes & M. S. Halliday (eds), *Inhibition and Learning*, pp. 301–336. London, Academic Press.

Wagner, M. K. & Cauthen, N. R. (1968). A comparison of reciprocal inhibition and operant conditioning in the systematic desensitization of a fear of snakes. *Behav. Res. Ther.*, **6**, 225–227.

Wahler, R. G. (1967). Infant social attachments: A reinforcement theory interpretation and investigation. *Child Dev.*, **4**, 1079–1088.

Wahlsten, D. L. & Cole, M. (1972). Classical and avoidance training of leg flexion in the dog. In A. H. Black & W. F. Prokasy (eds), *Classical Conditioning. II: Current Research and Theory*. New York, Appleton–Century–Crofts.

Walk, R. D. (1965). The study of visual depth and distance perception in animals. *Adv. stud. Behav.*, **1**.

Walk, R. D. & Gibson, E. J. (1961). A comparative and analytical study of visual depth perception. *Psychol. Monogr.*, **75**, (15, whole No. 519).

Wall, M. (1965). Discrete-trials analysis of fixed-interval discrimination. *J. comp. physiol. Psychol.*, **60**, 70–75.

Wallace, R. F. & Mulder, D. W. (1973). Fixed-ratio responding with human subjects. *Bull. Psychon. Soc.*, **1**, 359–362.

Waller, T. G. (1968). Effects of magnitude of reward in spatial and brightness discrimination tasks. *J. comp. physiol. Psychol.*, **66**, 122–127.

Walters, G. C. & Glazer, R. D. (1971). Punishment of instinctive behaviour in the Mongolian gerbil. *J. comp. physiol. Psychol.*, **75**, 331–340.

Wann, T. W. (1964). *Behaviourism and Phenomenology.* Chicago, Univ. of Chicago Press.

Warden, C. J., Fjeld, H. A. & Koch, A. M. (1940). Imitative behaviour in cebus and rhesus monkeys. *J. genet. Psychol.*, **56**, 311–322.

Warner, L. H. (1932). The association span of the white rat. *J. genet. Psychol.*, **41**, 57–90.

Wasik, B. H., Senn, K., Welch, R. H. & Cooper, B. R. (1969). Behaviour modification with culturally deprived school children: two case studies. *J. app. Behav. Anal.*, **2**, 181–194.

Wasserman, E. A. (1973). Pavlovian conditioning with heat reinforcement produces stimulus-directed pecking in chicks. *Science*, **181**, 875–877.

Watson, J. B. (1913). Psychology as the behaviorist views it. *Psychol. Rev.*, **20**, 158–177.

Watson, J. B. (1914). *Behaviour, an Introduction to Comparative Psychology.* New York, Holt.

Watson, J. B. (1919). *Psychology from the Standpoint of a Behaviourist.* Philadelphia, Lippincott.

Watson, J. B. (1920). Is thinking merely the action of language mechanisms? *Br. J. Psychol.*, **11**, 87–104.

Watson, J. B. (1924). The unverbalized in human behaviour. *Psychol. Rev.*, **31**, 273–280.

Watson, J. B. (1925). *Behaviourism.* New York, Norton.

Watson, J. B. & Rayner, R. (1920). Conditioned emotional reactions. *J. exp. Psychol.*, **3**, 1–14.

Weidmann, U. (1956). Some experiments on the following and the flocking reaction of mallard ducklings. *Br. J. Anim. Behav.*, **4**, 78–79.

Weiner, H. (1962). Some effects of response cost upon human operant behaviour. *J. exp. Anal. Behav.*, **5**, 201–208.

Weiner, H. (1964). Conditioning history and human fixed-interval performance. *J. exp. Anal. Behav.*, **7**, 383–385.

Weiner, H. (1965). Conditioning history and maladaptive human operant performance. *Psychol. Rep.*, **17**, 935–942.

Weiner, H. (1969). Controlling human fixed-interval performance. *J. exp. Anal. Behav.*, **12**, 349–373.

Weiner, H. (1972). Controlling human fixed-interval performance with fixed-ratio responding or differential reinforcement of low-rate responding in mixed schedules. *Psychon. Sci.*, **26**, 191–192.

Weisberg, P. & Fink, E. (1966). Fixed ratio and extinction performance of infants in the second year of life. *J. exp. Anal. Behav.*, **9**, 105–109.

Weisinger, R. S., Parker, L. F. & Bolles, R. C. (1973). Effects of amount of reward on acquisition of a black-white discrimination. *Bull. Psychon. Soc.*, **2**, 27–28.

Weisman, R. G. (1969). Some determinants of inhibitory stimulus control. *J. exp. Anal. Behav.*, **12**, 443–450.

Weiss, B. & Laties, V. G. (1961). Changes in pain tolerance and other behaviour produced by salicylates. *J. Pharmac. exp. Ther.*, **131**, 120–129.

Weiss, S. J. (1964). Summation of response strengths instrumentally conditioned to stimuli in different sensory modalities. *J. exp. Psychol.*, **68**, 151–155.

Weiss, S. J. (1967). Free-operant compounding of variable-interval and low-rate discrimination stimuli. *J. exp. Anal. Behav.*, **10**, 535–540.

Weiss, S. J. (1969). Response distributions during free-operant compounding of high- and low-rate discriminative stimuli. *Proc. 77th Ann. Conf. Am. Psychol. Ass.*, **4**, 827–828.

Weiss, S. J. (1971). Discrimination training and stimulus compounding consideration of non-reinforcement and response differentiation consequences of S. *J. exp. Anal. Behav.*, **15**, 387–402.

Weiss, S. J. (1972). Stimulus compounding in free-operant and classical conditioning: A review and analysis. *Psychol. Bull.*, **78**, 189–208.

Weissman, A. (1962). Nondiscriminated avoidance behaviour in a large sample of rats. *Psychol. Rep.*, **10**, 591–600.

Welgan, P. R. (1974). Learned control of gastric acid secretions in ulcer patients. *Psychosom. Med.*, **36**, 411–419.

Welker, R. L., Hansen, G., Engberg, L. A. & Thomas, D. R. (1973). A reply to Gamzu *et al.*, *Science*, **181**, 368–369.

Wells, M. J. (1961). Weight discrimination by *Octopus*. *J. exp. Biol.*, **38**, 127–133.

Wells, M. J. (1968). *Lower Animals*. London, Weidenfeld and Nicolson.

Wenger, M. A. (1936). An investigation of conditioned responses in human infants. *Univ. Iowa Stud. Child Welf.*, **12**, No. 1.

Westbrook, R. F. (1973). Failure to obtain positive contrast when pigeons press a bar. *J. exp. Anal. Behav.*, **20**, 499–510.

Wheeler, H. (1973). *Beyond the Punitive Society*. London, Wildwood House.

Wheeler, L. & Davis, H. (1967). Social disruption of performance on a DRL schedule. *Psychon. Sci.*, **7**, 249–250.

White, K., Juhasz, J. B. & Wilson, P. J. (1973). Is man no more than this? Evaluative bias in interspecies comparison. *J. hist. behav. Sci.*, **9**, 203–212.

White, S. H. (1965). Evidence for a hierarchical arrangement of learning processes. In L. P. Lipsitt & C. C. Spiker (eds), *Advances in Child Development and Behaviour*, Vol. 2, pp. 187–220. New York, Academic Press.

Wickens, D. D. & Wickens, C. D. (1940). A study of conditioning in the neonate. *J. exp. Psychol.*, **26**, 94–102.

Wiesel, T. N. & Hubel, D. H. (1965). Comparison of the effects of unilateral and bilateral eye closure on cortical unit response in kittens. *J. Neurophysiol.*, **28**, 1029–1040.

Wiesel, T. N. & Hubel, D. H. (1974). Ordered arrangement of orientation columns in monkeys lacking visual experience. *J. comp. Neurol.*, **158**, 307–318.

Wilcove, W. G. & Miller, J. C. (1974). CS-UCS Presentations and a lever: human autoshaping. *J. exp. Psychol.*, **103**, 868–877.

Wilcoxon, H. C., Dragoin, W. B. & Kral, P. A. (1969). Differential conditioning to visual and gustatory cues in quail and rat: Illness induced aversion. *Psychon. Sci.*, **17**, 52.

Wilkins, W. (1972). Desensitization: Getting it together with Davison and Wilson, *Psychol. Bull.*, **78**, 32–36.

Willems, E. P. (1974). Behavioural technology and behavioural ecology. *J. app. Anal. Behav.*, **7**, 151–165.

Williams, C. D. (1959). The elimination of tantrum behaviours by extinction procedures. *J. abnorm. soc. Psychol.*, **59**, 269.

Williams, D. R. (1965). Classical conditioning and incentive motivation. In W. F.

Prokasy (ed.), *Classical Conditioning: A Symposium*, pp. 340–357. New York, Appleton–Century–Crofts.

Williams, D. R. (1968). The structure of response rate. *J. exp. Anal. Behav.*, **11**, 251–258.

Williams, D. R. & Williams, H. (1969). Automaintenance in the pigeon: Sustained pecking despite contingent nonreinforcement. *J. exp. Anal. Behav.*, **12**, 511–520.

Wilson, A. & Smith, F. J. (1968). Counter-conditioning therapy using free association: A pilot study. *J. abnorm. Psychol.*, **73**, 474–478.

Wilson, M. P. & Keller, F. J. (1953). On the selective reinforcement of spaced responses. *J. comp. physiol. Psychol.*, **46**, 190–193.

Wilson, P. D. & Riesen, A. H. (1966). Visual deprivation in rhesus monkeys neonatally deprived of patterned light. *J. comp. physiol. Psychol.*, **61**, 87–95.

Wilson, W. A. & Rollin, A. (1959). Two-choice behaviour of rhesus monkeys in a noncontingent situation. *J. exp. Psychol.*, **58**, 174–180.

Wilsoncroft, W. E. (1969). Babies by bar-press: maternal behaviour in the rat. *Behav. Res. Meth. Instrum.*, **1**, 229–230.

Wilton, R. N. & Clements, R. O. (1971). Behavioural contrast as a function of the duration of an immediately preceding period of extinction. *J. exp. Anal. Behav.*, **16**, 425–428.

Wilton, R. N. & Clements, R. O. (1972). A failure to demonstrate behavioural contrast when the S + and S − components of a discrimination schedule are separated by about 23 hours. *Psychon. Sci.*, **28**, 137–139.

Winograd, E. (1965). Escape behaviour under different FR and shock intensities. *J. exp. Anal. Behav.*, **8**, 117–124.

Winograd, E. (1971). Some issues relating animal memory to human memory. In W. K. Honig & P. H. R. James (eds), *Animal Memory*. New York, Academic Press.

Wodinsky, J., Varley, M. A. & Bitterman, M. E. (1953). The solution of oddity-problems by the rat. *Am. J. Psychol.*, **66**, 137–140.

Wolf, M. M. (1963). Some effects of combined $S^D$s. *J. exp. Anal. Behav.*, **6**, 343–347.

Wolfe, J. B. (1934). The effect of delayed reward upon learning in the white rat. *J. comp. Psychol.*, **17**, 1–21.

Wolff, P. H. (1966). The causes, controls and organization of behaviour in the neonate. *Psychol. Issues*, **5**, (1, whole No. 17).

Wolpe, J. (1958). *Psychotherapy by Reciprocal Inhibition*. Stanford, Stanford Univ. Press.

Wolpe, J. (1962). Isolation of a conditioning procedure as the crucial psychotherapeutic factor: a case study. *J. Nerv. Ment. Dis.*, **134**, 316–329.

Wolpe, J. & Lazarus, A. A. (1966). *Behaviour Therapy Techniques: A Guide to the Treatment of Neuroses*. New York, Pergamon Press.

Wolpin, M. & Raines, W. (1966). Visual imagery, expected roles and extinction as possible factors in reducing fear and avoidance behaviour. *Behav. Res. Ther.*, **4**, 25–37.

Worsham, R. W. (1975). Temporal discrimination factors in the delayed matching-to-sample task in monkeys. *Anim. Learn. Behav.*, **3**, 93–97.

Yerkes, R. M. & Yerkes, A. W. (1929). *The Great Apes*. New Haven, Yale.

Zahorik, D. M. & Maier, S. F. (1969). Appetitive conditioning with recovery from thiamine deficiency as the unconditioned stimulus. *Psychon. Sci.*, **17**, 309–310.

Zajonc, R. B. (1969). *Animal Social Psychology*. New York, Wiley.

Zamble, E. (1967). Classical conditioning of excitement anticipatory to food reward. *J. comp. physiol. Psychol.*, **63**, 526–529.

Zamble, E. (1969). Conditioned motivational patterns in instrumental responding of rats. *J. comp. physiol. Psychol.*, **69**, 536–543.

Zbrozyna, A. W. (1958). On the conditioned reflex of the cessation of the act of eating.

I. Establishment of the conditioned cessation reflex. *Acta Biol. Exp.*, **18**, 137–162.

Zeaman, D. (1949). Response latency as a function of the amount of reinforcement. *J. exp. Psychol.*, **39**, 446–483.

Zeaman, D. & House, B. J. (1967). The relation of I.Q. and learning. In R. M. Gague (ed.), *Learning and Individual Differences*, pp. 192–212. Columbus, Ohio, Merrill.

Zeiler, M. D. (1972a). Fixed interval behaviour: effects of percentage reinforcement. *J. exp. Anal. Behav.*, **17**, 177–189.

Zeiler, M. D. (1972b). Superstitious behaviour in children: an experimental analysis. *Adv. Child Dev. Behav.*, **7**, 1–29.

Zener, K. (1937). The significance of behaviour accompanying conditioned salivary secretion for theories of the conditioned response. *Am. J. Psychol.*, **50**, 384–403.

Zentall, T. R. (1973). Memory in the pigeon: retroactive inhibition in a delayed matching task. *Bull. Psychon. Soc.*, **1**, 126–128.

Zentall, T. R. & Hogan, D. E. (1976). Imitation and social facilitation in the pigeon. *Anim. Learn. Behav.*, **4**, 427–430.

Zuriff, G. E. (1969). Collateral responding during differential reinforcement of low rates. *J. exp. Anal. Behav.*, **12**, 971–976.

# Glossary

Procedures and apparatus commonly used in studies of animal learning.

*Active Avoidance*
An avoidance procedure where the subject has to make a designated 'active' response in order to avoid shock (for example, to jump a barrier, run an alleyway). This contrasts with passive avoidance procedures.

*Autoshaping Procedures*
A classical conditioning procedure for developing CS oriented behaviour in animals. See p. 180ff.

*Backward Conditioning*
In classical conditioning, presenting UCS before the CS, see p. 23.

*CER* – see Conditional Suppression

*Chain Schedule*
In operant conditioning, responding under one stimulus produces a stimulus change indicating the operation of a second schedule which terminates with positive reinforcement. For example, chain FR20 : FI 1-min; the 20th response under one stimulus produces a stimulus change, under this second stimulus the animal is reinforced for the first response after one minute. See p. 46.

*Changeover Delay (COD)*
A feature which is often incorporated into a concurrent schedule of reinforcement in order to minimize the chances of superstitiously reinforcing alternation between choices. Typically, a changeover delay provides that no responses can be reinforced within a specified period of time following the changeover from one choice to the other.

*Concurrent Chain Schedule*
In operant conditioning, a procedure for assessing a subject's 'choice' of reinforcement schedules. Two response manipulanda are available to the subject: responses on one of these will produce a stimulus change indicating the operation of a terminal component on that manipulandum. The subject can then only obtain reinforcement on that manipulandum according to the schedule in operation on that manipulandum. See p. 99.

*Concurrent Schedules (conc)*
In operant conditioning, two or more schedules of reinforcement which are arranged independently but are operating simultaneously such that reinforcements are available for both. See p. 99.

*Conditioned Suppression/Conditioned Anxiety/Conditioned Emotional Response (CER)*
A procedure first developed by Estes and Skinner (1941) as an analogue for 'anxiety'. See p. 142ff.

A stimulus (CS) which has been paired with unavoidable shock is presented during operant responding with the result that rate of operant responding is suppressed during the CS. The CS can be paired with shock at the same time as the animal is responding for food (on-the-baseline conditioning) or in a different apparatus to that in which food is acquired (off-the-baseline conditioning).

*Conjunctive Schedules (conj)*
In operant conditioning, a schedule in which two or more contingencies must be met before the reinforcer is delivered. For example, conj. FR10; FI 1-min – the subject gets food for the first response after one minute providing that 10 responses have occurred during that time.

*Continuous Reinforcement (crf)*
In operant conditioning, presenting a reinforcer for every occurrence of the operant response. See p. 43.

*Cumulative Recorder/Cumulative Recording*
A recording device for plotting rate and patterning of operant responses. See p. 40 and figure 2.9.

*Delay Conditioning*
In classical conditioning, a procedure where the CS is presented for some time before the UCS, but CS and UCS offset are simultaneous. See p. 23.

*Delayed Matching to Sample (DMTS)*
A procedure for the study of short-term retention in animals. See p. 286.

*Delayed Response Problems*
A retention test in which one of a number of sites is baited with food in view of the subject. At some later time the subject is returned to the experimental situation and required to identify the baited food site.

*Differential Reinforcement of High Rate (DRH)*
A schedule for developing very high rates of responding. The subject is only reinforced if he emits a certain number of responses in a short specified period of time.

*Differential Reinforcement of Low Rate (DRL)*
A schedule of reinforcement which develops a steady low rate of operant responding. See p. 79.

*Differential Reinforcement of Other Behaviour (DRO)*
A schedule which reinforces behaviour other than some selected response. It is usually used as a procedure for eliminating responding. A DRO 20-s schedule delivers a reinforcer every 20 s if no response has occurred in that time. If a response does occur, the timer is reset to zero and the interval starts again without reinforcement occurring.

*Discrete-Trial Probability Learning*
A choice procedure in which one response is randomly reinforced on a proportion of trials while the alternative is reinforced on the remaining trials. See p. 98.

*Discriminated Avoidance*
A procedure in which a predesignated operant response avoids an aversive event. The aversive event is signalled by an exteroceptive CS. See p. 132.

*DMTS –* see Delayed Matching To Sample

*Double-Alleyway*
An apparatus designed by Abram Amsel for studying the effects of reinforcement omission on rate of subsequent responding. See p. 95.

*DRH* – See Differential Reinforcement of High Rate

*DRL* – see Differential Reinforcement of Low Rate

*DRO* – see Differential Reinforcement of Other Behaviour

*Errorless Discrimination Learning*
A procedure for developing good operant discrimination by minimising errors (that is, responses to S −). See p. 106.

*Escape*
A negative reinforcement procedure in which an operant response terminates an aversive event. See p. 130.

*Extinction*
In classical and operant conditioning, the removal of the contingency between CS and UCS, and between response and consequence. See pp. 26 and 43.

*Fixed-Interval Schedule (FI)*
A schedule of reinforcement which develops a characteristic positively accelerated response rate between reinforcers (the fixed-interval *scallop*). See p. 63.

*Fixed-Ratio Schedule (FR)*
A schedule of reinforcement which develops a characteristic *break-and-run* pattern of operant responding. The subject has to complete a fixed number of responses before being reinforced. See p. 70.

*Fixed-Time Schedule (FT)*
A schedule which delivers some event (usually food or electric shock) at fixed time intervals independently of the subject's behaviour. See p. 87.

*Free-reinforcement* – see Noncontingent Reinforcement

*GO-NOGO Discrimination*
A discrimination procedure in which S + requires the subject to make a defined response in order to acquire reinforcers (go) while responses in the presence of S − go unreinforced (nogo). Responding to S + and not responding to S − indicates discrimination, for example, *mult VI : EXT*.

*Heterogeneous Chained Schedule*
In operant conditioning, a chain schedule where the response is topographically different in each component. See p. 48.

*Higher-Order Conditioning*
In classical conditioning, a $CS_1$ is paired with a given UCS and then subsequently $CS_2$ is paired with $CS_1$. The experimenter then tests for the occurrence of a conditioned response to $CS_2$.

*Homogeneous Chained Schedule*
In operant conditioning a chain schedule where the response is topographically similar in each component. See p. 48.

*Interdimensional Discrimination Training*
A discrimination procedure in which S + and S − are on different dimensions, for example, S + is a tone and S − a light.

*Inter-response Time (IRT)*
The time recorded between successive responses. When collected into a frequency

distribution, inter-response time histograms provide a fine-grain analysis of the temporal distribution of responses during an experimental session, this is a response measure frequently used with DRL schedules (see also p. 81).

*Intradimensional Discrimination Training*
A discrimination procedure in which S + and S − are on the same dimension. For example, S + is a tone of 1000 Hz while S − is a tone of 850 Hz.

*IRT Criterion*
On differential reinforcement of low-rate schedules, this is the minimum inter-response time that can precede reinforcement. For instance, on a DRC 20-s schedule only responses spaced apart by 20 s or more are followed by food.

*Lashley Jumping Stand*
An apparatus for rapidly developing discriminative responding. The subject is placed on a pedestal within jumping distance of two distinctively patterned doors (the S + and S −). If the subject jumps to S + the door opens and he receives food. If the subject jumps to S −, he finds the door locked and he falls some distance into a safety net, thus punishing him for jumping towards S −.

*Latent Learning*
A procedure in which the subject (usually a rat) is pre-exposed to a maze without food or water present. Subsequent learning of the maze with rewards present is usually faster by pre-exposed animals than by naive animals.

*Learned Helplessness Procedure*
A procedure in which the subject is given initial exposure to unavoidable electric shocks; following this the subject is taught to avoid shock. Avoidance Learning usually proceeds more slowly in such animals as compared with non-pretrained control animals. See p. 150.

*Limited-Hold Contingency (LH)*
On DRL schedules, a procedure for sharpening the subject's discrimination of the temporal parameters of the schedule. See p. 82.

*Mixed Schedule (Mix)*
In operant conditioning, reinforcement is programmed by two or more schedules which alternate randomly for periods of time. No exteroceptive stimuli are correlated with the schedules.

*Multiple Schedule (mult)*
In operant conditioning, reinforcement is programmed according to two different schedules which alternate randomly and are signalled by distinctive $S^D$s, for example, mult FI 1-min : FR20.

*Non-contingent Reinforcement/Response-Independent Reinforcement/Free Reinforcement*
A procedure whereby reinforcers are presented independently of the subject's behaviour. See p. 87.

*Non-discriminated Avoidance ('Sidman' Avoidance)*
An avoidance procedure first used by Murray Sidman in which shock is not preceded by an exteroceptive stimulus. Brief shocks are scheduled at regular intervals and responses can delay these shocks for further periods of time. See p. 133.

*Observational Learning*
A procedure for assessing the effect of observing an experienced conspecific demonstrating the required response on learning in a naive subject. See p. 267.

*Omission Training*
In classical conditioning, a control procedure to test for the influence of operant reinforcement on conditioned responding. Conditioned responding is developed using the normal classical conditioning procedure of pairing a CS with a UCS. The importance of this pairing procedure in developing CRs can be assessed by imposing an omission contingency such that occurrence of the CR results in the withholding of the UCS on that trial. See p. 5.

*Paced Responding*
In operant conditioning a procedure for altering the overall rate of responding without affecting reinforcement frequency, for example, reinforcement may be scheduled on a VI 1-min schedule as long as the response prior to reinforcement meets a specified IRT criterion – the larger the IRT criterion, the lower will be the overall rate of responding.

*Passive Avoidance*
An avoidance procedure where the aversive event is avoided only if the subject remains immobile or in some designated area of the conditioning chamber.

*Positive Conditioned Suppression (PCS)*
Presenting a stimulus which signals free food (an appetitive CS) during on-going operant responding for food. This procedure usually suppresses appetitive operant responding. See p. 230.

*Post-reinforcement Pause (PRP)*
On schedules of reinforcement, the time from the delivery of food up to the first response following food.

*Reactivation Treatment*
A procedure used to alleviate warm-up effects. Aspects of the learning situation are presented to the subject (for example, the reinforcer CSs) prior to entry into the experimental situation. See p. 230.

*Response-Independent Reinforcement* – see Non-contingent Reinforcement

*Response Shaping (successive approximations)*
A method for hastening the acquisition of an operant response. See p. 40.

*Reversal Learning*
A discrimination procedure in which the contingencies of reinforcement programmed for two external stimuli are reversed. For example, if a subject has learnt a left-right discrimination with left as S + and right as S −, the contingencies can be reversed so that left becomes S − and right S +. The contingencies can be reversed on several occasions until a 'learning set' may be formed. See p. 309.

*Second-Order Schedules*
In operant conditioning, a composite schedule of reinforcement in which one schedule is programmed as if it were the operant on some different schedule. For example, second-order FR20 (FI-1 min). The subject here has to complete 20 FI 1 min before he receives food.

*Sensory Preconditioning*
In classical conditioning, two neutral stimuli, $CS_1$ and $CS_2$ are first presented together

on a number of occasions. After this, $CS_1$ is paired with a UCS. Finally, CRs are now measured to $CS_2$.

*Shuttle-box*
An apparatus for developing avoidance learning. See p. 132.

*'Sidman' Avoidance* – see Non-discriminated Avoidance

*Simultaneous Conditioning*
In classical conditioning, a procedure where CS and UCS are presented at the same time. See p. 23.

*Simultaneous Discrimination*
A discrimination procedure in which S + and S − are presented concurrently. See p. 49.

*Skinner-Box*
A pigeon conditioning chamber developed for free-operant learning studies by B. F. Skinner. See p. 38 and figure 2.7.

*Stimulus Compounding*
In operant conditioning, the simultaneous presentation of two independently established $S^D$s. See p. 108ff.

*Successive Approximations* – see Response Shaping

*Successive Discrimination*
A discrimination procedure in which S + and S − are presented to the subject individually on successive trials, or for alternating periods of time. See p. 38.

*Summation Test/Combined Cue Test*
In classical conditioning the simultaneous presentation of two independently established CSs in order to assess the excitatory or inhibitory effects of those CSs.

*T-Maze*
A T-shaped alleyway with start-box at the base of the T and goal-boxes at either end of the head of the T. Animals are usually taught either to turn to one particular goal-box (a left-right discrimination), or to approach a distinctively coloured goal-box which can be randomly alternated from left to right on successive trials. The goal-box is usually baited with food.

*Tandem Schedule (tand)*
In operant conditioning a reinforcer is delivered when the subject completes *two* schedule requirements in succession. No stimuli are correlated with the schedules. For example, tand FR1 FI 1-min – the subject receives the reinforcer on having made one response (FR1) followed immediately by completion of the FI requirement.

*Temporal Conditioning*
In classical conditioning the UCS is presented periodically at regular intervals. Time-since-last-UCS-delivery therefore acts as the CS. See p. 23.

*Time-Out (TO)*
A period of time during which the subject is prevented from acquiring reinforcers. In animal studies this is usually achieved by turning off all the lights in the experimental chamber and making the response-manipulanda inoperative.

*Titration*
A procedure for establishing a subject's threshold level for certain events, particularly electric shock. See p. 131.

### Trace Conditioning
In classical conditioning, a procedure in which the onset and offset of the CS occur some fixed time before delivery of the UCS. See p. 23.

### Truly Random Control (TRC)
In classical conditioning, a control procedure in which CS and UCS are presented with the same frequency as each other but independently of each other. Thus conditions are identical to a classical conditioning procedure except the contingency between CS and UCS is missing (Rescorla, 1967).

### Variable-Interval Schedule (VI)
A schedule of reinforcement which develops a steady, constant rate of responding. See p. 74.

### Variable-Ratio Schedule (VR)
A schedule of reinforcement which develops a relatively high, steady rate of responding. See p. 79.

### Variable-Time Schedule (VT)
A schedule which delivers some event (usually food or electric shock) at variable time intervals independently of the subject's behaviour. See p. 87.

### Wisconsin General Test Apparatus (WGTA)
An apparatus developed by Harlow and Harlow (1962) to study perceptual learning in primates. The subject is positioned opposite a horizontal tray in which there are a number of depressions. Suitable rewards can be placed in the depressions and are then covered by various objects. This is frequently used for studies of concept formation, 'learning sets' and memory in primates.

### Yoked Control
An experimental procedure where subjects are run in pairs. One member of each pair receives the experimental treatment, which usually involves an operant contingency of some kind (for example, omission training): the yoked partner receives reinforcers at the same time as the experimental subject, but his behaviour has no influence on the delivery of reinforcers. This is normally used to assess the relative influence of operant and classical contingencies on conditioned responding. See p. 239.

# Author Index

Numbers in italic refer to the reference list.

# Subject Index

skeletal responses, classical
      conditioning of    57ff
Skinner-box    11, 38, 467
social facilitation    268, 272
song-learning in birds    260
species-specific defence reactions
      (SSDRs)    192ff, 328
species-specific reactions    154ff
spontaneous recovery    27, 295ff
state-dependent learning    290ff,
      292
stimulus
   associations with reinforcers
      228ff
   associations with response
      216ff
   nature of, in constraints on
         conditioning    159ff
stimulus compounds, during operant
      conditioning    108ff, 467
stimulus intensity    111ff
stimulus substitution    188, 219ff
successive approximations    41
   see also response shaping
sucking response, in neonate
      conditioning    324ff, 334
superstitious reinforcement    42ff
   and interim and terminal
         behaviours    163ff
   and IRT reinforcement    78
   on DRL schedules    83ff
   see also adventitious reinforcement
symptom substitition    366, 389ff
systematic densensitisation    376ff

T-maze    467
tandem schedule    467
temporal conditioning    23, 25,
      326, 467

temporal discrimination    65, 69,
      73, 84, 133
   in theories of forgetting    297ff
terminal behaviours    165ff, 180
thinking
   in animals    318ff
   in man    396ff
time-out (TO)    382ff, 467
titration    131, 467
token economy    378ff
trace conditioning    23ff, 468
trace decay    293ff
trial and error learning    20, 309
two-factor theory of avoidance
      learning    135ff

unconditioned reflexes    19
unconditioned stimulus    21
   see also classical conditioning
unconditioned response    21
   see also classical conditioning

variable-interval schedule    74ff,
      336, 468
   human performance on    352
   local rate of reinforcement in
      75, 77
variable-ratio schedule    79, 468
   human performance on    352
visual cliff    249, 250
visual experience, and cortical
      development    250ff

warm-up decrement    289ff, 292
wheel running    171

yawning    200
yoked controls    239, 468